环境暴露与人群健康丛书

电子垃圾污染与健康管控

安太成 郭 杰 等 编著

科学出版社
北京

内 容 简 介

本书主要介绍电子垃圾拆解处理过程中的污染物排放特征与健康风险消减,以及与电子垃圾污染相关的暴露特征、毒性效应及控制技术等方面的研究内容。全书共9章。第1章概述电子垃圾的来源组成、污染特征及对人体健康效应,并简述我国电子垃圾的污染特征;第2章介绍电子垃圾相关的重金属污染特征、迁移转化及暴露特征;第3章介绍电子垃圾拆解处理排放挥发性有机物的污染特征和暴露风险;第4章对电子垃圾拆解处理产生持久性有机物污染进行详细介绍;第5章对电子垃圾产生污染物的人体暴露特征、健康风险及人体生物监测方法进行阐述;第6章对电子垃圾相关的重金属和有机污染物的毒性效应、流行病学案例进行介绍;第7章对电子垃圾拆解处理排放水体和沉积物中有机污染物的分布特征、转化机制和健康风险消减进行介绍;第8章介绍电子垃圾拆解处理排放大气污染物(颗粒物、挥发性有机物和持久性毒害有机物等)的控制净化、风险消减;第9章介绍电子垃圾相关的全球治理现状和经验,并提出电子垃圾健康风险管控的发展趋势和挑战。

本书可供高等院校环境科学、环境地学、毒理学、环境健康和暴露等专业的研究生、高年级本科生以及相关领域的科研工作者及管理人员阅读参考。

图书在版编目(CIP)数据

电子垃圾污染与健康管控 / 安太成等编著. —北京:科学出版社,2023.11
(环境暴露与人群健康丛书)
ISBN 978-7-03-076929-9

Ⅰ. ①电… Ⅱ. ①安… Ⅲ. ①电子产品-重金属污染-影响-健康-污染控制 Ⅳ. ①X503.1

中国国家版本馆 CIP 数据核字(2023)第 216011 号

责任编辑:杨 震 刘 冉 / 责任校对:杜子昂
责任印制:徐晓晨 / 封面设计:北京图阅盛世

科学出版社 出版
北京东黄城根北街 16 号
邮政编码:100717
www.sciencep.com

涿州市般润文化传播有限公司 印刷
科学出版社发行 各地新华书店经销

*

2023 年 11 月第 一 版 开本:720×1000 1/16
2023 年 11 月第一次印刷 印张:26 3/4
字数:540 000
定价:150.00 元
(如有印装质量问题,我社负责调换)

丛书编委会

顾　　问：魏复盛　陶　澍　赵进才　吴丰昌
总 主 编：于云江
编　　委：（以姓氏汉语拼音为序）
　　　　　安太成　陈景文　董光辉　段小丽　郭　杰
　　　　　郭　庶　李　辉　李桂英　李雪花　麦碧娴
　　　　　向明灯　于云江　于志强　曾晓雯　张效伟
　　　　　郑　晶
丛书秘书：李宗睿

《电子垃圾污染与健康管控》编著者名单

安太成　郭　杰　马盛韬　李桂英　陈江耀
余应新　高艳蓬　熊举坤　刘冉冉　林美卿
吴颖君　梁志梳　王　梅　唐　僭

丛 书 序

近几十年来，越来越多的证据表明环境暴露与人类多种不良健康结局之间存在关联。2021年《细胞》杂志发表的研究文章指出，环境污染可通过氧化应激和炎症、基因组改变和突变、表观遗传改变、线粒体功能障碍、内分泌紊乱、细胞间通信改变、微生物组群落改变和神经系统功能受损等多种途径影响人体健康。《柳叶刀》污染与健康委员会发表的研究报告显示，2019年全球约有900万人的过早死亡归因于污染，相当于全球死亡人数的1/6。根据世界银行和世界卫生组织有关统计数据，全球70%的疾病与环境污染因素有关，如心血管疾病、呼吸系统疾病、免疫系统疾病以及癌症等均已被证明与环境暴露密切相关。我国与环境污染相关的疾病近年来呈现上升态势。据全球疾病负担风险因素协作组统计，我国居民疾病负担20%由环境污染因素造成，高于全球平均水平。环境污染所导致的健康危害已经成为影响全球人类发展的重大问题。

欧美发达国家自20世纪60年代就成立了专门机构开展环境健康研究。2004年，欧洲委员会通过《欧洲环境与健康行动计划》，旨在加强成员国在环境健康领域的研究合作，推动环境风险因素与疾病的因果关系研究。美国国家研究理事会（NRC）于2007年发布《21世纪毒性测试：远景与策略》，通过科学导向，开展系统的毒性通路研究，揭示毒性作用模式。美国国家环境健康科学研究所（NIEHS）发布的《发展科学，改善健康：环境健康研究计划》重点关注暴露、暴露组学、表观遗传改变以及靶点与通路等问题；2007年我国卫生部、环保部等18个部委联合制订了《国家环境与健康行动计划》。2012年，环保部和卫生部联合开展"全国重点地区环境与健康专项调查"项目，针对环境污染、人群暴露特征、健康效应以及环境污染健康风险进行了摸底调查。2016年，党中央、国务院印发了《"健康中国2030"规划纲要》，我国的环境健康工作日益受到重视。

环境健康研究的目标是揭示环境因素影响人体健康的潜在规律，进而通过改善生态环境保障公众健康。研究领域主要包括环境暴露、污染物毒性、健康效应以及风险评估与管控等。在环境暴露评估方面，随着质谱等大型先进分析仪器的有效利用，对环境污染物的高通量筛查分析能力大幅提升，实现了多污染物环境暴露的综合分析，特别是近年来暴露组学技术的快速发展，对体内外暴露水平进行动态监测，揭示混合暴露的全生命周期健康效应。针对环境污染低剂量长期暴露开展暴露评估模型和精细化暴露评估也成为该领域的新的研究方向；在环境污染物毒理学方面，高通量、低成本、预测能力强的替代毒理学快速发展，采用低等动物、体外试验和非生物手段的毒性试验替代方法成为毒性测试的重要方面，解析污染物毒性作用通路、确定生物暴露标志物正成为该领域研究热点，通过这些研究可以大幅提高污染物毒性的筛查和识别能力；在环境健康效应方面，近年来基因组学、转录组学、代谢组学和表观遗传学等的快速发展为探索易感效应生物标志物提供了技术支撑，有助于理解污染物暴露导致健康效应的分子机制，探寻环境暴露与健康、疾病终点之间的生物学关联；在环境健康风险防控方面，针对不同暴露场景开展环境介质-暴露-人群的深入调查，实现暴露人群健康风险的精细化评估是近年来健康风险评估的重要研究方向；同时针对重点流域、重点区域、重点行业、重点污染物开展环境健康风险监测，采用风险分区分级等措施有效管控环境风险也成为风险管理技术的重要方面。

环境健康问题高度复杂，是多学科交叉的前沿研究领域。本丛书针对当前环境健康领域的热点问题，围绕方法学、重点污染物、主要暴露类型等进行了系统的梳理和总结。方法学方面，介绍了现代环境流行病学与环境健康暴露评价技术等传统方法的最新研究进展与实际应用，梳理了计算毒理学和毒理基因组学等新方法的理论及其在化学品毒性预测评估和化学物质暴露的潜在有害健康结局等方面的内容，针对有毒有害污染物，系统研究了毒性参数的遴选、收集、评价和整编的技术方法；重点污染物方面，介绍了大气颗粒物、挥发性有机污染物以及阻燃剂和增塑剂等新污染物的暴露评估技术方法和主要健康效应；针对典型暴露场景，介绍了我国电子垃圾拆解活动污染物的排放特征、暴露途径、健康危害和健康风险管控措施，系统总结了污染场地土壤和地下水的环境健康风险防控技术方面的创新性成果。

近年来环境健康相关学科快速发展，重要研究成果不断涌现，亟须开展从环境暴露、毒理、健康效应到风险防控的全链条系统梳理，这正是本丛书编撰出版的初衷。"环境暴露与人群健康丛书"以科技部、国家自然科学基金委员会、生态环境部、卫生健康委员会、教育部、中国科学院等重点支持项目研究为基础，汇集了来自我国科研院所和高校环境健康相关学科专家学者的集体智慧，系统总结了环境暴露与人群健康的新理论、新技术、新方法和应用实践。其成果非常丰富，可喜可贺。我们深切感谢丛书作者们的辛勤付出。冀望本丛书能使读者系统了解和认识环境健康研究的基本原理和最新前沿动态，为广大科研人员、研究生和环境管理人员提供借鉴与参考。

2022 年 10 月

前　言

随着电子信息技术的发展和电子电器产品使用寿命的缩短，大量的电子电器产品被淘汰、废弃，从而产生了数量巨大、组分复杂的电子垃圾。根据《2020年全球电子垃圾监测报告》的有关数据，2019年全球产生了5360万吨电子垃圾，并以每年近200万吨的速度增长。电子垃圾蕴含非常高的回收价值，估计2019年全球电子垃圾中原材料的价值约为570亿美元，铁、铜和金等金属是其价值的主要来源。电子垃圾的资源属性催生了电子垃圾回收处理行业，该行业甚至成为某些区域的经济支柱和特色产业，并形成了一定规模的从业人员群体。

万事万物具有两面性。在对电子垃圾进行资源回收利用的同时，一定量的重金属和有机污染物等毒害物质，将从电子垃圾的材料源头释放出来，加上电子垃圾回收处理工艺产生的废水、废气及废渣等污染物，一并形成电子垃圾的污染源。电子垃圾拆解回收产生的污染物，通过多个扩散途径迁移进入大气、水体、沉积物及土壤等环境介质中，发生复杂的物化、生化等转化过程，并能在环境介质中长期存在，再通过多种暴露途径或食物链进入人体并产生健康危害。早期由于对电子垃圾缺乏足够认识和科学管理，只注重对电子垃圾中有价值资源的回收利用，而忽视了电子垃圾中毒害物质的污染控制及排放消减，从而给电子垃圾拆解场地造成了严重的生态破坏和健康风险。

我国生态文明建设已经进入资源循环和减污降碳的新发展阶段，电子垃圾作为具有资源回收价值和环境污染特性的典型代表，对其进行回收利用符合国家的发展战略需求。随着电子垃圾相关的法律条例、管理制度及治理体系的不断完善，我国的电子垃圾处理模式已经从分散无序的家庭作坊式回收转变为有资质企业或园区的集中拆解处理，并且对处理过程排放污染物的管控也更加严格、规范。但是，由于电子垃圾组分的复杂性及含有潜在毒害物质，对其进行拆解回收处理仍然存在污染物的排放及对相关人群的暴露风险。

电子垃圾拆解回收处理导致的环境污染问题及健康风险是一项复杂的研究课题。本书全面系统地介绍了电子垃圾拆解处理相关的环境污染特征、暴露途径和健康风险消减，以及近年来该研究领域的最新进展和管理制度。本书由9章组成，主要包括以下研究内容：第1章概述电子垃圾的来源组成、污染特征及对人体健康效应，并简述我国电子垃圾的污染特征；第2章介绍电子垃圾相关的重金属污染特征、迁移转化及暴露特征；第3章介绍电子垃圾拆解处理排放挥发性有机物的污染特征和暴露风险；第4章对电子垃圾拆解处理产生持久性有机物污染进行详细介绍；第 5 章对电子垃圾产生污染物的人体暴露特征、健康风险及人体生物监测方法进行阐述；第 6 章对电子垃圾相关的重金属和有机污染物的毒性效应、流行病学案例进行介绍；第 7 章对电子垃圾拆解处理产生水体和沉积物中有机污染物的分布特征、转化机制和健康风险消减进行介绍；第 8 章介绍电子垃圾拆解处理排放大气污染物（颗粒物、挥发性有机物和持久性毒害有机物等）的控制净化、风险消减；第 9 章介绍电子垃圾相关的全球治理现状和经验，并提出电子垃圾健康风险管控的发展趋势和挑战。

本书由安太成和郭杰策划、统稿、修改及定稿，内容包括作者十余年的主要研究成果总结，这些成果是集体智慧的结晶。本书的撰写任务安排如下：第 1 章由安太成、郭杰、马盛韬撰写，第 2 章由李桂英、吴颖君、安太成撰写，第 3 章由刘冉冉、李桂英、安太成撰写，第 4 章由马盛韬、余应新、郭杰、安太成撰写，第 5 章由林美卿、马盛韬、唐僭、安太成撰写，第 6 章由高艳蓬、王梅、李桂英、安太成撰写，第 7 章由熊举坤、梁志梳、李桂英、安太成撰写，第 8 章由陈江耀、郭杰、安太成撰写，第 9 章由安太成、郭杰撰写。

本书的研究工作是在国家自然科学基金重点项目（41731279）、广东省战略性新兴产业核心技术攻关项目（2012A032300017）和粤港关键领域重点突破项目（2009A030902003）等持续资助下完成的。作者衷心感谢"环境暴露与人群健康丛书"顾问魏复盛院士、陶澍院士、赵进才院士、吴丰昌院士，丛书总主编于云江研究员及丛书编委们在本书撰写过程中给予的指导、帮助和鼓励。感谢科学出版社刘冉编辑耐心、细致的编校工作。本书的大部分数据和结果来自公开发表的期刊论文和研究生学位论文，感谢所有参与相关研究工作的老师、学生的辛勤付出。同时也向关心、支持我们完成相关工作的领导、同事表示衷心感谢。没有你们的帮助、支持，就没有本书的出版。

由于作者水平有限，书中疏漏之处在所难免，恳请读者批评指正。

著 者

2023 年 10 月

目 录

丛书序
前言

第1章 电子垃圾发展现状及其排放污染物的源与汇 ·········· 1
  1.1 电子垃圾的定义与分类 ·········· 1
  1.2 电子垃圾的来源与组成 ·········· 2
  1.3 电子垃圾污染现状及其健康风险 ·········· 4
    1.3.1 电子垃圾的污染概述 ·········· 4
    1.3.2 电子垃圾对生态环境的影响 ·········· 6
    1.3.3 电子垃圾对人体健康的影响 ·········· 8
    1.3.4 我国电子垃圾的污染特征 ·········· 9
  参考文献 ·········· 12

第2章 电子垃圾拆解过程重金属污染 ·········· 15
  2.1 重金属污染概述 ·········· 15
  2.2 电子垃圾中的重金属 ·········· 15
  2.3 电子垃圾拆解区重金属的环境分布 ·········· 17
    2.3.1 大气颗粒物 ·········· 17
    2.3.2 水和沉积物 ·········· 19
    2.3.3 灰尘和土壤 ·········· 22
    2.3.4 重金属在环境中的迁移转化 ·········· 23
  2.4 电子垃圾的重金属暴露 ·········· 24
    2.4.1 外暴露 ·········· 24
    2.4.2 内暴露 ·········· 29
  参考文献 ·········· 33

第3章 电子垃圾拆解排放挥发性有机物污染 ·········· 38
  3.1 挥发性有机物概述 ·········· 38
  3.2 电子垃圾拆解排放 VOCs 污染特征 ·········· 38
    3.2.1 芳香烃类 VOCs ·········· 45
    3.2.2 脂肪烃类 VOCs ·········· 50
    3.2.3 卤代烃类 VOCs ·········· 52

  3.2.4 其他 VOCs ·················54
 3.3 挥发性有机物的暴露风险 ·················54
 参考文献 ·················59
第 4 章 电子垃圾拆解排放持久性有机物污染 ·················63
 4.1 颗粒物污染 ·················63
  4.1.1 颗粒物的来源 ·················64
  4.1.2 颗粒物的组成 ·················65
  4.1.3 颗粒物的粒径分布 ·················67
 4.2 持久性有机污染物污染 ·················76
  4.2.1 多溴联苯醚 ·················78
  4.2.2 多氯联苯 ·················93
  4.2.3 四溴双酚 A ·················96
  4.2.4 得克隆 ·················103
  4.2.5 有机磷阻燃剂 ·················106
  4.2.6 邻苯二甲酸酯 ·················111
  4.2.7 多环芳烃 ·················116
  4.2.8 其他持久性有机污染物 ·················156
 参考文献 ·················159
第 5 章 电子垃圾污染的人体暴露特征及其风险 ·················176
 5.1 呼吸暴露及风险 ·················176
  5.1.1 车间工人的暴露 ·················176
  5.1.2 周边居民的暴露 ·················177
  5.1.3 个体暴露监测 ·················178
 5.2 摄食暴露及风险 ·················178
  5.2.1 饮用水暴露 ·················178
  5.2.2 食物暴露 ·················179
 5.3 皮肤接触暴露风险 ·················181
 5.4 电子垃圾拆解人群的人体生物监测方法 ·················182
  5.4.1 血液样品 ·················183
  5.4.2 尿液样品 ·················187
  5.4.3 头发和指甲样品 ·················188
  5.4.4 皮肤擦拭样品 ·················190
 5.5 拆解工人的暴露特征 ·················191
  5.5.1 阻燃剂暴露特征 ·················191
  5.5.2 塑料添加剂暴露特征 ·················195

  5.5.3 多环芳烃暴露 195
  5.5.4 其他持久性有机污染物暴露 196
 参考文献 197

第6章 电子垃圾排放污染物的毒理学与健康危害 204
 6.1 毒性来源与效应概述 204
  6.1.1 电子垃圾产生毒性效应的原因 204
  6.1.2 重金属的毒性效应 205
  6.1.3 有机污染物的毒性效应 206
  6.1.4 电子垃圾污染物的毒性分类 208
 6.2 重金属的毒性效应 208
  6.2.1 神经毒性效应 208
  6.2.2 遗传毒性效应 209
  6.2.3 肝肾毒性效应 211
  6.2.4 心血管系统毒性效应 212
  6.2.5 肠胃系统毒性效应 213
  6.2.6 呼吸系统毒性效应 214
 6.3 有机污染物的毒性效应 215
  6.3.1 内分泌干扰效应 215
  6.3.2 神经毒性效应 217
  6.3.3 免疫毒性效应 218
  6.3.4 生殖系统毒性 218
  6.3.5 疾病 218
 6.4 流行病学证据 219
  6.4.1 重金属暴露引发健康疾病风险的研究案例 219
  6.4.2 持久性有机污染物暴露引发健康疾病风险的研究案例 226
  6.4.3 阻燃剂暴露引发健康疾病风险的研究案例 227
  6.4.4 细颗粒物暴露引发健康疾病风险的研究案例 228
  6.4.5 多环芳烃暴露引发健康疾病风险的研究案例 228
 参考文献 229

第7章 电子垃圾拆解排放水体和沉积物中有机污染物的转化与风险消减 238
 7.1 污染物在水体和沉积物中分布特征 238
  7.1.1 阻燃剂 238
  7.1.2 塑料其他助剂 244
  7.1.3 多环芳烃 245
  7.1.4 其他持久性有机污染 247

7.2 污染物在水体和沉积物中的转化机制 249
  7.2.1 生物转化机制 249
  7.2.2 水解机制 267
  7.2.3 光化学转化机制 268
7.3 污染物在水体和沉积物中健康风险消减 272
  7.3.1 阻燃剂的健康风险 274
  7.3.2 阻燃剂健康风险的消减方法 278
参考文献 290

第 8 章 电子垃圾拆解排放大气污染物的控制与健康风险消减 308
8.1 颗粒物的控制与健康风险消减 308
  8.1.1 颗粒物的控制与减排 308
  8.1.2 颗粒物健康风险消减 319
8.2 VOCs 的控制与健康风险消减 323
  8.2.1 VOCs 的控制与减排 323
  8.2.2 VOCs 的健康风险消减 357
8.3 持久性毒害有机物的控制与健康风险消减 376
  8.3.1 持久性毒害有机物的排放 377
  8.3.2 持久性毒害有机物的控制 380
  8.3.3 持久性毒害有机物的健康风险消减 383
参考文献 387

第 9 章 电子垃圾健康风险管控：发展趋势与挑战 393
9.1 全球的政策 393
9.2 各大洲的电子垃圾管控 395
  9.2.1 欧美地区 397
  9.2.2 非洲地区 400
  9.2.3 拉美地区 402
  9.2.4 亚洲地区 405
9.3 发展趋势及健康风险挑战 411
  9.3.1 发展趋势 411
  9.3.2 健康风险挑战 412
参考文献 413

第 1 章 电子垃圾发展现状及其排放污染物的源与汇

随着电气信息时代的快速发展，琳琅满目的电器电子产品在丰富和提升人民生活水平的同时，也产生了废弃电器电子产品(俗称电子垃圾)的环境污染及处理处置问题。由于电子垃圾具有资源和污染的双重特性，在对电子垃圾进行资源化回收利用的同时，若采用不恰当的处理方式，容易导致电子垃圾中毒害物质对环境介质的释放，造成生态环境系统的污染，并对人体健康造成风险。本章阐述了电子垃圾的定义与分类、来源组成，并对电子垃圾的污染特征、源汇排放及人体健康危害等内容进行了概述。

1.1 电子垃圾的定义与分类

"电子垃圾"又称"电子废弃物"或"废弃电器电子产品"，其英文表述有"electronic waste"、"e-waste"、"waste electrical and electronic equipment(WEEE)"和"e-scrap"等。"废弃电器电子产品"属于正式的书面用语，我国《废弃家用电器与电子产品污染防治技术政策》(环发〔2006〕115 号)将"废弃家用电器与电子产品"定义为：已经失去使用价值或因使用价值不能满足要求而被丢弃的家用电器与电子产品，以及其元器件、零部件和耗材。现行的《废弃电器电子产品处理目录(2014 年版)》主要包括电冰箱、空气调节器、吸油烟机、洗衣机、电热水器、燃气热水器、打印机、复印机、传真机、电视机、监视器、微型计算机、移动通信手持机和电话单机等 14 大类废弃电器电子产品。根据我国《废弃电器电子产品回收处理管理条例》(国务院令第 551 号)，将"废弃电器电子产品的处理活动"定义为：对列入目录的废弃电器电子产品进行拆解，从中提取物质作为原材料或者燃料，用改变废弃电器电子产品物理、化学特性的方法减少已产生的废弃电器电子产品数量，减少或者消除其危害成分，以及将其最终置于符合环境保护要求的填埋场的活动，不包括产品维修、翻新以及经维修、翻新后作为旧货再使用的活动。相比于"废弃电器电子产品"，"电子垃圾"是个泛称，其包括的物品种类更多、范围更广，如废覆铜板、电路板产品边角料、废电缆电线、废塑料等与电器电子产品相关的废弃物。为此，本书采用"电子垃圾"这个泛称术语。

根据欧盟的 WEEE 指令(2012/19/EU)第 3(1)条款，将电子电气设备(EEE)定义为"依赖于电流或电磁场才能正常工作的设备，以及产生、传输和测量这些电流和电磁场的设备，且其设计使用额定电压不超过交流电 1000 V 或直流电 1500

V."因此，欧盟定义上述 EEE 被丢弃时，即为 WEEE。自 2018 年 8 月 15 日起，欧盟将 EEE 分为六大类产品(2012/19/EU 附录 3)，并采用开放式范围管理，即 2012/19/EU 附录 4 所列清单并非详尽清单，未列的产品亦属 WEEE 范围(除太空设备等规定排除设备外)。EEE 具体包括：①温度交换设备，如冰箱、冰柜、空调设备、除湿设备、热泵、含油的散热器和其他使用非水流体进行温度交换的温度交换设备；②屏幕、显示器以及包含屏幕的面积大于 100 cm² 的设备，如电视、液晶相框、显示器、笔记本电脑等；③灯具，如各种荧光灯、高压钠灯和金属卤化物灯等高压放电灯、低压钠灯、发光二极管等；④大型设备，如洗衣机、干衣机、洗碗机、炊具、电炉、照明设备、声音或图像再现设备、大型计算机主机、大型印刷机、大型医疗设备、大型监控仪器、自动投币设备、光伏电池板等；⑤小型设备，如吸尘器、地毯清扫机、缝纫用具、微波炉、通风设备、熨斗、烤面包机、电动刀、电动水壶、钟表、电动剃须刀、磅秤、头发和身体护理用具、计算器、收音机、摄像机、录像机、高保真音响设备、乐器、声音或图像复制设备、电器和电子玩具、运动器材等用电脑、烟雾探测器、加热调节器、恒温器、小型电气和电子工具、小型医疗设备、小型监测和控制仪器、自动输送产品的小型电器、集成光伏板等小型设备；⑥小型信息技术和电信设备(外形尺寸不超过 50 cm)，如移动电话、全球定位系统(GPS)设备、袖珍计算器、路由器、个人计算机、打印机、电话等。由上可知，欧洲绝大多数类型和种类的 EEE 淘汰废弃后，都列为 WEEE 指令的管理范畴，而 EEE 的六大类与其废物管理特点密切相关，例如，根据欧盟 WEEE 指令的要求，对不同类型的电子垃圾具有不同的收集回收率目标。

1.2 电子垃圾的来源与组成

20 世纪中后期，电子垃圾被认为是增长速度最快的一类固体废物[1]，随着阴极射线管(CRT)电视机、台式计算机等电器电子产品进入百姓家庭，尤其是发达国家对电器电子产品的需求量更大，导致废弃淘汰的家电数量也较多。而发展中国家，如亚洲的中国、印度和巴基斯坦等国，由于劳动力廉价、缺乏相应的环境标准和监管，成为美国等发达国家电子垃圾的主要出口国。据估计，美国所收集回收的 50%~80% 电子垃圾被出口到发展中国家[2]，而我国早期也承担了全球 70% 电子垃圾的处理活动，这说明全球电子垃圾的产生和回收处置存在严重的区域不平衡问题[3]。

根据英特尔联合创始人戈登·摩尔提出的"摩尔定律"，对于消费类的数码电子产品，当价格不变时，集成电路上可容纳的元器件的数目，约每隔 18~24 个月便会增加 1 倍，性能也将提升 1 倍。过去十多年中，智能手机和互联网的使用量急剧增加。根据 Digital 2020 报告，截至 2020 年 7 月，全球有 51.5 亿手机用户和 45.7 亿互联网用户，约 60% 的人口正在使用手机和互联网，且手机用户和互联网用户数分别比上年增长了 2.4%和 8%。随着我国经济的迅速发展,社会消费水平的不断提高,电子垃圾的数量也呈迅速增长态势。据央视财经消息,2021 年我国以手机为主的移动通信终端社会保有量已达 18.56 亿部,且废旧手机的产生量和闲置量逐年增长,预计"十四五"期间闲置手机总量累计将达到 60 亿部。由此可见,随着信息和通信技术的高速发展,EEE 的种类和数量日益增多,其淘汰周期逐渐缩短,从而导致全球 WEEE 的年产生量不断攀升。

根据《2020 年全球电子垃圾监测报告》,全球电子垃圾的产生量由 2014 年的 4440 万吨增长到 2019 年的 5360 万吨(人均 7.3 kg),并预计 2030 年和 2050 年将分别达到 7470 万吨和 1.1 亿吨[4,5]。表 1-1 给出了 2019 年全球五大洲(亚洲、美洲、欧洲、非洲、大洋洲)的电子电器产品市场保有量和电子垃圾产生量、正式收集及跨国流通量的相关数据信息[4-6],从中可以看出,电子垃圾的产生量因时间、空间、人口、生活方式和社会经济发展水平而异。亚洲以 2490 万吨(人均 5.6 kg)的电子垃圾产生量位居各大洲之首,其次是美洲 1310 万吨(人均 13.3 kg)、欧洲 1200 万吨(人均 16.2 kg)、非洲 290 万吨(人均 2.5 kg)和大洋洲 70 万吨(人均 16.1 kg)。若电子垃圾按照国家相关法律条例要求,经零售商、市政企业收集点或收集服务等途径进入专门处理企业,采用环境可控方式回收其中有价值材料,并以环境无害化管理有害物质,称之为"正式收集"。2019 年,全球电子垃圾得到正式收集的量为 929 万吨,仅占比 17%。其中,正式收集率最高的为欧洲(42%),其次是亚洲(12%),然后为美洲(9.2%)、大洋洲(8.6%),而非洲的仅为 1.0%。2019 全球约有 510 万吨的电子垃圾在不同国家间流通,跨国流通后的电子垃圾去向和环境影响在不同区域大相径庭。在高收入国家,废物回收基础设施比较完善,一方面,约 8%的电子垃圾(以小型设备为主)被丢弃在垃圾箱中,随后被填埋或焚烧。另一方面,废旧电器电子产品被翻新为二手产品或者被非法出口到低收入国家。而对于中低收入国家,电子垃圾相关法律和基础设施不完善,电子垃圾主要由非正规部门回收管理。

表 1-1　2019 年全球各大洲电子电器产品保有量和电子垃圾的产生量、正式收集及跨国流通量

地区	市场上的 EEE		电子垃圾产生量		电子垃圾正式收集量 (万 t,回收率)		跨国流通量 (万 t)	
	万 t	kg/人	万 t	kg/人	电子垃圾	废电路板	进口	出口
亚洲	4210	9.5	2490	5.6	290(12%)	10(17%)	288.9	253.7
美洲	1620	16.5	1310	13.3	120(9.2%)	10(44%)	39.3	54.7
欧洲	1360	18.1	1200	16.2	510(42%)	20(61%)	124.8	185.0
非洲	440	3.8	290	2.5	3(1.0%)	1(13%)	54.6	13.2
大洋洲	80	19.7	70	16.1	6(8.6%)	0.5(31%)	0	2.1
全球	7710	10.5	5360	7.3	929(17%)	41.5(34%)	507.6	508.7

电子垃圾具有资源和污染双重属性。一方面，电子垃圾是座不折不扣的"城市矿山"。电子垃圾中的主要组分包括黑色金属、有色金属、玻璃、塑料等。2019年电子垃圾中的原材料价值估计可达570亿美元，铁、铜、金是这一价值的主要来源。按17%的正规收集率计算，以环保方式可以从电子垃圾中回收的原材料价值约为100亿美元，且可回收400万吨可再利用原材料。另一方面，电子垃圾是潜在污染源。电子垃圾回收处理产生的毒害物质主要有两大类：一类是汞、铅、镉、铬等重金属，另一类是氟氯烃、多环芳烃（PAHs）、多溴联苯醚（PBDEs）、多氯联苯（PCBs）、多氯二苯并二噁英及呋喃（PCDD/Fs）和多溴二苯并二噁英及呋喃（PBDD/Fs）等有机污染物。2019年全球未记录在案的4430万吨（83%）电子垃圾的去向不明，其中含有的毒害物质将成为潜在污染源，若采用不适当的处理方式，将对生态环境释放大量毒害物质[7]。

1.3 电子垃圾污染现状及其健康风险

1.3.1 电子垃圾的污染概述

早期电子垃圾的非正规处理是造成电子垃圾拆解场地环境污染的重要原因。非正规处理活动，包括通过氰化物浸取或者硝酸-汞混合物从废电路板中回收黄金、加热拆解或露天燃烧废电路板、电缆组件分离或焊锡回收、墨粉清洗、塑料破碎-熔融再造粒、燃烧电线回收铜、手工拆解阴极射线管显示器等[8]，将会造成电子垃圾中毒害物质的释放及迁移转化，并进入大气、水体、土壤等环境介质中，且多种污染物在环境中持久存在，从而造成环境污染和健康风险[9]，早期典型电子垃圾处理方式的污染物排放特征和潜在健康风险如表1-2所示[7]。

表1-2 典型电子垃圾处理过程及其潜在危害

电子垃圾元器件	处理过程	潜在的健康危险	潜在的环境危害
阴极射线管（CRT）	打破、取出铜轭、倾倒废弃	1.硅化物；2.吸入或接触镉和其他金属的磷光体	铅、钡等重金属污染地下水；有毒磷光体污染空气；碎玻璃污染环境
电路板	熔化焊锡及电脑晶片去除	1.吸入锡、铅、铍、镉、汞；2.可能吸入PBDD/Fs、BFRs	同类物质在空气中的挥发
拆除过的电路板	露天燃烧去除晶片收集残余金属	1.吸入PBDD/Fs、BFRs、锡、铅、铍、镉、汞；2.呼吸道刺激	1.锡、铅等污染地表水和地下水；2.PBDD/Fs、铍、镉及汞污染空气
晶片及其他含金元件	用王水在河岸边提取金	酸、氯气和二氧化硫气体对眼睛、皮肤、呼吸道的腐蚀	1.酸化地表水及地下水；2.直接倾倒严重污染河流

续表

电子垃圾元器件	处理过程	潜在的健康危险	潜在的环境危害
电脑及辅助设备的塑料部分	破碎或低温熔化，作低质塑料利用	接触吸收 PBDD/Fs 等毒害有机物和重金属	PBDD/Fs、重金属及有毒碳氢化物污染空气
电线	露天燃烧回收铜	接触 PBDD/Fs、PCDD/Fs、BFRs 和 PAHs 等致癌物质	PCDD/Fs 和 PBDD/Fs 及 PAHs 对空气、水和土壤造成严重污染
橡胶或塑料中的电脑元件如钢轴	露天燃烧回收钢铁和其他金属	接触并吸收包括二噁英、PAHs 的有机物	分别对空气、水和土壤造成严重污染
墨盒	没有任何保护措施下用刷子回收油墨	1.呼吸道和肺刺激；2.油墨粉尘是致癌物	油墨粉尘污染空气，丢弃的黑色塑料污染环境
钢铁和铜及其他贵金属的冶炼	用炉子从废物中再生钢或铜	呼吸吸收或皮肤接触二噁英和重金属	重金属及二噁英挥发污染空气

据估计，美国 1997~2007 年间淘汰产生了 5 亿台废旧电脑，其中含有约 287 万吨塑料（塑料中含有阻燃剂等毒害物质）、71.7 万吨铅(Pb)、1361 吨镉(Cd)、862 吨铬(Cr)和 287 吨汞(Hg)[2]。电子垃圾中的典型毒害物质及健康危害如下[7]：

铅 阴极射线管(CRT)显示器的玻璃面板、印刷电路板和焊料中含有铅。铅在环境中容易积累，对植物、动物和微生物具有急性和慢性效应，对人体的中枢和末梢神经系统、血液系统、肾脏、生殖系统和内分泌系统造成损害，且对儿童大脑发育有严重负面影响。

镉 芯片电阻、红外探测器、半导体芯片、CRT 显示器及塑料稳定剂等零部件中含有镉。镉化合物是有毒的，容易在人体，特别在肾脏中富集，从而对人体健康产生不可逆转的影响。

六价铬 六价铬常被用作未处理和镀锌钢板的防腐保护，并作为装饰或钢材外壳的硬化剂。六价铬具有很强的毒性，很容易穿过细胞膜而被吸收，并在被污染的细胞中产生各种毒性作用，如造成 DNA 损伤等。

汞 汞常用于恒温器、传感器、继电器、开关、医疗设备、灯具、移动电话和电池。汞会对各种器官造成损伤，包括大脑和肾脏等。当母体接触汞时，发育中的胎儿极易受到影响。当无机汞在水中扩散时，它会在底部沉积物中转化为甲基化汞。甲基化汞很容易在生物体内积累，通过食物链富集，并进入人体产生毒害效应。

聚氯乙烯(PVC)等塑料 一台老式台式电脑的塑料质量为 6.3 kg。PVC 由于阻燃性能好而被广泛应用于电缆和计算机外壳，与其他含氯化合物一样，PVC 塑料的燃烧是形成二噁英的主要原因之一。

溴化阻燃剂(BFRs) BFRs 被广泛应用于电子设备的塑料外壳和电路板，四溴双酚 A(TBBPA)、PBDEs、六溴环十二烷(HBCD)是早期使用较多的 BFRs，在老旧电子产品中的检出率很高。随着 PBDEs 和 HBCD 被列入《关于持久性有机污染物的斯德哥尔摩公约》的名录清单,两种 BFRs 的生产和使用已经被禁止或受到限制。

墨粉 废弃打印机墨盒中含有黑色炭黑和彩色墨粉。国际癌症研究机构将炭黑列为 2B 类致癌物,而彩色墨粉中含有重金属。吸入是墨粉的主要暴露途径,急性暴露可导致呼吸道刺激。

荧光粉和添加剂 荧光粉是一种无机化合物,用作 CRT 显示器内部的涂层。荧光粉中含有镉、锌、钒及其他稀土金属,这些金属及其化合物毒性很大,对手动拆解 CRT 的工人具有潜在危险。

1.3.2 电子垃圾对生态环境的影响

电子垃圾的影响不仅局限于拆解场地或倾倒场所,它还会延伸到处理场所以外,对区域生态系统(如土壤、水、空气和其他生物圈等)产生不利影响[10]。据估计,从电子垃圾中每年可回收 82 万吨铜,但每年仍有 5000 吨铜释放到环境中[11]。PBDEs 大都作为添加型阻燃剂用于塑料等产品中,即 PBDEs 与塑料树脂基体间没有化学键作用,在拆解处理过程中,具有半挥发性的 PBDEs 容易从塑料中释放出来,并进入环境介质,从而对人类和生态系统造成影响[12]。

1. 对土壤的影响

电子垃圾的处理活动是影响土壤质量的重要原因之一,尤其是早期落后的处理方式会释放高浓度的重金属和有机污染物等。电子垃圾的随意丢弃或倾倒在垃圾填埋场,其中所含有的铅、镉、铬、锰、砷等有毒重金属将随着雨水的渗透而进入土壤当中,给土壤和地下水造成污染[13]。同时,农作物可能吸收、富集土壤中的重金属等毒害物质,给土壤生态环境造成影响。对印度德里电子垃圾非正规拆解场地表层土壤中重金属的研究表明[14],重金属 As(17.08 mg/kg)、Cd(1.29 mg/kg)、 Cu(115.50 mg/kg)、 Pb(2645.31 mg/kg)、 Se(12.67 mg/kg)、 Zn(776.84 mg/kg)的浓度均存在超标现象,且在当地狗牙根(*Cynodon dactylon*)植物样本中观察到重金属的高度积累。2004 年对广东某电子垃圾拆解场地附近表层土壤和燃烧残留物进行了分析检测,得出塑料碎片和电缆燃烧残渣、酸浸场土壤和打印机硒鼓倾倒场土壤中的 PBDEs 干重浓度分别为 33 000~97 400 ng/g、2720~4250 ng/g 和 593~2890 ng/g,酸浸场土壤和燃烧残渣中 PCDD/Fs 的干重浓度分别高达 12 500~89 800 pg/g(203~1100 pg WHO-TEQ/g)和 13 500~25 300 pg/g(84.3~174 pg WHO-TEQ/g)[15]。2007 年广东某电子垃圾露天焚烧场地土壤中 Cd、Cu、Pb 和 Zn 的浓度分别为 17.1 mg/kg、11 140 mg/kg、4500 mg/kg 和 3690 mg/kg,

表明电路板或其他金属芯片的燃烧排放已造成周边土壤及附近农田的重金属污染。重金属在土壤中具有不同的形态分布，残渣态和碳酸盐可交换态的比例较高。可交换态重金属组分在电子垃圾污染土壤中所占比例相对较低，而在稻田和其他农业土壤中的比例较高，从而对水环境和生物圈等生态系统构成威胁。例如，Ni、Cd和Zn等金属的可交换态组分远远高于其他金属组分，从而这些金属的迁移率远远高于其他金属，对环境的威胁也更高[16,17]。

2. 对水生生态系统的影响

废水是电子垃圾污染水生系统的主要输送途径之一。研究表明，电子垃圾拆解场地的水，其pH值、总溶解固体、硬度、氯化物和电导率等参数指标均存在超标现象。电子垃圾中PCDD/Fs、PAHs、PCBs、PBBs等持久性有机物和重金属的存在与电子垃圾渗滤液含有的几种基因毒性和诱变物质直接相关，从而使水生生物处于危险之中[18]。除了电子垃圾渗滤液对水生生态系统的直接污染外，电子垃圾回收或倾倒场所排放的灰尘颗粒沉积也会污染地表水[12]。早期酸洗回收电路板中贵金属的拆解场所主要位于河流或池塘附近，因为回收工艺需要水且产生的废水便于向河流排放，导致拆解场地附近的地表水和沉积物中发现较高浓度的重金属。广东南阳河是我国最早受到电子垃圾拆解回收过程污染的河流之一，其沉积物中PBDEs的浓度高达16 000 ng/g[19]，且河流的鲤鱼也有766 ng/g的PBDEs积累[20]。在印度德里电子垃圾拆解场地附近的溪流中发现了Pb、Ag、Cr和Se等重金属，其金属浓度是饮用水的4倍或更多，其Cr、Cu、Cd、Fe、Pb、Zn和Al浓度分别为0.60 mg/L、0.70 mg/L、0.05 mg/L、0.46 mg/L、0.04 mg/L、1.89 mg/L和3.67 mg/L，明显高于距离电子垃圾拆解场地500 m居民区水样的0.02 mg/L、0.05 mg/L、0.002 mg/L、0.32 mg/L、0.002 mg/L、1.46 mg/L和0.06 mg/L[14]。上述研究证实了电子垃圾落后处理方式导致了水生生态系统的污染。

3. 对大气的影响

电子垃圾的露天焚烧、电路板加热烤板等早期落后处理技术是向大气排放各种污染物的主要原因，而拆解场地将作为点源向大气中排放重金属、颗粒物、二噁英和其他毒害物质[21,22]。大气中污染物的类型与该地区所从事的电子垃圾拆解工作类型有一定关系。例如，从事废电路板拆解回收区域倾向于向大气中释放大量的铅和铜，而电池车间大气中检测到了高浓度的铅和镉[23]。2004年，对我国广东某电子垃圾拆解场地附近的大气颗粒物进行测试分析，得出总悬浮颗粒物（TSP）中铬、锌、铜、铅、锰和砷的浓度分别为1.161 $\mu g/m^3$、1.038 $\mu g/m^3$、0.483 $\mu g/m^3$、0.444 $\mu g/m^3$、0.0606 $\mu g/m^3$和0.010 $\mu g/m^3$，这些重金属大多具有致癌性[24]。随着电子垃圾的燃烧，会产生浓烟，而重金属和其他化合物被排放到大气中或附着在颗粒物中，在空气中停留时间较长，并运输传送到其他区域，通过干湿沉降作用而沉积在土壤、地表水、植物等表面上，从而对周边区域生态环境造成影响。例如，随着大气中重金属在植物上沉积，可能发生叶片吸收，研究已证实铅进入炼锌厂附近玉米叶片和籽粒组织的主要途径是叶片对大气铅的吸收[25]。另外，在电子垃圾回收处理或倾倒区附近生长的植物可食用叶状部位的金属浓度高于可食用根状部位[16]。

1.3.3 电子垃圾对人体健康的影响

电子垃圾的不恰当处理是对人体健康造成伤害的主要原因，特别是对从事电子垃圾拆解回收的从业人员及周边人群。由于大多数非正规拆解场地的工人不了解电子垃圾的相关知识和缺乏相应的防护措施，容易造成拆解工人对毒害物质的多途径接触暴露，包括不同的污染源、暴露途径、暴露时间等复杂的暴露场景，即电子垃圾含有的毒害物质的数量和类型，回收处理的方法和环保管理措施等，从业人员的年龄、性别、身体状况及保护措施等多种因素影响着电子垃圾回收排放污染物对人体的健康风险[26,27]。

人体接触毒害物质的暴露途径包括呼吸吸入、经口摄入和皮肤接触三种。重金属的主要途径之一是经口摄入的土壤农作物。我国早期非正规电子垃圾拆解场地大多数位于农村地区，而且这些农村地区存在一定规模的农作物生产，如水稻大米等。大米是包括中国在内的大多数亚洲人的主食，这可能是某些特定地区人群重金属摄入的主要来源。Fu 等[17]的研究表明，从电子垃圾拆解场地附近农田采集的大米样品，Hg、Cd 和 Pb 的重金属含量分别超过国家标准的 15.3%、31%和 100%。动物类食物也是人体污染负荷的重要来源，研究发现淡水鱼、海鲜、肉类和鸡蛋等食品中 PBDEs 的浓度较高[8,21]，且非食草性鱼类含有更高水平的 PBDEs。由于 PBDEs 等持久性有机污染物可以在动植物体内富集积累较长时间，当作为食物被人体摄入后，导致有机污染物进入人体，并在人体组织中积累[28]。对广东某电子垃圾拆解地的 9 种食物中 PBDEs 的含量进行检测分析，得到 PBDEs 的估计每日摄入量为 931 ng/(kg·d)，是美国环保署膳食摄入 PBDEs 参考剂量（100 ng/(kg·d)）的 9 倍多[29]。对我国两个电子垃圾拆解区域进行暴露风险评价，得出浙江台州的鸭蛋和广东清远的鸡蛋中 PBDEs 的估计日均摄入量分别为 104 ng/(kg·d) 和 200 ng/(kg·d) [30,31]。更令人担心的是，由于母亲摄入了受污染的食物，婴儿也将处于暴露风险中。在电子垃圾回收和倾倒地区对母乳喂养的母亲进行的研究表明，作为 6 个月婴儿唯一食物的母乳中 PBDEs 的估计日均摄入量为 572 ng/(kg·d)，比对照区高 57 倍[32]。除饮食摄入外，呼吸吸入途径也是电子垃圾污染物污染人体的主要原因之一，特别是对拆解工人而言。在我国进行的一项研究表明，通过呼吸吸入的 PCBs 的平均浓度比对照区高 5 倍[28]。当含毒害物质的颗粒物沉积在人体呼吸系统时，可产生急性和慢性毒理学影响[33]。对电子垃圾的拆解工人来说，其 PCDD/Fs 的每日估计摄入量为 3.43 pg TEQ/(kg·d)，包括饮食摄入量为 0.53 pg TEQ/(kg·d)，呼吸吸入估计摄入量为 2.54 pg TEQ/(kg·d)，通过土壤/灰尘摄入和皮肤接触的摄入量为 0.363 pg TEQ/(kg·d)，即非饮食摄入对 PCDD/Fs 每日总摄入量的贡献为 85%，高于普通人群的 30%[27]。

人体不同身体部位的污染物浓度水平可以作为人体暴露评价的重要指标[34-36]，可以一定程度显示污染物对人体的暴露负荷水平及健康风险。据报道，婴儿出生后收集的胎盘含有较高浓度的污染物，如 Cd、Pb、Ni、PCDD/Fs 和 PBDEs[37]。例如，从我国广东某电子垃圾拆解地采集的胎盘中铅的含量为 301.4 ng/g，是对照地 (165.8 ng/g)的两倍。在我国台州收集的胎盘中 PBDEs 浓度是对照地点的 19 倍(分别为 19.5 ng/g 和 1.02 ng/g)[32]。这些毒害物质积累的最主要原因可能是母亲在怀孕期间和怀孕前参与了电子垃圾回收活动，或者居住在电子垃圾拆解场地附近。研究发现电子垃圾拆解场地区域儿童血铅水平升高，且随着年龄的增长，血铅水平也在增加，对年龄较大的儿童造成的健康负担比年龄较小的儿童更大[38]。除了重金属，PCBs、PBDEs 也在不同的人群血液中被检出，且从事电子垃圾回收作业的职业人群面临更大的危险[39]。除了血液，人体尿液中污染物的浓度也会升高，Zhang 等[40]研究表明，居住在电子垃圾拆解场地附近居民的尿液中积累了双酚类物质，这与人体内氧化应激升高有关。人的头发可以一定程度指示污染物在人体的污染水平[32,41]。研究发现，从电子垃圾回收中心和附近收集的工人和居民的头发样本中含有 Cd、Cu、Ni、Cr、Mn、As、PBDEs 和 PCDD/Fs 等，其浓度高于对照组人群[42,43]。儿童和新生儿是接触电子垃圾而对身体负担最敏感的群体，因为他们有多种摄入途径，如通过母乳和胎盘接触、手接触口腔活动和呼吸吸入相对较大的空气量和较低的毒害物质消除率。

1.3.4 我国电子垃圾的污染特征

早在 2002 年，美国的巴塞尔行动网络(Basel Action Network，BAN)和硅谷防止有毒物质联盟(Silicon Valley Toxics Coalition，SVTC)联合发表了一份调研报告——《输出危害：流向亚洲的高科技垃圾》("Exporting Harm — The High-Tech Trashing in Asia")[2]，对我国广东某地的电子垃圾拆解活动及污染现状进行了实地调研，并拍摄了令人触目惊心的电子垃圾拆解处理画面，包括露天焚烧产生滚滚黑烟、简易煤炉烘烤电路板、简易酸洗炼金，残渣、废酸等随意丢弃排放，当地居民包括儿童等暴露在受污染的环境中。报道显示，广东某地多年来的电子垃圾拆解回收技术原始落后，给当地造成严重的大气、土壤、水等环境污染，给当地居民尤其是拆解工人及儿童造成巨大的健康危害。由此揭开了国际社会对全球电子垃圾拆解之都的关注，并引起环境科学等领域的学者对该地进行大量的研究工作。结合国家对电子垃圾的管理进展，可以将我国电子垃圾处理过程分为4个不同阶段，每个阶段具有不同的特征[44]。

(1) 20世纪90年代至2008年，由于对电子垃圾的回收处理缺乏管理，大部分的电子垃圾在家庭小作坊进行手工拆解、露天焚烧、简易酸洗等落后方法处理，对电子垃圾拆解场地及周边环境的污染特征研究也证实，电子垃圾的非正规拆解处理造成了严重的环境污染[7]，也严重危害了拆解工人和当地居民的健康。在2004~2005年间，对电子垃圾拆解场地附近的环境样品进行采样分析得出：①沉积物。河流沉积物中重金属Cd(0~10.3 mg/kg)、Cu(17.0~4 540 mg/kg)、Ni(12.4~543 mg/kg)、Pb(28.6~590 mg/kg)、Zn(51.3~324 mg/kg)的浓度远高于其他地区[45]。②大气。大气中总悬浮颗粒物(TSP)和$PM_{2.5}$的质量浓度分别为124 $\mu g/m^3$和62.1 $\mu g/m^3$，其中22种PBDEs同系物的浓度分别为21.5 ng/m^3和16.6 ng/m^3，是其他地方大气中PBDEs浓度的58~691倍，这主要是由含PBDEs塑料等电子垃圾的加热处理或露天焚烧导致的[46]。大气中PCDD/Fs浓度和毒性当量(TEQ)值分别为64.9~2365 pg/m^3和0.909~48.9 pg W-TEQ $/m^3$，是当时世界上环境空气中发现的最高纪录值。大气中PBDD/Fs的浓度和毒性当量分别为8.124~61 pg/m^3和1.6~2104 pg I-TEQ$/m^3$，也处于高污染水平，呼吸风险评估得出该地居民的二噁英日暴露量远超世界卫生组织的摄入量限值[47]。③土壤。露天焚烧地区土壤样品中PBDEs(2906~44 473 ng/g)、PCDD/Fs(30 948~967 500 pg/g)和PAHs(820~3206 μg/kg)等有机污染物的干重浓度很高，其中PBDEs浓度是背景地区的7200倍，而PAHs以萘、菲、荧蒽为主[7,48,49]。

(2) 2008~2011年，电子垃圾规范化处理处置的初期探索阶段。2008年2月1日起施行的《电子废物污染环境防治管理办法》(国家环境保护总令第40号)对我国境内电子垃圾的拆解利用处置、相关方责任、罚则等环境管理内容进行了规定。国务院第551号令《废弃电器电子产品回收处理管理条例》自2011年1月1日起施行，从国家顶层设计方面规范了电子垃圾的回收处理活动。2009年6月至2011年期间，国家采用家电"以旧换新"的政策，极大促进了废电视机、电冰箱、洗衣机、空气调节器和微型计算机五类(简称"四机一脑")等大型废旧家电的淘汰收集，为有资质处理企业的规范拆解处理提供了大量拆解原料。同时，国家建立了一批电子垃圾定点拆解处理试点企业，尝试探索电子垃圾的标准化和规范化的处理技术和工艺。

(3) 2012年1月至2020年8月，随着《废弃电器电子产品回收管理条例》的颁布实施，我国逐步形成以废弃电器电子产品处理基金为核心内容的生产者责任延伸制度，完善了废弃电器电子产品回收处理相关的规划、资格许可、处理名录、基金征收与补贴等规范制度，并初步建成了符合我国国情的电子垃圾环境综合管理体系。2012年，国家设立废弃电器电子产品处理基金，向电器电子产品生产者征收费用，对有处理资格许可的企业开展规范回收处理予以补贴，简称基金制度。目前，针对"四机一脑"开展补贴。2016年，国家发展改革委等六部委发布《废弃电器电子产品处理目录（2014年版）》，将"四机一脑"等5类扩充到14类，包含了复印机、电热水器、吸油烟机、移动通信手持机等。2017年，原环境保护部联合其他五部委联合印发《关于联合开展电子废物、废轮胎、废塑料、废旧衣服、废家电拆解等再生利用行业清理整顿的通知》，集中解决了一系列电子垃圾回收处理等再生资源行业的突出环境问题，依法取缔一大批无经营资质、无环保措施的电子垃圾非法拆解作坊。以"电子垃圾拆解第一镇"为例，在中央和地方生态环境等相关职能部门的强力推进下，累计投入财政资金16.54亿元用于循环经济产业园的建设，实现了"家家拆解，户户冒烟；酸液排河，黑云蔽天"的散乱污作坊模式向"统一规划、统一建设、统一运营、统一治污、统一监管"的集约绿色发展模式转变。根据生态环境部废弃电器电子产品处理信息系统的相关数据，我国对"四机一脑"废旧家电的年实际拆解量由2012年的1010万台快速增长至2014年的7000万台，2015年至2020年稳步增长至年拆解量8000多万台。2012年至2020年，约6亿台"四机一脑"进入正规处理企业进行拆解处理，得到1456.27万吨拆解产物，包括CRT玻璃580.57万吨、塑料290.24万吨、铁及其合金264.26万吨、压缩机74.66万吨、保温材料46.85万吨、电动机45.15万吨、印刷电路板78.85万吨、铜及其合金22.19万吨，累计发放219亿元处理基金用于补贴处理企业[50]。

（4）2020年9月1日，修订后的《中华人民共和国固体废物污染环境防治法》开始生效，其中第六十六条、六十七条对我国电子垃圾相关的管理工作提出新的要求。2020年底，生态环境部、商务部、发展改革委、海关总署联合发布《关于全面禁止进口固体废物有关事项的公告》，公告要求自2021年1月1日起，禁止以任何方式进口固体废物，禁止我国境外的固体废物进境倾倒、堆放、处置。但是，我国仍然是电子信息及家电产品的产销大国，工业和信息化部的数据显示[51,52]，全国2021年共生产了移动通信手持机166151.6万台、微型计算机设备46692万台、集成电路3594.3亿块、彩色电视机18496.5万台、电冰箱8992.1万台、家用洗衣机8618.5万台及房间空气调节器21835.7万台。由此可见，虽然我国从法律层面杜绝了各类电子垃圾等固体废物的入境问题，可以一定程度减少跨境电子垃圾的处理量，但我国作为世界上最大的电器电子产品生产国和消费国，仍将面临国内大量电器电子产品的淘汰更新、处理处置的问题。

（安太成　郭　杰　马盛韬）

参 考 文 献

[1] Ogunseitan O A, Schoenung J M, Saphores J D M, et al. The electronics revolution: From e-wonderland to e-wasteland [J]. Science, 2009, 326(5953): 670-671.
[2] Puckett J, Smith T. Exporting harm: The high-tech trashing of Asia. The Basel Action Network. Seattle: Silicon Valley Toxics Coalition; 2002[R].
[3] Wang Z H, Zhang B, Guan D B. Take responsibility for electronic-waste disposal [J]. Nature, 2016, 536(7614): 23-25.
[4] Forti V, Baldé C P, Kuehr R, et al. The Global E-waste Monitor 2020: Quantities, flows and the circular economy potential. United Nations University (UNU)/United Nations Institute for Training and Research (UNITAR) – co-hosted SCYCLE Programme, International Telecommunication Union (ITU) & International Solid Waste Association (ISWA), Bonn/Geneva/Rotterdam[R].
[5] Baldé C P, D'Angelo E, Luda V, et al. Global transboundary e-waste flows monitor — 2022, United Nations Institute for Training and Research (UNITAR), Bonn, Germany[R].
[6] Rautela R, Arya S, Vishwakarma S, et al. E-waste management and its effects on the environment and human health [J]. Science of the Total Environment, 2021, 773: 145623.
[7] 彭平安, 盛国英, 傅家谟. 电子垃圾的污染问题 [J]. 化学进展, 2009, 21(Z1): 550-557.
[8] Song Q B, Li J H. A systematic review of the human body burden of e-waste exposure in China [J]. Environment International, 2014, 68: 82-93.
[9] Orlins S, Guan D B. China's toxic informal e-waste recycling: local approaches to a global environmental problem [J]. Journal of Cleaner Production, 2016, 114: 71-80.
[10] 傅建捷, 王亚韡, 周麟佳, 等. 我国典型电子垃圾拆解地持久性有毒化学污染物污染现状 [J]. 化学进展, 2011, 23(8): 1755-1768.
[11] Bertram M, Graedel T E, Rechberger H, et al. The contemporary European copper cycle: Waste management subsystem [J]. Ecological Economics, 2002, 42(1-2): 43-57.
[12] Robinson B H. E-waste: An assessment of global production and environmental impacts [J]. Science of the Total Environment, 2009, 408(2): 183-191.
[13] 林文杰, 吴荣华, 郑泽纯, 等. 贵屿电子垃圾处理对河流底泥及土壤重金属污染 [J]. 生态环境学报, 2011, 20(1): 160-163.
[14] Pradhan J K, Kumar S. Informal e-waste recycling: Environmental risk assessment of heavy metal contamination in Mandoli industrial area, Delhi, India [J]. Environmental Science and Pollution Research, 2014, 21(13): 7913-7928.
[15] Leung A O W, Luksemburg W J, Wong A S, et al. Spatial distribution of polybrominated diphenyl ethers and polychlorinated dibenzo-p-dioxins and dibenzofurans in soil and combusted residue at Guiyu, an electronic waste recycling site in southeast China [J]. Environmental Science & Technology, 2007, 41(8): 2730-2737.
[16] Luo C, Liu C, Wang Y, et al. Heavy metal contamination in soils and vegetables near an e-waste processing site, South China [J]. Journal of Hazardous Materials, 2011, 186(1): 481-490.
[17] Fu J J, Zhou Q F, Liu J M, et al. High levels of heavy metals in rice (*Oryza sativa* L.) from a typical E-waste recycling area in southeast China and its potential risk to human health [J]. Chemosphere, 2008, 71(7): 1269-1275.
[18] Alabi A O, Bakare A A. Genetic damage induced by electronic waste leachates and contaminated underground water in two prokaryotic systems [J]. Toxicology Mechanisms and Methods, 2017, 27(9): 657-665.
[19] Luo Q, Wong M H, Cai Z W. Determination of polybrominated diphenyl ethers in freshwater fishes from a river polluted by e-wastes [J]. Talanta, 2007, 72(5): 1644-1649.

[20] Luo Q, Cai Z W, Wong M H. Polybrominated diphenyl ethers in fish and sediment from river polluted by electronic waste [J]. Science of the Total Environment, 2007, 383(1-3): 115-127.

[21] Chan J K Y, Wong M H. A review of environmental fate, body burdens, and human health risk assessment of PCDD/Fs at two typical electronic waste recycling sites in China [J]. Science of the Total Environment, 2013, 463: 1111-1123.

[22] 余莉萍, 李会茹, 孟祥周, 等. 电子垃圾焚烧排放的二噁英对周围大气环境的影响 [J]. 环境污染与防治, 2008, (2): 8-11+28.

[23] Brigden K, Labunska I, Santillo D, et al. Recycling of electronic wastes in China and India: Workplace and environmental contamination[R]. Greenpeace Research Laboratories, 2005.

[24] Deng W J, Louie P K K, Liu W K, et al. Atmospheric levels and cytotoxicity of PAHs and heavy metals in TSP and $PM_{2.5}$ at an electronic waste recycling site in southeast China [J]. Atmospheric Environment, 2006, 40(36): 6945-6955.

[25] Bi X Y, Feng X B, Yang Y G, et al. Allocation and source attribution of lead and cadmium in maize (*Zea mays* L.) impacted by smelting emissions [J]. Environmental Pollution, 2009, 157(3): 834-839.

[26] 孙朋, 于云江, 李定龙, 等. 电子垃圾对环境与健康的影响研究进展 [J]. 环境与健康杂志, 2008, (5): 452-455.

[27] Ma J, Kannan K, Cheng J, et al. Concentrations, profiles, and estimated human exposures for polychlorinated dibenzo-*p*-dioxins and dibenzofurans from electronic waste recycling facilities and a chemical industrial complex in Eastern China [J]. Environmental Science & Technology, 2008, 42(22): 8252-8259.

[28] Xing G H, Chan J K Y, Leung A O W, et al. Environmental impact and human exposure to PCBs in Guiyu, an electronic waste recycling site in China [J]. Environment International, 2009, 35(1): 76-82.

[29] Chan J K Y, Man Y B, Wu S C, et al. Dietary intake of PBDEs of residents at two major electronic waste recycling sites in China [J]. Science of the Total Environment, 2013, 463: 1138-1146.

[30] Labunska I, Harrad S, Santillo D, et al. Domestic duck eggs: An important pathway of human exposure to PBDEs around e-waste and scrap metal processing areas in Eastern China [J]. Environmental Science & Technology, 2013, 47(16): 9258-9266.

[31] Zheng X B, Wu J P, Luo X J, et al. Halogenated flame retardants in home-produced eggs from an electronic waste recycling region in South China: Levels, composition profiles, and human dietary exposure assessment [J]. Environment International, 2012, 45: 122-128.

[32] Leung A O W, Chan J K Y, Xing G H, et al. Body burdens of polybrominated diphenyl ethers in childbearing-aged women at an intensive electronic-waste recycling site in China [J]. Environmental Science and Pollution Research, 2010, 17(7): 1300-1313.

[33] Zheng X B, Xu X J, Yekeen T A, et al. Ambient air heavy metals in $PM_{2.5}$ and potential human health risk assessment in an informal electronic-waste recycling site of China [J]. Aerosol and Air Quality Research, 2016, 16(2): 388-397.

[34] Guo Y Y, Huo X, Li Y, et al. Monitoring of lead, cadmium, chromium and nickel in placenta from an e-waste recycling town in China [J]. Science of the Total Environment, 2010, 408(16): 3113-3117.

[35] Ni W Q, Huang Y, Wang X L, et al. Associations of neonatal lead, cadmium, chromium and nickel co-exposure with DNA oxidative damage in an electronic waste recycling town [J]. Science of the Total Environment, 2014, 472: 354-362.

[36] Zhao Y X, Ruan X L, Li Y Y, et al. Polybrominated diphenyl ethers (PBDEs) in aborted human fetuses and placental transfer during the first trimester of pregnancy [J]. Environmental Science & Technology, 2013, 47(11): 5939-5946.

[37] Zhang X Y, Ruan X L, Yan M C, et al. Polybrominated diphenyl ether (PBDE) in blood from children (age 9-12) in Taizhou, China [J]. Journal of Environmental Sciences, 2011, 23(7): 1199-1204.

[38] Zheng L K, Wu K S, Li Y, et al. Blood lead and cadmium levels and relevant factors among children from an

e-waste recycling town in China [J]. Environmental Research, 2008, 108(1): 15-20.
[39] Ren G F, Yu Z Q, Ma S T, et al. Determination of dechlorane plus in serum from electronics dismantling workers in south China [J]. Environmental Science & Technology, 2009, 43(24): 9453-9457.
[40] Zhang T, Xue J C, Gao C Z, et al. Urinary concentrations of bisphenols and their association with biomarkers of oxidative stress in people living near e-waste recycling facilities in China [J]. Environmental Science & Technology, 2016, 50(7): 4045-4053.
[41] Zheng J, Luo X J, Yuan J G, et al. Heavy metals in hair of residents in an e-waste recycling area, south China: Contents and assessment of bodily state [J]. Archives of Environmental Contamination and Toxicology, 2011, 61(4): 696-703.
[42] Zheng J, Luo X J, Yuan J G, et al. Levels and sources of brominated flame retardants in human hair from urban, e-waste, and rural areas in South China [J]. Environmental Pollution, 2011, 159(12): 3706-3713.
[43] Ma J, Cheng J P, Wang W H, et al. Elevated concentrations of polychlorinated dibenzo-p-dioxins and polychlorinated dibenzofurans and polybrominated diphenyl ethers in hair from workers at an electronic waste recycling facility in Eastern China [J]. Journal of Hazardous Materials, 2011, 186(2-3): 1966-1971.
[44] Zeng X, Li J, Stevels A L N, et al. Perspective of electronic waste management in China based on a legislation comparison between China and the EU [J]. Journal of Cleaner Production, 2013, 51: 80-87.
[45] Wong C S C, Wu S C, Duzgoren-Aydin N S, et al. Trace metal contamination of sediments in an e-waste processing village in China [J]. Environmental Pollution, 2007, 145(2): 434-442.
[46] Deng W J, Zheng J S, Bi X H, et al. Distribution of PBDEs in air particles from an electronic waste recycling site compared with Guangzhou and Hong Kong, South China [J]. Environment International, 2007, 33(8): 1063-1069.
[47] Li H R, Yu L P, Sheng G Y, et al. Severe PCDD/F and PBDQ/F pollution in air around an electronic waste dismantling area in China [J]. Environmental Science & Technology, 2007, 41(16): 5641-5646.
[48] Wong M H, Wu S C, Deng W J, et al. Export of toxic chemicals—A review of the case of uncontrolled electronic-waste recycling [J]. Environmental Pollution, 2007, 149(2): 131-140.
[49] Yu X Z, Gao Y, Wu S C, et al. Distribution of polycyclic aromatic hydrocarbons in soils at Guiyu area of China, affected by recycling of electronic waste using primitive technologies [J]. Chemosphere, 2006, 65(9): 1500-1509.
[50] 中国电子废物环境综合管理(2012—2021) [R]. 生态环境部固体废物与化学品管理技术中心, 2021.
[51] 工业和信息化部运行监测协调局. 2021年电子信息制造业年度统计数据. https://www.miit.gov.cn/dznj2021/hlw.html [Z].
[52] 国家统计局. 2021年家电行业生产情况. https://www.miit.gov.cn/gxsj/tjfx/xfpgy/jd/art/2022/art_47407bc485a64efbbc8a37fbb0542508.html [Z].

第 2 章 电子垃圾拆解过程重金属污染

电子垃圾中含有的重金属组分是一类重要的环境污染物。电子垃圾的拆解回收处理，主要是通过手工拆解、机械破碎、加热处理等方法对其中的有用金属材料进行回收及资源化，在此过程中会不可避免地存在重金属对环境介质的释放，尤其是非正规拆解处理过程所导致的重金属污染更为严重，而受重金属污染的环境介质进一步对相关暴露人群造成健康风险。本章主要阐述关于电子垃圾拆解过程中所排放重金属的时空污染特征、典型重金属的来源解析及其在环境各介质中的归趋，进而探讨电子垃圾拆解处理所排放重金属的主要暴露途径和暴露结局。

2.1 重金属污染概述

重金属广泛用于各种电子产品的制造，例如电路板中的铅和镉、计算机电池中的镉、电线中的铜等等。早期非正规拆解所使用的工艺，如强酸酸洗和露天焚烧等粗犷、原始的回收操作过程所产生的废气、废液和固体废物往往未得到妥善处理，而是直接排放并进入到周围的大气、水体、土壤等环境介质中[1]。这些电子垃圾拆解操作产生的重金属会在环境中持续存在，重金属能够扩散到空气中，沉积在表土上，最后通过雨水渗透到底土和地下水中[2]，通过生物积累和食物链的传递，在低浓度下就导致生物机体中毒[3]。以往的研究表明，电子垃圾拆解活动导致土壤、大气和灰尘受到重金属铅、镉、铬、铜等的污染[4]。根据文献报道，当前使用的规范化处理电子垃圾的工艺如下：小口径和大口径电吹风机用于剥离手机或者电脑硬盘元器件、电热炉拆解电视机、旋转焚烧炉拆解电视机和微型电机等、电烙铁剥离电脑主板元器件、机械切割电脑主板等。尽管这些工艺所产生的污染物的排放减少了，但其无组织排放的粉尘、废气依然危害着拆解工人及周边居民的身体健康[5]。

2.2 电子垃圾中的重金属

在没有规范管控下，人们为了回收电子垃圾中的贵金属，采用了一些原始的回收方法：从电子垃圾中手动拆卸印刷电路板，使用电吹风/蜂窝炉/电热板等工具加热熔融焊料，从线路板上取下贵金属含量较高的电子元器件，通过使用王水浸泡以提炼贵金属，剩余的裸线路板则使用蜂窝煤炉焙烤以去除线路板的塑料后用于炼铜等。在这些过程中，废气、废水向环境中无组织地排放。随着电子垃圾的逐渐规范化拆解处理，如使用尾气收集和处理装置，使用电热板代替蜂窝炉，增设回收塑料的企业，在园区内建设污水处理设备等，这些措施可以减轻污染的无组织排放。

印刷电路板是所有电气电子设备系统的重要组成部分，它们通常含有各种金属，如铜、铁、铅、锌、金、银、铂等。然而，其他功能性但有毒的金属也存在于印刷电路板中，Rai等[6]整理了文献报道的印刷电路板金属含量，如表2-1所示。

表 2-1 印刷电路板中金属的种类及含量(w/w) [6]

金属	均值(%)	误差	金属	均值(‰)	误差
铜	20.0	5.8	银	1.239	1.654
铝	3.2	1.7	金	0.344	0.377
铅	2.1	0.8	镉	1.183	N/A
锌	1.0	0.9	钾	0.180	N/A
镍	1.0	1.4	钢	0.500	N/A
铁	3.7	3.0	锰	4.594	4.837
锡	3.1	1.4	硒	0.021	N/A
锑	0.05	0.007	砷	0.011	N/A
铬	0.18	0.2	镁	0.750	0.354
钠	0.48	N/A	钯	0.101	0.103
钙	1.4	0.2	钴	0.350	0.071
			钛	0.400	N/A

注：w/w表示质量分数

其中，有毒有害的金属因无经济价值，也没有特殊的工艺去回收，这些金属在电子垃圾中的分布如下：铅分布在电路板、阴极射线管、焊料和电池中，铅在典型个人电脑的占比为 6.3%(w/w)，在显示器玻璃（用于辐射屏蔽）中占20%(w/w)，在印刷线路板上的锡铅焊料占 37%(w/w)[7]；在铅蓄电池中，铅作为电极材料含量尤其高[8]。镉存在于电阻器、红外线发生器、半导体、充电镍镉电池等，同时作为塑料的固化剂，被应用于作为某一些开关元件和电子产品的焊接点，以及旧式阴极射线管中的荧光粉涂层中[9]。虽然电子垃圾中的铜大部分被回收，但由于其含量非常高，随着电子垃圾拆解也有部分被释放到环境中。铜在电子垃圾中主要存在于电真空器件如高频和超高频发射管、渡导管、磁控管等，作为印刷电路基板导电材料等[7]。汞主要用于电子设备中的荧光灯、继电器和温度开关[10]。汞和砷还存在于液晶显示器中[11,12]。六价铬化合物则常用于抑制金属外壳腐蚀，镍是不锈钢等合金的主要成分[13,14]。锑用于半导体元件和电子设备内电路板的阻燃剂[15]。

不过，尽管印刷电路板被认为是电子垃圾处理过程中重金属排放的主要来源，但其他类型的电子垃圾，如从电子垃圾中分离出来的塑料，也含有一定量的重金属和其他有毒物质[16]。Dimitrakakis 等[17]分析了电子垃圾的塑料中重金属的含量，铅为 17.4 mg/kg，镉为 5.7 mg/kg，铬为 8.4 mg/kg，表明塑料也是整个电子垃圾中产生重金属污染的重要组件。此外，在阴极射线管禁用后，新型的平板电脑、液晶电视等也含有大量的重金属元素，在回收过程中也可能进一步产生新的重金属污染问题，如铍、钒、钼等鲜有被报道的金属[18]。

2.3 电子垃圾拆解区重金属的环境分布

2.3.1 大气颗粒物

大气颗粒物是电子垃圾拆解地区的大气重金属的主要载体，非正规的电子垃圾拆解工艺向环境释放重金属。由于早前分析技术的问题，在大气中研究的重金属元素数量少、缺乏对毒性较大的痕量元素的定量分析，如砷、锑、镍等。后来随着电感耦合等离子体质谱仪和光谱仪的普遍使用，同一大气样品中大多数元素可以被同时分析，获得更多金属元素的污染特征等方面的信息。

表 2-2 总结了我国几个典型的电子垃圾拆解和回收区大气颗粒物重金属的污染水平。通过对比不同地区及不同年份的数据，发现电子垃圾拆解区大气中重金属具有明显的时空差异性。在从 2004 年到 2016 年的相关研究中，发现大气颗粒物上毒害重金属元素主要包含十种，分别为镉、铜、镍、铅、锌、砷、锰、锑、铬、铁。不同地区的电子垃圾拆解区的大气中不同重金属污染水平和特征有差异，其中在 2007 年得到的广东某电子垃圾拆解地的大气重金属含量非常高，特别是镉、镍、铅和锌的污染水平远大于其他研究地区，其可能的原因是这个采样点设置在"蜂窝煤炉烧废旧电路板"车间内。

表 2-2 不同省市的电子垃圾拆解和回收区大气颗粒物重金属污染水平 (ng/m³)

地点	采样时间	粒径	镉	铜	镍	铅	锌	砷	锰	锑	铬	铁	文献
	2004 年	PM$_{2.5}$	7.3	126	7.19	392	924	6.04	25.4	—	1152		[19]
		TSP	7.3	483	9.93	444	1038	10.2	60.6	—	1161		
汕头市	2005 年	TSP	3.7	274		63.1	397		83.3	—		3682	[20]
	2007 年	TSP	80	570	80	4420	3320			150	—		[21]
	2013 年	PM$_{2.5}$	5.58			152.96			22.10		6.49		[22]

续表

地点	采样时间	粒径	镉	铜	镍	铅	锌	砷	锰	锑	铬	铁	文献
清远市	2012 年	TSP	6.1	95	11	170	880	22	76	15	23	—	[23]
	2015 年	PM$_{10}$	1.10	32.1	4.28	32.9	79.5	—	—	—	—	—	[24]
江苏省	2011 年	TSP	28	1220		1400							[25]
兰州市	2016 年	PM$_{1.0}$	8	89	28	157	890				31		[26]
		PM$_{2.5}$	12	103	29	196	1174	—			38	—	
		PM$_{10}$	12	110	38	240	1245				41		

由于在 2016 年前，广东某地主要是分散的家庭作坊式拆解电子垃圾，因此可能会造成不同的采样点不一样的结果。根据 Deng 等[19]在该地区 2004 年的研究结果，铬（1.161 μg/m³）和锌（1.038 μg/m³）在 TSP 中含量较高，其次是铜（0.483 μg/m³）、铅（0.444 μg/m³）、锰（0.0606 μg/m³）和砷（0.010 μg/m³），这些金属在 PM$_{2.5}$ 中也显示出同样的趋势。除了锰（44%）和铜（25%）之外，所有金属都在细颗粒部分中占主导地位（大于 78%的质量浓度分布在 PM$_{2.5}$）。与其他非电子垃圾拆解区 TSP 中重金属对比，铬、铜、锌的浓度高出 4~33 倍。而在 2007 年，对我国广东某地的一家进行废印刷电路板回收利用的典型车间的空气颗粒样品进行了化学成分分析，结果发现无机元素以磷为主，其次是地壳元素和金属元素铅、锌、锡，再次为铜、锑、锰、镍、钡、镉。具体来说，相比 2004 年在室外采集到的大气颗粒物上的重金属浓度，废印刷电路板车间内 TSP 上的镉、镍和铅的浓度高十倍。因而，印刷电路板的回收被认为是广东省汕头市电子垃圾拆解区重金属污染的重要因素，尤其是镉、铅和镍，这种电子垃圾的非正规回收方法是电子垃圾回收场所周围大气中颗粒物的重要来源[21]。

Xue 等[25]从江苏某正规的废电路板回收车间收集空气样本，发现该车间 TSP 中重金属（铜除外）含量均低于广东某非正规拆解车间。曹红梅[26]对甘肃省兰州市一家规模化电子垃圾拆解厂空气重金属的污染特征进行研究，发现拆解车间外重金属在不同粒径颗粒物中的含量有着较大差异，重金属在 PM$_{10}$ 和 PM$_{1.0}$ 中含量分别最高和最低，说明环境空气的重金属在车间外趋于吸附在较粗的 PM$_{10}$ 上，不过这可能是自然扬尘和车间内向外释放重金属的共同结果。黄春莉[23]对广东省清远市的电子垃圾拆解区 TSP 的重金属进行研究，发现该场地 TSP 中的砷浓度尤其高。此外，在此场地开展了不同采样高度中分段粒径颗粒物的重金属污染研究，如表 2-3 所示。发现 15 种颗粒结合态重金属的总浓度范围为 670~5200 ng/m³，平均值和中值分别为 2800 ng/m³ 和 2200 ng/m³。在所有目标金属中，铁含量最高，平均浓度为 1400 ng/m³，而钴含量最低[27]。此外，清远地区铜、铅、铬、镉的平均浓度(95 ng/m³、170 ng/m³、23 ng/m³、6.1 ng/m³)远低于广东某电子垃圾拆解地三层楼房顶上大气颗粒物中重金属浓度(483 ng/m³、444 ng/m³、1161 ng/m³、7.3 ng/m³)[28]。这一结论部分归因于 MOUDI 采样器的颗粒捕获效率低于大流量空气采样器。但是距离地面不同高度大气颗粒态重金属的浓度并不存在显著差异，这表明从电子垃圾拆解作坊排放出来的重金属在近地面处就已经混合均匀，然后迁移至高处，同时也说明大气颗粒物的重金属的迁移扩散能力较强，若遇到降雨沉降等情况，也会间接导致周边土壤受到重金属污染。

表 2-3 电子垃圾拆解区距离地面不同高度大气颗粒态重金属污染水平 (ng/m³)[27]

元素	不同高度的浓度：平均值(范围)			不同高度浓度的比值		
	1.5 m	5 m	20 m	1.5∶5	1.5∶20	5∶20
钒	11 (1.3~24)	12 (1.4~25)	12 (1.4~24)	0.972	0.934	0.965
铬	21 (12~38)	23 (2.2~40)	25 (11~42)	0.879	0.707	0.852
锰	74 (20~130)	73 (21~150)	81 (30~160)	0.979	0.838	0.839
铁	1500 (320~2200)	1200 (460~2300)	1400 (590~2500)	0.654	0.855	0.791
钴	0.83 (0.25~1.6)	0.92 (0.18~1.8)	0.91 (0.26~1.9)	0.836	0.855	0.986
镍	10 (3.5~16)	12 (4.6~18)	12 (6.0~19)	0.590	0.576	0.983
铜	80 (9.4~200)	90 (13~240)	110 (21~230)	0.867	0.560	0.705
锌	820 (210~1400)	870 (160~1500)	940 (380~1600)	0.884	0.729	0.845
砷	22 (14~30)	20 (9.2~33)	24 (16~34)	0.656	0.669	0.426
硒	11 (3.8~1.6)	10 (3.9~16)	11 (4.2~15)	0.827	0.931	0.876
镉	5.8 (2.5~12)	5.8 (1.5~13)	6.8 (2.9~14)	0.988	0.702	0.733
铅	150 (45~240)	160 (48-260)	190 (73~280)	0.976	0.504	0.572
钛	130 (21~230)	140 (41~250)	140 (42~270)	0.876	0.770	0.900
钼	2.7 (0.76~5.0)	3.2 (1.1~5.2)	3.8 (1.7~6.9)	0.737	0.427	0.633
锑	13 (4.9~27)	14 (3.1~30)	17 (5.8~33)	0.903	0.547	0.642
总和	2800 (670~4200)	2700 (780~4700)	3000 (1200~5200)	0.876	0.896	0.790

2.3.2 水和沉积物

电子垃圾拆解对水体的污染包括水体中自身的重金属污染和沉积物的重金属污染。金属污染物经过各种途径进入河流等水体环境，会沉积下来并富集在沉积物中。当外界环境条件发生改变时，容易引起水体或沉积物的二次污染。因此，同时研究水体中和沉积物的重金属污染水平可更准确地判断重金属的近期或累积污染程度。

非正规电子垃圾拆解使用强酸浸泡含贵金属的芯片后，废水直接排往开放水域当中，造成当地池塘、湖泊、河流等水体重金属污染[29,30]。2006 年，广东省汕头的电子垃圾拆解区处于练江中游，Wong 等[31]在 2006 年采集练江两个支流水体样品并对其中的溶解重金属浓度进行分析(原文标记为练江和南阳河)，结果显示这两个水体中的溶解性金属浓度明显高于该地区的水库。其中，在靠近电子垃圾拆解区的支流水体(南阳河靠近酸浸提点)含有明显浓度升高的溶解性银、铍、镉、钴、铜、镍、铅，而和另一离拆解区稍远(南阳河)的水体则含有溶解性砷、铬、锂、钼、锑和硒，结果如表 2-4 所示。这些金属的时间分布在练江和南阳河之间也存在显著差异，说明两条支流中溶解金属的水平受不同因素和来源的控制，并且这些金属的时间变化可能受到河流环境中新的金属输入影响，因为溶解的金属通常反映最近的输入。

表 2-4 电子垃圾拆解区两个支流水体溶解性重金属的水平[31]

元素	单位	练江(n=8)		南阳河(n=8)		南阳河靠近酸浸提(n=2)	水库(n=2)
		均值±标准差	范围	均值±标准差	范围	范围	范围
铝	mg/L	0.052 ± 0.008	0.033~0.058	0.510 ± 0.067	0.382~0.580	0.476~0.519	0.459~0.527
钙	mg/L	20.7 ± 0.6	20.1~21.8	18.7 ± 0.4	18.0~19.3	18.4~185	2.43~2.50
铁	mg/L	0.100 ± 0.005	0.092~0.106	0.318 ± 0.044	0.235~0.358	0.297~0.322	0.126~0.182
镁	mg/L	5.28 ± 0.45	4.69~5.80	2.65 ± 013	2.52~2.91	2.75~2.76	0.653~0.683
锰	mg/L	0.225 ± 0.014	0.206~0.246	0.199 ± 0.010	0.188~0.217	0.216~0.217	0.002~0.006
锶	mg/L	0.182 ± 0.001	0.181~0.184	0.196 ± 0.003	0.193~0.200	0.198~0.199	0.041~0.042
银	µg/L	0.030 ± 0.004	0.024~0.036	0.161 ± 0.097	0.059~0.303	0.335~0.354	0.006~0.014
砷	µg/L	6.46 ± 0.42	5.95~7.12	4.74 ± 0.43	4.28~5.51	4 36~4 80	0.939~1.13
铍	µg/L	0.025 ± 0.004	0.020~0.031	0.134 ± 0.019	0.109~0.162	0.159~0.165	0.019~0.029
镉	µg/L	0.091 ± 0.010	0.073~0.101	0.315 ± 0.032	0.260~0.362	0.540~0.554	＜0.002
钴	µg/L	0.860 ± 0.096	0.744~1.00	3.62 ± 0.82	2.79~4.94	4.98~5.08	0.033~0.048
铬	µg/L	2.20 ± 0.49	1.65~2.85	1.21 ± 0.10	1.07~1.35	1.08~1.14	0.572~0.616
铜	µg/L	7.80 ± 1.70	5.92~9.97	50.8 ± 10.0	39.7~67.3	85.5~89.7	0.460~0.728
锂	µg/L	7.30 ± 0.30	7.05~8.00	3.22 ± 0.09	3.11~3.33	3.38~3.44	0.497~0.613
钼	µg/L	3.26 ± 0.15	3.04~3.45	1.75 ± 0.40	1.39~2.27	2.32~2.35	0.247~0.422
镍	µg/L	36.6 ± 6.2	29.8~47.9	52.4 ± 7.6	43.3~66.0	92.0~94.0	0.849~1.23
铅	µg/L	1.48 ± 0.09	1.35~1.60	1.81 ± 0.30	1.33~2.24	1.73~1.87	0.151~0.221
锑	µg/L	31.1 ± 1.7	28.7~33.1	18.7 ± 2.3	16.3~22.1	21.1~21.4	0.179~0.221

续表

元素	单位	练江(n=8)		南阳河(n=8)		南阳河靠近酸浸提(n=2)	水库(n=2)
		均值±标准差	范围	均值±标准差	范围	范围	范围
硒	μg/L	7.76 ± 0.43	7.07~8.36	2.32 ± 0.46	1.77~2.93	3.04~3.14	0.429~0.470
钛	μg/L	158 ± 9	148~176	161 ± 8	153~175	155~158	10.4~149
钒	μg/L	1.05 ± 0.06	0.991~1.18	1.26 ± 0.07	1.17~1.40	1.13	0.260~0.313
锌	μg/L	30.6 ± 4.2	24.0~343	106 ± 10	89.9~117	122~128	2.93~3.46

一般来说，与沉积物结合的金属相对稳定，表明金属在给定水生系统中的历史积累。Wong 等[32]在对练江沉积物金属含量进行了研究，发现镉(n.d.~10.3 mg/kg)、铜(17.0~4540 mg/kg)、镍(12.4~543 mg/kg)、铅(28.6~590 mg/kg)和锌(51.3~324 mg/kg)污染严重，已经导致电子垃圾拆解区当地的水生环境质量的严重下降。一项在 2018 年开展的练江(原非正规电子垃圾拆解点)沉积物研究发现，沉积物仍然受到银、镉、铜、铅、锑、锡和锌污染，结果如表 2-5 所示。除了电子垃圾拆解直接影响的水域，其支流和干流表层沉积物中金属均存在不同程度的富集。支流表层沉积物锑的含量范围为 13.9~937 mg/kg，其平均值是广东省土壤背景值的 383.5 倍；银、镉、铜、铅、锡和锌的平均值分别是背景值的 239.7、116.9、194.2、28.5、17.7 和 25.1 倍。干流表层沉积物锑的含量范围为 3.1~63.6 mg/kg，其平均值为广东省土壤背景值的 63.0 倍；银、镉、铜、铅、锡和锌的平均值分别为广东省土壤背景值的 19.9、21.8、15.6、3.1、2.6 和 9.3 倍。总体来看，支流表层沉积物中的金属污染比干流更为严重[33]。研究表明，早期非正规的电子垃圾拆解活动给当地的水土环境造成了严重且持久的金属污染问题。

表 2-5 电子垃圾拆解区附近河流表层沉积物中金属含量(mg/kg)[33]

元素	支流(T1~T13)		干流	
	平均值±标准差	范围	平均值±标准差	范围
银	23.73 ± 28.15	0.45~81	1.97 ± 4.28	0.29~13.35
镉	4.77 ± 4.83	0.32~16.95	0.89 ± 1.13	0.22~3.86
钴	16.02 ± 7.81	8.7~33.3	10.31 ± 4.29	7.6~21.6
铬	93.62 ± 54.49	40~215	67.44 ± 14.16	47~92
铜	2038.96 ± 2814.22	102~10000	163.30 ± 155.63	23.1~547
锰	615.08 ± 235.9	341~1160	445.33 ± 167.65	306~785
镍	194.35 ± 278.98	25~1080	40.24 ± 36.24	17.9~132
铅	849.22 ± 740.37	81.7~2240	92.41 ± 23.82	62.7~141.5

续表

元素	支流(T1~T13)		干流	
	平均值±标准差	范围	平均值±标准差	范围
锑	157.25 ± 263.18	13.9~937	25.82 ± 18.24	3.1~63.6
锡	318.40 ± 211.46	28.4~500	47.42 ± 29.99	16.1~118
铊	1.08 ± 0.26	0.78~1.59	1.40 ± 0.16	1.14~1.55
锌	911.15 ± 747.63	164~2100	337.11 ± 150.33	136~676

2.3.3 灰尘和土壤

人为活动产生的重金属污染可以向大气排放，沉降到地面形成灰尘，大气颗粒物沉降过程形成的灰尘是能够提供关于其污染水平、分布特征和存在于地球表面的环境污染物的最终目的地信息的显著环境介质[15,34,35]。灰尘可以表明近期通过大气沉降传输的重金属污染，而土壤则可以表明是经大气、灰尘、水共同传输长时间的重金属污染情况[36]。

Leung 等[37]于 2004 年在废旧电路板拆解车间内的粉尘中检测出高浓度的铅、铜、锌和镍污染，其中铅浓度在 22 900~206 000 mg/kg，以及镍是室内灰尘很典型的元素；Bi 等[15]于 2010 年在 13 个位于电子垃圾拆解区的村子居民房屋室内采集灰尘样品，分析发现所有村庄灰尘中的锑浓度(6.1~232 mg/kg)显著高于非电子垃圾拆解区，是非电子垃圾拆解区的 3.9~147 倍，表明电子垃圾拆解是灰尘锑污染的重要来源。Wu 等[38]于 2008 年在浙江省台州市的电子垃圾拆解区采集了室内粉尘，分析发现其中铅、铬、砷、镉的平均浓度分别为 399 mg/kg、151 mg/kg、48.13 mg/kg 和 5.85 mg/kg，室外粉尘中的平均浓度分别为 328 mg/kg、191 mg/kg、17.59 mg/kg 和 4.07 mg/kg，且随着离电子垃圾拆解点的距离越远，这些重金属的浓度越低，说明电子垃圾拆解是这些重金属的主要来源。He 等[39]对清远电子垃圾拆解区的 2013~2014 年室内灰尘分析，发现电子垃圾拆解区的镉、铅、铜和铬远高于其他非电子垃圾区，且灰尘的铅是电子垃圾拆解区最为特征的重金属污染。

Luo 等[2]于 2007 年在汕头市原电子垃圾焚烧场的土壤中检测出高浓度的镉、铜、铅和锌，平均值分别为 17.1 mg/kg、11 140 mg/kg、4500 mg/kg 和 3690 mg/kg，附近的稻田和菜园的土壤中镉和铜的浓度也相对较高。这还可能是使用受污染的水源对农田灌溉，与大气沉降共同造成土壤的重金属污染。我国浙江台州电子垃圾拆解区域周边土壤也受到多种重金属污染，其中汞、铜、镉、铅和锌是环境样品中含量较高的金属，汞是该地区最严重的金属污染物[40]。位于广东省清远市的电子垃圾拆解区，其中的四个拆解地土壤受到了重金属污染，特别是镉、铜、铅和锑的浓度分别比当地背景浓度高 2.83~2306、2.17~1880、0.96~1971 和 9.28~5607 倍。废弃场地土壤中的镉、铬、汞、铅和锑浓度显著低于酸浸、拆解和焚烧场地，通过其来源解析表明：不同地点的金属浓度差异主要是由不同的电子垃圾拆解工艺造成的，但是电子垃圾的非正规拆解是土壤重金属污染的重要来源[41]。对相关电子垃圾拆解区周边土壤和灰尘中重金属污染水平的文献进行归纳总结，见表 2-6。

表 2-6 电子垃圾拆解区周边土壤和灰尘中重金属的污染水平 (mg/kg)

地点	时间	采样点	铬	锰	镉	铜	镍	铅	锌	汞	文献
汕头市	2007 年	菜园土壤	9.66~19	—	0.26~1.17	210~450	7.04~10.3	73.3~134	92.4~142	—	[42]
		稻田土壤	10.5~24.1	—	0.04~1.43	40.1~260	10.8~66	48.1~97	62.1~252		
		焚化厂土壤	23.6~122	—	3.05~46.8	1500~21400	12.2~132	629~7720	682~8970	—	
汕头市	2013 年	倾倒场土壤	22.5±1.96	111±9.45	0.14±0.05	513±43.8	14.5±1.22	198±19.7	179±26.4		[43]
		燃烧现场土壤	30.9±3.27	123±19.9	0.52±0.05	1981±252	16.6±2.55	206±25.8	194±21.8		
		酸浸现场土壤	58.3±1.52	173±14.5	0.39±0.03	2981±32.0	135±16.6	131±7.88	316±30.9		
		稻田土壤	46.8±4.48	127±9.65	0.62±0.04	329±38.4	14.8±2.37	66.7±6.63	182±6.06		
		植被土壤	24.5±3.90	61.4±4.08	0.11±0.01	134±2.98	13.6±0.80	14.3±0.65	46.9±6.08		
清远市	2013~2014 年	电子垃圾拆解室内灰尘	134		40.4	1200		1380	212		[39]
台州市	2008 年	电子垃圾拆解厂灰尘	21.0~269.1	—	1.7~7.4	77.9~296.3	11.7~83.8	81.3~501.9	177.7~660.8	0.2~0.9	[2]
		金回收厂灰尘	88.6~91.3		0.6~9.6	272.3~576.7	207.4~120.4	102.8~184.4	143.8~263.0	0.4~3.2	
		简易家庭式作坊灰尘	65.9~125.9		2.8~7.9	222.5~1641.3	53.3~71.0	200.1~2374.1	221.2~518.7	1.7~654.1	

2.3.4 重金属在环境中的迁移转化

电子垃圾拆解活动排放的重金属对大气、水体、土壤等环境造成重金属污染，通过在不同介质之间的相互转化，使重金属在多种介质中发生迁移。大气中的金属污染物可以通过气体吸收、挥发和干湿沉降等作用在土壤、沉积物和植被中运输和积累；对于陆地生态系统，土壤是大气污染物的最初接收者，也是重金属污染物进入农业和野生动物食物链的主要途径；而表层土壤中的微量金属会通过径流和垂直输送对地表水和地下水造成污染[42,44,45]；水体中的重金属一部分继续赋存在水体中,而另一部分则进一步沉积在底泥中;当用受到重金属污染的水进行农田灌溉时,不仅导致农田土壤受到重金属污染,还使得农作物中重金属含量增加,食用这些受污染的农作物也将增加人类健康风险[46,47]。此外,草本植物已被用作生物指标来研究重金属污染物的远距离迁移[48]。

电子垃圾拆解区,特别是我国早期的非正规电子垃圾拆解场所,通常位于农业用地附近。生长在受污染农田及附近的作物通过根部吸收重金属,然后将这些元素转移到地上部分,储存在可食用部分[49]。植物也可以直接吸收微量金属并在根际土壤中引入有机碳和其他化合物,引入根际的有机酸可以与铁/锰吸附或碳酸盐中的微量金属进行螯合,从而有助于微量金属的释放[50,51],有研究表明在电子垃圾拆解污染土壤中的蔬菜和野生植物中观察到高水平的铜和锌。这是由于在金属回收过程中,高浓度的重金属将随着废渣和废水被排放到露天区域和排水沟中,并进入到表层土壤中。Fu 等[52]于 2006~2010 年对我国浙江台州电子垃圾拆解区和非电子垃圾拆解区的水稻样品进行检测得出,电子垃圾拆解区大米样品中砷、镉、铜和铅的几何均值分别为 111 ng/g、217 ng/g、4676 ng/g 和 237 ng/g。铅含量在采样期内呈显著下降趋势,而其他 3 种金属含量保持相对稳定或呈上升趋势。电子垃圾拆解区中大米含铅、铜、镉浓度显著高于非电子垃圾拆解区($p<0.05$),说明电子垃圾拆解活动与铅、铜、镉的高浓度密切相关。Yu 等[24]报道了广东省清远市的电子垃圾拆解正规化后,尽管环境中的重金属污染有所下降,但是土壤中的重金属仍有一定程度的污染,使得农作物蔬菜和稻米中的重金属含量较高,成为当地成人及儿童致癌和非致癌风险中主要的暴露来源。

2.4 电子垃圾的重金属暴露

2.4.1 外暴露

由于重金属在多种环境介质中迁移,可以通过多种直接或间接暴露途径进入到人体中。重金属的外暴露一般可以通过饮食(包括灰尘和土壤)的经口摄入、呼吸吸入、皮肤接触等方式暴露,表 2-7 总结了部分重金属在环境介质中的累积以及可能的暴露途径及健康影响。可以看出,重金属的外暴露途径有吸入空气、意外摄入灰尘或土壤、饮用受污染的水、食用植物和鱼类,重金属包含铅、铬、镉、汞、锌、镍、锑、砷、钴、铜和硒等。

外暴露的健康评估方法参考美国环保署推荐的健康评估暴露模型,通过环境介质中的重金属含量,计算其经过不同途径的每日摄入量,使用参考剂量/参考浓度和单位风险值(IUR)来计算的致癌风险(LCR)和非致癌风险商(HQ)。简化的计算公式如下[53,54]：

$$CDI = \frac{C_{HM} \times ET \times EF \times ED \times ADAF \times IR}{AT \times BW} \quad (2.1)$$

$$HQ = \frac{CDI}{RfD} \quad (2.2)$$

$$LCR = CDI \times SF \quad (2.3)$$

式中，CDI 为每日摄入量；C_{HM}(ng/m³)是某种金属的浓度，包括人类呼吸系统中所有尺寸部分的总浓度和沉积浓度；ET 是暴露时间(h/d)；EF 为暴露频率(d/a)；ED 为暴露持续时间；AT 为平均时间(以年为单位的寿命 × 365 d/a × 24 h/d)；BW 是体重(一般成人平均体重：70 kg；儿童平均体重：15 kg)；ADAF 被定义为与年龄相关的调整因子(成人：1；儿童：3)；IR 为吸入空气的速率(m³/h)；RfD 是参考剂量，用美国环保署提供的参考浓度(RfC，mg/m³)计算，即

$$RfD = \frac{RfC \times IR}{BW} \quad (2.4)$$

SF 是吸入斜率因子(kg·d/mg)，根据吸入单位风险值(IUR，(m³/mg)⁻¹)计算，即

$$SF = \frac{IUR \times BW}{IR} \quad (2.5)$$

表 2-7　典型重金属在环境中的累积介质、可能的暴露途径及其健康影响

重金属	暴露的生态源	暴露途径	健康影响	文献
铅	空气、灰尘、水和土壤	饮食、呼吸或皮肤接触	儿童的神经行为发展、贫血、肾损害、慢性神经毒性	[55]
铬或六价铬	空气、灰尘、水和土壤	饮食、呼吸	致癌性，生殖功能，内分泌功能，卵毒性	[55-57]
镉	空气、灰尘、水、土壤和食物(尤其是大米和蔬菜)	饮食、呼吸	肾损害，肾毒性，骨病(骨软化症和骨质疏松症)，可能有生殖损伤和肺气肿	[58]
汞	空气、灰尘、水、土壤和食物(鱼体内的生物累积性)	饮食、呼吸或皮肤接触	儿童的神经行为发育(尤其是甲基汞)，贫血，肾损害，慢性神经毒性	[58]
锌	空气、灰尘、水和土壤	饮食、呼吸	铜缺乏的风险增加(贫血、神经系统异常)	
镍	空气、水、土壤和食物(植物)	饮食、呼吸或皮肤接触以及经胎盘	致癌、肺栓塞、呼吸衰竭、出生缺陷、哮喘和慢性支气管炎	[59]
锑	空气、水和土壤	饮食、呼吸或皮肤接触	损害肺、心、肝和肾，眼睛刺激和脱发	

续表

重金属	暴露的生态源	暴露途径	健康影响	文献
砷	空气、土壤、水和食物	饮食、呼吸或皮肤接触	皮肤改变,神经传导减弱,患糖尿病和癌症(皮肤和其他组织)的风险增加	[60]
钴	空气、灰尘、水、土壤和食物	饮食、呼吸或皮肤接触	哮喘和肺炎,呕吐和恶心、视力问题、心脏问题、甲状腺损伤、脱发	
铜	空气、灰尘、水和土壤	饮食、呼吸或皮肤接触	长期接触铜会刺激鼻子、嘴巴和眼睛,并导致头痛、胃痛、头晕、呕吐和腹泻	[59]
硒	空气、灰尘、水和土壤	饮食、呼吸或皮肤接触	脱发、指甲脆性、心血管、肾脏和神经系统异常	

此外,由于不同粒径的颗粒物能够深入到肺部不同深度的位置,因而大气颗粒物的重金属在肺部不同位置的沉积效率和通量不同,得到不同程度的健康风险。基于国际防辐射委员会提供的 ICRP 模型计算人体呼吸道中粒径分级颗粒的沉积分数和通量。沉积在人体呼吸道三个主要部分的吸入颗粒的部分,即头部气道(HA,包括鼻、口、咽和喉)、气管支气管区(TB)和肺泡区(AR)。特别是特定尺寸颗粒在头部气道($DF_{HA,i}$)、气管支气管区($DF_{TB,i}$)和肺泡区($DF_{AR,i}$)的沉积效率,简化的计算公式如下[61]:

$$DF_{HA,i} = IF_i \times \left(\frac{1}{1+\exp(6.84+1.183\ln D_{p,i})} + \frac{1}{1+\exp(0.924-1.885\ln D_{p,i})} \right) \quad (2.6)$$

$$DF_{TB,i} = \frac{0.00352}{D_{p,i}} \times \left(\exp\left(-0.234\left(\ln D_{p,i}+3.40\right)^2\right) \right. \\ \left. +63.9\exp\left(-0.819\left(\ln D_{p,i}-1.61\right)^2\right) \right) \quad (2.7)$$

$$DF_{AR,i} = \frac{0.0155}{D_{p,i}} \times \left(\exp\left(-0.416\left(\ln D_{p,i}+2.84\right)^2\right) \right. \\ \left. +19.11\exp\left(-0.482\left(\ln D_{p,i}-1.362\right)^2\right) \right) \quad (2.8)$$

其中 IF 为颗粒物的可吸入部分,由下面的公式推算:

$$IF = 1-0.5\left(1-\frac{1}{1+0.00076 D_{p,i}^{2.8}}\right) \quad (2.9)$$

而呼吸道中颗粒结合金属的总沉积通量(D，ng/h)由下式推算：

$$D = \sum (\mathrm{DF}_i \times C_i) \times \mathrm{IR} \tag{2.10}$$

Huang 等[23, 27]使用上述方法，评估中国广东省清远的电子垃圾拆解区大气分段粒径颗粒物的重金属经呼吸的暴露风险，结果如表 2-8 所示。成人对重金属的呼吸摄入量是 530 ng/(kg·d)（范围为：130~1000 ng/(kg·d)），儿童对重金属的呼吸摄入量为 3700 ng/(kg·d)（范围为：890~7000 ng/(kg·d)），儿童对颗粒态重金属的呼吸摄入量近乎是成人的 7 倍。而且发现电子垃圾拆解区颗粒态重金属在肺部不同位置的沉积通量存在显著性差异(t 检验，p = 0.038)：肺泡部位的沉积通量(成人：5050 ng/d ± 2700 ng/d，儿童：2500 ng/d ± 1400 ng/d)低于鼻腔咽喉部(成人：30100 ng/d ± 17 200 ng/d，儿童：15 050 ng/d ± 8600 ng/d)，但明显高于气管支气管部(成人：1950 ng/d ± 1000 ng/d，儿童：980 ng/d ± 500 ng/d)。此结果表明，大多数(17%)进入鼻腔咽喉部的颗粒态重金属能穿过气管支气管进入到肺泡部位，导致当地居民将面临更高的健康风险。

表 2-8 电子垃圾拆解区 1.5 m 处成人及儿童通过呼吸吸入大气颗粒态重金属的每日摄入量(CDI，ng/kg) [27]

	成人				儿童			
	$\mathrm{CDI}_{\mathrm{C-total}}{}^a$	$\mathrm{CDI}_{\mathrm{C-lung}}{}^b$	$\mathrm{CDI}_{\mathrm{D-total}}{}^c$	$\mathrm{CDI}_{\mathrm{D-AR}}{}^d$	$\mathrm{CDI}_{\mathrm{C-total}}{}^a$	$\mathrm{CDI}_{\mathrm{C-lung}}{}^b$	$\mathrm{CDI}_{\mathrm{D-total}}{}^c$	$\mathrm{CDI}_{\mathrm{D-AR}}{}^d$
钒	3.3±2.7	2.5±2.0	2.0±1.6	0.29±0.24	23±19	17±14	14±11	2.0±1.7
锰	21±13	16±9.9	14±8.4	1.7±1.0	105±92	110±69	96±59	12±7.2
铁	430±220	320±160	295±150	36±29	3000±1500	2200±1100	2100±1000	250±220
铜	23±23	17±17	14±13	1.9±2.0	160±160	120±120	97±91	13.4±14
锌	230±150	180±110	140±90	24±15	1600±1040	1200±780	990±630	170±100
硒	3.1±1.4	2.3±1.1	1.2±0.61	0.30±0.13	22±10	16±7.5	8.7±4.3	2.1±0.91
铅	44±23	33±17	22±11	4.2±2.1	310±160	230±120	150±80	30±15
铊	36±25	27±18	30±20	2.3±1.5	250±170	190±130	210±140	16±10.4
钼	0.78±0.54	0.59±0.40	0.44±0.27	0.06±0.04	5.5±3.8	4.1±2.8	3.1±1.9	0.41±0.29
锑	3.6±2.8	2.7±2.1	1.7±1.2	0.34±0.26	25±20	19±15	12±8.5	2.4±1.8
铬	6.1±3.3	4.6±2.5	4.1±2.3	0.49±0.24	43±23	32±18	28±16	3.4±1.7

续表

	成人				儿童			
	$CDI_{C\text{-}total}^a$	$CDI_{C\text{-}lung}^b$	$CDI_{D\text{-}total}^c$	$CDI_{D\text{-}AR}^d$	$CDI_{C\text{-}total}^a$	$CDI_{C\text{-}lung}^b$	$CDI_{D\text{-}total}^c$	$CDI_{D\text{-}AR}^d$
钴	0.24±0.17	0.18±0.13	0.17±0.13	0.02±0.01	1.7±1.2	1.3±0.89	1.2±0.91	0.12±0.08
镍	2.9±1.4	2.1±1.1	1.8±0.93	0.23±0.11	20±10	15±7.5	12±6.5	1.6±0.77

a. 按颗粒态金属元素的总浓度计算
b. 按75%浓度吸入肺部计算
c. 按粒径分布的沉积通量计算
d. 按肺泡区沉积通量计算

根据重金属的沉积浓度估计了居民通过暴露于空气中颗粒结合的重金属的非癌症风险和 LCR。成人和儿童总金属（包括锰、砷、镍和镉等）的 HQ 值分别为 2.7（95% CI：1.0~5.5）和 8.0（95% CI：3.0~17）。对于致癌风险，成人（1.3×10^{-3}）和儿童（3.9×10^{-3}）的铬、钴、镍、砷和镉的总 LCR 远高于美国环保署规定的可接受的癌症风险阈值 10^{-6}[53]。此外，儿童的总 LCR 是成人的 3 倍，儿童的铬癌症风险还超过了不可接受的风险值 10^{-4}[53]。因此，电子垃圾拆解区居民因吸入空气中的颗粒结合重金属而面临的健康风险很高，尤其是对儿童而言[27]。

除了经呼吸暴露，灰尘或土壤也会造成有意识或无意识地被摄入。由于儿童的身高比成人矮，儿童无意识接触灰尘或土壤，使得他们更容易暴露于灰尘或土壤的重金属[62,63]。Yang 等[64]在中国汕头市的电子垃圾拆解区，通过元素示踪研究儿童暴露于灰尘重金属的健康风险，发现儿童通过手/物到口摄入途径接触重金属对儿童的整体健康风险很高。虽然非致癌风险在相关指南规定的最低范围内，但铬和砷的致癌风险超过了 10^{-6}，且该地区有超过 75%的儿童存在致癌健康风险。

食用在电子垃圾区种植的蔬菜，可导致重金属的摄入。Liu 等[65]评估摄入电子垃圾区生产的 11 种常见蔬菜的重金属暴露风险，结果发现对成年人来说，只有红薯中铬的 HQ 超过 1。但考虑到综合风险，部分蔬菜类型（芸豆、辣椒、生菜、茄子、蚕豆和红薯）的 HI（HQ 的总和）高于 1，表明食用这些蔬菜可能会导致不利的健康风险。对于儿童来说，莴苣中镉和铬的 HQ 和甘薯中铬的 HQ 均超过 1，表明食用这些蔬菜可能对儿童的健康构成不利风险。考虑到所有有害元素，除卷心菜外，其他所有蔬菜类型的 HI 值均超过 1，生菜、辣椒和甘薯的 HI 值居前三位。这一结果表明，受污染的蔬菜对当地儿童构成了巨大的健康风险。对于致癌风险，成年人的致癌风险从芸豆中镉的 2.61×10^{-7} 到茄子中镉的 7.55×10^{-4} 不等。最高的致癌风险是镉，其次是铅。蔬菜的致癌风险值从大到小依次为：生菜（1.92×10^{-3}）> 甘薯（9.02×10^{-4}）> 茄子（8.64×10^{-4}）> 辣椒（7.25×10^{-4}）> 大白菜

($5.98×10^{-4}$) > 蒜苗 ($4.51×10^{-4}$) > 番茄 ($4.22×10^{-4}$) > 蚕豆 ($2.62×10^{-4}$) > 白菜 ($2.45×10^{-4}$) > 芸豆 ($2.06×10^{-4}$) > 豇豆 ($1.93×10^{-4}$)。总之，与多叶蔬菜、根茎类和茄科蔬菜相比，豆类对成年人的健康风险较低。而儿童致癌风险相比成人较低，红薯铬值最高 ($2.62×10^{-4}$)，豇豆铬值最低 ($1.47×10^{-8}$)。铬比镉和铅具有更严重的致癌风险。所研究的 11 种蔬菜的致癌风险值排名（从高到低）依次为：红薯 ($2.64×10^{-4}$) > 生菜 ($2.19×10^{-4}$) > 辣椒 ($1.43×10^{-4}$) > 蒜苗 ($1.28×10^{-4}$) > 芸豆 ($8.47×10^{-5}$) > 豇豆 ($6.45×10^{-5}$) > 蚕豆 ($5.33×10^{-5}$) > 沼泽白菜 ($5.05×10^{-5}$) > 茄子 ($4.52×10^{-5}$) > 番茄 ($4.02×10^{-5}$) > 卷心菜 ($2.54×10^{-5}$)。这一结果表明根茎类蔬菜可能比其他三种蔬菜更不安全。

2.4.2 内暴露

人类接触电子垃圾，通过吸入和皮肤接触等途径直接暴露于电子垃圾中的重金属[4,66-68]。目前在电子垃圾拆解区附近人群中的重金属负荷标志物浓度明显高于非电子垃圾拆解区[66]。正是由于早期人们的环保意识和自我防护意识比较薄弱，工人在拆解过程中缺乏必要的职业防护。除工人容易受到各种有害物质的侵害外，周边居民区的孕妇、儿童和新生儿也是对电子垃圾暴露最敏感的人群。

重金属暴露对儿童健康有不利影响，包括肛门生殖器距离缩短、阿氏评分（Apgar 评分）降低、体重降低、肺功能降低、较低的乙型肝炎表面抗体水平、较高的注意力缺陷/多动障碍患病率，以及较高的 DNA 和染色体损伤。重金属影响许多不同的系统和器官，对儿童的健康造成急性和慢性影响，从轻微的上呼吸道刺激到慢性呼吸道、心血管、神经、泌尿和生殖疾病，以及加重原有症状和疾病[67]。

关于电子垃圾重金属的人体负荷，铅被广泛研究，因为其对儿童智力发育有很大的影响[68]。此外，毒性较高的致癌金属如镉和铬等也有较多的研究。铅在血液和骨骼中的半衰期分别约为 30 天和 30 年，吸入的无机铅有一半可能会被肺部吸收，而儿童还可能会通过胃肠道吸收经口摄入的 50%以上的铅，当铅进入到血液中，超过 95%的血铅与红细胞结合[69]。铅排出体外的方式主要通过尿液，但大约 90%的铅会沉积在骨骼中，其余 10%的铅在血液中循环并进一步分布于各种组织和器官中，当血液中的铅浓度下降到一定水平以下时，铅会随着时间的推移从骨骼中渗出到血液中[70]。

1. 血液

由于儿童主要受到铅的影响，霍霞等在中国广东省汕头市的电子垃圾拆解区，多年研究儿童血铅、血镉及其他金属与健康的关系。表 2-9 列举了儿童全血铅的水平,从早期到近期的时间上来看,儿童血铅是有所下降的。其中,在 2016 年中,儿童血铅较其他年份低,可能是因为采样对象为某小学的儿童,而不是电子垃圾拆解从业人员的子女[22]。根据 2008 年的报道,生活在非正规电子垃圾拆解区的 1~6 岁儿童平均血铅水平为 15.3 µg/dL,远高于非电子垃圾回收区的儿童。血铅水平超过 10 µg/dL 的儿童比例约为 81.8%,远高于非电子垃圾回收和拆解区域的 37.7%,并且血铅水平随着年龄的增长而持续上升[71]。当与同一时间生活在附近城镇的儿童(血铅:3.63 µg/dL ± 0.24 µg/dL)相比,电子垃圾拆解区儿童的血铅(5.29 µg/dL ± 0.29 µg/dL)显著增加,并且他们相关的认知能力降低[71,72]。可见,电子垃圾拆解对儿童产生了很大的影响。除了铅以外,发现电子垃圾拆解区儿童的血镉也更高,不过根据不同年份的文献报道,血镉随着时间有下降的趋势,从 2008 年的 1.58 µg/L 降到 2012 年的 0.12 µg/L[73]。

母体血液重金属水平将直接影响到胎儿的发育,Kim 等[74]还报道了中国广东省汕头市的电子垃圾拆解区母体血铅为 6.66 µg/dL,范围在 1.87~27.09 µg/dL,并且发现电子垃圾拆解区域这些孕妇的健康风险增加。

除了儿童和孕妇,生活在电子垃圾拆解区周边居民的血液重金属也有所上升。其中,居民血液的铅、镍、钴和汞是显著高于对照组人群,中位数浓度分别为 33.49 ng/mL、2.11 ng/mL、0.33 ng/mL 和 2.00 ng/mL,并且发现较高的钴暴露造成了居民肺部发生纤维化的风险[75]。

表 2-9 我国广东汕头市电子垃圾拆解区儿童的血铅(µg/dL)和镉(µg/L)

研究对象	时间*	均值/中位数(范围)		参考文献
		铅	镉	
儿童全血	2007 年	15.3 (4.40~32.67)		[76]
儿童全血	2008 年	13.17 (4.14~37.78)	1.58 (0.00~9.72)	[71]
儿童全血	2016 年	6.24 (4.55~8.00)	0.576 (0.442~0.779)	[22]
儿童全血	2017 年**	9.43 (6.84~13.13)	0.12 (0.11~0.29)	[73]
儿童全血	2017 年	9.40 (7.00~31.00)		[77]

*部分研究没有提供采样时间,表中年份为文献发表年份
**采样时间为 2012 年

2. 尿液

由于尿液样品相比血液样品更易获得,尿重金属也作为人体重金属负荷的典型生物样品。一项在电子垃圾拆解区开展的暴露途径研究,暴露区人群的尿

重金属（镉：2.12 ng/mL；铅：4.98 ng/mL；铜：22.2 ng/mL；锑：0.20 ng/mL）升高与人类氧化应激的生物标志物 8-OHdG 显著相关[3]。一项在我国汕头市电子垃圾拆解区孕妇尿镉的研究发现，男婴和女婴孕妇的尿镉水平分别为 (1.38 ± 0.74) μg/g creatinine 和 (1.59 ± 0.92) μg/g creatinine，是相邻非电子垃圾拆解区村镇孕妇尿镉的 1.84 倍和 2.09 倍，并发现女婴的新生儿 Apgar 评分与母体尿镉负相关，但在男婴中没有显著相关，镉暴露及其代谢可能有性别差异[78]。此外，Kuang 等[79]在该电子垃圾拆解区中连续四年分析儿童和成人尿液重金属暴露水平（表 2-10），随着中国对非法拆解电子垃圾的管控，从 2016 年至 2019 年，儿童尿液重金属含量呈下降趋势（图 2-1），说明这些儿童在年幼时重金属暴露水平更高；并且结果表明儿童的重金属暴露水平高于成年人，其中尿铜、钼和镉与氧化应激呈现正相关。

表 2-10 中国汕头市电子垃圾回收区儿童和成人尿液污染物 (μg/g creatinine)[79]

	儿童(*n*=275)					成人(*n*=485)				
	几何均值	25th 分位数	中位数	75th 分位数	95th 分位数	几何均值	25th 分位数	中位数	75th 分位数	95th 分位数
铬	3.57	1.23	3.08	7.67	53.6	3.69	1.23	2.70	10.4	42.3
锰	3.01	1.29	2.48	4.53	43.0	2.15	1.25	1.80	2.93	8.98
钴	0.598	0.348	0.590	1.05	3.23	0.332	0.149	0.263	0.780	2.16
镍	6.34	2.65	6.54	11.9	35.8	4.21	2.11	3.89	7.30	19.1
铜	17.4	10.9	16.4	26.8	87.0	11.8	7.92	11.0	18.1	37.6
钼	131	83.1	136	203	347	54.5	37.7	54.0	80.4	151
镉	0.348	0.227	0.454	0.695	1.41	0.828	0.495	0.840	1.60	3.13
锡	3.50	1.64	3.11	8.08	31.5	2.20	1.11	2.05	4.02	11.2
钡	5.83	2.91	3.44	9.81	65.1	4.09	2.79	3.17	4.63	17.3
铅	5.23	1.55	4.69	15.4	74.2	3.56	1.65	3.50	6.91	25.0

总之，电子垃圾拆解处理活动所排放废水、废气及废渣中含有大量重金属，随后进入环境产生重金属污染问题，并可能对相关人群造成暴露风险和健康危害。例如，电子垃圾暴露区人群的血液重金属和尿液重金属含量均高于非电子垃圾拆解区人群的内暴露，可能导致暴露区的儿童生长发育受阻、新生儿健康风险以及人群肺部疾病的发生。

(李桂英 吴颖君 安太成)

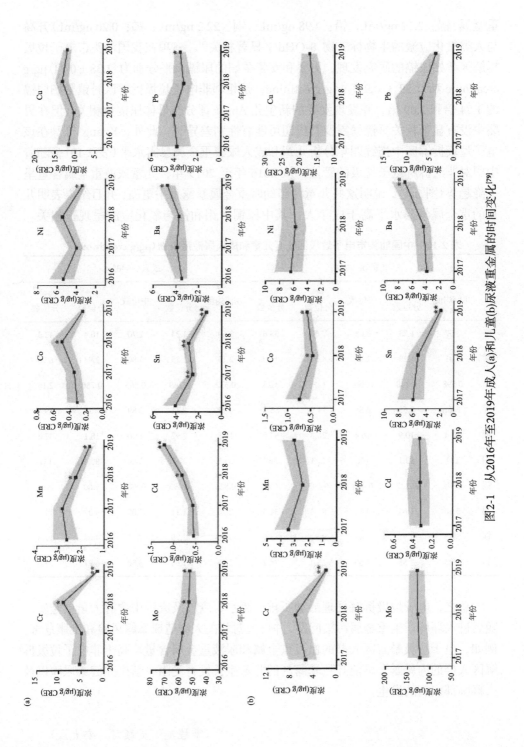

图2-1 从2016年至2019年成人(a)和儿童(b)尿液重金属的时间变化[79]

参 考 文 献

[1] Ladou J, Lovegrove S. Export of electronics equipment waste [J]. International Journal of Occupational and Environmental Health, 2008, 14(1): 1-10.

[2] Luo C L, Liu C P, Wang Y, et al. Heavy metal contamination in soils and vegetables near an e-waste processing site, south China [J]. Journal of Hazardous Materials, 2011, 186(1): 481-490.

[3] Zhang T, Ruan J J, Zhang B, et al. Heavy metals in human urine, foods and drinking water from an e-waste dismantling area: Identification of exposure sources and metal-induced health risk [J]. Ecotoxicology and Environmental Safety, 2019, 169: 707-713.

[4] Chen H J, Ma S T, Yu Y X, et al. Seasonal profiles of atmospheric PAHs in an e-waste dismantling area and their associated health risk considering bioaccessible PAHs in the human lung [J]. Science of the Total Environment, 2019, 683: 371-379.

[5] An T C, Huang Y, Li G Y, et al. Pollution profiles and health risk assessment of VOCs emitted during e-waste dismantling processes associated with different dismantling methods [J]. Environment International, 2014, 73: 186-194.

[6] Rai V, Liu D B, Xia D, et al. Electrochemical approaches for the recovery of metals from electronic waste: A critical review [J]. Recycling, 2021, 6(3): 53.

[7] Robinson B H. E-waste: An assessment of global production and environmental impacts [J]. Science of the Total Environment, 2009, 408(2): 183-191.

[8] Shen H R, Peters T M, Casuccio G S, et al. Elevated concentrations of lead in particulate matter on the neighborhood-scale in Delhi, India as determined by single particle analysis [J]. Environmental Science & Technology, 2016, 50(10): 4961-4970.

[9] Akesson A, Lundh T, Vahter M, et al. Tubular and glomerular kidney effects in Swedish women with low environmental cadmium exposure [J]. Environmental health perspectives, 2005, 113(11): 1627-1631.

[10] Ni W, Chen Y, Huang Y, et al. Hair mercury concentrations and associated factors in an electronic waste recycling area, Guiyu, China [J]. Environmental Research, 2014, 128: 84-91.

[11] Bakhiyi B, Gravel S, Ceballos D, et al. Has the question of e-waste opened a Pandora's box? An overview of unpredictable issues and challenges [J]. Environment International, 2018, 110: 173-192.

[12] Woo S H, Lee D S, Lim S R. Potential resource and toxicity impacts from metals in waste electronic devices [J]. Integrated Environmental Assessment and Management, 2016, 12(2): 364-370.

[13] Peng P, Sheng G, Fu J. The pollution by electronic and electric wastes [J]. Progress in Chemistry, 2009, 21(0203): 550-557.

[14] Priya A, Hait S. Toxicity characterization of metals from various waste printed circuit boards [J]. Process Safety and Environmental Protection, 2018, 116: 74-81.

[15] Bi X Y, Li Z G, Zhuang X C, et al. High levels of antimony in dust from e-waste recycling in southeastern China [J]. Science of the Total Environment, 2011, 409(23): 5126-5128.

[16] Li J, Duan H, Shi P. Heavy metal contamination of surface soil in electronic waste dismantling area: Site investigation and source-apportionment analysis [J]. Waste Management and Research, 2011, 29(7): 727-738.

[17] Dimitrakakis E, Janz A, Bilitewski B, et al. Determination of heavy metals and halogens in plastics from electric and electronic waste [J]. Waste Management, 2009, 29(10): 2700-2706.

[18] Lim S R, Schoenung J M. Human health and ecological toxicity potentials due to heavy metal content in waste electronic devices with flat panel displays [J]. Journal of Hazardous Materials, 2010, 177(1-3): 251-259.

[19] Deng W J, Louie P K K, Liu W K, et al. Atmospheric levels and cytotoxicity of PAHs and heavy metals in TSP and PM2.5 at an electronic waste recycling site in southeast China [J]. Atmospheric Environment, 2006,

40(36): 6945-6955.

[20] 陈多宏, 高博, 毕新慧, 等. 典型电子垃圾拆解区大气颗粒物中元素污染的季节变化特征 [J]. 环境监测管理与技术, 2010, 22(04): 19-22.

[21] Bi X H, Simoneit B R T, Wang Z Z, et al. The major components of particles emitted during recycling of waste printed circuit boards in a typical e-waste workshop of south China [J]. Atmospheric Environment, 2010, 44(35): 4440-4445.

[22] Zeng X, Xu X J, Zheng X B, et al. Heavy metals in $PM_{2.5}$ and in blood, and children's respiratory symptoms and asthma from an e-waste recycling area [J]. Environmental Pollution, 2016, 210: 346-353.

[23] 黄春莉. 电子设备的生产和回收过程:大气颗粒态重金属的风险暴露与疏水性有机物在沉积物中的特征[D]. 广州: 中国科学院大学(中国科学院广州地球化学研究所), 2016.

[24] Yu Y, Zhu X, Li L, et al. Health implication of heavy metals exposure via multiple pathways for residents living near a former e-waste recycling area in China: A comparative study [J]. Ecotoxicology and Environmental Safety, 2019, 169: 178-184.

[25] Xue M Q, Yang Y C, Ruan J J, et al. Assessment of noise and heavy metals (Cr, Cu, Cd, Pb) in the ambience of the production line for recycling waste printed circuit boards [J]. Environmental Science & Technology, 2012, 46(1): 494-499.

[26] 曹红梅. 西北干旱区某规模化电子垃圾拆解厂空气环境中多溴联苯醚与重金属的污染特征及职业呼吸暴露风险 [D]. 兰州: 兰州大学, 2019.

[27] Huang C L, Bao L J, Luo P, et al. Potential health risk for residents around a typical e-waste recycling zone via inhalation of size-fractionated particle-bound heavy metals [J]. Journal of Hazardous Materials, 2016, 317: 449-456.

[28] Zheng X B, Xu X J, Yekeen T A, et al. Ambient air heavy metals in $PM_{2.5}$ and potential human health risk assessment in an informal electronic-waste recycling site of China [J]. Aerosol and Air Quality Research, 2016, 16(2): 388-397.

[29] Chapman P M, Wang F Y, Janssen C, et al. Ecotoxicology of metals in aquatic sediments: Binding and release, bioavailability, risk assessment, and remediation [J]. Canadian Journal of Fisheries and Aquatic Sciences, 1998, 55(10): 2221-2243.

[30] Gao Y, Kan A T, Tomson M B. Critical evaluation of desorption phenomena of heavy metals from natural sediments [J]. Environmental Science & Technology, 2003, 37(24): 5566-5573.

[31] Wong C S C, Duzgoren-Aydin N S, Aydin A, et al. Evidence of excessive releases of metals from primitive e-waste processing in Guiyu, China [J]. Environmental Pollution, 2007, 148(1): 62-72.

[32] Wong C S C, Wu S C, Duzgoren-Aydin N S, et al. Trace metal contamination of sediments in an e-waste processing village in China [J]. Environmental Pollution, 2007, 145(2): 434-442.

[33] Du Y M, Wu Q H, Kong D G, et al. Accumulation and translocation of heavy metals in water hyacinth: Maximising the use of green resources to remediate sites impacted by e-waste recycling activities [J]. Ecological Indicators, 2020, 115: 106384.

[34] Labunsk I, Harrad S, Santillo D, et al. Levels and distribution of polybrominated diphenyl ethers in soil, sediment and dust samples collected from various electronic waste recycling sites within Guiyu town, southern China [J]. Environmental Science-Processes & Impacts, 2013, 15(2): 503-511.

[35] Ma J, Addink R, Yun S H, et al. Polybrominated dibenzo-p-dioxins/dibenzofurans and polybrominated diphenyl ethers in soil, vegetation, workshop-floor dust, and electronic shredder residue from an electronic waste recycling facility and in soils from a chemical industrial complex in eastern China [J]. Environmental Science & Technology, 2009, 43(19): 7350-7356.

[36] Fujimori T, Takigami H. Pollution distribution of heavy metals in surface soil at an informal electronic-waste recycling site [J]. Environmental Geochemistry and Health, 2014, 36(1): 159-168.

[37] Leung A O W, Duzgoren-Aydin N S, Cheung K C, et al. Heavy metals concentrations of surface dust from

e-waste recycling and its human health implications in southeast China [J]. Environmental Science & Technology, 2008, 42(7): 2674-2680.

[38] Wu Y Y, Li Y Y, Kang D, et al. Tetrabromobisphenol A and heavy metal exposure via dust ingestion in an e-waste recycling region in southeast China [J]. Science of the Total Environment, 2016, 541: 356-364.

[39] He C T, Zheng X B, Yan X, et al. Organic contaminants and heavy metals in indoor dust from e-waste recycling, rural, and urban areas in south China: Spatial characteristics and implications for human exposure [J]. Ecotoxicology and Environmental Safety, 2017, 140: 109-115.

[40] Tang X J, Shen C F, Shi D Z, et al. Heavy metal and persistent organic compound contamination in soil from Wenling: An emerging e-waste recycling city in Taizhou area, China [J]. Journal of Hazardous Materials, 2010, 173(1-3): 653-660.

[41] Han Y, Tang Z W, Sun J Z, et al. Heavy metals in soil contaminated through e-waste processing activities in a recycling area: Implications for risk management [J]. Process Safety and Environmental Protection, 2019, 125: 189-196.

[42] Wu C F, Luo Y M, Deng S P, et al. Spatial characteristics of cadmium in topsoils in a typical e-waste recycling area in southeast China and its potential threat to shallow groundwater [J]. Science of the Total Environment, 2014, 472: 556-561.

[43] Wu Q H, Leung J Y S, Geng X H, et al. Heavy metal contamination of soil and water in the vicinity of an abandoned e-waste recycling site: Implications for dissemination of heavy metals [J]. Science of the Total Environment, 2015, 506: 217-225.

[44] Arya S, Rautela R, Chavan D, et al. Evaluation of soil contamination due to crude e-waste recycling activities in the capital city of India [J]. Process Safety and Environmental Protection, 2021, 152: 641-653.

[45] Zhang W H, Wu Y X, Simonnot M O. Soil contamination due to e-waste disposal and recycling activities: A review with special focus on China [J]. Pedosphere, 2012, 22(4): 434-455.

[46] Rattan R K, Datta S P, Chhonkar P K, et al. Long-term impact of irrigation with sewage effluents on heavy metal content in soils, crops and groundwater: A case study [J]. Agriculture Ecosystems & Environment, 2005, 109(3-4): 310-322.

[47] Chen G Q, Zeng G M, Du C Y, et al. Transfer of heavy metals from compost to red soil and groundwater under simulated rainfall conditions [J]. Journal of Hazardous Materials, 2010, 181(1-3): 211-216.

[48] Hassanin A, Johnston A E, Thomas G O, et al. Time trends of atmospheric PBDEs inferred from archived U.K. herbage [J]. Environmental Science & Technology, 2005, 39(8): 2436-2441.

[49] Jolly Y N, Islam A, Akbar S. Transfer of metals from soil to vegetables and possible health risk assessment [J]. Springerplus, 2013, 2: 385.

[50] Xu W H, Liu H, Ma Q F, et al. Root exudates, rhizosphere Zn fractions, and Zn accumulation of ryegrass at different soil Zn levels [J]. Pedosphere, 2007, 17(3): 389-396.

[51] Hernandez-Soriano M C, Jimenez-Lopez J C. Effects of soil water content and organic matter addition on the speciation and bioavailability of heavy metals [J]. Science of the Total Environment, 2012, 423: 55-61.

[52] Fu J J, Zhang A Q, Wang T, et al. Influence of e-waste dismantling and its regulations: Temporal trend, spatial distribution of heavy metals in rice grains, and its potential health risk [J]. Environmental Science & Technology, 2013, 47(13): 7437-7445.

[53] U.S. Environmental Protection Agency (EPA). Guidelines for Exposure Assessment [R]. Risk Assessment Forum, Washington, DC, 1992.

[54] U.S. Environmental Protection Agency (EPA). A framework for assessing health risks of environmental exposures to children [R]. National Center for Environmental Assessment Office of Research and Development, Washington, DC, 2006.

[55] Wu Q H, Leung J Y S, Du Y M, et al. Trace metals in e-waste lead to serious health risk through consumption of rice growing near an abandoned e-waste recycling site: Comparisons with PBDEs and AHFRs [J].

Environmental Pollution, 2019, 247: 46-54.

[56] Banu S K, Samuel J B, Arosh J A, et al. Lactational exposure to hexavalent chromium delays puberty by impairing ovarian development, steroidogenesis and pituitary hormone synthesis in developing Wistar rats [J]. Toxicology and Applied Pharmacology, 2008, 232 (2): 180-189.

[57] Quinteros F A, Machiavelli L I, Miler E A, et al. Mechanisms of chromium (VI)-induced apoptosis in anterior pituitary cells [J]. Toxicology, 2008, 249 (2-3): 109-115.

[58] Gangwar C, Choudhari R, Chauhan A, et al. Assessment of air pollution caused by illegal e-waste burning to evaluate the human health risk [J]. Environment International, 2019, 125: 191-199.

[59] Li W L, Achal V. Environmental and health impacts due to e-waste disposal in China — A review [J]. Science of the Total Environment, 2020, 737: 139745.

[60] Gamble A V, Givens A K, Sparks D L. Arsenic speciation and availability in orchard soils historically contaminated with lead arsenate [J]. Journal of Environmental Quality, 2018, 47 (1): 121-128.

[61] ICRP. Human respiratory model for radiological protection [J]. Ann ICRP, 1994, 24 (1-3): 1-3.

[62] Wu S, Peng S Q, Zhang X X, et al. Levels and health risk assessments of heavy metals in urban soils in Dongguan, China [J]. Journal of Geochemical Exploration, 2015, 148: 71-78.

[63] Zheng N, Liu J H, Wang Q C, et al. Health risk assessment of heavy metal exposure to street dust in the zinc smelting district, northeast of China [J]. Science of the Total Environment, 2010, 408 (4): 726-733.

[64] Yang Y, Zhang M D, Chen H J, et al. Estimation of children's soil and dust ingestion rates and health risk at e-waste dismantling area [J]. International Journal of Environmental Research and Public Health, 2022, 19 (12): 7332.

[65] Liu X M, Gu S B, Yang S Y, et al. Heavy metals in soil-vegetable system around e-waste site and the health risk assessment [J]. Science of the Total Environment, 2021, 779: 146438.

[66] Song Q, Li J. A systematic review of the human body burden of e-waste exposure in China [J]. Environment International, 2014, 68 (4): 82-93.

[67] Zeng X, Xu X, Boezen H M, et al. Children with health impairments by heavy metals in an e-waste recycling area [J]. Chemosphere, 2016, 148: 408-415.

[68] Zhang Y H, Hou D Y, O'Connor D, et al. Lead contamination in Chinese surface soils: Source identification, spatial-temporal distribution and associated health risks [J]. Critical Reviews in Environmental Science and Technology, 2019, 49 (15): 1386-1423.

[69] Huo X, Dai Y F, Yang T, et al. Decreased erythrocyte CD44 and CD58 expression link e-waste Pb toxicity to changes in erythrocyte immunity in preschool children [J]. Science of the Total Environment, 2019, 664: 690-697.

[70] Zeng X, Huo X, Xu X J, et al. E-waste lead exposure and children's health in China [J]. Science of the Total Environment, 2020, 734: 139286.

[71] Zheng L K, Wu K S, Li Y, et al. Blood lead and cadmium levels and relevant factors among children from an e-waste recycling town in China [J]. Environmental Research, 2008, 108 (1): 15-20.

[72] Li Y, Xu X J, Wu K S, et al. Monitoring of lead load and its effect on neonatal behavioral neurological assessment scores in Guiyu, an electronic waste recycling town in China [J]. Journal of Environmental Monitoring, 2008, 10 (10): 1233-1238.

[73] Lin X J, Xu X J, Zeng X, et al. Decreased vaccine antibody titers following exposure to multiple metals and metalloids in e-waste-exposed preschool children [J]. Environmental Pollution, 2017, 220: 354-363.

[74] Kim S, Xu X J, Zhang Y L, et al. Metal concentrations in pregnant women and neonates from informal electronic waste recycling [J]. Journal of Exposure Science and Environmental Epidemiology, 2019, 29 (3): 406-415.

[75] Xue K B, Qian Y, Wang Z Y, et al. Cobalt exposure increases the risk of fibrosis of people living near e-waste recycling area [J]. Ecotoxicology and Environmental Safety, 2021, 215: 112145.

[76] Huo X, Peng L, Xu X J, et al. Elevated blood lead levels of children in Guiyu, an electronic waste recycling town in China [J]. Environmental Health Perspectives, 2007, 115 (7): 1113-1117.

[77] Zhang B, Huo X, Xu L, et al. Elevated lead levels from e-waste exposure are linked to decreased olfactory memory in children [J]. Environmental Pollution, 2017, 231: 1112-1121.

[78] Zhang Y L, Xu X J, Chen A M, et al. Maternal urinary cadmium levels during pregnancy associated with risk of sex-dependent birth outcomes from an e-waste pollution site in China [J]. Reproductive Toxicology, 2018, 75: 49-55.

[79] Kuang H X, Li Y H, Li L Z, et al. Four-year population exposure study: Implications for the effectiveness of e-waste control and biomarkers of e-waste pollution [J]. Science of the Total Environment, 2022, 842: 156595.

第3章 电子垃圾拆解排放挥发性有机物污染

集约式的电子垃圾拆解工艺流程包括物理机械拆解、熔融焊锡和高温焚烧等基本步骤[1],上述过程涉及多种高温加热方式,加之分离过程较为剧烈的空气扰动,电子垃圾中的有机组分受热容易生成并排放出高浓度的有机废气,并成为多种阻燃剂、多环芳烃、重金属等毒害性物质的载体[2-4]。电子垃圾加热处理过程释放的有毒废气中,挥发性有机物(volatile organic compounds,VOCs)是不可忽视的一类重要污染物,导致电子垃圾拆解工作场所存在严重的 VOCs 职业暴露风险。本章通过对电子垃圾拆解处理过程产生废气中 VOCs 的排放水平、污染特征和暴露风险进行研究,进而为降低 VOCs 等毒害性污染物对职业工人的暴露提供数据参考,同时为电子垃圾的绿色拆解提供科学指导。

3.1 挥发性有机物概述

VOCs 作为环境中普遍存在的一类重要的大气污染物,在空气、水和食物等环境介质中频繁检出。大气中 VOCs 种类繁多、成分复杂,根据其化学结构可分为脂肪烃类、芳香烃类、卤代烃类、含氧类化合物(如酯类、醛类、酮类、有机酸类等)和含 N/S 类化合物(如硫醇、硫醚)等五类[5]。一般而言,含氧含氮类统称为其他类 VOCs。VOCs 作为大气对流层中重要的组成部分,可发生一系列反应,生成大气对流层臭氧、硝酸过氧乙酰基和二次有机气溶胶等污染物,直接或间接地对人体健康构成严重威胁[6]。

大气中的 VOCs 来源非常广泛,可分为自然来源和人为来源两类。从全球范围来看,VOCs 排放以异戊二烯为主,约占生物源排放的 65%和 VOCs 总排放量的 40%[7]。在城市地区,VOCs 主要源自人为排放,而人为排放源相对比较复杂,包括工业过程[8]、生物质燃烧[9]、化石燃料燃烧、溶剂使用和机动车尾气排放[10]等五种主要的 VOCs 人为释放源,具体可分为石油化工、包装印刷、涂料喷涂、纺织和塑料制品等多种工业场所的固定源,以及机动车、飞机和轮船等交通工具的流动源和溶剂挥发、生物质燃烧等无组织排放源。

3.2 电子垃圾拆解排放 VOCs 污染特征

电子垃圾中含有大量的塑料和金属,如 1 吨电脑板卡中大约含有 272 kg 塑料以及大量的铜、铁、铅、锡、金、银、锑等金属和钯、铂等贵金属[11]。塑料在高温下容易热解释放 VOCs，而高含量的金属可能对塑料的热解产生催化作用，导致热解加速或产生毒性更强的 VOCs。其可能的形成机制为：电子垃圾中比重较大的有机树脂基体的断链和自由基($\cdot$Br、$\cdot$Cl 和 $\cdot$CH$_3$ 等)或前驱体分子的重排，生成芳香烃类 VOCs 和各种开环产物[12]。表 3-1 为废电路板(WPCB)等电子垃圾处理车间的主要拆解工艺的详细情况。

表 3-1 电子垃圾资源化过程车间情况[1]

拆解工艺	车间简称	车间	车间情况
电吹风工艺	SEBMP	小口径电吹风处理手机 WPCB 车间	室内
	LEBMP	大口径电吹风处理手机 WPCB 车间	室内
	LEBHD	大口径电吹风处理硬盘 WPCB 车间	室内
电烤锡炉工艺	EHF-TV	电烤锡炉处理电视机 WPCB 车间	室内，电烤锡炉正上方有集气罩收集释放废气
	EHF-PS	电烤锡炉处理电源 WPCB 车间	
旋转灰化炉工艺	RI-TV	旋转灰化炉处理电视机 WPCB 车间	室内，在密闭的焚化炉内进行拆解，正上方有集气罩收集产生废气
	RI-HD	旋转灰化炉处理硬盘 WPCB 车间	
	RI-MM	旋转灰化炉处理马达车间	
电烙铁工艺	ESI	电烙铁处理电脑 WPCB 车间	室内
机械切割工艺	MC	机械切割处理电脑 WPCB 车间	室内

如图 3-1 至图 3-3 所示，SEBMP、LEBMP 和 LEBHD 车间内 VOCs 的浓度分别为 $(179.96 \pm 5.62)\mu g/m^3$、$(172.55 \pm 1.97)\mu g/m^3$ 和 $(234.51 \pm 2.41)\mu g/m^3$，略微增加电吹风口径对车间内产生 VOCs 含量影响不大。但是，用相同的旋转灰化炉工艺拆解不同的线路板时产生 VOCs 浓度差别比较明显。其中 RI-TV 内 VOCs 的浓度最高 $(32\,507.26 \pm 1363.14)\mu g/m^3$，其次为 RI-MM $(10\,543.37 \pm 257.28)\mu g/m^3$，RI-HD 内 VOCs 浓度为最低 $(3307.10 \pm 15.96)\mu g/m^3$，仅为 RI-TV 内的 VOCs 浓度的 1/10，RI-MM 的 1/3。Molto 等[13]研究发现：在热解和焚烧手机板过程中产生的轻芳香烃类含量远低于热解和焚烧手机外壳及外壳与手机板混合物中产生的轻芳香烃类。VOCs 浓度的巨大差异可能是由以下原因引起的：RI-TV 内拆解电子垃圾中含电视机外壳且其电视机线路板中的许多元器件为热塑性塑料材料组成，RI-MM 内马达中的电线也是由热塑性塑料制成，这些热塑性塑料热稳定性差，在热解时释放出的 VOCs 远比热固性塑料的要多，而 RI-HD 拆解的硬盘板的主要成分为热固性的环氧树脂。因此，RI-TV 和 RI-MM 车间内 VOCs 浓度远高

于 RI-HD 车间。电烤锡炉处理电视机车间内环境大气 VOCs 的浓度分别为 795.63 μg/m³，比电烤锡炉加热处理电视机线路板排放烟气中 VOCs 浓度水平（1600~6700 μg/m³）低 1 个数量级[14]，而煤炉加热方式拆解回收电视机线路板大气中 VOCs 浓度为 48~920 μg/m³[15]，即拆解工艺的改变和不同年份的线路板均可影响 VOCs 的排放通量。

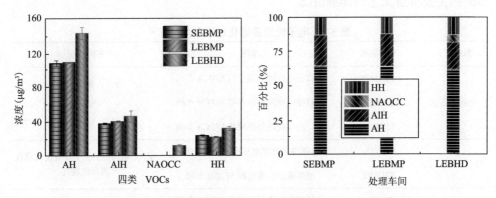

图 3-1 电吹风拆解不同电子垃圾车间内四类 VOCs 浓度及其占比
AH：芳香类 VOCs；AlH：脂肪烃类 VOCs；NAOCC：含氮含氧类 VOCs；HH：卤代烃类 VOCs
请扫描封底二维码查看本书彩图

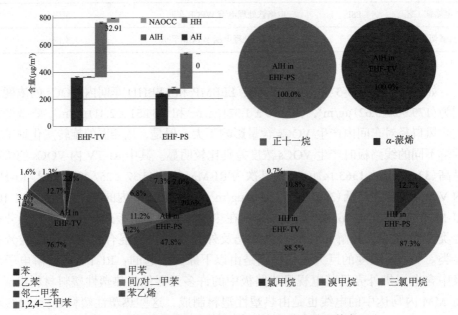

图 3-2 电烤锡炉拆解车间内 VOCs 各组分含量及其占比

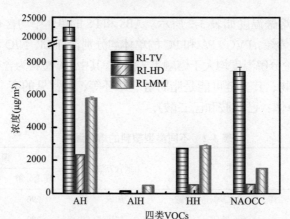

图 3-3 旋转灰化炉拆解不同电子垃圾车间内四类 VOCs 浓度

综上所述，采用不同的拆解工艺线路板产生的 VOCs 浓度迥异。一般而言，电子垃圾拆解过程需要经历机械切割(应用切割机)、熔焊锡(采用电烙铁、电吹风或电烤锡炉)和焚烧(应用旋转灰化炉)三个阶段。因此，为更好地获得不同电子垃圾拆解阶段产生 VOCs 的污染特征，针对上述三个阶段，选取 5 个具有代表性的车间，分别是机械切割车间(MC)、电烙铁车间(ESI)、LEBHD、EHF-TV 和 RI-TV，进一步开展了不同拆解工艺车间内 VOCs 浓度的对比研究。从图 3-4 中可知，RI-TV 内 VOCs 的浓度最高，其浓度分别为 MC、ESI、LEBHD 和 EHF-TV 的 190、180、139 和 40 倍。即在拆解电子垃圾时随着拆解温度的升高和处理面积的扩大，产生的 VOCs 越多。在拆解电子垃圾车间中产生 VOCs 的浓度按大小排列如下：LEBHD＜EHF-TV≪RI-TV。由于在 MC、ESI 和 LEBHD 车间内处理温度较低，处理面积较小，因此，这三个车间内产生 VOCs 含量变化不大。

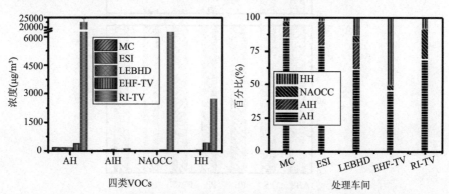

图 3-4 五个不同工艺电子垃圾拆解车间中四类 VOCs 含量及占比

塑料再生工艺一般包含废旧塑料回收、预处理除掉杂质、进入加热熔融、塑化挤出、冷却过水和切割造粒等六个过程，仅有进入加热熔融塑化才会产生有机废气，不同塑料熔融温度如表 3-2 所示。ABS 和 PS 的单体中均含有苯乙烯，PE 和 PP 均属于聚直链烯烃，PVC、PA 和 PC 的单体均分别含有 Cl、N/O 和 O 有机物，而且七类塑料的理论分解温度均大于熔融工艺温度。其中 PA 塑料为含有（—NHCO—）重复单元的高聚物，其支链可能是脂肪链、苯环等，最常见的有 PA6（单体：己内酰胺）和 PA66（单体：己二胺和己二酸）。

表 3-2 不同类型塑料的熔融温度

塑料类型	单体	聚合方法	温度(℃)	
			理论分解	熔融工艺
ABS	苯乙烯，1,3-丁二烯，丙烯腈	加成	290	200~300
PS	苯乙烯	加成	290	200~260
PE	乙烯	加成	350	150~250
PP	丙烯	加成	350	150~250
PVC	氯乙烯	加成	210	150~200
PA	二胺，二羧酸	加成/缩合	300~355	200~230
PC	双酚 A，碳酰氯	缩合	>350	180~230

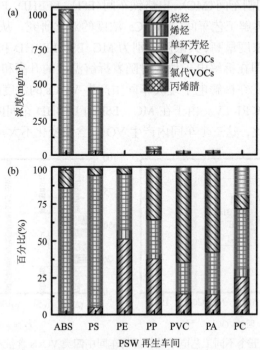

图 3-5 七种塑料再生车间塑化熔融工艺排放的六类 VOCs 的浓度及比例[16]
ABS：丙烯腈-丁二烯-苯乙烯共聚物；PS：聚苯乙烯；PE：聚乙烯；PP：聚丙烯；PVC：聚氯乙烯；
PA：聚酰胺；PC：聚碳酸酯

七种废旧塑料加热熔融过程 VOCs 主要包括烷烃类、烯烃类、单环芳烃类、含氧 VOCs、氯代 VOCs 和丙烯腈等组分(图 3-5)。总的来说，塑料再生塑化熔融工艺中排放 VOCs 浓度的大小顺序为：ABS＞PS＞PP＞PA＞PVC＞PE＞PC。相比于其他车间，ABS 塑料再生车间 VOCs 浓度最高为 $(1.0 \pm 0.4 \times 10^3)\,\text{mg/m}^3$，远高于 PS 的 $(4.7 \pm 1.0 \times 10^2)\,\text{mg/m}^3$。PE 和 PP 其单体分别为乙烯和丙烯，其 VOCs 的总浓度分别是 $(2.8 \pm 2.4)\,\text{mg/m}^3$ 和 $(59.0 \pm 14.0)\,\text{mg/m}^3$，仅仅为 PS 塑料再生车间的 0.6%和 12.4%。由表 3-2 可知，由于纯的 PE 和 PP 的理论分解温度高于 350℃，而塑化熔融工艺的温度仅有 150~250℃，导致塑化过程排放的 VOCs 浓度不高。虽然如此，由于废旧塑料的使用时间长，且长期热暴露老化以及和添加剂等相互作用等原因，还是会释放出少量 VOCs。对于 PVC，其排放的 VOC 浓度为 $(25 \pm 1)\,\text{mg/m}^3$，PA 车间中排放的 VOCs 浓度为 $(26.0 \pm 12.0)\,\text{mg/m}^3$，与 PVC 塑化工艺中浓度相似；而 PC 塑料再生车间排放的 VOCs 浓度最低，为 $(0.99 \pm 0.21)\,\text{mg/m}^3$，仅为 ABS 车间的 1/1000。这可能是因为 PC 的理论分解温度大于 350℃，然而在塑化阶段的温度为 180~230℃，所以其结构很稳定，难以分解。Jang 等[17]同样发现 PC 具有良好的热稳定性，其高温降解产物一般是双酚 A 和苯酚等较难挥发的物质。此外，大部分的 PC 是用于光盘的使用，因此其表面含有一层金属薄膜和有机颜料，在塑化时需要用水冲洗，这也是导致 PC 难以分解的主要原因之一。

早期的塑料再生主要以家庭作坊式回收为主，这些处理车间存在于居民楼之间，甚至在同一栋楼里面，排放的 VOCs 会直接对居民造成室内污染。对塑料再生作坊的室内 VOCs 进行检测分析得出，VOCs 的浓度大小顺序如下：PS(3.9 mg/m³)＞PA(2.1 mg/m³)＞PVC(1.9 mg/m³)＞ABS(1.8 mg/m³)＞PP(1.2 mg/m³)＞PE(1.1 mg/m³)＞PC(0.60 mg/m³)。Huang 等[18]研究发现聚碳酸酯-丙烯腈-丁二烯-苯乙烯(PC-ABS)和苯乙烯-丁二烯共聚物(K 树脂)塑料造粒过程车间大气 VOCs 浓度水平分别为 3.7 mg/m³ 和 3.2 mg/m³，与作坊车间 VOCs 浓度水平相当。比较塑料再生车间作坊室内 VOCs 与塑料加热熔融过程排放 VOCs 的浓度及相关性得出(图 3-6)，车间作坊内的 VOCs 主要来自于塑化熔融工艺的排放，部分来自于其他邻近的车间或自然源。而 ABS 作坊中的通风系统比较好，能够很好地将 VOCs 排出室外，导致室内的 VOCs 浓度有所较低。

大量实验室规模的模拟研究结果表明，电子垃圾热处理过程(热解为主)将产生 VOCs 的排放及 VOCs 污染。加热拆解过程释放 VOCs 的种类随着热处理的电子元器件的类型和工艺条件不同而发生变化。电脑显示器外壳主要由 ABS 和 PVC 组成，热解时可产生大量的苯乙烯、甲苯和乙苯[19]；电脑显卡等热解过程则释放出丙酮、苯酚和溴代甲苯[20]；热解 WPCB 过程会产生较高浓度的苯酚及酚类衍生物和苯乙烯等芳香烃类 VOCs[12,21,22]。低密度聚乙烯、高密度聚乙烯和 PP 等聚烯

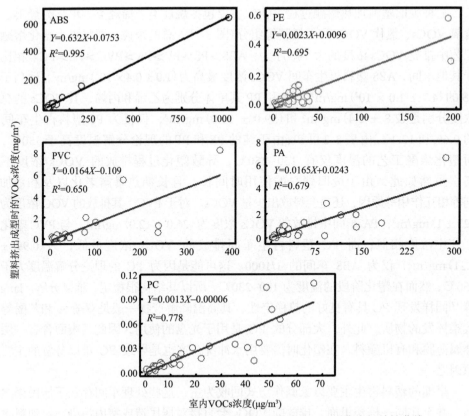

图 3-6 塑料再生作坊内与相应塑化熔融工艺排放的 VOCs 浓度相关性曲线

烃塑料热解废气中主要以氢气、烷烃和烯烃类 VOCs 为主；PS 塑料热解主要生成苯乙烯、芳香烃及多环芳香烃等组成的油；PVC 热解主要产物则为氯化氢和芳香烃类焦油；聚对苯二甲酸乙酯则热解产生大量的二氧化碳和一氧化碳以及大量的焦炭；当塑料混合之后，气态产物随着热解温度的升高而增加，焦油产量呈现先上升后下降的趋势，焦炭含量因裂解反应和焦化过程持续增加[23]。丙烯和甲苯为热解手机线路板的主要组分，而手机外壳与线路板混合物中甲苯为主要成分[24]。此外，VOCs 的种类和含量还会因热解温度而发生变化。当温度低于 300℃时，热解 WPCB 产生的废气以芳香烃 VOCs 为主，随着热解温度的升高芳香烃 VOCs 生成的比例不断降低，高于 400℃时则有 Br-VOCs 生成[25]。线路板在 600℃热解时，气相产物中苯和甲苯的产量分别为 1.06 g/kg 和 1.05 g/kg，当温度升到 850℃，其产量增加了 10 倍(13.8 g/kg)和 3 倍(3.35 g/kg)之多[22]。当热解多种混合塑料时，苯、甲苯和苯乙烯等芳香烃受温度影响较大(温度：650~800℃)，其在 719℃生成量达到最高[26]。

此外，VOCs 容易从电子垃圾拆解过程扩散到周边环境大气中[16]。受塑料再生作坊的影响，塑料再生地区大气 VOCs 的浓度水平高达 213.98 μg/m³，比非工业区的高出 3.4 倍。塑料再生地区塑料塑化阶段所排放的特征 VOCs 包括：异戊烷、3-甲基戊烷、3-甲基己烷、正壬烷、正癸烷、正丙苯、3-乙基甲苯、2-乙基甲苯、1,2,3-三甲苯和对二乙基苯，其浓度分别为 12.49 μg/m³、1.17 μg/m³、0.78 μg/m³、3.45 μg/m³、13.24 μg/m³、11.00 μg/m³、11.33 μg/m³、17.79 μg/m³、10.43 μg/m³ 和 35.21 μg/m³。对电子垃圾拆解工业园区及周边地区进行 9 km × 9 km 网格化季节采样，发现整个区域浓度最高的 VOCs 组分为异戊烷、正戊烷、BTEX（苯、甲苯、乙苯、二甲苯）、一氟三氯甲烷、二氟二氯甲烷、1,2-二氯乙烷和 1,2-二氯丙烷。时空分布结果表明，电子垃圾拆解园区是 BTEX、1,2-二氯乙烷和 1,2-二氯丙烷的热点，预示这些物质可能是电子垃圾排放的特征污染物。源解析结果发现汽车尾气（28%~56%）是园区外区域 VOCs 最主要来源，电子垃圾拆解地区 VOCs 有 20%来自于电子垃圾拆解贡献[27]。

3.2.1 芳香烃类 VOCs

如图 3-1 至图 3-4 所示，各电子垃圾拆解回收过程有机废气中芳香烃类 VOCs(AHs) 的浓度水平差异较大，较低浓度的 AHs 在电吹风工艺拆解线路板车间大气样品中被检出 (108.5~143.4 μg/m³)，该水平与机械切割车间 (145.1 μg/m³) 和电烙铁车间 (142.5 μg/m³) 大气中 AHs 的浓度水平相当[1]；略低于 EHF-TV (360.0 μg/m³) 和 EHF-PS (236.2 μg/m³) 车间[1]大气中 AHs 的观测结果；比旋转灰化炉拆解线路板车间 (>2299.1 μg/m³)[1]和塑料造粒车间 (PC-ABS：2969.6 μg/m³ 和 K 树脂：2463.6 μg/m³)[18]大气中 AHs 的观测值低 1~3 个数量级。此外，AHs 在旋转灰化炉处理线路板烟道气中浓度为 (1700 ± 57) ~ (25000 ± 840) μg/m³[28]。电子垃圾线路板拆解烟道气中 AHs 的浓度呈现出与相应的拆解车间大气中的污染水平相当。

电吹风工艺拆解电子垃圾线路板车间内 AHs 的含量最高，在 VOCs 所占的比例均超过了 61.1%。高含量的 AHs 主要因为无论是手机还是硬盘的印制电路板的主要成分均为环氧树脂，这些环氧树脂的单体为双酚 A；由于苯环相对比较稳定，在加热过程中双酚 A 中苯环与碳原子间的碳碳单键会断裂或者被其他自由基取代，最终释放出大量的芳香烃。此外，AHs 的含量在旋转灰化炉车间内排放的 VOCs 中也最高（占 VOCs 比例>54%）。这可能是因为旋转灰化炉内温度很高（>400℃），此时有机物被分解，释放出大量的芳香烃和脂肪烃[29]。然而，由于脂肪烃相对较易被氧化而芳香烃中的苯环比较稳定，导致最终检测到的脂肪烃类浓度远低于芳香烃类。RI-TV 内 AHs 的浓度远高于 RI-MM 和 RI-HD 内，分别是 RI-MM

和 RI-HD 内的 3.9 和 9.8 倍。与此同时，AHs 在 MC 和 ESI 车间内高达 84.9%和 79.0%（图 3-4）。EHF-TV 中试研究排放有机废气中同时也发现了高含量的 AHs（30.5%）[14]。

不同电子垃圾拆解车间中 AHs 不同组分的分布特征存在差异。如图 3-2 及图 3-7 至图 3-9 所示，甲苯为电吹风拆解线路板、EHF-TV、EHF-PS 和 RI-HD 车间的主要贡献组分，分别贡献 AHs 总浓度的 21.6%~27.2%、76.7%、47.9%和 86.3%。RI-TV

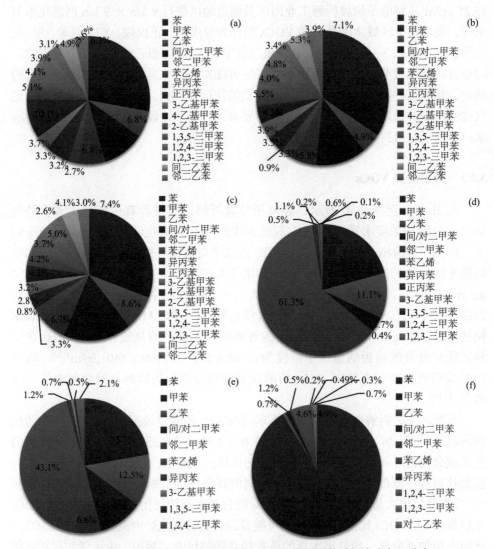

图 3-7　电吹风和旋转灰化炉拆解不同电子垃圾车间内芳香烃类中各组分占比
(a) SEBMP；(b) LEBMP；(c) LEBHD；(d) RI-TV；(e) RI-MM；(f) RI-HD

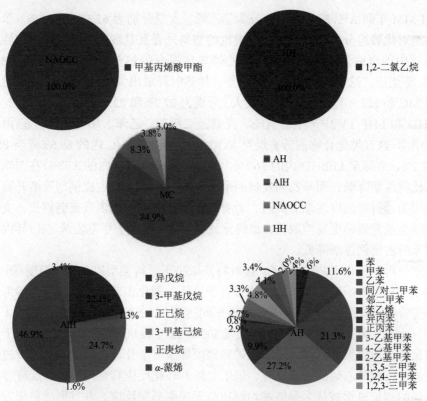

图 3-8 机械切割电子垃圾车间中四类 VOCs 及其主要成分占比图

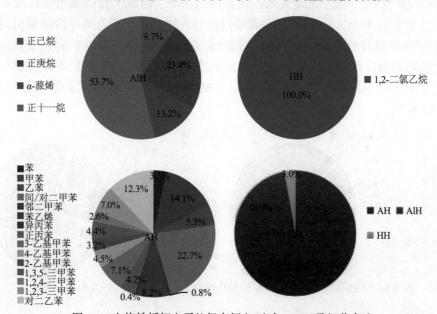

图 3-9 电烙铁拆解电子垃圾车间(ESI)内 VOCs 及组分占比

和 RI-MM 车间 AHs 中最主要成分为苯乙烯，含量分别为 61.3%和 43.1%。苯乙烯作为绝对优势组分的原因如下：电视机的塑料外壳及其线路板中大量热塑性元件与马达中的电线及其他元件均由苯乙烯单体聚合的热塑性塑料中的如 ABS 或 HIPS 等组成，这些材料热稳定性较差，加热时释放出大量的苯乙烯。间/对二甲苯对 MC 和 ESI 车间 AHs 的贡献最大，分别为 27.2%和 22.7%。此外，在 MC、ESI、LEBHD 和 EHF-TV 中产生的 AHs，尤其是二甲苯、乙苯、甲基乙苯、正/丙基苯及三甲苯（这五类化合物占芳香烃类 VOCs 比例依次由 MC 内的 69.5%降至 ESI 内的 57.2%，再降至 LEBHD 内的 36.1%，最后降至 EHF-TV 内的 3.53%）在 VOCs 中所占比例逐渐降低；而苯乙烯在这四个车间内的 VOCs 所占比例却逐渐升高。可能是因为当拆解温度逐渐升高时，芳香烃中苯环上侧链越多或支链越长，支链越容易断裂或者侧链更易被取代，最终导致二甲苯、乙苯、甲基乙苯、正/异丙苯及三甲苯所占比例逐渐降低。

  塑料再生塑化工艺中 VOCs 排放特征取决于塑料类型以及操作温度（图 3-10 至图 3-12）。ABS 和 PS 的单体中均含有苯乙烯，塑化工艺中排放以苯乙烯为主的单环芳香烃，苯乙烯浓度在 ABS 和 PS 车间达到 $(6.3 \pm 2.1 \times 10^2)$ mg/m³ 和 $(3.1 \pm 0.7 \times 10^2)$ mg/m³，分别贡献 ABS 和 PS 大气 VOCs 浓度的 60.9%和 66.4%[16]。由于 ABS 是共聚物，包含聚苯乙烯、聚丁二烯和聚丙烯腈三个片段，因此其热分解过程包括三部分：聚苯乙烯的解聚反应、聚丁二烯的环化反应和聚丙烯腈的反应[30]。聚苯乙烯的解聚过程被认为是链端均裂和链无规则断裂反应，并且形成易挥发性共轭结构的产物[31]。因此，随着聚苯乙烯片段的解聚，形成大量的苯乙烯以及单环芳香烃为主的 VOCs。由于 PS 的降解过程与 ABS 中的聚苯乙烯片段解聚机理一样，因此导致排放的 VOCs 也包含了大部分的以苯乙烯为主的单环芳香烃。但是由于塑化熔融的温度低于 ABS，因此 VOCs 的浓度相对较低。甲苯是塑料造粒车间大气 VOCs 的主要贡献者，分别占 PC-ABS 和 K 树脂车间 VOCs 含量的 33.4%和 31.5%[18]。

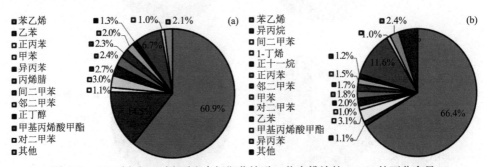

图 3-10 ABS(a)和 PS(b)再生车间塑化熔融工艺中排放的 VOCs 的百分含量

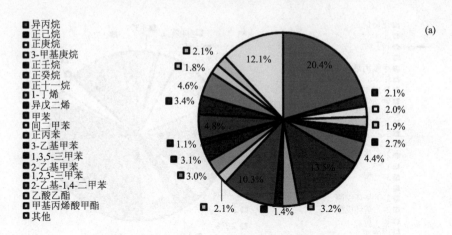

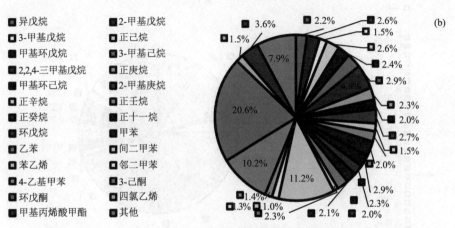

图 3-11　PE(a)和 PP(b)再生车间塑化熔融工艺中排放的 VOCs 的百分含量

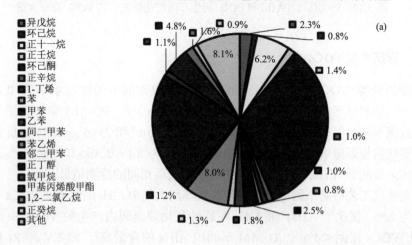

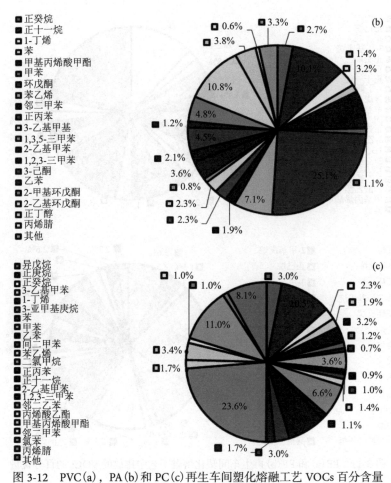

图 3-12　PVC(a), PA(b) 和 PC(c) 再生车间塑化熔融工艺 VOCs 百分含量

3.2.2　脂肪烃类 VOCs

脂肪烃类 VOCs(AlHs) 在不同 WPCB 拆解车间内的浓度水平为 0.76~440 μg/m³[1], 明显低于塑料熔融过程(320~28 600 μg/m³)[16]。PC-ABS 和 K 树脂等塑料造粒过程车间大气 AlHs 浓度水平分别为 752.9 μg/m³ 和 759.6 μg/m³[18]。旋转灰化炉拆解线路板烟道气中 AlHs 的浓度为 $(55 \pm 4.8) \sim (8.4 \pm 0.20) \times 10^2$ μg/m³[28]。RI-HD 车间内未检测到 AlHs, EHF-TV 烟道气中也发现相同的检测结果[14]。

电吹风工艺拆解不同电子垃圾线路板车间大气中 AlHs 的含量在 VOCs 中的比例约为 20%, 仅次于 AHs; 而 AlHs 在旋转灰化炉车间内产生的 VOCs 中含量最低(占 VOCs 比例<4%)。RI-MM 车间内 AlHs 的含量高于 RI-TV 和 RI-HD 车间内, 可能原因是拆解的不同电子垃圾的组成不同所致。此外, EHF-TV 和

RI-TV 内 AlHs 占比非常低(≤0.3%)，表明高温和大范围加热对 AlHs 的生成无显著影响。

不同电子垃圾拆解车间大气中 AlHs 的成分相对比较简单，且分布特征存在明显差异(图 3-2，图 3-8，图 3-9，图 3-13 和图 3-14)。电吹风工艺拆解不同电子垃圾线路板车间大气中 AlHs 含量最高的是异丙烷(约占 AlHs 的 50%)，另外两种主要组分分别是正戊烷(约占 22%)和正庚烷(约占 20%)。EHF-TV 内仅检测到少量的 α-蒎烯(0.76μg/m³± 0.01 μg/m³)；EHF-PS 车间内仅检测到少量的正十一烷(39.88 μg/m³± 9.80 μg/m³)。RI-TV 中只检测到 2 种脂肪烃——1-正己烯(86.4%)和环戊烯(13.6%)；RI-MM 中检测到 6 种 AlHs，其中含量最高的分别为 1-丁烯(25.9%)、1-己烯(20.5%)和 3-甲基庚烷(21.5%)。MC 内产生的 AlHs 中正庚烷比例几乎达一半(46.9%)，另一半则为正己烷(24.7%)和异戊烷(22.1%)。ESI 内 AlHs 中，正十一烷含量最高(占 AlHs 的 53.7%)，其次为正庚烷(占 AlHs 的 23.4%)，还有少量的蒎烯和正己烷。

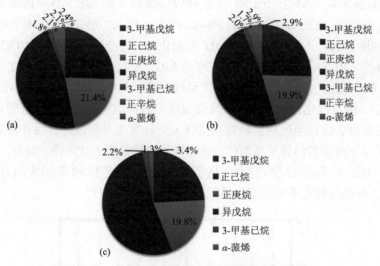

图 3-13 不同电吹风拆解车间内脂肪烃各组分占比
(a) SEBMP；(b) LEBMP；(c) LEBHD

AlHs 在塑料回收过程也没有较为明显的分布规律(图 3-10 至图 3-12)。仅 ABS 过程大气中 AlHs 的组成比较简单，检测到 6 种 AlHs。正十一烷是 PS、PVC 和 PA 车间内主要的贡献者，分别占 AlHs 总含量的 32.52%、40.47%和 59.60%。1-丁烯(40.38%)、异戊烷(38.22%)、2,2,4-三甲基庚烷(11.67%)和正庚烷(31.25%)分别是 ABS、PE、PP 和 PC 车间大气中 AlHs 的主要成分。PC-ABS 和 K 树脂等塑料造粒过程车间大气测到高含量的三甲基壬烷[18]。

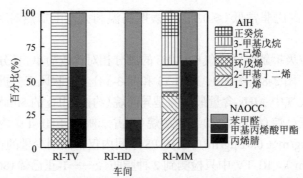

图 3-14 旋转灰化炉不同车间内脂肪烃和含氮含氧类 VOCs 各组分含量

3.2.3 卤代烃类 VOCs

卤代烃类 VOCs(HHs)在机械切割、电烙铁、电吹风(3 个车间)、电烤锡炉(2 个车间)和旋转灰化炉(3 个车间)工艺回收线路板车间(表 3-1)中的浓度分别占 VOCs 的 3%、3%、12.8%~13.8%、48.3%~50.4%和 8.3%~27.0%(图 3-2,图 3-8,图 3-9,图 3-15 和图 3-16)。MC、ESI、LEBHD 和 EHF-TV 内产生的 HHs 在 VOCs 所占比例逐渐升高。RI-TV 车间内 HHs 的浓度是 MC、ESI、LEBHD 和 EHF-TV 的 538、509、84 和 7 倍,与所采用的工艺温度呈现一致的趋势(RI-TV＞EHF＞LEBHD＞ESI＞MC),即氯代化合物含量随着焚烧温度的升高也明显升高。这可能是因为随着温度的升高,卤代阻燃剂会释放到空气中并且发生分解反应,另外烃类中的氢原子在高温下可能被卤原子所取代,最终导致卤代烃类占比升高。此外,EHF-TV 烟气中 HHs 的百分含量(67.4%)高于前期相同工艺拆解车间大气中的含量(50.5%),可能与得克隆等新型含氯阻燃剂的使用有关[32,33]。

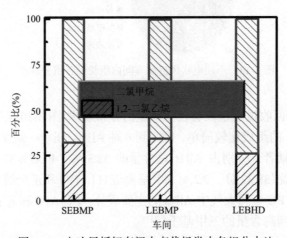

图 3-15 电吹风拆解车间内卤代烃类中各组分占比

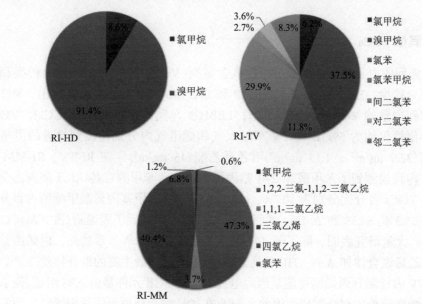

图 3-16　旋转灰化炉拆解不同电子垃圾车间卤代烃类中各组分占比

与 AlHs 观测结果类似，HHs 在不同电子垃圾拆解车间大气中成分相对比较简单，且分布特征存在明显差异(图 3-2，图 3-9，图 3-10，图 3-15 和图 3-16)。电吹风工艺检测到二氯甲烷和 1,2-二氯乙烷两种成分，最主要成分二氯甲烷在 SEBMP、LEBMP 和 LEBHD 车间内的浓度分别为 $(15.94 \pm 1.41)\,\mu g/m^3$、$(14.60 \pm 0.61)\,\mu g/m^3$ 和 $(23.97 \pm 2.09)\,\mu g/m^3$；EHF-TV 和 EHF-PS 内一溴甲烷是最主要的组分，分别占 HHs 的 88.5%和 87.3%；MC 和 ESI 车间内检测到 100%的 1,2-二氯乙烷。旋转灰化炉车间内 HHs 成分非常复杂并且变化剧烈。在 RI-TV 内，最主要的成分是一溴甲烷(占 HHs 的 37.5%)及苄基氯(29.9%)，接下来依次是对/间/邻-二氯甲苯(14.6%)、氯苯(11.8%)和一氯甲烷(6.2%)。与 RI-TV 内不同的是，在 RI-MM 内 HHs 不同成分含量的大小顺序为 1,1,2-三氟-1,2,2-三氯乙烷(47.3%)、三氯乙烯(40.4%)、四氯乙烯(6.8%)、1,1,1-三氯乙烷(3.7%)及少量的邻二氯苯(1.2%)。而 RI-HD 内仅检测到了两种卤代烃类 VOCs，大量的一溴甲烷(91.36%)和少量的一氯甲烷(8.64%)。ESI 和 MC 内仅检测到一种卤代烃——1,2-二氯乙烷。

如图 3-10 至图 3-12 所示，塑料熔融过程 HHs 含量较低。虽然 PVC 中含有 Cl 原子，但 PVC 热降解第一步反应是脱氯反应，Cl 原子直接以 HCl 的形式脱去，而 PVC 则转化成多烯自由基[34,35]，这些多烯自由基容易与空气中的氧气发生氧化反应以及自身的环化反应等生成一系列的含氧 VOC(OVOC)，单环芳香烃以及其他烷烯烃，只产生少量的 HHs(3.7%)。

3.2.4 其他 VOCs

低温回收电子垃圾线路板过程含氧含氮类 VOCs(NAOCC)的检出种类和含量都相对较少。ESI、SEBMP、LEBMP 和 EHF-PS 内没有 NAOCC 检出；MC 内仅检测到少量的 2-甲基丙烯酸甲酯；LEBHD 车间内释放的 NAOCC 在 VOCs 中所占比例仅约为 5%。EHF-TV 车间大气和烟道气内分别检测到少量的甲基异丁基酮(32.9 μg/m³ ± 1.13 μg/m³)和乙酸乙酯(45 μg/m³)。在 RI-TV、RI-MM 和 RI-HD 内均检测到了苯甲醛和 2-甲基丙烯酸甲酯，苯甲醛在各 RI 车间内占含氮含氧类 VOCs 百分比分别为 40.7%、35.5%和 79.9%，2-甲基丙烯酸甲酯所占百分比分别为 38.3%、64.5%和 20.1%。然而仅有 RI-TV 内检测到了丙烯腈(占 NAOCC 的 21.0%)。大量研究表明，除了苯、甲苯、苯乙烯及氯苯等苯系物外，丙烯腈是在热解苯乙烯聚合体如 ABS、HIPS 及 SAN 等中产生的最主要的组分物质[19, 36]。仅在 RI-TV 内检测到丙烯腈可能是因为电视机的 WPCB 元件是由 ABS 组成，而 ABS 是由丙烯腈等单体聚合而成，因此，ABS 在加热时会释放出大量丙烯腈，而马达中的电线是由其他热塑性塑料制成，硬盘板则是由环氧树脂制成，这些都导致 RI-MM 和 RI-HD 内无丙烯腈检出。

NAOCC 在塑料熔融工艺过程有不同浓度水平的检出，以 ABS 车间浓度最高。研究发现，一些特殊用途的 ABS 中含有聚甲基丙烯酸甲酯，比如电子产品[37]，从而导致 ABS 过程产生高含量的甲基丙烯酸甲酯(70.0 mg/m³ ± 33.0 mg/m³)。NAOCC 在 PVC 和 PA 工艺内含量分别高达 60.56%和 56.39%，明显高于其他 VOCs 组分。对于 PVC，其中含量最高的是 OVOC，占总含量的 61.2%，其中含量最高的 OVOC 主要包括环戊酮、正丁醇和甲基丙烯酸甲酯，其浓度分别是(8.2 ± 0.2) mg/m³、(5.5 ± 0.1) mg/m³ 和 (1.2 ± 0.0) mg/m³，对应占总含量的 33.1%、22.1% 和 4.8%。塑料熔融过程比线路板回收过程排放更多的酮类化合物。环戊酮是 PP 和 PA 塑化工艺含量最高的 NAOCC。Ballistreri 等[38]在热分解 PA66 时同样发现以环戊酮为主要的热降解产物，他们认为 PA 容易发生热氧化反应，其中己二酸片段中的 CO 容易热分解失去而形成环戊酮。PP 比 PE 多出的支链—CH_3 致使热降解时形成较稳定的自由基，首先键断裂形成分子量较大的多聚物或大分子化合物，通过氧化反应、歧化反应和裂解反应等将中间产物氧化成分子量较小的化合物，包括烷烃、OVOC 和芳香烃等[39,40]。

3.3 挥发性有机物的暴露风险

工业有机废气的持续排放，必然会导致附近空气中 VOCs 含量不断增加，加之多数 VOCs 具有高挥发性，可通过呼吸、皮肤摄入等多种途径暴露渗透到身体内部，造成更为严重的负面影响。因此，评价人体通过呼吸暴露职业环境中 VOCs 的风险具有十分重要的意义。

电子垃圾拆解回收排放大气 VOCs 的呼吸暴露风险研究中，每种目标分析物的非致癌风险计算式为 $HR_i = \dfrac{C_i}{RfC_i}$，即用日常浓度（$C_i$，μg/m³）除以目标物对应的参考浓度（$RfC_i$，μg/m³），VOCs 各组分参考浓度均摘自文献[31]。呼吸暴露 VOCs 的非致癌风险指数如图 3-17 所示。若 HI 大于 1.0，说明此浓度的 VOCs 会对健康造成不利影响[41]。RI-TV 呼吸暴露 VOCs 的总危险指数（HI）值最高（1788.85），分别是 MC、ESI、LEBHD 和 EHF-TV 总危险指数的 213、223、278 和 24 倍。在 RI-TV 内，12 种 VOCs 的危险指数超过 1.0，4 种 VOCs 的指数介于 0.1 与 1.0 之间，表明它们均超过了潜在危害浓度[42]。需要特别指出的是 RI-TV 内产生的 2-甲基丙烯酸甲酯的危险指数最高，高达 1415，占总危险指数的 79%。另外，RI-TV 内 HHs 的总危险指数也远高于其他 4 个车间内，分别是 MC、ESI、LEBHD 和 EHF-TV 的 100 332、95 293、4806 和 3 倍。因此，需要特别注意旋转灰化炉拆解电子垃圾车间，加强旋转灰化炉的密封性，收集产生废气以便后续深度净化处理。

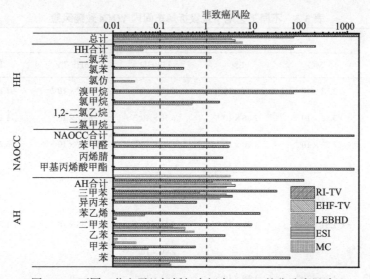

图 3-17　不同工艺电子垃圾拆解车间内 VOCs 的非致癌风险[1]

在所有产生的 VOCs 中，三甲苯是一种比较危险的污染物，在 MC（1.92），ESI（3.58），LEBHD（2.66）和 RI-TV（31.64）的危险指数均超过 1.0，在 EHF-TV 的危险指数也高达 0.68。除三甲苯外，在 MC，ESI，LEBHD 和 EHF-TV 产生的 VOCs 的危险指数几乎都没有超过 1.0，只有 EHT-TV 的一溴甲烷（71.15）和 MC 的 2-甲基丙烯酸甲酯(3.26)超过 1.0。苯和二甲苯是除 RI-TV 外的其他四个车间内的两种重要的潜在危害物，它们的危险指数介于 0.1 和 1.0 之间，表明苯和二甲苯也可能对车间工人造成潜在危害。

根据文献[41]描述，当化合物的致癌风险值 $>10^{-4}$ 被认定为其有确定致癌风险；当其介于 10^{-5} 和 10^{-4} 被认定为很可能有致癌风险；当其介于 10^{-5} 和 10^{-6} 认定为可能有致癌风险。如表 3-3 所示，RI-TV 车间内致癌风险(CR)非常高，该车间内已检测到的 VOCs 中丙烯腈(1.06×10^{-1})、苯(1.10×10^{-2})、乙苯(6.24×10^{-3})和对二氯苯(1.07×10^{-3})均具有确定致癌风险(1.0×10^{-4})。在 RI-TV 内总致癌风险(LCR)最高，其值高达 1.24×10^{-1}，分别是 MC，ESI，LEBHD 和 EHF-TV 致癌风险的 519、666、400 和 900 倍，是美国环保署制定的可接受致癌风险值(1.0×10^{-6})的 124 000 倍，是国际防辐射委员会推荐的最大可接受值(5×10^{-5})的 2480 倍。尽管 MC，ESI，LEBHD 和 EHF-TV 的致癌风险比 RI-TV 低 3 个数量级，分别为 2.4×10^{-4}，1.86×10^{-4}，3.11×10^{-4} 和 1.36×10^{-4}，但其 LCR 均超过了确定致癌风险值 10^{-4}。MC、ESI 和 LEBHD 车间大气中 1,2-二氯乙烷的致癌风险值超过了 10^{-4}，是最重要的致癌物。另外，MC、ESI、LEBHD 和 EHF-TV 的乙苯也有可能的致癌风险(LCR $>10^{-5}$)。

表 3-3　不同工艺电子垃圾拆解车间内 VOCs 致癌风险

化合物	LCR				
	MC	ESI	LEBHD	EHF-TV	RI-TV
AH	3.12×10^{-5}	2.86×10^{-5}	6.38×10^{-5}	6.07×10^{-5}	1.10×10^{-2}
苯	3.12×10^{-5}	2.86×10^{-5}	6.38×10^{-5}	6.07×10^{-5}	1.10×10^{-2}
乙苯	7.72×10^{-5}	1.90×10^{-5}	3.08×10^{-5}	1.17×10^{-5}	6.24×10^{-3}
NAOCC	—	—	—	—	1.06×10^{-1}
丙烯腈	—	—	—	—	1.06×10^{-1}
HH	1.31×10^{-4}	1.39×10^{-4}	2.16×10^{-4}	6.56×10^{-5}	1.07×10^{-3}
二氯甲烷	—	—	2.40×10^{-7}	—	—
1,2-二氯乙烷	1.31×10^{-4}	1.39×10^{-4}	2.16×10^{-4}	0	—
氯仿	—	—	—	6.56×10^{-5}	—
对二氯苯	—	—	—	—	1.07×10^{-3}
ΣLCR_i	2.40×10^{-4}	1.86×10^{-4}	3.11×10^{-4}	1.38×10^{-4}	1.24×10^{-1}

需要特别指出的是 EHT-TV 产生的 VOCs 含量远比 MC、ESI 和 LEBHD 高，而得到的总致癌风险值却反而更低，这可能是由于该车间内产生的绝对优势组分——一溴甲烷(占 VOCs 的 44.7%)没有确切的呼吸致癌单位风险而未能计算其致癌效应，另外一主要组分甲苯，由于缺乏致癌风险参考值未能计入总致癌风险。总之，在这五个车间尤其是 RI-TV 产生的 VOCs 的致癌及非致癌风险值均很高。所有车间内的 ΣHRs 均超过 1.0，ΣLCR_i 也超过了 10^{-4}，表明车间内释放的 VOCs 对车间工人有致癌和非致癌的危害。

如图 3-18 和图 3-19 所示，采用美国政府工业卫生专家协会制定方法计算的职业暴露风险值在 RI-TV、ABS(13) 和 PS(4.2) 车间内超过 1.0，对操作工人会造成一定的伤害，应采取一定的防护措施。具体来说，在 ABS 车间中，苯乙烯和丙烯腈对工人危害最大，其暴露风险值分别是 6.8 和 5.3，占总职业暴露的 51.1% 和 39.8%；PS 车间中，苯乙烯的暴露风险值高达 3.4，占了总暴露风险的 72.1%；RI-TV 车间内，苯(1.72)的暴露风险值超过 1.0，其他主要有害物质如丙烯腈(5.37×10^{-1})、一溴甲烷(3.93×10^{-1})和苯乙烯(2.42×10^{-1})均被划定为不能认定为致癌物的 A2 或 A3 组，它们的浓度也均超过了 0.1，表明对车间内工人可能存在潜在非致癌方面危害。除了 EHF-TV 车间的总职业暴露风险值(1.56×10^{-1})高于 0.1 外，MC、ESI 和 LEBHD 车间内的总职业暴露风险值均小于 0.1，其值分别为 6.01×10^{-3}、5.72×10^{-3} 和 1.18×10^{-2}，表明如果这些车间内的工人做好安全防护措施，车间内产生的 VOCs 不会对他们的身体健康造成明显伤害。

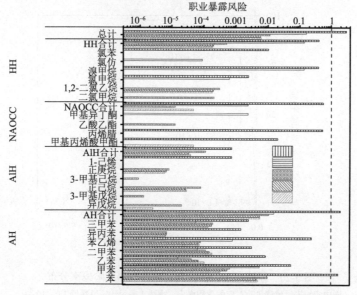

图 3-18 不同工艺电子垃圾拆解车间内 VOCs 的职业暴露风险

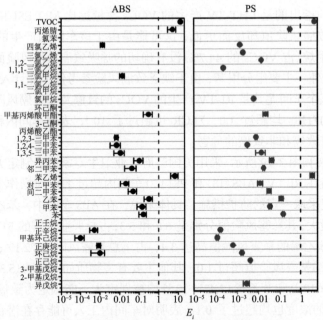

图 3-19　不同塑料塑化工艺产生 VOCs 职业暴露评价

由于个体代谢差异，进一步根据尿液中 BTEX 代谢产物的浓度水平评估了园区内职业工人的暴露风险。基于目前 BTEX 的代谢转化系数，仅对尿液中致癌性苯的代谢产物 S-PMA 进行了评估（图 3-20）。苯所导致的内暴露致癌风险平均值为

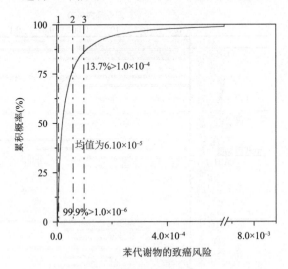

图 3-20　尿液中苯的代谢物造成的致癌风险概率分布

图中实线表示致癌风险对应的累积概率，虚线 1 表示致癌风险值 1.0×10^{-6}，虚线 2 表示致癌风险概率平均值 6.10×10^{-5}，虚线 3 表示致癌风险值 1.0×10^{-4}

6.10×10^{-5}（95%置信区间：$2.61 \times 10^{-6} \sim 2.23 \times 10^{-4}$），约为美国环保署规定的可接受值的 60 倍，而且 99.9%概率分布超过 1.0×10^{-6}。这进一步证实了电子垃圾拆解过程除多环芳烃及重金属等污染物之外，BTEX 的暴露不容小觑[43-45]。与仅通过呼吸暴露评估对比发现，基于内暴露评估的风险值要高出很多。这说明在职业场所暴露环境下，由于 BTEX 的亲脂特性，由于皮肤等暴露途径对内暴露剂量贡献很大。虽然这一结果尚未得到证实，但是前期的研究成果均表明皮肤暴露的可能性。研究发现，通过尿液中苯乙烯代谢产物可预测热塑性板厂内职业暴露于苯乙烯的皮肤暴露贡献[46]。此外，Caro 和 Gallego 发现[47]，游泳 2 h 比泳池工作 4 h 摄入更多的三氯甲烷，即除呼吸暴露和偶然的经口摄入外，皮肤暴露对 VOCs 内暴露的贡献不容忽视。因此，对于职业暴露场所，除环境大气和生物监测外，皮肤暴露评估将呈现重要的研究价值。

VOCs 除直接危害人体健康外，还可能造成拆解工人嗅觉损伤。Cheng 等[48]发现，与对照组比较，塑模工人暴露 1 个工作日后，其嗅觉有所降低，暴露组与对照组呈现显著的统计学差异。Tsai 等[49]调查了不同塑料回收车间内 VOCs 的嗅觉浓度（odor concentration，OC），计算发现 PVC 塑料回收厂、PE/塑料回收厂及 PP 塑料回收厂的嗅觉浓度分别为 447~1315 OC，0~100 OC 和 0~300 OC，绝大多数生产线 VOCs 的嗅觉浓度超过阈值 50 OC，而且不同生产线释放的乙苯嗅觉浓度均超过其阈值 13 OC。

综上所述，目前电子垃圾拆解过程大气毒害有机污染物的排放特征与相应的人体呼吸健康研究已经取得一定的进展，但是以下几个关键的环境问题还需在今后的研究中重点关注：①尽管针对拆解过程中典型高污染的 VOCs 的浓度水平进行了相对较多的研究和调查，但是其降解转化机理、转化产物污染水平及环境归趋的深入研究比较匮乏；②电子垃圾拆解过程中排放的 VOCs 对人体的暴露途径主要有经呼吸暴露和皮肤接触暴露，但是到底哪种途径对人体的健康危害更大目前还存在一定的争议；③拆解过程中大气毒害性物质的有效去除研究还比较有限，虽然一些联用技术取得了一定的进展，但是在降解去除过程中是否会生成某些不确定的中间产物，是否对环境造成二次污染等问题仍有待商榷。

（刘冉冉　李桂英　安太成）

参 考 文 献

[1] An T, Huang Y, Li G, et al. Pollution profiles and health risk assessment of VOCs emitted during e-waste dismantling processes associated with different dismantling methods [J]. Environment International, 2014, 73: 186-194.

[2] Liu R R, Ma S T, Yu Y Y, et al. Field study of PAHs with their derivatives emitted from e-waste dismantling

processes and their comprehensive human exposure implications [J]. Environment International, 2020, 144.

[3] Liu R, Ma S, Li G, et al. Comparing pollution patterns and human exposure to atmospheric PBDEs and PCBs emitted from different e-waste dismantling processes [J]. Journal of Hazardous Materials, 2019, 369: 142-149.

[4] Gangwar C, Choudhari R, Chauhan A, et al. Assessment of air pollution caused by illegal e-waste burning to evaluate the human health risk [J]. Environment International, 2019, 125: 191-199.

[5] Yang C, Miao G, Pi Y, et al. Abatement of various types of VOCs by adsorption/catalytic oxidation: A review [J]. Chemical Engineering Journal, 2019, 370: 1128-1153.

[6] He C, Cheng J, Zhang X, et al. Recent advances in the catalytic oxidation of volatile organic compounds: A review based on pollutant sorts and sources [J]. Chemical Reviews, 2019, 119(7): 4471-4568.

[7] Bon D M, Ulbrich I M, de Gouw J A, et al. Measurements of volatile organic compounds at a suburban ground site (T1) in Mexico City during the MILAGRO 2006 campaign: measurement comparison, emission ratios, and source attribution [J]. Atmospheric Chemistry and Physics, 2011, 11(6): 2399-2421.

[8] Wu R, Xie S. Spatial distribution of ozone formation in China derived from emissions of speciated volatile organic compounds [J]. Environmental Science and Technology, 2017, 51(5): 2574-2583.

[9] Wang H, Lou S, Huang C, et al. Source profiles of volatile organic compounds from biomass burning in Yangtze River Delta, China [J]. Aerosol and Air Quality Research, 2014, 14(3): 818-828.

[10] Kumar A, Singh D, Kumar K, et al. Distribution of VOCs in urban and rural atmospheres of subtropical India: Temporal variation, source attribution, ratios, OFP and risk assessment [J]. Science of the Total Environment, 2018, 613-614: 492-501.

[11] 陈烈强, 谢明权. 废印刷电路板回收处理技术的研究进展 [J]. 广东化工, 2008, (9): 100-103.

[12] Guo J, Luo X, Tan S, et al. Thermal degradation and pollutant emission from waste printed circuit boards mounted with electronic components [J]. Journal of Hazardous Materials, 2020, 382: 121038.

[13] Molto J, Egea S, Conesa J A, et al. Thermal decomposition of electronic wastes: Mobile phone case and other parts [J]. Waste Management, 2011, 31(12): 2546-2552.

[14] Liu R, Chen J, Li G, et al. Using an integrated decontamination technique to remove VOCs and attenuate health risks from an e-waste dismantling workshop [J]. Chemical Engineering Journal, 2017, 318: 57-63.

[15] Chen J, Zhang D, Li G, et al. The health risk attenuation by simultaneous elimination of atmospheric VOCs and POPs from an e-waste dismantling workshop by an integrated de-dusting with decontamination technique [J]. Chemical Engineering Journal, 2016, 301: 299-305.

[16] He Z, Li G, Chen J, et al. Pollution characteristics and health risk assessment of volatile organic compounds emitted from different plastic solid waste recycling workshops [J]. Environment International, 2015, 77: 85-94.

[17] Jang B N, Wilkie C A. A TGA/FTIR and mass spectral study on the thermal degradation of bisphenol A polycarbonate [J]. Polymer Degradation and Stability, 2004, 86(3): 419-430.

[18] Huang D Y, Zhou S G, Hong W, et al. Pollution characteristics of volatile organic compounds, polycyclic aromatic hydrocarbons and phthalate esters emitted from plastic wastes recycling granulation plants in Xingtan Town, South China [J]. Atmospheric Environment, 2013, 71: 327-334.

[19] Hall W J, Williams P T. Fast pyrolysis of halogenated plastics recovered from waste computers [J]. Energy & Fuels, 2006, 20(4): 1536-1549.

[20] Duan H, Li J. Thermal degradation behavior of waste video cards using thermogravimetric analysis and pyrolysis gas chromatography/mass spectrometry techniques [J]. Journal of the Air & Waste Management Association, 2012, 60(5): 540-547.

[21] Quan C, Li A, Gao N, et al. Characterization of products recycling from PCB waste pyrolysis [J]. Journal of Analytical and Applied Pyrolysis, 2010, 89(1): 102-106.

[22] Ortuno N, Conesa J A, Molto J, et al. Pollutant emissions during pyrolysis and combustion of waste printed circuit boards, before and after metal removal [J]. Science of the Total Environment, 2014, 499: 27-35.

[23] Williams P T, Williams E A. Interaction of plastics in mixed-plastics pyrolysis [J]. Energy & Fuels, 1999, 13(1):

188-196.

[24] Moltó J, Font R, Gálvez A, et al. Pyrolysis and combustion of electronic wastes [J]. Journal of Analytical and Applied Pyrolysis, 2009, 84(1): 68-78.

[25] Chiang H L, Lin K H. Exhaust constituent emission factors of printed circuit board pyrolysis processes and its exhaust control [J]. Journal of Hazardous materials, 2014, 264: 545-551.

[26] Cho M H, Jung S H, Kim J S. Pyrolysis of mixed plastic wastes for the recovery of benzene, toluene, and xylene (BTX) aromatics in a fluidized bed and chlorine removal by applying various additives [J]. Energy & Fuels, 2010, 24(2): 1389-1395.

[27] Chen D, Liu R, Lin Q, et al. Volatile organic compounds in an e-waste dismantling region: From spatial-seasonal variation to human health impact [J]. Chemosphere, 2021, 275.

[28] Chen J, Huang Y, Li G, et al. VOCs elimination and health risk reduction in e-waste dismantling workshop using integrated techniques of electrostatic precipitation with advanced oxidation technologies [J]. Journal of Hazardous materials, 2016, 302: 395-403.

[29] Blazso M, Czégény Z, Csoma C. Pyrolysis and debromination of flame retarded polymers of electronic scrap studied by analytical pyrolysis [J]. Journal of Analytical and Applied Pyrolysis, 2002, 64(2): 249-261.

[30] Wang Y, Zhang J. Thermal stabilities of drops of burning thermoplastics under the UL 94 vertical test conditions [J]. Journal of Hazardous materials, 2013, 246: 103-109.

[31] Demirbas A. Pyrolysis of municipal plastic wastes for recovery of gasoline-range hydrocarbons [J]. Journal of Analytical and Applied Pyrolysis, 2004, 72(1): 97-102.

[32] Qiu X, Marvin C H, Hites R A. Dechlorane plus and other flame retardants in a sediment core from Lake Ontario [J]. Environmental science & technology, 2007, 41(17): 6014-6019.

[33] Xiao K, Wang P, Zhang H D, et al. Levels and profiles of dechlorane plus in a major E-waste dismantling area in China [J]. Environmental Geochemistry and Health, 2013, 35(5): 625-631.

[34] Jafari A J, Donaldson J D. Determination of HCl and VOC emission from thermal degradation of PVC in the absence and presence of copper, copper(II) oxide and copper(II) chloride [J]. E-Journal of Chemistry, 2009, 6(3): 685-692.

[35] Kim S. Pyrolysis kinetics of waste PVC pipe [J]. Waste Management, 2001, 21(7): 609-616.

[36] Karaduman A, Simsek E H, Cicek B, et al. Flash pyrolysis of polystyrene wastes in a free-fall reactor under vacuum [J]. Journal of Analytical and Applied Pyrolysis, 2001, 60(2): 179-186.

[37] An J, Kim C, Choi B H, et al. Characterization of acrylonitrile-butadiene-styrene (ABS) copolymer blends with foreign polymers using fracture mechanism maps [J]. Polymer Engineering and Science, 2014, 54(12): 2791-2798.

[38] Ballistreri A, Garozzo D, Giuffrida M, et al. Mechanism of thermal decomposition of nylon 66 [J]. Macromolecules, 2002, 20(12): 2991-2997.

[39] Canevarolo S V. Chain scission distribution function for polypropylene degradation during multiple extrusions [J]. Polymer Degradation and Stability, 2000, 70(1): 71-76.

[40] Hinsken H, Moss S, Pauquet J R, et al. Degradation of polyolefins during melt processing [J]. Polymer Degradation and Stability, 1991, 34(1-3): 279-293.

[41] Ramirez N, Cuadras A, Rovira E, et al. Chronic risk assessment of exposure to volatile organic compounds in the atmosphere near the largest Mediterranean industrial site [J]. Environment International, 2012, 39(1): 200-209.

[42] Li G Y, Zhang Z Y, Sun H W, et al. Pollution profiles, health risk of VOCs and biohazards emitted from municipal solid waste transfer station and elimination by an integrated biological-photocatalytic flow system: A pilot-scale investigation [J]. Journal of Hazardous materials, 2013, 250: 147-154.

[43] Zhang B, Zhang T, Duan Y, et al. Human exposure to phthalate esters associated with e-waste dismantling: Exposure levels, sources, and risk assessment [J]. Environment International, 2019, 124: 1-9.

[44] Lin M, Tang J, Ma S, et al. Insights into biomonitoring of human exposure to polycyclic aromatic hydrocarbons with hair analysis: A case study in e-waste recycling area [J]. Environment International, 2020, 136: 105432.

[45] Wang H, Han M, Yang S, et al. Urinary heavy metal levels and relevant factors among people exposed to e-waste dismantling [J]. Environment International, 2011, 37(1): 80-85.

[46] Creta M, Moldovan H, Poels K, et al. Integrated evaluation of solvent exposure in an occupational setting: air, dermal and bio-monitoring [J]. Toxicology Letters, 2018, 298: 150-157.

[47] Caro J, Gallego M. Assessment of exposure of workers and swimmers to trihalomethanes in an indoor swimming pool [J]. Environmental Science and Technology, 2007, 41(13): 4793-4798.

[48] Cheng S F, Chen M L, Hung P C, et al. Olfactory loss in poly (acrylonitrile-butadiene-styrene) plastic injection-moulding workers [J]. Occupational Medicine-Oxford, 2004, 54(7): 469-474.

[49] Tsai C J, Chen M L, Chang K F, et al. The pollution characteristics of odor, volatile organochlorinated compounds and polycyclic aromatic hydrocarbons emitted from plastic waste recycling plants [J]. Chemosphere, 2009, 74(8): 1104-1110.

第4章 电子垃圾拆解排放持久性有机物污染

电子垃圾拆解处理过程包括机械破碎、人工敲打、加热焚烧等步骤，拆解处理活动将使电子垃圾内部聚集的灰尘逸出扩散，机械切割、摩擦、撞击等作用力可以产生新的粉尘颗粒物，而对有机组分的加热焚烧也可以形成新的颗粒污染物。在拆解处理过程造成颗粒物污染的同时，具有半挥发特性的有机污染物将以气态或颗粒态的形式进入周边环境，并通过吸附吸收、干湿沉降、大气或水体迁移等环境地球化学行为对局部、区域甚至全球的持久性有机物污染造成影响。本章首先阐述了电子垃圾拆解处理过程产生的颗粒物污染，然后对电子垃圾拆解处理所排放持久性有机污染物的污染特征进行了全面论述，重点讨论比较不同处理方式下多溴联苯醚(PBDEs)、多环芳烃(PAHs)及衍生物等典型持久性有机污染物的排放特征，并概述了电子垃圾拆解处理产生持久性有机污染物的相关研究进展。

4.1 颗粒物污染

颗粒物(particulate matter, PM)是大气中均匀分散的各种粒径大小不一的固体颗粒或液体微粒混合物的总称。通常按照它们的空气动力学粒径(D_p)进行分级，常见的包括 $PM_{2.5}$(空气动力学当量直径小于或等于 2.5 μm)、PM_{10}(空气动力学当量直径小于或等于 10 μm)和总悬浮颗粒物(TSP)等。相比于 TSP 和 PM_{10}，粒径更细的 $PM_{2.5}$ 对人体具有更高的健康风险[1]，因为 $PM_{2.5}$ 能被吸入人体的支气管和肺泡，在人体肺部具有更大的沉积系数[2]，且容易附着微生物、病毒和重金属等物质；当携带有害物质的 $PM_{2.5}$ 通过呼吸系统进入人体后，会损害人体健康，诱发各种疾病，引起或加重呼吸系统的疾病，甚至导致癌症[3,4]。环境流行病学研究证实，电子垃圾拆解工人直接接触 PM 易导致肺功能损伤，并易患哮喘和慢性阻塞性肺疾病(COPD)等小气道功能障碍性疾病[5]；孕期暴露于电子垃圾拆解活动产生的大气 $PM_{2.5}$ 影响甲状腺激素(THs)的基因 DNA 甲基化，从而可能增加新生儿大脑发育异常风险[6]；暴露于电子垃圾拆解活动释放的高浓度 $PM_{2.5}$ 显著增加特定位点的全局 DNA 甲基化水平[7]；从电子垃圾拆解地采集的 $PM_{2.5}$ 可能导致炎症反应、氧化应激和 DNA 损伤，且 $PM_{2.5}$ 中有机可溶性组分危害更大[8]。此外，粗放式的电子垃圾活动造成工人因心脏自主神经功能障碍且患心血管疾病的风险增加，如心率变异性降低和心率加快[9]。

4.1.1 颗粒物的来源

我国早期粗放式的电子垃圾拆解活动排放了大量的颗粒物,包括用烧蜂窝煤的炉子或电烤炉烘烤电路板回收电子元器件,用酸洗的方式提取贵金属,剩余的边角料将被焚烧,用过的废酸直接排放到河流及田地等,这些粗放式的电子垃圾拆解活动对环境破坏性极大。Chen 等[10]模拟了印刷线路板和塑料外壳热解过程颗粒物的生成特征,结果显示粒径为 0.4~2.1 μm 细颗粒物分别占线路板和塑料外壳热解排放 PM 的 78.9%和 89.3%,相应的排放因子分别为(9.68 ± 4.81)g/kg 和(18.49 ± 7.2)g/kg。与其他排放源(如煤炭、生物质和交通废气)相比,电子垃圾拆解排放颗粒物的排放因子相对较高;其中,线路板中碳和氮的初始含量与 PM 排放因子呈显著正相关($p<0.05$),而塑料外壳的相关关系不显著。在泰国某电子垃圾拆解活动现场观察到的数据显示,$PM_{2.5}$ 浓度增加与焚烧电子垃圾有关,而粗颗粒物 $PM_{2.5\sim10}$ 则主要来源于清扫和整理电子垃圾拆解活动现场向周围的大气中排放[11]。An 等[12]于 2007 年对我国华南地区某电子垃圾不同回收工艺车间的 TSP 进行了分析,结果显示正在作业的塑料回收车间、电子垃圾焚烧车间及电子垃圾拆解车间大气中 TSP 的浓度分别为 1.78 mg/m³、1.94 mg/m³ 及 2.21 mg/m³,该浓度 2.4~3.3 倍于非作业期间,明显超过我国《环境空气质量标准》(GB 3095—2012)关于 TSP 的标准(二级浓度限值 0.3 mg/m³)。Guo 等[13]于 2013 年对上海某一电子垃圾正规处理企业不同车间大气颗粒物(TSP、PM_{10} 和 $PM_{2.5}$)进行监测得出,4 个车间中(废电视机人工拆解车间、废电路板加热预处理车间、废电路板破碎分选车间和废塑料破碎车间)$PM_{2.5}$、PM_{10} 和 TSP 的平均浓度分别为 103~230 μg/m³、335~802 μg/m³ 和 442~1512 μg/m³,这表明车间内的环境空气受到电子垃圾回收活动的污染。车间内空气中的颗粒物排放与几个因素有关,例如电视机内部灰尘含量、使用的回收方法和设备以及车间内的空气循环等。由于收集的废电视机已经使用或存放了多年,大量的灰尘聚集在电视机内外,该粉尘在回收过程中被转移到环境空气中。电视机拆解车间的 PM_{10} 和 TSP 浓度最高,平均值分别为 802 μg/m³ 和 1512 μg/m³,但是电视机拆解车间的 $PM_{2.5}/PM_{10}$(0.27)最低,表明颗粒物主要以从电视机外部和内部收集的灰尘排放的粗颗粒物的形式存在。研究发现,$PM_{2.5}$ 的最高值出现在废电路板加热去除元器件车间(230 μg/m³)。这可以解释为,由于使用了电热炉,且废电路板的加热过程将释放出较多的细颗粒物,虽然操作台配备了负压罩,但人工加热脱焊过程仍会造成烟气的泄漏逸出到车间环境,废电路板加热去除元器件过程的污染控制将在第 8 章介绍。

4.1.2 颗粒物的组成

PM$_{2.5}$ 主要由离子组分、元素组分以及碳组分构成[14],其中碳组分包括有机碳(OC)和元素碳(EC)。OC 的来源较为复杂,可能是污染源直接排放的 PAHs 等一次有机碳(POC),也可能是气态前体物通过光化学反应生成的二次有机碳(SOC)。OC 中的有机含碳物质通常对环境和人体具有致癌性和致突变性等毒害效应。EC 的来源较为稳定,通常为来源于石油燃料和生物质的不完全燃烧,由污染源直接排放到大气中,其性质较为稳定,吸附能力较强,可以吸附其他有害物质。EC 通常与 OC 一起由生物质或者煤炭燃烧产生,极少单独生成。空气质量和气候变化均会受空气颗粒物中的 OC 和 EC 等组分的影响[15]。对废旧线路板回收作坊内空气颗粒物 PM、OC 和 EC、有机物(OM)及重金属的浓度进行了分析(表 4-1),PM 的浓度范围为 1129~1688 μg/m^3,平均为(1430 ± 200.8)μg/m^3。PM 的主要成分是碳质颗粒物,平均含量占 PM 的 50.9%,其中主要有机物(OM)占 48.4%[16]。OM 的主要有机成分是有机磷酸酯类,包括磷酸三苯酯(TPP)及其甲基取代物、十六酸甲酯、十八酸甲酯、左旋葡聚糖和双酚 A(BPA)等,EC 对 PM 的贡献较小。有机磷酸酯及 BPA 广泛用作印刷线路板的阻燃剂和增塑剂,而十六酸甲酯、十八酸甲酯、左旋葡聚糖可能来源于煤炭和生物质燃烧。

表 4-1 线路板回收车间大气颗粒物中 PM、OC 和 EC、主要有机物(OM)的污染水平[16]

项目	平均浓度(μg/m^3)	占 PM 比例(%)
PM	1430±200.8	—
OC	493.6±67.4	34.5
EC	36.2±6.3	2.5
OC+EC	529.8±72.3	37.0
OM	691.0±94.4	48.4
左旋葡聚糖	6.79±1.90	0.48
十六酸甲酯	20.9±5.72	1.47
十六酸	1.34±0.45	0.09
十八酸甲酯	9.24±2.84	0.65
十八酸	0.69±0.33	0.048
二十酸甲酯	1.42±0.63	0.099
双酚 A	1.11±0.14	0.078
磷酸三苯酯	46.6±16.3	3.26
ΣC_1-TPP	66.5±19.7	4.65
ΣC_2-TPP	47.7±11.6	3.34
ΣC_3-TPP	59.9±18.4	4.19

通过网格化布点，Chen 等[17,18]采集分析了我国广东某典型电子垃圾拆解地及周边地区的大气颗粒物，包括 TSP、PM_{10} 和 $PM_{2.5}$，结果如图 4-1 所示，在电子垃圾拆解地区，TSP 浓度范围是 69.9~473 μg/m³，其中园区的 TSP 浓度最高，为 190~473 μg/m³。在园区内，春季的 TSP 浓度最高，其次是冬季和秋季(平均浓度分别是 290 μg/m³ 和 269 μg/m³)，夏季的 TSP 浓度最低，是 191 μg/m³。

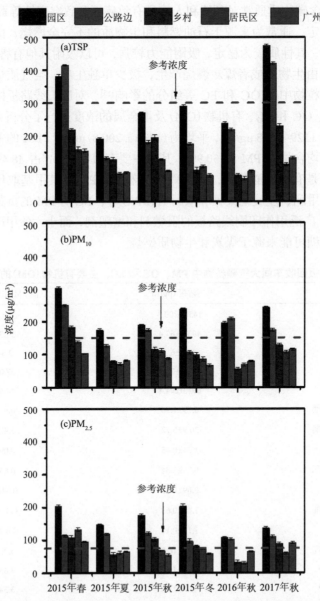

图 4-1　不同采样点 TSP (a)、PM_{10} (b) 和 $PM_{2.5}$ (c) 浓度的时间变化和空间分布[18]

除了 2015 年春季和 2017 年秋季的园区内和公路旁以外,其他采样点的 TSP 浓度都低于《环境空气质量标准》(GB 3095—2012)的 TSP 二类区域参考浓度 (300 μg/m³)。此外,园区采样点的 TSP 平均浓度为 280 μg/m³,相比稍微高于清远(109~314 μg/m³)[19]和早期电子垃圾拆解地区的浓度(30.2~201 μg/m³)[20]。然而,园区周边地区的 TSP 浓度(乡村和居民区采样点 TSP 的平均浓度分别为 162 μg/m³ 和 110 μg/m³)低于广东典型电子垃圾拆解地区的浓度。从年变化趋势而言,TSP 总体上呈现先下降后上升趋势,在 2017 年达到最高。例如,园区采样点在 2015 年、2016 年和 2017 年的 TSP 平均浓度分别是 269 μg/m³、237 μg/m³ 和 473 μg/m³。这些采样点的污染物浓度受到气象条件的影响[18],采样点在夏季时经常下雨,颗粒物容易发生湿沉降,导致颗粒物浓度在夏季明显下降。

电子垃圾拆解地区 PM_{10} 和 $PM_{2.5}$ 浓度范围分别是 55.6~302 μg/m³ 和 30.8~204 μg/m³。其中,在园区内,春季的 $PM_{2.5}$ 浓度最高,平均浓度是 204 μg/m³,其次是冬季、秋季和夏季,平均浓度分别是 203 μg/m³、174 μg/m³ 和 146 μg/m³。在春冬季节,大气边界层高度较低,不利于污染物扩散,导致颗粒物在大气中累积,浓度偏高。在夏季时雨量充沛,颗粒物容易发生湿沉降,导致颗粒物浓度在夏季明显下降。大多数采样点的 $PM_{2.5}$ 浓度超过我国《环境空气质量标准》(GB 3095—2012)的 $PM_{2.5}$ 二类区域参考浓度(75 μg/m³)。值得一提的是,自 2016 年起,乡村和居民区的大气颗粒物浓度与广州城区的大气颗粒物浓度相当甚至更低,说明当地政府将拆解作业集中在工业园区内进行,对周边地区大气颗粒物的污染控制起到了积极作用。

4.1.3 颗粒物的粒径分布

1. 排放特征

荷电低压撞击器(electrical low pressure impactor,ELPI,芬兰 Dekati 公司)可以实时分级测量直径在 0.03~10 μm 范围内大气颗粒物的粒径分布,其工作原理主要基于颗粒荷电,颗粒物在层叠撞击器中的惯性分级以及气溶胶颗粒的电荷检测。大气样品进入 ELPI 后,颗粒物先被带电离子荷电,荷电后的颗粒被低压撞击器分级捕获,且带电颗粒撞击到每一级上产生的瞬时电流被测量记录,电流值越大,对应的颗粒数量越多。ELPI 的空气采样速率为 10 L/min,电流值每秒记录一次。ELPI 根据撞击器的切割粒径(0.03 μm、0.06 μm、0.108 μm、0.17 μm、0.26 μm、0.4 μm、0.65 μm、1.0 μm、1.6 μm、2.5 μm、4.4 μm、6.8 μm、10 μm)将颗粒物分为空气动力学粒径几何均值为 0.039 μm、0.071 μm、0.12 μm、0.20 μm、

0.31 μm、0.48 μm、0.76 μm、1.22 μm、1.94 μm、3.06 μm、5.12 μm、8.08 μm 的 12 个通道，然后分别测出不同粒径通道颗粒物的质量浓度(单位是 μg/m³)、数量浓度(单位是个/立方厘米，count/cm³)和肺沉积表面积(lung deposited surface area，LDSA，单位是 μm²/cm³)。

利用 ELPI 对上海某一正规电子垃圾回收处理企业进行了不同车间环境中颗粒物污染的现场监测研究[21]，分别在废电视机人工拆解车间(TV-D)、废阴极射线管拆解车间(CRT-D)、废电路板加热去除元器件车间(PCB-H)、废电路板破碎-分选车间(PCB-C)、拆解产物存储仓库(WH)和办公楼室内(Office)进行实时在线监测，其中 PCB-C 车间分别在废电路板进料口(PCB-C-I)和出料口(PCB-C-O) 2 个点位进行了现场监测(图 4-2)。

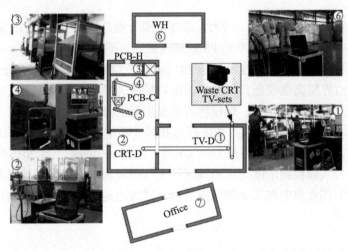

图 4-2　不同车间大气颗粒物的 ELPI 采样示意图

不同采样点大气颗粒物的质量浓度、数量浓度和 LDSA 如表 4-2 所示，不同采样点颗粒物的荷电电流值、LDSA 及数量浓度粒径分布等值线随采样时间的变化规律如图 4-3 所示。

表 4-2　大气颗粒物的质量浓度(PM)、数量浓度(PN)和肺沉积表面积(LDSA)

采样点	PM_{10}	$PM_{10-2.5}$	$PM_{2.5}$	$PM_{0.1}$	PN_{10}	$PN_{0.1}$	LDSA 平均值	10 百分位	90 百分位
	μg/m³				count/cm³		μm²/cm³		
Office	25.4	—	25.4	0.7	$1.09×10^4$	$6.79×10^3$	39.3	38.2	40.5
TV-D	379	278	101	8.9	$1.04×10^5$	$8.71×10^4$	197	156	230
CRT-D	302	219	83.4	4.2	$5.96×10^4$	$4.56×10^4$	143	104	188
PCB-H	498	270	228	4.8	$7.04×10^4$	$4.67×10^4$	237	202	273
PCB-C-I	235	148	86.8	2.5	$3.35×10^4$	$2.43×10^4$	90.6	74.2	107

续表

采样点	PM_{10}	$PM_{10-2.5}$	$PM_{2.5}$	$PM_{0.1}$	PN_{10}	$PN_{0.1}$	LDSA 平均值	10 百分位	90 百分位
		$\mu g/m^3$			count/cm^3		$\mu m^2/cm^3$		
PCB-C-O	243	137	106	5.2	7.08×10^4	5.29×10^4	173	149	189
WH-1	99.7	55	44.7	1.1	1.70×10^4	1.15×10^4	53.9	50.1	58.7
WH-2	248	181	67.2	1.6	2.41×10^4	1.49×10^4	83.4	58.6	120

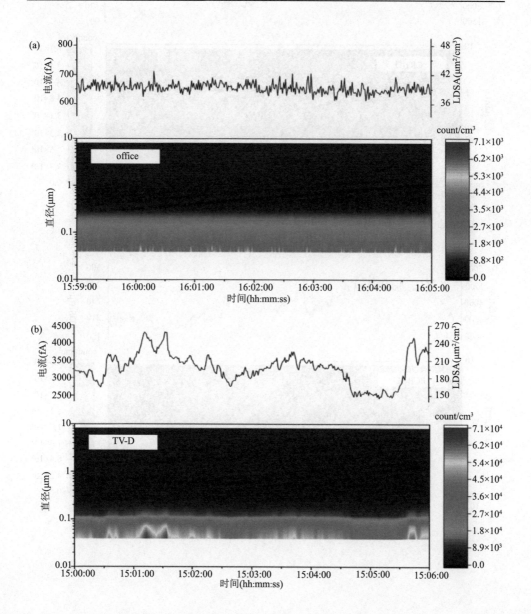

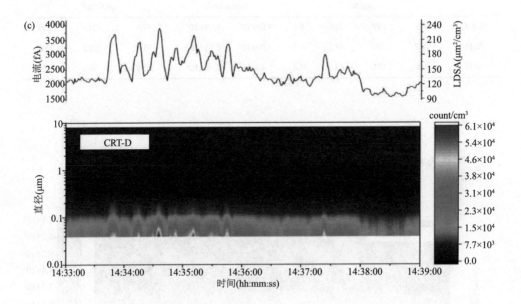

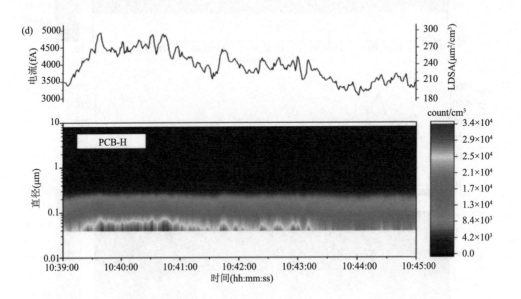

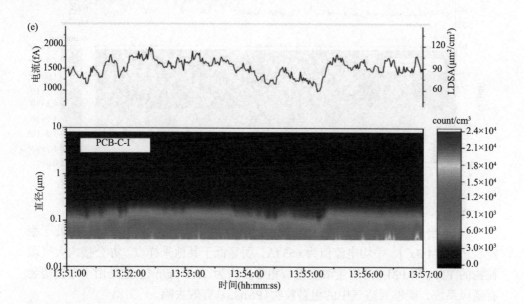

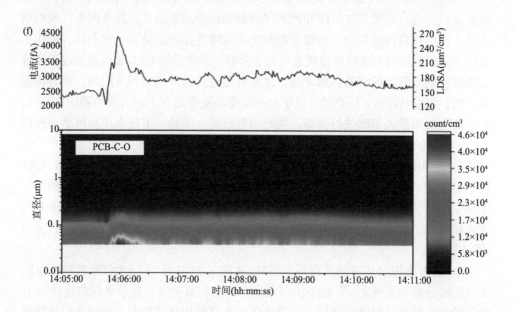

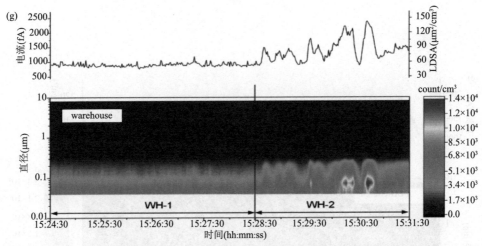

图 4-3　大气颗粒物随采样时间的电流值、肺沉积表面积及数量浓度的粒径分布

(1) 办公楼室内 (office)：办公楼内大气颗粒物的电流值和粒径分布均处于稳定状态[图 (4-3a)]，平均电流值为 655 fA，明显低于其他采样点。办公楼内大气颗粒物的 PM_{10} 和 $PM_{2.5}$ 的质量浓度近似相等，均为 25.4 $\mu g/m^3$。这是由于办公楼装有新风系统，能够对空气中的粗颗粒物 ($PM_{10\sim2.5}$) 有效去除。

(2) 废电视机人工拆解车间 (TV-D)：图 4-3(b) 显示了采样时间内颗粒物电流值的强烈波动，表明了 TV-D 车间大气颗粒物污染的复杂性，其车间大气颗粒物主要来自电视机内的灰尘、拆解过程和叉车运输等操作活动。工作期间，操作工人将废阴极射线管 (CRT) 电视机进行人工拆解，拆解后的 CRT 显示器通过滚轮输送机转移到废阴极射线管拆解车间，废塑料通过传送带传送到车间外，而其他拆解产物，如废电路板、扬声器、电子枪和电线等被收集在不同的收纳箱中。其间，不定期有叉车进入车间进行拆解产物的物料搬运。因此，上述多项拆解活动形成了 TV-D 车间颗粒物污染的复杂特性。

(3) 废阴极射线管拆解车间 (CRT-D)：操作工人首先使用切割机切断 CRT 显示器包裹的金属支撑带，然后采用加热电阻丝法加热焊缝实现显示器的屏、锥玻璃的分离，接着工人在封闭的负压操作室通过真空吸尘装置将屏玻璃上的荧光粉吸除。因此，颗粒物电流值的波动是由切割机的间歇式切割过程引起的[图 4-3(c)]。

(4) 废电路板加热去除元器件车间 (PCB-H)：PCB-H 车间大气颗粒物数量浓度的粒径分布等值线图[图 4-3(d)]与其他车间有明显差异，其超细颗粒物 ($PM_{0.1}$) 的数量浓度较高且保持时间较长，这表明废电路板加热过程中超细颗粒物的持续排放。前期研究也证实，在加热拆解废电路板上电子元器件时，由于人工拔除电子元器件和敲击电路板基板等操作过程将造成加热电路板产生烟气逸出负压集气罩，从而导致 PCB-H 车间的细颗粒物污染[22]。

（5）废电路板破碎-分选车间（PCB-C）：废电路板破碎-分选生产线采用多级破碎和风力分选相结合的方法进行回收，生产线配备旋风除尘器和袋式除尘器，可以去除大部分的细颗粒物料粉尘。该车间设置了两个采样点，废电路板进料口和分选后物料出料口。操作工人首先将废电路板等物料倾倒入生产线进料口，然后通过传送皮带将废电路板转移到破碎机，再经多级破碎和风力分选得到铜颗粒和废电路板树脂粉两类物料，在出料口有工人进行物料的打包及搬运。图 4-3(e) 和 (f) 显示了 PCB-C 车间中两个采样点大气颗粒物的 LDSA 和数量浓度粒径分布等值线图。从中可以得出，出料口处（PCB-C-O）的 LDSA（173 $\mu m^2/cm^3$）和颗粒物数量浓度平均值（7.08×10^4 count/cm^3）要高于进料口处（PCB-C-I）的 LDSA（90.6 $\mu m^2/cm^3$）和颗粒物数量浓度（3.35×10^4 count/cm^3），这主要由于破碎分选等机械处理过程产生了较多的物料颗粒及粉末，在出料口处铜颗粒表面附着粉尘将造成一定程度的颗粒物污染。

（6）拆解产物存储仓库：拆解处理得到的拆解产物通过叉车转移到仓库。ELPI 装置放在仓库保管员附近，旁边有个磅秤，用于入库物品的称重记录。在采样期间，实时数据可分为前 4 min 的 WH-1 段和后 3 min 的 WH-2 段。在 WH-1 期间，仓库处于静止状态，没有发生操作活动，所以 WH-1 的电流值较低且稳定。与 WH-1 的稳态相比，WH-2 的电流值存在波动，这主要由于叉车运输拆解产物进库造成的。在 WH-2 期间，仓库保管员指挥叉车司机给拆解产物称重、入库。此时，叉车燃烧柴油所排放的尾气及地面扬尘被 ELPI 的采样泵吸入收集，从而在图 4-3(g) 的 WH-2 段可明显地观察到细颗粒物数量浓度的增加。数据显示，叉车称重入库操作使仓库车间大气颗粒物的平均数量浓度从 1.70×10^4 count/cm^3（WH-1）增加到 2.41×10^4 count/cm^3（WH-2），PN_{10} 和 $PN_{0.1}$ 分别增加了 41.8%和 29.6%。在 0.071~0.20 μm 的粒径范围内，颗粒物的数量浓度增加较大，空气动力学粒径几何均值为 0.071 μm、0.12 μm 和 0.2 μm 的粒径通道内，其颗粒物的数量浓度分别从 4506 count/cm^3、2823 count/cm^3、2074 count/cm^3 增加到 7278 count/cm^3、4883 count/cm^3、3476 count/cm^3，分别增加了 61.5%、73.0%和 67.6%。另外，叉车的移动也会导致地面灰尘的再悬浮。

从表 4-2 可以得出，不同处理车间的 PM_{10} 的范围在 99.7~498 μg/m^3 之间，$PM_{0.1}$ 的范围在 0.7~8.9 μg/m^3 之间。PM_{10} 最高的是 PCB-H 车间（498 μg/m^3），其次是 TV-D、CRT-D 和 PCB-C，均高于办公室 PM_{10} 的质量浓度（25.4 μg/m^3）。而 TV-D 车间的 $PM_{0.1}$ 浓度最高（8.9 μg/m^3），其次是 PCB-C-O、PCB-H 和 CRT-D。对于不同粒径颗粒物的质量分布，$PM_{0.1}$ 仅占 $PM_{2.5}$ 质量的 2.1%~8.8%，而 $PM_{2.5}$ 质量占 PM_{10} 的 27%~46%。ELPI 检测得出到的颗粒物质量浓度与大气颗粒物采样器检测到的 PM_{10} 结果具有可比性。利用中流量采样器（100 L/min，采样时间为 8 h）

检测相同处理车间的 PM_{10} 和 $PM_{2.5}$ 发现[13],其值分别为 TV-D 车间的 802 μg/m³ 和 218 μg/m³、PCB-H 车间的 397 μg/m³ 和 230 μg/m³、PCB-C 车间的 353 μg/m³ 和 103μg/m³、仓库内的 70 μg/m³ 和 42 μg/m³,与 ELPI 所监测的 PM_{10} 和 $PM_{2.5}$ 的质量浓度相当。

2. 数量浓度

图 4-4 是不同采样点大气颗粒物的数量浓度(PN)的箱线图,可以看出,TV-D 内 PN_{10} 的平均值最高(1.04×10^5 count/cm³),然后依次是 PCB-C-O(7.08×10^4 count/cm³)、PCB-H(7.04×10^4 count/cm³)、CRT-D(5.96×10^4 count/cm³)和 PCB-C-I(3.35×10^4 count/cm³)。办公楼的 PN_{10} 平均值(1.09×10^4 count/cm³)明显低于其他采样点。超细颗粒物(UFPs,粒径小于 0.1 μm $PN_{0.1}$)占颗粒物数量个数的大部分,其 $PN_{0.1}$ 与 PN_{10} 的比值在 0.62~0.84 范围内。

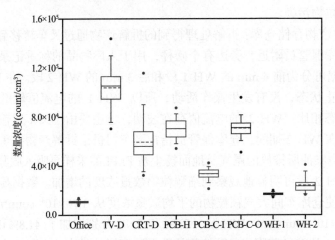

图 4-4 不同采样点大气颗粒物的数量浓度的箱线图
每个方框内的实线和虚线分别表示中位数和平均浓度,框表示第 25 和第 75 百分位,上下限表示第 5 和第 95 百分位,实心圆点表示最大值和最小值

一般地,粗颗粒的质量占比较大,但细颗粒物的数量浓度更大。不同采样点的颗粒物数量浓度的粒径分布如图 4-5 所示。颗粒物数量浓度随粒径的减小而增加,且当颗粒物粒径均小于 0.2 μm 时,数量浓度值均有明显增加。对于 PCB-H 车间,当粒径小于 0.5 μm 时,其颗粒物数量浓度值开始增加;对于 TV-D 车间,当粒径小于 0.1 μm 时,其颗粒物数量浓度值显著增加,最小粒径段(0.037 μm)颗粒物的数量浓度最高,接近 5.0×10^4 count/cm³。

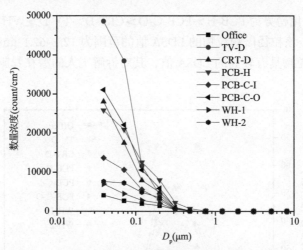

图 4-5 不同采样点的大气颗粒物数量浓度的粒径分布

3. 肺沉积表面积

LDSA 结合了肺沉积和呼吸系统中颗粒的表面化学性质，可以作为表征颗粒物暴露的一个评价指标。对于一定质量的颗粒物，其表面积随着粒径的减小而增大。虽然人体吸入的 UFPs 的质量浓度和质量占比很低，但 UFPs 具有较高的数量浓度，且容易沉积在肺泡区域。另外，细颗粒物，特别是 UFPs，由于单位质量的表面积大，可作为其他毒害污染物的载体进入肺泡区甚至人体血液循环，其作用效应与颗粒的表面化学性质等因素有关，并对人体健康造成影响。目前关于电子垃圾拆解处理车间的 LDSA 数据还比较少，表 4-2 显示了不同采样点的 LDSA 平均值及第 10 和 90 百分位的 LDSA 值，不同采样点颗粒物的 LDSA 平均值范围为 39.3~237 μm²/cm³，办公楼内 LDSA 值最低，为 39.3 μm²/cm³，明显低于车间内的浓度。办公室的 LDSA 值与巴塞罗那城市背景环境中的 37 μm²/cm³[23]和葡萄牙里斯本城市环境中的 35~89.2 μm²/cm³ 相当[24]。处理车间内 LDSA 值约为办公楼浓度的 3~6 倍，其中 PCB-H 车间的 LDSA 值最高 (237 μm²/cm³)。从图 4-6 所示的 LDSA 粒径分布，可以得出 LDSA 属于单峰分布模式，当粒径大于 2 μm 时，LDSA 值很低，这是由于粗颗粒物能够被人体鼻腔和气管去除，且粗颗粒很难进入肺泡区域，从而沉积在肺部的粗颗粒物较少。随着粒径的减小，LDSA 值先增大后减小。TV-D 和 PCB-H 车间中 LDSA 的粒径分布与其他采样点的不同。TV-D 车间的 LDSA 峰值 (67.3 μm²/cm³) 出现在 0.07 μm 粒径附近，其他采样点的 LDSA 峰值出现在 0.2 μm 的粒径附近。在 PCB-H 车间中，当粒径小于 2 μm 时，LDSA 值急剧增加，而其他 LDSA 值在粒径小于 0.5 μm 时，LDSA 值才开始增加。换句话说，在 0.2~2 μm 的粒径范围内，PCB-H 的 LDSA 值远高于其他车间。当颗粒物粒径为 0.2 μm 时，

LDSA 峰值的变化趋势为 PCB-H＞PCB-C-O＞CRT-D＞TV-D＞WH-2＞PCB-C-I＞WH-1＞Office，该粒径段颗粒物的 LDSA 值的范围为 12.1~67.1 μm²/cm³。因此，PCB-H 车间颗粒物具有较高的 LDSA 值，其对拆解工人的潜在暴露风险较大，应该特别关注。

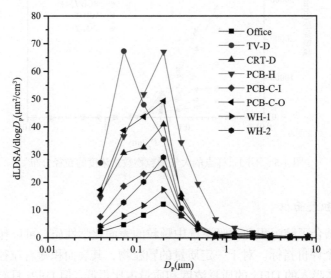

图 4-6　不同采样点大气颗粒物肺沉积表面积(LDSA)的粒径分布

根据我国生态环境部《废电器电子产品拆卸运行和生产管理指南》（2015 年版），电子垃圾正规处理企业可以采用人工拆解、机械处理、热处理相结合的处理方法，其车间颗粒物的排放源包括许多方面：电子垃圾内部的聚集灰尘、物料机械破碎产生的粉尘、PCB 加热释放的废气以及叉车的废气排放。通过不同车间现场的实时监测发现，不同处理车间的颗粒物污染状况有所不同，其颗粒物的排放与电子垃圾材料组分和处理工艺等因素有关。废电视机人工拆解车间产生的 TSP 浓度最高，而废电路板加热过程导致严重的 $PM_{2.5}$ 污染。另外，由于叉车在不同车间的物料运送中被广泛使用，叉车的尾气排放成为车间大气污染的一个移动源。

4.2　持久性有机污染物污染

持久性有机污染物(persistent organic pollutants, POPs)是指通过各种环境介质（大气、水、生物体等）能够长距离迁移并长期存在于环境，具有长期残留性、生物蓄积性、半挥发性和高毒性，对人类健康和环境具有严重危害的天然或人工合成的有机污染物质。2001 年 5 月 22 日，联合国环境规划署(UNEP)在瑞典斯德哥尔摩主持通过了《关于持久性有机污染物的斯德哥尔摩公约》(简称《斯德哥尔摩公约》），旨在保护人类健康和环境免受 POPs 影响。《斯德哥尔摩公约》于 2004 年 5 月 17 日正式生效，最初提出淘汰包括滴滴涕(DDT)、氯丹、灭蚁灵、六氯苯、毒杀芬、艾氏剂、狄氏剂、异狄氏剂、七氯、多氯联苯、二噁英、多氯代二苯并呋喃等在内的 12 种 POPs。截至 2022 年 5 月，又有 18 种(类)新 POPs 物质被增列入公约管制物质名单，包括 α-六氯环己烷、β-六氯环己烷、林丹、十氯酮、五氯苯、五氯苯酚及其盐类和酯类、六氯丁二烯、多氯萘、短链氯化石蜡、三氯杀螨醇、硫丹及其相关异构体、六溴联苯、六溴环十二烷(HBCD)、四溴联苯醚和五溴联苯醚、六溴联苯醚和七溴联苯醚、全氟辛基磺酸及其盐类和全氟辛基磺酰氟、全氟辛酸(PFOA)及其盐类和相关化合物等。事实上，符合 POPs 定义的化学物质还远远不止上面所提到的 12 种或 18 种，名单也在不断地更新之中，一些污染物如 PAHs 里面的部分化合物也符合 POPs 的特性，被称为广义上的 POPs 而受到关注。

粗放式的电子垃圾拆解活动已经被证实是 POPs 的一个重要排放源。早期粗放式的电子垃圾拆解活动导致多种有毒害重金属及有机污染物释放到环境中，造成严重的生态危害，也对电子垃圾拆解从业人员及拆解区周边居民的健康造成了损伤。除了第 2 章阐述的重金属污染物外，粗放式拆解各种电子垃圾废塑料含有聚氯乙烯(PVC)、聚乙烯(PE)、聚丙烯(PP)、聚乙烯对苯二甲酸酯(PET)等，以及各种塑化剂、溴化阻燃剂(BFRs)、调色剂、表面涂层等多种有害物质，这些物质在高温或加热作用下从塑料基体中释放出，且可能发生转化形成新的污染物，从而释放到环境介质中，其中第 3 章对排放 VOCs 的污染特征进行了阐述。表 4-3 对电子垃圾拆解地区最常见的 POPs 进行了总结，这些 POPs 主要分为两大类，其中多氯联苯、短链氯化石蜡、六溴联苯、HBCD、四溴联苯醚和五溴联苯醚、六溴联苯醚和七溴联苯醚、全氟辛基磺酸及其盐类和全氟辛基磺酰氟、全氟辛酸(PFOA)及其盐类和相关化合物等主要在拆解时直接排放到环境中，而 PAHs、多氯萘、二噁英、多氯代二苯并呋喃等则是粗放式电子垃圾拆解活动(焚烧或热解)无意产生的副产物。

表 4-3　粗放式电子垃圾拆解活动产生的 POPs

序号	化学品名称	主要来源
1	多环芳烃(PAHs)	来自电子垃圾元器件拆解焚烧以及各种煤和生物质能源用于烘烤电路板
2	多氯萘(PCNs)	来自电子垃圾拆解焚烧以及废旧电路板二次提炼贵金属过程
3	二噁英、多氯代二苯并呋喃(PCDD/Fs)	来自电子垃圾拆解焚烧以及废旧电路板二次提炼贵金属过程
4	多氯联苯(PCBs)	拆解含多氯联苯电力装置及其废物

续表

序号	化学品名称	主要来源
5	短链氯化石蜡(SCCPs)	含短链氯化石蜡阻燃剂的电子垃圾拆解排放
6	六溴联苯(hexabromobiphenyl)	含六溴联苯阻燃剂的电子垃圾拆解排放
7	六溴环十二烷(HBCD)	含HBCD阻燃剂的电子垃圾拆解排放
8	四溴联苯醚和五溴联苯醚(tetra-PBDEs; penta-PBDEs)	含工业品五溴联苯醚阻燃剂的电子垃圾拆解排放
9	六溴联苯醚和七溴联苯醚(hexa-PBDEs; hepta-PBDEs)	含工业品八溴联苯醚阻燃剂的电子垃圾拆解排放
10	全氟辛基磺酸(PFOS)及其盐类和全氟辛基磺酰氟	含全氟辛基磺酸类添加剂的电子垃圾拆解排放
11	全氟辛酸(PFOA)及其盐类和相关化合物	含全氟辛酸类添加剂的电子垃圾拆解排放

4.2.1 多溴联苯醚

多溴联苯醚(polybrominated diphenyl ethers，PBDEs)属于含溴阻燃剂的一种，具有阻燃效率高、热稳定性好、价格便宜及对材料性能影响小等特点，被广泛地应用在电子、电器、化工、交通、建材、纺织等领域中，家用电器、计算机、室内装潢中的泡沫塑料、地毯和布料中约有 5%~30%的成分是 PBDEs[25]。PBDEs 曾是世界上使用量第二大的溴代阻燃剂，根据取代溴原子个数和位置的不同，理论上 PBDEs 共有 209 种同系物，常见 PBDEs 同系物的物理化学性质见表 4-4[26]。PBDEs 和多氯联苯(polychlorinated biphenyls, PCBs)一样都按国际纯粹与应用化学联合会(IUPAC)编号系统命名。PBDEs 在室温下具有环境稳定性、脂溶性高、不易降解等特性，熔点为 310~425℃，在水中溶解度小。PBDEs 具有相当稳定的化学结构，很难通过物理、化学或生物方法降解[27]。

表 4-4　PBDEs 的基本理化性质

名称	辛醇-空气分配系数 ($\log K_{OA}$) (15~45℃)	辛醇-水分配系数 ($\log K_{OW}$)	蒸气压(Pa) 25℃	水溶性(μg/L) 25℃
BDE-28	8.71~9.99	5.98	2.19×10^{-3}	0.07
BDE-47	9.43~11.13	6.55	1.86×10^{-4}	1.5×10^{-2}
BDE-66	9.67~11.52	6.73	1.22×10^{-4}	1.8×10^{-2}
BDE-77	9.84~11.49	6.96	6.79×10^{-5}	6×10^{-3}
BDE-85	10.54~12.31	7.03	9.86×10^{-6}	6×10^{-3}
BDE-99	10.26~11.85	7.13	1.76×10^{-5}	9.4×10^{-3}
BDE-100	9.99~11.76	6.86	2.86×10^{-5}	0.04

续表

名称	辛醇-空气分配系数 ($\log K_{OA}$) (15~45℃)	辛醇-水分配系数 ($\log K_{OW}$)	蒸气压(Pa) 25℃	水溶性(μg/L) 25℃
BDE-138	—	7.91	1.58×10^{-6}	—
BDE-153	10.53~12.32	7.62	2.09×10^{-6}	8.7×10^{-4}
BDE-154	10.79~12.46	7.39	3.8×10^{-6}	8.7×10^{-4}
BDE-183	10.98~11.96	7.14	4.68×10^{-7}	1.5×10^{-3}
BDE-209	—	9.97	2.95×10^{-9}	4.17×10^{-6}

根据含溴原子数的不同，商用 PBDEs 产品可大致分为三类：五溴联苯醚(penta-BDEs)、八溴联苯醚(octa-BDEs)和十溴联苯醚(deca-BDEs)(表 4-5)[28]。其中，penta-BDEs 和 octa-BDEs 是由多种 PBDEs 同系物组成的混合物，其组成成分随厂商的不同而有所变化，而 deca-BDEs 则以十溴联苯醚为主(BDE-209 占比超过97%)，还含有少量九溴联苯醚。

表 4-5 商用多溴联苯醚中 PBDEs 同系物组成特征

商用 PBDEs	五溴联苯醚	八溴联苯醚	十溴联苯醚
四溴联苯醚	24%~38%(BDE-47)		
五溴联苯醚	50%~60%(BDE-99、BDE-100)		
六溴联苯醚	4%~8%	10%~12%(BDE-153)	
七溴联苯醚		44%(BDE-183)	
八溴联苯醚		31%~35%(BDE-196、BDE-197)	
九溴联苯醚		10%~11%(BDE-207)	<3%(BDE-206、BDE-207)
十溴联苯醚		<1%	97%~98%(BDE-209)

由于 PBDEs 在早期产品中的大量使用，在环境介质和生物体内均检测到 PBDEs 的存在，如大气、沉积物或污泥、鱼类、人体血液和脂肪组织、母乳等[29-31]，PBDEs 单体对人体神经系统、内分泌系统和生殖系统等具有较大危害，从而引起了公众、政府部门和学者的广泛关注。欧盟于 2003 年首次禁止 penta-BDEs 和 octa-BDEs 的使用[32]，并于 2008 年禁止在电子电气设备中使用 deca-BDEs[33]。加拿大环保组织在 2006 年发布了 penta-BDEs 和 octa-BDEs 的毒性效应研究，于 2009 年禁止在新产品中使用 penta-BDEs 和 octa-BDEs，同时政府呼吁在新产品中采取措施限制 deca-BDEs 的释放[34]。21 世纪初，美国局部地区(华盛顿、缅因州)禁止使用 PBDEs，到 2004 年全国禁止使用 penta-BDEs 和 octa-BDEs，直到 2013 年全

面禁止使用 deca-BDEs。2009 年 5 月，联合国环境规划署正式将四溴联苯醚和五溴联苯醚、六溴联苯醚和七溴联苯醚列入《斯德哥尔摩公约》，十溴联苯醚也于 2019 年被正式列入《斯德哥尔摩公约》管控名单，虽然 PBDEs 在全球大部分地区已被禁用，但由于早期大量使用，使得环境中有较高的 PBDEs 赋存量，可通过食物链传递富集进入哺乳动物和人体。另外，只要含 PBDEs 的产品一直在使用，则 PBDEs 对人群的暴露一直存在，尤其对于 PBDEs 污染严重的电子垃圾拆解地区，沉积物、土壤和室内灰尘中都报道了全球较高水平的 PBDEs，且相应暴露人群的血液中也有高浓度的 PBDEs 检出，这将给拆解区生态环境和人体健康带来巨大的潜在风险。

1. 大气中的 PBDEs

1)电子垃圾拆解场地

PBDEs 广泛应用于不同的电子电气设备中，在电子垃圾回收或拆解活动中容易被释放出来。在电子垃圾拆解过程中会排放有机废气，其中含有高浓度的 PBDEs 等污染物[35,36]。对我国广东某典型电子垃圾拆解工业园区及其周边地区的大气样品中 PBDEs 进行研究[18]，采样地点共有五个(图 4-7)，分别是电子垃圾拆解工业园区(EP)和周边地区包括公路旁边(RS)、乡村(VS)、居民区对照点(RA)，以及代表典型城市生活的对照点——距离电子垃圾拆解地区大约 400 公里的广州市(GZ)。使用流量为 0.3 m³/min 的大气采样器采集 $PM_{2.5}$、PM_{10} 和 TSP 以及对应的气相样品，采样时间为 8 小时，采样高度距离地面 1.5~2 m。颗粒物使用石英纤维滤膜采集，气相污染物使用聚氨基甲酸酯泡沫塞(PUF)吸附。在 2015 年四个季节和 2016 年、2017 年秋季，每个季节采集 15 对样品(大气颗粒物及对应的气相样品)，一共采集 180 个样品。

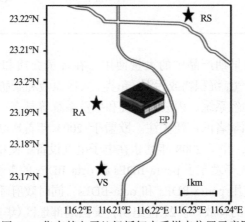

图 4-7 华南某电子垃圾拆解点采样点位置示意图

不同采样点大气样品中 PBDEs 的浓度水平如图 4-8 所示。在电子垃圾拆解地区，气相和 TSP 中 PBDEs 的总浓度范围为 43.7~211 ng/m³。PBDEs 在 EP 点的浓度最高，为 211 ng/m³，其次是 RS 点、VS 点、GZ 点和 RA 点，分别是 89.5 ng/m³、88.3 ng/m³、80.2 ng/m³ 和 43.67 ng/m³，EP 点的 PBDEs 浓度明显高于其他采样点（Mann-Whitney U test，$P<0.05$）。这些结果说明该地区电子垃圾拆解过程是 PBDEs 的重要来源。在电子垃圾拆解地区，TSP、PM_{10} 和 $PM_{2.5}$ 上的 PBDEs 浓度分别为 2.7~72.7 ng/m³、2.4~70.0 ng/m³ 和 1.4~67.9 ng/m³，62.3%~89.9%的 PBDEs 附着在 $PM_{2.5}$ 上。此外气固相分配特征显示，64.8%~96.7%的 PBDEs 存在于气相中。对照点广州市也有类似的结果，93.6%的 PBDEs 存在于气相中，表明大气中绝大部分的 PBDEs 存在于气相中和 $PM_{2.5}$ 上。在电子垃圾拆解地区，$PM_{2.5}$ 上 PBDEs 浓度的中位值为 67.9 ng/m³，远低于电子垃圾拆解车间内的颗粒物上 PBDEs 浓度（254~2656 ng/m³[37]和 26.7~11 800 ng/m³[35]）。然而，相比于文献中报道电子垃圾拆解地区户外大气的结果，电子垃圾拆解园区大气中 PBDEs 浓度比 2005 年广东某地[38]、2007 年台州[39]和 2011 年清远[40]电子垃圾拆解地区的浓度（分别是 5.3~39.0 ng/m³、0.51~1.66 ng/m³ 和 0.013~0.234 ng/m³）都高。这说明电子垃圾的集中拆解，使得工业园区大气中的 PBDEs 污染更加集中，还需配套相应的控制措施减少污染排放。

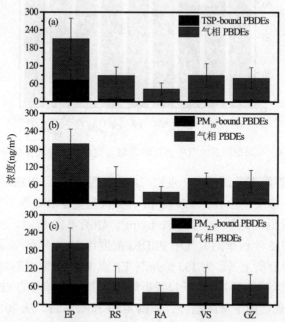

图 4-8 不同采样点颗粒相[TSP(a)、PM_{10}(b) 和 $PM_{2.5}$(c)]和气相中 PBDEs 浓度空间分布

气固相分布方面，三溴联苯醚大部分存在于气相中，四溴联苯醚在气相和颗

粒相中均有分布，而五溴联苯醚则大部分存在于颗粒相中。如图4-9所示，三溴联苯醚、四溴联苯醚和五溴联苯醚占总PBDEs的比重最高。在气相中，三溴联苯醚和四溴联苯醚浓度最高，占了总PBDEs的92.9%~98.5%；在颗粒相中，四溴联苯醚和五溴联苯醚浓度最高，占了总PBDEs的75.6%~92.8%。在其他电子垃圾拆解地区，也有类似的结果[35, 38]。PBDEs同系物的辛醇-空气分配系数（K_{OA}）随着溴原子数增加而增加[35]。因此，低溴联苯醚倾向于分布在气相中，而高溴联苯醚则倾向于分布在颗粒相中，尤其在$PM_{2.5}$上。各采样点大气中PBDEs同系物的组分构成与DE-71和Bromakl 70-5DE阻燃剂商业品最为接近，而电子电气设备的材料中往往会添加这些商业品来提高材料的阻燃性能。因此，该地区大气中PBDEs主要来源于这些PBDEs商业品。但大气中PBDEs同系物中的低溴组分比例更高，除了与PBDEs阻燃剂的组成有关外，还可能由于电子垃圾拆解过程中PBDEs发生脱溴降解，或是挥发出来的高溴PBDEs在大气中发生光化学脱溴反应形成。

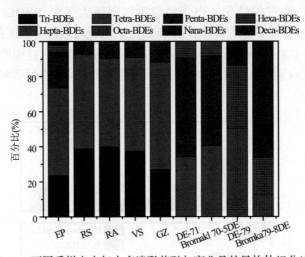

图4-9 不同采样点大气中多溴联苯醚与商业品的异构体组分比较

在电子垃圾拆解工业园区（EP点），2015年四季中，PBDEs的浓度（TSP加气相）春季时最高，浓度中位数为274 ng/m³，其次是夏季和冬季，分别为212 ng/m³和201 ng/m³，最低的是秋季，为132 ng/m³，如图4-10所示。2015~2017年的秋季中，EP点的2016年秋季的大气中PBDEs浓度比2015年秋季低，为93.7 ng/m³，到2017年秋季时有所上升，为135 ng/m³。EP点PBDEs浓度随时间的变化，主要与其电子垃圾的拆解量有关。电子垃圾拆解地区其他采样点的PBDEs浓度随时间变化，总体上呈现逐渐下降的趋势，如图4-10和图4-11。从2015年春季到2017年秋季，RA点PBDEs浓度从82.8 ng/m³下降至8.2 ng/m³，VS点PBDEs浓度从165 ng/m³下降至3.4 ng/m³，RS点PBDEs浓度从159 ng/m³下降至2.5 ng/m³，浓度下降了 1~2 个数量级，说明这个地区的 PBDEs 污染正在减轻。电子垃圾拆解地区和对照点都位于亚热带地区，在一定程度上，这些采样点的大气污染物浓度受到亚热带季风气候的影响。一般来说，在夏季和秋季，由于受到来自海洋的夏季风的影响，大气污染物会被稀释和扩散；而在冬季和春季，由于受到来自大陆的冬季风的影响，中国北方的大气污染物会被转运到中国南方地区，大气污染物会有所增加[41]。采样点大气中 PBDEs 浓度的季节变化符合一般大气污染物在季风气候影响下的变化规律。

人体通过呼吸对 PBDEs 的"中等"和"高"日摄入量(EDI)，分别通过 $PM_{2.5}$ 和气相中 PBDEs 浓度的中位数和第 95 百分位数计算获得。各采样点 PBDEs 的"中等"和"高"日摄入量分别为 3.5~133.8 ng/(kg·d) 和 10.9~332.4 ng/(kg·d)；考虑 PBDEs 在人体肺部的生物有效性，PBDEs 相应"中等"和"高"日吸收量(EDU)分别降低至 2.6~94.6 ng/(kg·d) 和 8.1~241.0 ng/(kg·d)；EP 点人们的 PBDEs 日吸收量最高，非致癌风险也最高，"高"风险指数 HI 为 1.1(大于 1.0)，说明 EP 点大

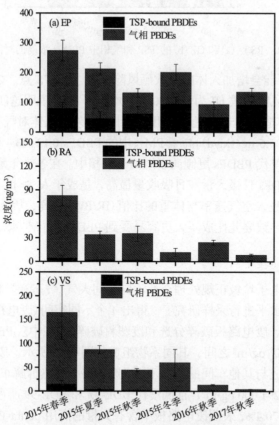

图 4-10　EP 点(a)、RA 点(b) 和 VS 点(c) 的 TSP 和气相中 PBDEs 浓度的时间变化

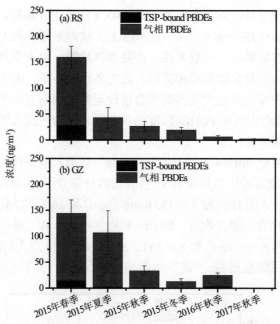

图 4-11　RS 点(a)和 GZ(b)的 TSP 和气相中 PBDEs 浓度的时间变化

气中的 PBDEs 可能会增加人体的非致癌风险。RS 点、VS 点、GZ 点和 RA 点的 PBDEs 的"高"风险指数 HI 为 0.11~0.43，均小于 1.0，说明这四个采样点大气中的 PBDEs 对人体造成的非致癌风险相对较小。此外，人体对气相中的 PBDEs 的日吸收量[2.45~197.6 ng/(kg·d)]高于 $PM_{2.5}$ 中的 PBDEs[0.12~43.4 ng/(kg·d)]，这是因为高毒性的低溴代 PBDEs 更倾向于游离于气相中。在各个年龄段的居民中，婴儿(<1 岁)的 PBDEs 日摄入量和日吸收量最高，是成年人(≥18 岁)的 2~3 倍。这主要是由于婴儿吸入空气速率与体重的比值(IR/BW)较高，因为他们需要更大的氧气量[42]，从而导致婴儿比成年人更容易受到 PBDEs 污染带来的潜在健康影响。

2) 电子垃圾处理车间

对上海某一电子垃圾正规处理企业不同车间大气(气态和颗粒态)中 PBDEs 的化学组成和浓度水平进行采样研究[13]，得出 4 个不同车间(废电视机人工拆解、废电路板加热拆解、废电路板破碎分选和废塑料破碎)大气中 Σ_{12}PBDEs 的平均浓度在 6780~2.28 × 10⁶ pg/m³ 之间，其同系物组成如图 4-12 所示。废电路板加热拆解车间的同系物组成与其他车间的明显不同，废电路板加热拆解车间气态 PBDEs 的浓度最高，达到 291 000 pg/m³，且低溴代 BDE-47 和 BDE-99 所占比例最高，其两者之和占比达到 70.4%，表明废电路板中含有大量的商用 penta-PBDEs 产品，而其他车间大气中主要以 BDE-209 为主，其中废电视机拆解和塑料破碎车间大气中

BDE-209 的贡献率超过了 98%，这说明废塑料和废电视机内含有一定数量的商用 deca-PBDEs 产品。4 个车间颗粒物 $PM_{2.5}$ 和 PM_{10} 中Σ_{12}PBDEs 浓度分别为 6.8~6670 µg/g 和 32.6~6790 µg/g。废电路板加热预处理车间内 $PM_{2.5}$ 和 PM_{10} 中 PBDEs 的浓度分别为 3740 µg/g 和 1220 µg/g，其中 BDE-47、BDE-99 等低溴代联苯醚所占比例较高，而 BDE-209 的所占比例小于 10%。这主要因为废电视机拆解得到电路板主要以单层纸基酚醛树脂板为主，其中含有大量的添加型商用 penta-PBDEs 产品，容易在加热条件下从基板中释放出来并吸附在颗粒物中，从而污染车间大气环境。在废电路板破碎-分选车间，$PM_{2.5}$ 中低溴代联苯醚的所占比例要高于 PM_{10} 中的含量。塑料粉碎车间 $PM_{2.5}$ 和 PM_{10} 中 PBDEs 的浓度分别为 6670 µg/g 和 6790 µg/g，含有 BDE-206、BDE-207、BDE-208 和 BDE-209 等高溴代联苯醚，其中 BDE-209 是其主要同系物，占Σ_{12}PBDEs 的 95%，且颗粒物中 PBDEs 同系物组成与商用 deca-PBDEs 的组成类似。Die 等[43]对 3 家电子垃圾正规拆解处理企业不同操作区域进行了大气样品采集，得出大气中 PBDEs 的浓度范围为 $0.58\sim2.89\times10^3$ ng/m^3，其主要分布在颗粒相中(90.7%~99.9%)，且 BDE-209 的贡献率达到 84.0%~97.9%。电子垃圾正规处理车间大气中 PBDEs 污染主要与废电视机各材料组分、表面积灰及处理工艺等因素有关。例如，废电视机拆解处理导致内部积灰的释放迁移，电路板基板和废塑料等含 PBDEs 产品中 PBDEs 受热向空气中挥发释放，还有废电路板破碎分选过程产生含 PBDEs 的细小微粒等。

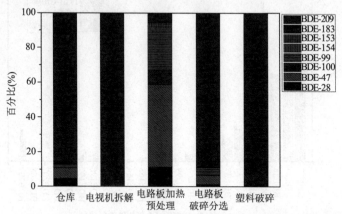

图 4-12　不同车间大气(气态和颗粒态)中 PBDEs 的同系物组成分布

3) 不同处理车间大气 PBDEs 的气粒分配和粒径分布

Chen 等[44]利用安德森八级撞击式采样器(Tisch TE-20-800，美国)分别在广东某循环经济产业园区两个车间(小型电子垃圾产品人工拆解车间和废电路板加热-人工拆解元器件车间，分别简写为 SE-DisW 和 PCB-TherW)和上海某电子垃圾正规处理企业两个车间(废电视机人工拆解车间和废电路板机械破碎-分选车间,分别简写为 TV-DisW 和 PCB-CruW),进行了车间工作运行状态下,大气样品颗粒物的分级采集和气态样品的现场监测研究。研究得出,TV-DisW、PCB-CruW、SE-DisW 和 PCB-TherW 四个车间大气气态Σ_{25}PBDEs 的浓度分别为 2.19 ng/m³、1.33 ng/m³、40.5 ng/m³ 和 10.6 ng/m³,即上海某正规电子垃圾拆解处理企业两个车间气态 PBDEs 浓度低于广东某循环经济产业园内处理车间。这与电子垃圾的拆解处理活动及车间的通风状况有一定关系,广东某地的 SE-DisW 车间进行人工拆解小型电子垃圾产品,车间通风条件差,而其他三个车间在进行拆解处理作业时,均配备负压通风装置,使车间内的气体得到较好的循环换气,一定程度上避免了气态 PBDEs 浓度的不断升高。图 4-13 显示了不同车间气态 PBDEs 的同系物所占比例,可以发现四个车间在气态 PBDEs 同系物组成上没有太大差异,主要以低溴代联苯醚同系物为主,其中三溴、四溴联苯醚同系物(BDE-17、BDE-25、BDE-28、BDE-32、BDE-35 和 BDE-37 为三溴联苯醚同系物,BDE-47、BDE-49/71 和 BDE-66 为四溴同系物)占总气态 PBDEs 的含量超过 85%,其中 TV-DisW、PCB-CruW、SE-DisW 和 PCB-TherW 车间中三溴、四溴联苯醚同系物占总气态 PBDEs 的 88%、91%、96% 和 95%。

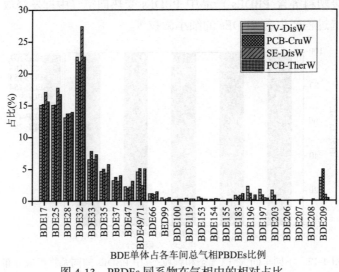

图 4-13 PBDEs 同系物在气相中的相对占比

在 TV-DisW、PCB-CruW 和 SE-DisW 中,除粒径<0.4 μm 颗粒物外,颗粒态 PBDEs 的浓度随粒径的减小而减小,而在 PCB-TherW 车间中,颗粒态 PBDEs 的浓度随粒径的减小而增大(图 4-14),这与车间中颗粒物的生成规律及粒径分布有一定关系。在 TV-DisW、PCB-CruW 和 SE-DisW 三个车间中,颗粒物主要由人工

拆解或机械破碎等操作造成的粉尘或粗颗粒物生成，而在 PCB-TherW 车间，废电路板在热应力作用下，一方面释放产生了较多数量的小粒径颗粒物，另一方面释放出大量的低溴代联苯醚同系物，由于小颗粒物具有较大的比表面积，从而吸附更多的 PBDEs。研究结果显示，TV-DisW、PCB-CruW 和 SE-DisW 车间大气中颗粒态Σ_{25}PBDEs 的浓度分别为 56 ng/m³、14.3 ng/m³ 和 57.7 ng/m³，明显低于 PCB-TherW 车间大气颗粒态Σ_{25}PBDEs 浓度（156 ng/m³）。

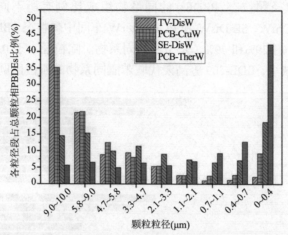

图 4-14　不同粒径颗粒中 PBDEs 在总颗粒态 PBDEs 中占比

将颗粒态 PBDEs 的质量浓度单位（ng/m³）转换为 μg/g，用于表征颗粒物中 PBDEs 的质量分数和粒径分布。图 4-15 表征了四个车间不同粒径范围颗粒物中 Σ_{25}PBDEs 的质量浓度，可以得出，TV-DisW、PCB-CruW、SE-DisW 和 PCB-TherW 等车间大气中Σ_{25}PBDE 在不同粒径颗粒物中的浓度范围分别为 50.0~104 μg/g、

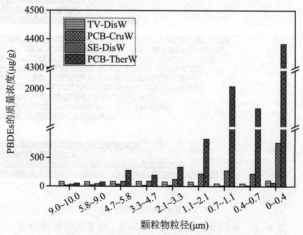

图 4-15　PBDEs 在不同粒径颗粒物中的质量浓度（μg/g）

12.3~64.2 μg/g、33.0~764 μg/g 和 51.6~4383 μg/g。在 SE-DisW 和 PCB-TherW 车间中，随着颗粒物粒径的减少，颗粒态 PBDEs 的质量浓度有明显的增加趋势，尤其在 PCB-TherW 车间，粒径小于 2.5 μm 的细颗粒物中，颗粒态 PBDEs 的质量浓度增加尤为显著。对于 TV-DisW 和 PCB-CruW 车间，不同粒径颗粒态 PBDEs 的质量浓度变化不大。

从不同车间颗粒态 PBDEs 同系物组成的粒径分布可以看出（图 4-16），四个车间大气中不同粒径颗粒态 PBDEs 的同系物组成比例有所不同，BDE-209 在 TV-DisW、PCB-CruW、SE-DisW 和 PCB-TherW 车间中颗粒态 PBDEs 总量的比重分别为 75%、65%、29%和 49%，是主要的同系物。除粒径小于 0.4 μm 的颗粒物外，随着粒径的减小，BDE-209 等高溴代联苯醚同系物的所占比例有下降的趋势。

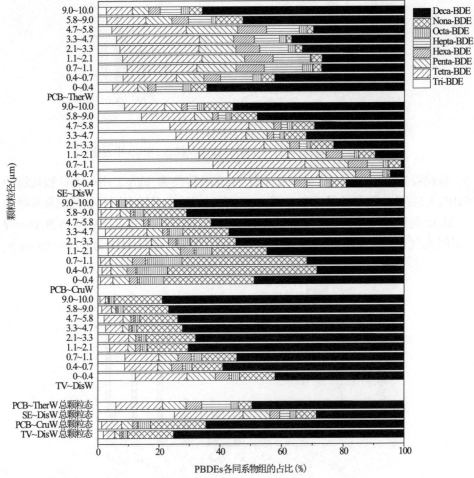

图 4-16 不同车间颗粒物中 PBDEs 同系物的粒径分布

图 4-17 表征了不同车间大气颗粒物中不同溴代联苯醚的质量中位数空气动力学直径(MMAD), 从图可以看出, 在 TV-DisW、PCB-CruW 和 SE-DisW 三个车间中, 不同溴代联苯醚的 MMAD 值随溴原子数量增加呈现增加的趋势, 这说明随着溴原子数量的增加, 高溴代联苯醚更趋向于在粒径大的颗粒物上聚集。在 TV-DisW 和 PCB-CruW, 不同溴代联苯醚的 MMAD 值均大于 2.5 μm, 表明 PBDEs 趋向于聚集在粗颗粒物中。对于 SE-DisW 车间, 三溴至八溴联苯醚的 MMAD 值位于 1~2.5 μm 之间, 而九溴、十溴及总 PBDEs 的 MMAD 值大于 2.5 μm。在 PCB-TherW 车间中各种溴代联苯醚的 MMAD 值均小于 2.5 μm, 甚至小于 1 μm, 即表明该车间颗粒态 PBDEs 主要富集在粒径较小的细颗粒物上。

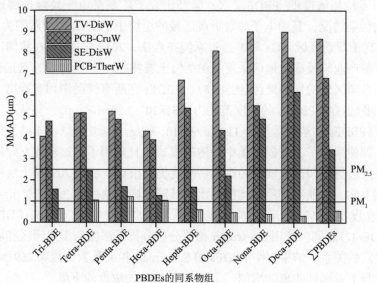

图 4-17 PBDEs 同系物的质量中位直径(MMAD)

2. 灰尘中的 PBDEs

Wang 等[45]在 2007 年采集了来自珠三角城市区域和靠近珠三角的电子垃圾回收区域的居民区室内灰尘和庭院灰尘, 发现在电子垃圾回收区 PBDEs 的浓度为 227~160 000 ng/g(中值浓度: 1441 ng/g), 而在市区为 530~44 000 ng/g(中值浓度: 4530 ng/g)。研究区域的灰尘样品中 BDE-202 的存在, BDE-197 与 BDE-201 的比率(电子垃圾区: 2.36; 市区: 1.81; 商业品: 28.5), 以及 nona-BDEs 与 deca-BDE 的比率(电子垃圾拆解区: 0.41; 市区: 0.10; 商业品: 0.06), 均表明它们可能来自 deca-BDE 的环境降解。Qiao 等[46]在 2015 年采集了华南某电子垃圾回收区的车间灰尘, 研究结果显示灰尘中 PBDEs 浓度为 36 180~418 746 ng/g(中值浓度:

126 944 ng/g），其中 BDE-209 是浓度最高（中值浓度：169 637 ng/g）和最丰富的 PBDEs（占比：97.6%）。Lin 等[47]在 2019 年采集了广东某电子垃圾拆解区和濠江对照区的居民区住宅灰尘，电子垃圾区住宅灰尘中 PBDEs 的浓度为 58.4~247 000 ng/g（中值浓度：7060 ng/g），约高出对照区浓度（中值浓度：78.3 ng/g，范围：44.7~16 400 ng/g）（$p<0.001$）两个数量级。

3. 土壤中的 PBDEs

PBDEs 具有较高的土壤/沉积物吸着系数（log K_{OC}），可与土壤中的有机碳等相结合，再加上土壤自身的低迁移特性，导致了电子垃圾拆解区污染点源附近土壤中的蓄积了较高浓度的 PBDEs。Xu 等[48]研究了广东某电子垃圾拆解地土壤中 PBDEs 的污染情况，其中电子垃圾拆解区域的土壤中 PBDEs 浓度范围为 13~1014 ng/g，远高于背景区域。Ge 等[49]通过系统布点法，2018 年 11 月在我国广东某电子垃圾拆解产业园及周边地区采集了 107 份土壤样品，其中 24 份来自园区内，其余样品来自园区外的周边地区（图 4-18）。PBDEs 在所有样品中均有检出，电子垃圾拆解工业园区内 PBDEs 的浓度范围为 $1.54×10^3$~$3.10×10^5$ ng/g，显著高于拆解园周边地区（PBDEs 的浓度范围为 11.6~$3.60×10^4$ ng/g；$p<0.05$）（表 4-6）。该研究结果表明，即使采用了工业园区集中拆解的模式，且采用了更加环境友好的拆解方式，电子垃圾拆解园区土壤中的污染物浓度仍然很高。BDE-209 是土壤中主要的 PBDEs 同系物，在所有样品中均能检出，其园区内的浓度为 $3.88×10^4$ ng/g，周边地区的浓度为 458 ng/g。而其他同系物的检出率在 38.4%（BDE-138）到 94.9%（BDE-47）之间。使用 Spearman 相关分析分析了浓度与工业园区和周边地区的距离之间的关系，结果表明 PBDEs 的 Spearman 相关系数值为 –0.269（$p<0.05$），表明随着与工业园区距离的增加，土壤中 PBDEs 的浓度均下降。

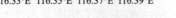

图 4-18　电子垃圾拆解园区、塑料回收站和采样点的位置信息

表 4-6 园区内与周边地区土壤中 PBDEs 的浓度

同系物	园区内(ng/g)					周边地区(ng/g)				
	Min	25%	Median	75%	Max	Min	25%	Median	75%	Max
BDE-17	n.d.	14.1	47.2	91.4	$1.10×10^3$	n.d.	n.d.	n.d.	1.21	30.4
BDE-28	n.d.	123	280	656	$4.67×10^3$	n.d.	n.d.	1.07	2.53	96.7
BDE-47	13.9	543	$1.19×10^3$	$2.28×10^3$	$1.82×10^4$	n.d.	1.63	5.00	15.0	532
BDE-66	4.30	127	372	938	$5.74×10^3$	n.d.	n.d.	1.21	3.90	168
BDE-71	6.59	162	399	781	$6.02×10^3$	n.d.	n.d.	n.d.	2.01	6.26	159
BDE-85	n.d.	20.4	56.0	99.4	971	n.d.	n.d.	n.d.	1.13	24.9
BDE-99	21.4	684	$1.23×10^3$	$2.16×10^3$	$1.73×10^4$	n.d.	1.21	4.74	17.2	580
BDE-100	n.d.	60.8	102	178	$2.36×10^3$	n.d.	n.d.	0.85	2.01	47.1
BDE-138	n.d.	13.3	37.8	63.8	466	n.d.	n.d.	n.d.	n.d.	19.8
BDE-153	5.98	226	394	589	$3.73×10^3$	n.d.	0.73	2.22	8.98	159
BDE-154	4.28	67.7	103	181	$1.31×10^3$	n.d.	n.d.	1.35	3.47	45.2
BDE-183	10.2	287	465	682	$2.75×10^3$	n.d.	0.17	2.65	17.0	208
BDE-190	n.d.	n.d.	19.4	40.3	209	n.d.	n.d.	n.d.	1.15	17.8
BDE-196	2.97	53.8	65.8	88.3	329	n.d.	n.d.	0.82	2.91	25.8
BDE-197	4.01	66.1	118	145	368	n.d.	n.d.	1.10	3.73	52.2
BDE-203	5.69	66.9	83.1	115	348	n.d.	n.d.	1.60	4.05	32.2
BDE-206	26.7	877	$1.10×10^3$	$1.60×10^3$	$6.35×10^3$	n.d.	n.d.	6.85	55.5	748
BDE-207	37.0	625	810	$1.01×10^3$	$3.03×10^3$	n.d.	3.35	10.9	50.8	465
BDE-208	24.8	347	473	610	$2.06×10^3$	n.d.	4.00	7.84	24.6	264
BDE-209	$1.35×10^3$	$3.19×10^4$	$3.88×10^4$	$4.98×10^4$	$2.34×10^5$	9.78	40.4	458	$3.15×10^3$	$3.38×10^4$
ΣPBDEs	$1.54×10^3$	$3.62×10^4$	$4.63×10^4$	$6.22×10^4$	$3.10×10^5$	11.6	61.8	575	$3.46×10^3$	$3.60×10^4$

和我国北方(天津，河北)、亚洲、大洋洲以及非洲一些国家相比，我国广东某典型电子垃圾拆解地土壤中 PBDEs 的浓度相对较高(表 4-7)。天津子牙循环经济园内土壤中 PBDEs 的浓度为 138 ng/g，其周边地区土壤中为 16 ng/g[50]。澳大利亚墨尔本两个电子垃圾拆解站周边地区土壤中 PBDEs 的浓度分别为 130 ng/g 和 160 ng/g，其背景点的浓度仅有 21 ng/g[51]。此外，广东和浙江等典型电子垃圾拆解地土壤中的污染程度相当[52]，可能是因为这两个地区都有着较长的电子垃圾拆解历史。此外，在电子垃圾不同的拆解功能区，其受污染的程度也是不一样的。Leung 等[53]在调查广东某典型电子垃圾拆解地土壤中的 PBDEs 时发现，酸浸池 ($2.72×10^3$~$4.25×10^3$ ng/g)＞打印机转毂存放区 (893~$2.89×10^3$ ng/g)＞鸭塘

(263~604 ng/g)＞稻田(34.7~70.9 ng/g)＞水库(2.0~6.2 ng/g)。Ohajinwa 等[54]也发现露天电子垃圾焚烧点的污染程度大于拆解站点，也大于背景点。

表 4-7　广东某典型电子垃圾拆解地不同年份和其他地区土壤中多溴联苯醚的研究

化合物	采样地/年份	采样点描述	浓度(ng/g)	文献
Σ_{20}PBDEs	广东/2018	电子垃圾拆解园区内和其周边地区	园区：1.54×10^3~3.10×10^5 周边地区：11.6~3.60×10^4	[49]
Σ_{13}PBDEs	广东/2004	8 个采样点从近到远分布于某电子垃圾露天焚烧站	219~1.42×10^3	[55]
Σ_{24}PBDEs	广东/2004	5 个采样点：水库，酸浸池，打印机滚筒存放区，稻田，和鸭塘	酸浸池：2.72×10^3~4.25×10^3 打印机滚筒存放区：893~2.89×10^3 鸭塘：263~604 稻田：34.7~70.9 水库：2.0~6.2	[53]
Σ_{37}PBDEs	广东/2009	3 个采样区域：背景点(无电子垃圾活动)，电子垃圾区域(露天焚烧和酸浸)，和邻近城镇(人工拆解和破碎)	电子垃圾区域：13~1.01×10^3 背景点：4.3~5.7 邻近城镇：8.7、2.6 和 10	[48]
Σ_{10}PBDEs	广东/2012	采样区域可分为居民区和农业区	居民区：168~6.54×10^3 农业区：10~4.45×10^3	[56]
Σ_{13}PBDEs(不包括 BDE-209)	广东/2004，2014	11 个采样点，包括 1 个背景点	2004 年：0.64~670 2014 年：12~2.10×10^3	[57]
Σ_{14}PBDEs	浙江/2005，2007，2009，2011	采样区域为路桥区，电子垃圾拆解区域	2005 年：489~1.49×10^4 2007 年：1.43×10^3~1.54×10^4 2009 年：628~9.64×10^3 2011 年：1.64×10^3~1.17×10^4	[52]
Σ_{14}PBDEs	天津/2015	采样区域为子牙电子垃圾拆解园区级周边地区	园区内：138(平均值) 周边地区：16(平均值) 总浓度范围：5.9~2.70×10^3	[58]
Σ_{13}PBDEs	Bui Dau/2012，2013，2014	采样区域为越南北部兴安省 Bui Dau 的露天焚烧站、拆解站和稻田	2012 年：0.1~9.20×10^3 2013 年：0.1~3.90×10^3 2014 年：0.1~2.90×10^3	[59]
Σ_{8}PBDEs	墨尔本/2017	采样区域为澳大利亚墨尔本 2 个电子垃圾拆解站的周边和 1 个背景点	拆解站 A：34~5.00×10^3 拆解站 B：8.3~9.80×10^4 背景点：0.10~44	[51]
Σ_{17}PBDEs	Lagos，Ibadan，Aba/2015	采样区域为尼日利亚 3 个城市中的露天焚烧站、拆解站和背景点	Lagos：4.67(背景点)，2.90×10^4(露天焚烧站)，1.58×10^3(拆解站) Ibadan：1.05×10^3(背景点)，6.97×10^3(露天焚烧站)，1.80×10^3(拆解站) Aba：77.7(背景点)，206(露天焚烧站)	[54]

PBDEs 是 BFRs 物美价廉的典型代表，从 20 世纪 70 年代开始在电子电器产品、日常生活用品等产品中广泛使用，导致电子垃圾材料中具有较大的赋存量。由于 PBDEs 主要作为添加型阻燃剂使用，其容易在电子垃圾拆解处理过程中释放迁移到环境中，并对当地生态环境和人群健康造成了不利影响。因此，电子垃圾相关的 PBDEs 环境污染及健康危害等问题一度成为环境领域的研究热点。随着 21 世纪《斯德哥尔摩公约》对 PBDEs 商业品的禁限用，PBDEs 的污染源头已逐渐消失，但还会出现非 PBDEs 类卤系阻燃剂替代品，如十溴二苯乙烷、六溴苯等，这些新型阻燃剂的环境风险还需引起更多关注。

4.2.2 多氯联苯

多氯联苯（PCBs）是联苯 1~10 位上的氢原子被不同数目的氯原子取代后形成的一系列氯代烃类化合物。PCBs 分子通式可表示为 $C_{12}H_{10-m-n}Cl_{m+n}$（$1 \leqslant m+n \leqslant 10$，其中 m，n 为正整数），分子量在 188.6~498.7 之间，结构式见图 4-19。根据联苯上氯取代的个数和其在苯环上取代位置的不同，PCBs 理论上有多达 209 种同系物。PCBs 于 1881 年由德国化学家 H. 施米特和 G. 舒尔茨首次合成，因其具有良好的化学稳定性、热稳定性、阻燃性、导热性、绝缘性以及合成工艺简单、成本低廉等特点，曾被广泛应用于电力工业、塑料加工业、化工和印刷业等领域。商品化 PCBs 由美国 Monsanto 公司于 1929 年最先开始生产，系列工业品主要包括 Aroclor 1016、Aroclor 1221、Aroclor 1232、Aroclor 1242、Aroclor 1254、Aroclor 1260、Aroclor 1262 和 Aroclor 1268 等，四位数字中前两位 12 表示碳原子数，后两位表示工业品中氯元素的百分含量。除 Aroclor 系列之外，工业用 PCBs 的商品名称还有日本生产的 Kanechlor（KC），德国生产的 Clophen，法国生产的 Phenochlor，苏联生产的 Sovols 等。

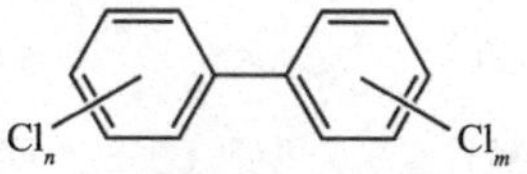

图 4-19 多氯联苯（PCBs）的化学结构

PCBs 的纯化合物为结晶态，混合物则为油状液体，无色或淡黄色，水溶性较差，易溶于动物脂肪和有机溶剂。PCBs 理化性质稳定，是一类半挥发性或不挥发性物质，具备生物毒性，难降解，易在生物体内蓄积，具有远距离迁移性等 POPs 的典型特征。已有研究表明，PCBs 具有致癌性，并可对免疫、生殖和神经系统等产生危害。由于 PCBs 具有高毒性，历史上曾发生过多次 PCBs 引起的重大环境事件。1968 年，日本发生了因食用受 PCBs 污染的米糠油而导致多人食物中毒的事故，共造成 1684 人中毒，其中 30 多人死亡。PCBs 对环境的影响和污染问题首次引起了人们的广泛关注。之后，1978 年在台湾彰化又发生类似的米糠油事件，近 2000 人中毒，53 人死亡。据估计，自 20 世纪 20 年代开始商业化生产 PCBs 至 70 年代全球范围内禁用 PCBs，全世界生产和使用的 PCBs 高达 2000 万吨，其中有大约 31%进入环境，对生态环境及人类健康造成危害。主要的几大类 PCBs 工业品混合物的组成成分及物理化学性质详见表 4-8 和表 4-9。

表 4-8 Aroclor 系列工业品中各种 PCBs 同族体的百分组成

	Aroclor 1016	Aroclor 1221	Aroclor 1232	Aroclor 1242	Aroclor 1254	Aroclor 1260	Aroclor 1262	Aroclor 1268
一氯联苯	0.7	60.1	27.5	0.7	0.02	0.02		
二氯联苯	17.5	33.4	26.8	15	0.09	0.1		0.1
三氯联苯	54.7	4.2	25.6	44.9	0.4	0.2		0.3
四氯联苯	22.1	1.1	10.6	20.2	4.9	0.3		4.4
五氯联苯	5.1	1.2	9.4	18.9	71.4	8.7	4.2	10.6
六氯联苯			0.2	0.3	22	43.3	30.9	40.7
七氯联苯			0.03		1.4	38.5	45.8	33.4
八氯联苯						8.3	17.7	9.4
九氯联苯					0.4	0.7	1.3	1.2
十氯联苯								0.02

表 4-9 多氯联苯各同族体的物理化学性质

异构体组	n	熔点 (℃)	蒸气压 (Pa)	溶解度 (g/m³)	log K_{ow}	BCF (鱼类)	蒸发速率 [g/(m²·h)] (25℃)
联苯	1	71	4.9	9.3	4.3	1000	0.92
一氯联苯	3	25-78	1.1	4	4.7	2500	0.25
二氯联苯	12	24-149	0.24	1.6	5.1	6300	0.065
三氯联苯	24	18-87	0.054	0.65	5.5	1.6×10^4	0.017
四氯联苯	42	47-180	0.012	0.26	5.9	4.0×10^4	4.2×10^{-3}
五氯联苯	46	76-124	2.6×10^{-3}	0.099	6.3	1.0×10^5	1.0×10^{-3}
六氯联苯	42	77-150	5.8×10^{-4}	0.038	6.7	2.5×10^5	2.5×10^{-4}
七氯联苯	24	122-149	1.3×10^{-4}	0.014	7.1	6.3×10^5	6.2×10^{-5}
八氯联苯	12	159-162	2.8×10^{-5}	5.5×10^{-3}	7.5	1.6×10^6	1.5×10^{-5}
九氯联苯	3	183-206	6.3×10^{-6}	2.0×10^{-3}	7.9	4.0×10^6	3.5×10^{-6}
十氯联苯	1	306	1.4×10^{-6}	7.6×10^{-3}	8.3	1.0×10^7	8.5×10^{-7}

虽然目前PCBs已经在全球范围内被禁止生产和使用,但早期生产的PCBs被封存或填埋,这些残留物依然对环境具有潜在的威胁。PCBs的理化性质稳定,不易降解,在生产和加工使用过程中因有意或无意排放泄漏到环境中,会长期存在,并对水体、大气、土壤等造成大规模的环境污染,并通过食物链的累积效应影响到多种生物,包括人类。因此,PCBs的危害仍然不容小视,而PCBs的环境污染问题一直是环境科学研究的热点。

我国从1965年开始工业化生产和使用PCBs,至1974年禁止,累计生产的PCBs接近1万吨,其中约1000吨作为油漆添加剂,通过开放性使用进入环境,9000吨作为电力变压器和电容器的电介质,目前已被废弃和淘汰。据查,废旧电容器的浸渍液中PCBs含量高于90%,而废弃的进口变压器的变压器油中PCBs的含量大于50%。我国PCBs的生产和使用量虽远不及发达国家(占全球产量的不到1%),但局部地区(如主要工业区、电子垃圾拆解回收地和废旧变压器存放地等)PCBs的含量仍然很高。与欧美日等发达地区比较,我国由于历史上PCBs使用较少,大气、土壤和水体沉积物中PCBs的浓度处于较低的水平,其中土壤中PCBs的浓度不到全球基准浓度的10%[60]。普通人群暴露调查数据也显示我国人体PCBs含量处在较低的水平[60-62]。然而,部分地区粗放式的电子垃圾拆解活动给当地生态环境造成严重的PCBs污染,部分环境介质中报道了较高水平的PCBs,电子垃圾从业人员和附近居民体内也蓄积了较高浓度的PCBs[63-69]。

早期大量国外电子垃圾通过非法途径进入我国,这些电子垃圾在一些地区(主要是农村或者城市郊区)进行粗放式的拆解,成为我国环境中PCBs的重要来源。进入环境中的PCBs由于受气候、生物、水文地质等因素的影响,在不同的环境介质间发生一系列的迁移转化,最终的贮存场所主要是土壤、河流和沿岸水体的底泥。释放到环境中的PCBs可通过大气长距离迁移,传输到偏远的极地地区,并通过沉降作用,成为土壤和水体中PCBs的主要来源之一。虽然在20世纪80年代,包括中国在内的多数国家已禁止生产和使用PCBs,其浓度呈现明显的下降趋势[70],但是在粗放式电子垃圾拆解地区仍可检测到高浓度的PCBs污染。环境中的PCBs主要来源于三个方面:PCBs产品的生产及使用、PCBs工业产品的处理焚烧和环境介质中PCBs的释放。

PCBs主要作为电机的导热体及变压器和电容器的绝缘液,在对其拆解处理过程中,PCBs将进入周围的大气,而电子垃圾拆解地区温和的气候条件可促进PCBs的二次排放,导致不同环境介质之间的界面交换和潜在的生物转化[71,72],并通过大气传输对海洋大气造成一定程度的污染[73]。在广东清远电子垃圾拆解地区,大气中PCBs的浓度(气相和颗粒相)为 25.6 ng/m^3(7.8~76.3 ng/m^3),比前期同一地区大气中PCBs的浓度(0.29~7.4 ng/m^3)高一个数量级,是台州的2倍之多[72,74,75]。同时在其对照区也发现了较高浓度的PCBs(2.2 ng/m^3),其浓度明显高于中国其他城市和农村地区，表明电子垃圾回收行为是 PCBs 的重要来源之一[76]。Han 等[75]对台州电子垃圾拆解活动集中治理后大气中Σ_{38}PCBs 进行了研究，冬季大气中 PCBs 的浓度为(12.4±9.6) ng/m³，显著低于先前报道的浓度(192~650 ng/m³ Σ_{17}PCBs)[77]，与同一时期其他研究的浓度相当(4.3~8.4 ng/m³ Σ_{17}PCBs；7.2 ng/m³ Σ_{18}PCBs)[78,79]。虽然集中治理后大气中的 PCBs 污染显著低于十年前的浓度水平，但仍然比对照区高 54 倍之多。Xing 等[80]研究了台州变压器拆解车间和居民区大气中的 PCBs 污染，拆解车间内大气中的 PCBs 浓度为 17.6 ng/m³，比居民区(3.37 ng/m³)高 5 倍，是对照区临安(0.46 ng/m³)的 38 倍，表明废旧变压器拆解车间工人面临着很高的 PCBs 暴露健康风险。其 PCBs 同系物以三氯和四氯为主，低氯代的 PCB-28 含量最高，这与它们的大量使用及易挥发等特性相一致。而另一项关于台州电子垃圾拆解的研究表明，工业园区车间内气相中 PCBs 的浓度高达 85.39 ng/m³，是工业区外浓度的 2 倍；颗粒相 PCBs 浓度(10.31 ng/m³)是工业区外浓度的 6 倍之多，其中三氯和四氯 PCBs 仍是主要的同系物[81]。通过对越南电子垃圾拆解车间的研究也发现，大气中 PCBs 的浓度高于非拆解车间，且大气中三氯 PCBs 是主要的同系物[82]。

4.2.3 四溴双酚 A

四溴双酚 A(tetrabromobisphenol A，TBBPA)是 BFRs 的一种，其化学结构式见图 4-20。TBBPA 主要以反应型阻燃剂被广泛应用于制造印刷线路板，也作为添加型阻燃剂用于电子电气设备外壳中。随着其他 BFRs，如 HBCD 和 PBDEs 等被列入《斯德哥尔摩公约》禁用名单之后，TBBPA 的产用量急剧攀升，造成的环境污染近年来也呈加剧态势。TBBPA 在室温下多呈现白色粉末状，具备较好的热稳定性、阻燃效率高、不易溶于水、耐腐蚀等特点，其理化性质见表 4-10[83,84]。由于 TBBPA 具有较低的水溶性、较高的辛醇水分配系数(log K_{OW})，且在土壤、水体等环境介质中均具有较长的半衰期，其容易在水体、土壤和沉积物等多种环境介质中长期、稳定存在。已有的数据表明，全球范围内大气和室内灰尘、土壤、水体和沉积物等环境介质都存在不同程度的 TBBPA 污染(表 4-11)[85,86]，并且 TBBPA 可在动物体和人体内蓄积，并被证明具有一定的毒性效应[87-89]。

TBBPA 既可作反应型又可用作添加型的 BFRs，其生产量约占 BFRs 生产总量的 50%以上[90]。早期 TBBPA 主要作为反应型阻燃剂添加入电路板基材中；作为 PBDEs 被禁后的理想代替品，TBBPA 近期也被广泛添加到电子电气设备外壳，在某些电子产品塑料组件中添加量高达 20%。在这些电子产品的拆解过程中，TBBPA 易以挥发、渗出等方式释放到大气，并容易与大气中颗粒物结合，通过干湿沉降等扩散至水环境中，也可以通过拆解过程产生的废液直接排放，以及垃圾填埋等方式进入水环境和土壤[91]。粗放式的电子垃圾拆解活动是部分地区环境中 TBBPA 的主要来源之一。在电子垃圾，尤其是废电路板机械拆解、粉碎、焚烧过程中，大量 TBBPA 进入环境并在生物体内蓄积[92]。在焚烧过程中，TBBPA 会产生大量烟雾和有毒气体如二噁英等[91]。电子垃圾拆解活动地区也成为研究 TBBPA 及其转化产物的环境行为及归趋的理想场地。

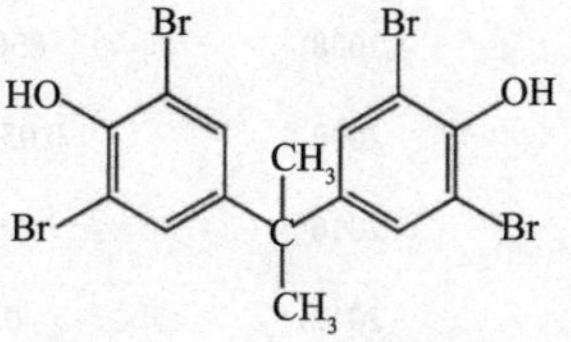

图 4-20 TBBPA 的化学结构

表 4-10 **TBBPA 的理化性质**

理化性质	TBBPA
CAS 号	79-94-7
分子式	$C_{15}H_{16}Br_4O_2$
分子量	543.87
熔点（℃）	178
沸点（℃）	316
密度（g/cm³）	2.12（20℃）
蒸气压（Pa）	1.9×10^{-5}（20℃）
辛醇水分配系数（log K_{OW}）	4.50~6.53
水溶性（S_W）（mg/L）	1.26（pH=7，25℃）
亨利系数（H_W）（Pa/mol·m³）	1.47×10^{-5}
生物积累系数（BAF）	9.56~22.64
半衰期（土壤）（d）	64
半衰期（水）（d）	6.6~80.7

表 4-11 **四溴双酚 A 在不同地区环境介质中的浓度**

地区	介质	时间	浓度范围	单位	文献
中国广东某地	大气	2007	66.01~95.04	ng/m³	[93]
瑞典斯德哥尔摩地区	大气	2000	30~40（厂区） 0.031~0.038（办公室）	ng/m³	[94]
韩国某地住宅区	室内灰尘	2016	78.87~463.81	ng/g	[95]
中国广东某地	室内灰尘	2019	5.5×10^3~2.38×10^4	ng/g	[96]

续表

地区	介质	时间	浓度范围	单位	文献
中国清远地区	土壤	2014	30~646	ng/g	[97]
中国广东某地	土壤	2018	82.6~9.83×10^4	ng/g	[98]
越南电子垃圾拆解厂	表层土壤	2014	5~2900	ng/g	[99]
中国巢湖	水体	2008	850~4870	ng/L	[100]
法国塞纳河支流	水体	2008	0.035~0.068	ng/L	[101]
中国渤海，黄河	水体	2016	56~607	ng/L	[102]
中国大亚湾	水体	2012	0.23~9	ng/L	[103]
中国巢湖	沉积物	2008	22.0~518	ng/g	[100]
瑞典某污水处理厂	污泥	2000	34~270	ng/g	[104]
日本名古屋	鱼类	2005	0.01~0.11	ng/g	[105]
中国巢湖	鱼类	2008	28.5~39.4	ng/g	[100]
韩国	人体头发	2017	n.d.~16.04	ng/g dw	[106]

1. 大气中的 TBBPA

在大气中，由于 TBBPA 的低蒸气压和高脂溶性特性，TBBPA 更容易与大气颗粒物中的碳黑结合，并通过干沉降进入室内灰尘和土壤，通过湿沉降进入水体。研究发现 TBBPA 在与电子产品相关的工业区大气中的浓度要远高于办公区浓度[94]。在室内灰尘中也有同样发现，Liu 等[96]在我国广东某电子垃圾拆解园区及附近住宅区设定 4 个采样点，分别是距离工业园区西南约 1 km 处的住宅区，工业园内办公楼以及两个园内车间，对各采样点灰尘中 TBBPA 进行分析后，发现所有的 TBBPA 脱溴产物及其溴氯混合转化产物均有检出（表 4-12），浓度范围在 $1.13×10^4$ ~ $2.06×10^5$ ng/g 之间。电子垃圾手工拆解车间灰尘中浓度最高（EW，$2.06×10^5$ ng/g），其次是电路板粉碎车间（RW，$1.38×10^5$ ng/g），这两个车间的浓度比高炉二次炼铜车间（BW，$3.24×10^4$ ng/g）高一个数量级，也远高于办公区（OA）和背景点（RA）的浓度，这表明粗放式的电子垃圾拆解活动是 TBBPA 及其转化产物的重要来源。高炉二次炼铜车间相对较低的浓度可能与较高的分解温度有关系。MoBBPA（$8.45×10^4$ ng/g）是电子垃圾拆解车间最主要的同系物，其次是 2,2'-DiBBPA（$3.74×10^4$ ng/g），TBBPA 的脱溴转化产物的浓度相比母体 TBBPA 的浓度更高，这些脱溴产物相比溴氯混合卤代产物浓度高了 2 个数量级，表明脱溴更容易发生。

表 4-12 电子垃圾不同拆解车间灰尘中 TBBPA 及其转化产物的浓度

化合物	EW ($n=5$)	RW ($n=3$)	BW ($n=3$)	OA ($n=2$)	RA ($n=2$)
BPA	$1.67×10^3 ± 1.15×10^3$	$1.08×10^3 ± 0.25×10^3$	$229 ± 6.0$	$196 ± 26.0$	$48.8 ± 17.0$
MoBBPA	$8.45×10^4 ± 6.35×10^4$	$5.68×10^4 ± 1.52×10^4$	$1.25×10^4 ± 0.47×10^4$	$3.27×10^3 ± 0.56×10^3$	$454 ± 191$
2-Cl-2′-MoBBPA	$8.15×10^3 ± 7.83×10^3$	$1.67×10^3 ± 0.4×10^3$	$275 ± 162$	$38.7 ± 0.36$	$14.6 ± 3.4$
2-Cl-6-MoBBPA	$192 ± 429$	$1.29×10^3 ± 0.27×10^3$	$81.7 ± 71.8$	$36.4 ± 10.1$	n.d. [a]
2,2′-DiBBPA	$3.74×10^4 ± 2.63×10^4$	$2.41×10^4 ± 0.55×10^4$	$4.83×10^3 ± 1.96×10^3$	$897 ± 165$	$145 ± 23.8$
2,6-DiBBPA	$1.87×10^4 ± 1.27×10^4$	$1.24×10^4 ± 0.30×10^4$	$2.03×10^3 ± 0.46×10^3$	$752 ± 126$	$115 ± 41.6$
2,2′-Cl$_2$-6-MoBBPA	$5.3 ± 7.6$	$2.6 ± 4.5$	n.d.	n.d.	n.d.
2-Cl-2′,6-DiBBPA [b]	$4.38×10^3 ± 3.33×10^3$	$2.18×10^3 ± 0.31×10^3$	$239 ± 15.8$	$53.9 ± 0.74$	$12.8 ± 0.04$
2,2′,6-TriBBPA	$2.31×10^4 ± 0.84×10^4$	$1.75×10^4 ± 0.16×10^4$	$4.69×10^3 ± 0.36×10^3$	$1.92×10^3 ± 0.18×10^3$	$1.16×10^3 ± 0.19×10^3$
2,2′-Cl$_2$-6,6′-DiBBPA	$3.5 ± 4.9$	$0.9 ± 1.6$	n.d.	n.d.	n.d.
2-Cl-2′,6,6′-triBBPA	$3.54×10^3 ± 2.07×10^3$	$1.53×10^3 ± 0.38×10^3$	$147 ± 50.1$	$44.5 ± 3.89$	$24.0 ± 6.26$
TBBPA	$2.38×10^4 ± 0.39×10^4$	$1.92×10^4 ± 0.04×10^4$	$7.40×10^3 ± 2.55×10^3$	$5.50×10^3 ± 0.65×10^3$	$9.35×10^3 ± 0.98×10^3$
Σ_7Cl-BBPA	$1.63×10^4 ± 1.30×10^4$	$6.67×10^3 ± 1.19×10^3$	$743 ± 79.0$	$174 ± 6.65$	$51.4 ± 9.71$
Σ_4Br-BBPA	$1.64×10^5 ± 1.11×10^5$	$1.12×10^5 ± 0.25×10^5$	$2.40×10^4 ± 0.64×10^4$	$6.83×10^3 ± 1.03×10^3$	$1.88×10^3 ± 0.45×10^3$
Σ_{all}X-BBPA	$2.06×10^5 ± 1.23×10^5$	$1.38×10^5 ± 0.26×10^5$	$3.24×10^4 ± 0.66×10^4$	$1.27×10^4 ± 0.17×10^4$	$1.13×10^4 ± 0.05×10^4$

a. 未检出；b. 2-Cl-2′,6-DiBBPA 和 2-Cl-2′,6′-DiBBPA 共溢出；n：样品数量；RW：电路板粉碎车间，EW：电子垃圾手工拆解车间，BW：高炉二次炼铜车间，OA：办公区，RA：背景点

从同系物组成来看(图 4-21)，MoBBPA 是最主要的同系物，贡献了脱溴产物的 51%，而 2-Cl-2′,6,6′-TriBBPA 则是最主要的溴氯混合卤代产物。这些脱溴产物的总浓度(Σ_4Br-BBPA)是溴氯混合卤代产物(Σ_7Cl$_x$-BBPA)的 8~42 倍。不同采样点 TBBPA 转化产物的同系物组成也有较大的差异。拆解车间脱溴产物，如一溴(MoBBPA)到三溴(TriBBPA)占主要，而背景点 TBBPA 则是主要的同系物(贡献超过 80%)，表明电子垃圾拆解活动是导致脱溴转化产物形成的主要原因。机械破碎拆解和其他拆解车间具有类似的同系物组成，表明机械破碎电路板也可能造成 TBBPA 转化产物的形成。

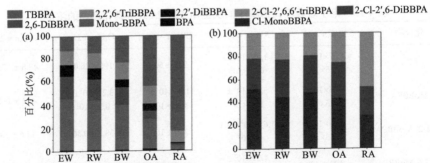

图 4-21 电子垃圾拆解地区灰尘中 TBBPA 及其转化产物的同系物组成特征
(a)脱溴转化产物；(b)溴氯混合卤代产物
RW：电路板粉碎车间，EW：电子垃圾手工拆解车间，BW：高炉二次炼铜车间，OA：办公区，RA：背景点

为了进一步确证这些 TBBPA 转化产物的来源，对不同类型的电子垃圾外壳和电路板进行粉碎，用丙酮和二氯甲烷混合溶剂进行超声萃取，并将萃取液进行 LC-MS/MS 分析，结果发现仅有少量的三溴转化产物 TriBBPA 在部分样品中有检出，占比不超过 3%。有文献表明工业品 TBBPA 中存在少量的 TriBBPA，因此这些 TriBBPA 可能来自工业品杂质，而 TBBPA 在生产和使用过程中并不会造成大量的转化产物的形成，从而更进一步证明粗放式的电子垃圾拆解活动是这些 TBBPA 转化产物的主要来源。

2. 土壤中的 TBBPA

在土壤中，TBBPA 由于具有较低的亨利系数和蒸气压，其易吸附在有机质或黏土颗粒上，从而在土壤中沉积，迁移能力弱[107]。TBBPA 主要以未经加热的电子垃圾或城市污水处理厂污泥填埋等方式进入土壤[104, 108]。研究表明，土壤中 TBBPA 的含量与污染源密切相关，典型污染地区如 BFRs 生产区以及电子垃圾拆解区土壤中的干重浓度分别高达 7758 ng/g[109]和 646.04 ng/g[110]，要远远高于非污染地区土壤如某农田土壤浓度仅为 5.6 ng/g[111]。

为了进一步探讨 TBBPA 及其转化产物的污染特征，Ge 等[49]以电子垃圾拆解地拆解园区为中心，划定一个 9 km × 9 km 的采样区域(采样点位置信息见图 4-18)，对拆解园区及其周边区域的表层土壤样品(0~10 cm)进行采样，其中拆解园区采集了 24 个样品，园区外周边区域采集了 83 个样品，对土壤中的 TBBPA 及其转化产物进行了分析。TBBPA 及其转化产物在电子垃圾拆解园及其周边地区土壤中广泛检出，其中拆解园区内土壤样品中 TBBPA 及其转化产物的浓度和检出率均高于周边地区($P<0.05$)。脱溴转化产物在拆解园区内土壤中普遍检出，周边地区土壤中检出率为 77.1%。溴氯混合卤代产物在园区内土壤中的检出率为 79.2%，而在周边地区土壤中检出率为 18.1%。TBBPA 是拆解园区及其周边区域土壤中浓

度最高的同系物，其浓度分别为 1.08×10^4 ng/g 和 293 ng/g(表 4-13)。一溴取代 MoBBPA 是拆解园区内第二高浓度的同系物(均值：3.86×10^3 ng/g)，但是周边区域土壤中三溴取代产物 TriBBPA 的浓度更高(均值：53.9 ng/g)。拆解园区土壤中总的脱溴转化产物浓度范围为 $131 \sim 2.49 \times 10^4$ ng/g，周边区域土壤中浓度范围为 n.d.$\sim 4.42 \times 10^3$ ng/g。此外，BPA 在园区内外也广泛检出，其在拆解园区及其周边区域土壤中浓度分别为 2.35×10^4 ng/g 和 491 ng/g。对于溴氯混合卤代产物而言，其在土壤中的浓度明显低于 TBBPA 及其脱溴产物。拆解园区及其周边区域土壤中七种溴氯混合卤代产物的总浓度(Σ_7X-BBPAs)范围分别为 n.d.$\sim 1.18 \times 10^3$ ng/g 以及 n.d.~ 268 ng/g。2-Cl-2',6,6'-TriBBPA 最浓度最高的同系物，其次是 2-Cl-2',6,6'-TriBBPA，2-Cl-2',6-DiBBPA，2-Cl-2'-MonoBBPA，2,2'-Cl$_2$-6,6'-DiBBPA，2-Cl-6-MonoBBPA 以及 2,2'-Cl$_2$-6-MonoBBPA。园区内外溴氯混合卤代产物的浓度顺序和组成特征也没有差异，表明周边地区土壤中的污染物主要来自拆解园区的扩散。

表 4-13 电子垃圾工业园及其周边地区土壤中 TBBPA 及其转化产物的干重浓度(ng/g dw)

化合物	电子垃圾拆解园区				园区周边			
	平均值	中位值	范围	检出频次	平均值	中位值	范围	检出频次
TBBPA	$1.08\times10^4\pm$ 2.26×10^4	527	$82.6\sim9.83\times10^4$	100%	$293\pm1.05\times10^3$	17.4	n.d.$\sim8.49\times10^3$	79.5%
TriBBPA	$1.62\times10^3\pm$ 2.19×10^3	337	$29.4\sim6.91\times10^3$	100%	53.9 ± 278	2.32	n.d.$\sim2.46\times10^3$	7.1%
2,2'-DiBBPA	$1.25\times10^3\pm$ 1.25×10^3	850	$28.8\sim4.13\times10^3$	100%	21.2 ± 98.7	n.d.	n.d.~834	36.1%
2,6-DiBBPA	512 ± 509	262	$3.78\sim1.42\times10^3$	100%	7.15 ± 37.1	n.d.	n.d.~324	28.9%
MonoBBPA	$3.86\times10^3\pm$ 4.49×10^3	2.77×10^3	$69.2\sim1.99\times10^4$	100%	39.1 ± 120	n.d.	n.d.~795	47.0%
Σ_4Br-BBPAs	$7.24\times10^3\pm$ 7.12×10^3	5.18×10^3	$131\sim2.49\times10^4$	100%	121 ± 515	5.68	n.d.$\sim4.42\times10^3$	77.1%
BPA	$2.35\times10^4\pm$ 2.28×10^4	1.87×10^4	$366\sim1.08\times10^5$	100%	491 ± 914	60.7	n.d.$\sim4.06\times10^3$	81.9%
2-Cl-2',6,6'-TriBBPA	103 ± 196	n.d.	n.d.~676	37.5%	2.75 ± 13.7	n.d.	n.d.~115	10.8%
2-Cl-2',6-DiBBPA [a]	46.6 ± 65.1	3.35	n.d.~200	62.5%	1.43 ± 9.37	n.d.	n.d.~85.1	9.6%
2-Cl-2'-MonoBBPA	43.5 ± 46.1	26.8	n.d.~130	75.0%	1.37 ± 9.12	n.d.	n.d.~80.2	6.0%
2-Cl-6-MonoBBPA	6.16 ± 9.23	2.15	n.d.~27.0	50.0%	0.11 ± 0.81	n.d.	n.d.~7.23	2.4%
2,2'-Cl$_2$-6,6'-DiBBPA	35.7 ± 58.3	n.d.	n.d.~233	45.8%	0.78 ± 4.03	n.d.	n.d.~35.4	8.4%
2,2'-Cl$_2$-6-MonoBBPA	2.67 ± 5.90	n.d.	n.d.~21.1	25.0%	0.05 ± 0.27	n.d.	n.d.~1.44	3.6%
Σ_7X-BBPAs	237 ± 328	79.0	n.d.$\sim1.18\times10^3$	79.2%	6.48 ± 32.8	n.d.	n.d.~268	18.1%

a. 2-Cl-2',6-DiBBPA 和 2-Cl-2',6'-DiBBPA 共馏出，以总浓度表示；n.d. 表示未检出

从同系物组成来看，园区内外土壤中污染物的同系物总体组成特征差异不大（图 4-22），细微差别主要在个别同系物组成上。在园区内土壤中，脱溴转化产物的贡献比例分别为 MonoBBPA（9.3%）＞TriBBPA（3.9%）＞2,2′-DiBBPA（3.0%）＞2,6-DiBBPA（1.2%），而在周边地区土壤中，三溴 TriBBPA 贡献的比例更高，其大小顺序为 TriBBPA（6.0%）＞MonoBBPA（4.3%）＞2,2′-DiBBPA（2.3%）＞2,6-DiBBPA（0.8%）。值得一提的是不论是园区内，还是园区周边，2,2′-DiBBPA 与 2,6-DiBBPA 的浓度之间有很好的相关性，且前后者的浓度比值为 2.4。理论上而言，2,2′-DiBBPA 与 2,6-DiBBPA 都是来自于 TriBBPA 的脱溴，如果脱溴加氢过程中三个溴原子取代位没有选择性的话，则两种产物的比例为 2∶1，这与研究观测到的 2.4 非常接近，因此，园区内土壤中 2,2′-DiBBPA 与 2,6-DiBBPA 的脱溴加氢过程可能属于非选择性反应，需要后续进一步的研究确证。

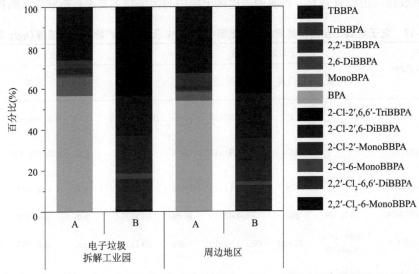

图 4-22 电子垃圾拆解工业园及其周边地区土壤中 TBBPA 及其转化产物的同系物组成特征
A：脱溴转化产物；B：溴氯混合卤代产物

根据电子垃圾拆解工业园区及其周边地区土壤中 TBBPA 及其转化产物的空间部分特征（图 4-23）可以得出，混合卤代产物集中在电子垃圾拆解园区内，而脱溴转化产物的扩散范围更大，浓度也更高。BPA 一方面可能来源于 TBBPA 的脱溴转化，另外也可能作为塑料添加剂广泛使用，因此 BPA 的扩散范围更广，且存在除了电子垃圾拆解园区以外的次中心。TBBPA 跟 BPA 的污染特征类似，也存在除了电子垃圾拆解园区以外的次中心，这与 TBBPA 和 BPA 作为塑料添加剂的广泛使用密切相关。

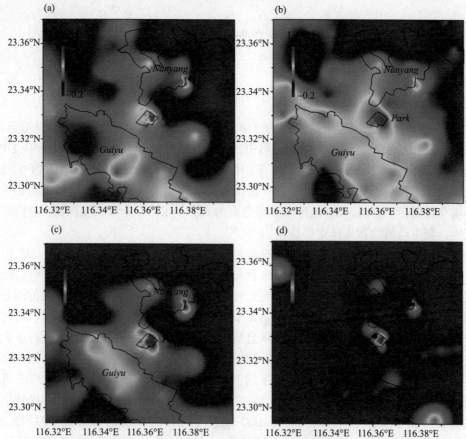

图 4-23 电子垃圾拆解工业园及其周边地区土壤中 TBBPA 及其转化产物的空间部分特征
(a)TBBPA；(b)BPA；(c)脱溴转化产物；(d)混合卤代产物

4.2.4 得克隆

得克隆(dechlorane plus，DP)，中文名又称为"敌可燃"。DP 的化学名称为双(六氯环戊二烯)环辛烷，分子式为 $C_{18}H_{12}Cl_{12}$。作为一种含氯阻燃剂，DP 被广泛用于电线和电缆的涂层、塑料屋顶材料、电视和计算机显示器的连接器添加剂，还作为非增塑阻燃剂用于尼龙和聚丙烯塑料等聚合物材料中。DP 具有较强的环境持久性及生物累积性，对生态环境及人体健康存在风险，已于 2018 年被欧洲化学品管理局(ECHA)确定为高关注度物质(SVHC)。工业品 DP 主要由 1,5-环辛二烯和六氯环戊二烯以 2∶1 的摩尔比通过 Diels-Alder 环化缩合反应制得，产物包括顺式 DP(*syn*-DP)和反式 DP(*anti*-DP)两种异构体(图 4-24)。理论上，*syn*-DP 和 *anti*-DP 的组成比例约为 1∶3，即 *anti*-DP 的百分比(f_{anti}=anti-DP/(anti-DP + syn-DP))为 0.75。

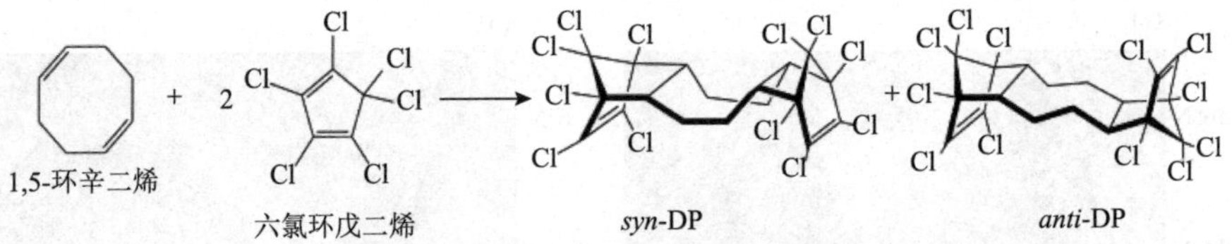

图 4-24　DP 工业品的合成路线

DP 主要用作添加型阻燃剂,由于 DP 自身与聚合物材料之间并无化学键键合,因此,在产品生产、加工和使用过程中,以及在废物处理和回收利用过程中 DP 会释放到环境中。DP 的挥发性低,亲脂性强,释放到环境中的 DP 易吸附于大气气溶胶、水体悬浮颗粒物,并通过干湿沉降等作用在大气灰尘、土壤及沉积物等环境介质赋存。目前,DP 已经在全球范围的大气、水体、土壤、沉积物、室内外灰尘、野生动植物和人体中广泛检出[49, 112]。

电子垃圾拆解区大气中通常具有非常高的 DP 污染[113,114]。总体而言,DP 由于分子体积较大挥发性低,主要存在于大气颗粒相中,贡献比超过 70%[115,116]。在我国华南地区清远某电子垃圾拆解地及对照地区大气颗粒相中 DP 的浓度范围分别为 13.1~1794 pg/m^3 和 0.47~35.7 pg/m^3,拆解区大气中 DP 浓度高于对照区约两个数量级;DP 在气相中检出率<40%且所占比例<2%。此外,电子垃圾拆解区大气 anti-DP 与 DP 总浓度的比值(f_{anti})平均值为 0.726 ± 0.037,与工业品相近(0.72),但显著高于($p = 0.036$)对照区大气(平均值为 0.701 ± 0.099),表明大气中的 DP 可能发生了异构体选择性迁移转化[113, 116]。在电子垃圾拆解区采集的桉树叶和松针中也检出了 DP,浓度范围分别为 0.45~16.7 ng/g 和 0.51~51.9 ng/g,显著高于对照区;植物中 DP 的 f_{anti} 值均显著低于对应的大气($p<0.03$),其中在电子垃圾拆解地区桉树叶与松针中 f_{anti} 值呈显著差异,分别为 0.699 ± 0.094 和 0.657 ± 0.047,这可能与不同植物选择性从大气中吸收 DP 异构体有关;此外,还在桉树叶及松针样品中检测到了 DP 的脱氯产物 anti-Cl$_{11}$DP[113]。在巴基斯坦卡拉奇地区某个电子垃圾拆解点附近的大气中也检出了较高浓度水平的 DP,其中拆解厂大气颗粒物中 DP 的平均浓度为 26.7 ng/m^3,显著高于我国华南地区电子垃圾拆解区域的浓度值[115-116],表明巴基斯坦等新兴电子垃圾拆解地区的 DP 污染也十分严重。

灰尘作为半挥发性有机物的蓄积库之一,其中 DP 的污染也一直备受关注。Wang 等[117]对采集自我国华南清远某电子垃圾拆解地不同室内灰尘中的 DP 进行了分析,发现在超过 85%的灰尘样品中都有 DP 检出,其浓度范围介于 n.d.~21000 ng/g(几何均值浓度为 604 ng/g),高出广州市区(浓度范围 2.78~70.4 ng/g,几何均值浓度为 14.5 ng/g)以及其他农村区(浓度范围 n.d.~27.1 ng/g,几何均值浓度为 2.89 ng/g) 1~2 个数量级,表明粗放式的电子垃圾拆解活动也是室内灰尘中 DP 的主要来源之一;城市区域灰尘中的 DP 则主要来源于工业活动和一些室内产品中的释放,而农村区域灰尘中的 DP 则可能受大气远距离迁移的影响。类似地,DP 也在我国北方最大的电子垃圾拆解地天津市子牙镇某拆解工业园区的室内灰尘中有检出(185 ng/g)[114]。Zheng 等[118]发现电子垃圾拆解车间室内灰尘中 DP 浓度为 342.79~4197 ng/g,高于电子垃圾拆解车间周边地区居民室内灰尘(45.22~1798 ng/g);该浓度也显著高于农村地区(32.62~118.26 ng/g)及城市(2.78~70.38 ng/g)对照区;电子垃圾拆解车间室内灰尘中 DP 的 f_{anti} 值(0.54)显著低于农村对照区(0.70)及城市对照区(0.76),也低于电子垃圾拆解区周边居民家庭室内灰尘(0.66)。在电子垃圾拆解车间,DP 脱氯产物 anti-Cl_{11}DP 与 anti-DP 的比值(0.014)比拆解区周边居民室内灰尘相应的比值(0.0012)高一个数量级,表明电子垃圾拆解活动中也存在 anti-DP 选择性转化。

粗放式的电子垃圾拆解活动也造成了拆解点土壤的污染,高浓度的 DP 也在拆解地土壤中广泛检出(表 4-14)。Ge 等[49]在我国华南地区某典型电子垃圾拆解园区内土壤中检出的 DP 浓度分别为 21.8~1.80×10^4 ng/g,在拆解园区周边地区 DP 的浓度为 1.76~4.05×10^3 ng/g。园区内 DP 的污染程度显著高于周边地区($p<0.05$)。Syn-DP 和 anti-DP 二者的检出率均为 100%,syn-DP 在园区内的浓度为 211 ng/g,要比在周边地区的 2.81 ng/g 高 2 个数量级;anti-DP 也有着相似的结果,其园区内和周边地区的浓度分别为 504 ng/g 和 8.52 ng/g。Xu 等[48]发现电子垃圾拆解区域土壤中的 DP 的浓度(13~1014 ng/g)也显著高于对照地区(5.7 ng/g);在华南清远某电子垃圾拆解区土壤中最高 DP 的浓度可达 3327 ng/g[119];在浙江台州某电子垃圾拆解区发现的土壤中 DP 的浓度在 0.17~1990 ng/g[120];在越南北部的某个电子垃圾拆解工厂附近土壤中 DPs 浓度为 1.6~1900 ng/g,另外该拆解地区 2012 年、2013 年及 2014 年沉积物中 DPs 的平均浓度分别为 1.9 ng/g、9.6 ng/g 和 41 ng/g,呈逐年上升趋势[59]。

表 4-14 电子垃圾拆解地区土壤中 DP 的浓度水平

采样时间	采样点描述	浓度(ng/g)	文献
2018 年	电子垃圾拆解园区内和其周边地区	园区:21.8~1.80×10^4 周边:1.76~4.05×10^3	[49]
2009 年	拆解区(露天焚烧和酸浸);邻近镇(人工拆解和破碎);背景点(无电子垃圾活动)	拆解区:0.57~146 背景点:0.21~0.47 邻近镇:0.32~6.6	[48]
2013 年	5 个采样区域采集的农田土壤	0~160	[121]
2014 年	11 个采样点,包括 1 个背景点	3.8~2.10×10^3	[57]
无	采样区域为电子垃圾回收点周边	n.d.~21.2	[119]
无	采样区域为电子垃圾回收点和周边地区	回收站点:3327 周边地区:0~47.4	[119]

由上可知，粗放式电子垃圾拆解活动造成了 DP 的普遍污染，高浓度的 DP 在电子垃圾拆解地区的大气、车间灰尘、周边土壤及河流沉积物中普遍检出，因此，DP 可能会进入生物体内，并发生环境-生物体内异构体选择性富集及迁移转化，后续研究需要关注 DP 在生物体内的潜在代谢产物及可能的毒性效应。

4.2.5 有机磷阻燃剂

有机磷酸酯类阻燃剂(OPFRs)是一类人工合成的化学品，因其具有良好的阻燃性能被广泛应用于塑料、纺织品、电子、家具和建筑材料中[122-124]。2015 年，全球 OPFRs 的产量估计为 68 万吨[125]，然而，OPFRs 大多是添加型阻燃剂，不会与材料发生化学键合，在相关产品的生产、应用和处置过程中容易释放到环境中造成污染[126]，进而危害人体健康。根据取代基的不同，OPFRs 可分成氯代烷基磷酸酯、非氯代烷基磷酸酯、芳基磷酸酯和其他取代基磷酸酯[127]，常见的 OPFRs 见表 4-15。关于 OPFRs 研究较多的是有机磷酸三酯(tri-OPEs)，除 tri-OPEs 外，近年来，有机磷二酯(di-OPEs)的环境污染问题也引起了人们的广泛关注。除环境污染问题外，tri-OPEs 和 di-OPEs 都表现出了多种不同的毒性效应，如发育毒性[128]、生殖毒性[129]、神经毒性等[130]。

表 4-15 常见的有机磷酸酯种类及理化性质

中文名称	缩写	分子式	CAS 号	log K_{OW}	BCF
磷酸三(2-氯丙基)酯	TCIPP	$C_9H_{18}Cl_3O_4P$	13674-84-5	2.89	3.268
磷酸三(1,3-二氯异丙基)酯	TDCIPP	$C_9H_{15}Cl_6O_4P$	13674-87-8	3.65	21.4
磷酸三(2-氯乙基)酯	TCEP	$C_6H_{12}Cl_3O_4P$	115-96-8	1.63	0.4254
磷酸三(2-丁氧基乙基)酯	TBOEP	$C_{18}H_{39}O_7P$	78-51-3	3	25.56
磷酸三甲酯	TMP	$C_3H_9O_4P$	512-56-1	—	—
磷酸三丁酯	TnBP	$C_{12}H_{27}O_4P$	126-73-8	4	39.81
磷酸三异丁酯	TiBP	$C_{12}H_{27}O_4P$	126-71-6	3.6	19.51
磷酸三乙酯	TEP	$C_6H_{15}O_4P$	78-40-0	0.87	3.162
磷酸三异丙酯	TiPrP	$C_9H_{21}O_4P$	513-02-0	—	—
磷酸三辛酯	TEHP	$C_{24}H_{51}O_4P$	78-42-2	9.49	1×10^6
磷酸三戊酯	TNP	$C_{15}H_{33}O_4P$	2528-38-3	—	—
磷酸三丙酯	TPrP	$C_9H_{21}O_4P$	513-08-6	—	—
磷酸三苯酯	TPHP	$C_{18}H_{15}O_4P$	115-86-6	4.7	113.3
2-乙基己基二苯基磷酸酯	EHDPP	$C_{20}H_{27}O_4P$	1241-94-7	6.3	855.3

电子垃圾拆解活动已被证实是环境中 OPFRs 的重要来源，但是目前关于电子垃圾拆解产生 tri-OPEs 及 di-OPEs 的研究还比较少。目前，已在大气[131]、灰尘[132]、沉积物[123]、土壤[49]、水体[127]等环境介质中检测到较高水平的 tri-OPEs，然而，关

于 di-OPEs 环境污染的研究主要集中在灰尘、污泥等方面[125,133]。Du 等[133]对中国南方电子垃圾拆解区灰尘中 di-OPEs 进行了报道，拆解车间灰尘中 di-OPEs 浓度范围为 2010~55600 ng/g，明显高于当地家庭灰尘中 di-OPEs 的浓度(186~4350 ng/g)，研究表明电子垃圾拆解同样是 di-OPEs 的重要来源。但是，目前关于电子垃圾园区 tri-OPEs 和 di-OPEs 污染情况的研究，一方面集中在正规产业园区和拆解企业建立之前，另一方面集中在某一个点位和区域，很少对电子垃圾拆解园区 tri-OPEs 和 di-OPEs 的迁移转化进行研究。因此，有必要对电子垃圾拆解区及其周围地区 tri-OPEs 和 di-OPEs 的污染特征、迁移转化、人体暴露风险等进行了解，从而为电子垃圾拆解区 OPFRs 的治理提供理论依据。

1. 大气中的 OPFRs

大气在 OPFRs 的迁移运输方面起着重要作用。例如，OPFRs 可以通过干湿沉降进入土壤和地表水，与颗粒物结合形成灰尘[131]。所以，研究大气中 OPFRs 至关重要。尽管 OPFRs 在大气中的半衰期很短，但 OPFRs 可以与空气颗粒结合，这大大增强了它们的持久性[134]。Bi 等[135]研究发现，有机磷酸盐是中国南方电子垃圾回收车间排放的颗粒物中的主要有机成分，其主要由 TPhP 组成。Wang 等[136]对中国南方的城市地区(包含 20 个工业站点)和农村地区(包含 4 个电子垃圾回收设施)收集的 $PM_{2.5}$ 进行了研究，在两个地区都发现了较高浓度的 tri-OPEs(中位值浓度分别为 2854 pg/m^3 和 3321 pg/m^3)。由于 OPFRs 工业品种类繁多且使用存在差异，其同系物组成特征也可能存在差异，Nguyen 等[137]对加拿大一个正式电子垃圾拆解区采集的大气样本进行了分析，研究发现磷酸三(2-氯乙基)酯(TCEP)是空气样品 tri-OPEs 中含量最高的，中值浓度为 59 ng/m^3，其次是中位数浓度为 50 ng/m^3 的 TCIPP。Liu 等[138]对中国十个城市大气 $PM_{2.5}$ 中 tri-OPEs 的浓度进行了报道，其中浓度较高的有磷酸三(2-氯乙基)酯(TCEP)、TCPP、磷酸三(1,3-二氯异丙基)酯(TDCPP)，其中 TCPP 的浓度最高。

Yue 等[139]对我国南方某电子垃圾拆解园区及其周围地区大气中的 tri-OPEs 和 di-OPEs 进行了深入研究。电子垃圾拆解园区及其周边地区总悬浮颗粒物(TSP)中 13 种 tri-OPEs 和 7 种 di-OPEs 的检出率(DFs)和浓度(气相+颗粒相)如表 4-16 所示[140]。所有这些目标化合物在电子垃圾拆解园区的大气样本中都能检测到，而周边地区这些化合物的检出率相对较低。三种有机磷酸三酯(p-TCP、m-TCP 和 o-TCP)的 DFs 范围为 44.5%~94.5%，对于 di-OPEs，除 BCIPP、DPhP 和 DBP 外的其他化合物的 DFs 范围为 72.2%~100%。2017 年、2018 年和 2019 年电子垃圾拆解园区 TSP 中 tri-OPEs 的总浓度(Σtri-OPEs)分别为 $1.30×10^8$ pg/m^3、$4.60×10^6$ pg/m^3、$4.01×10^7$ pg/m^3，分别是对应的周边地区中位浓度(分别为 $7.40×10^4$ pg/m^3、$2.37×10^4$ pg/m^3 和 $3.68×10^3$ pg/m^3)的 1757、194 和 10897 倍。对于 di-OPEs，电子垃圾拆解园区 di-OPEs 的总浓度

(Σdi-OPEs)分别为 1.14×10^3 pg/m³、1.10×10^3 pg/m³、0.35×10^3 pg/m³,而周边地区 di-OPEs 的总浓度(Σdi-OPEs)范围为 1.36~141 pg/m³,也明显低于电子垃圾拆解园区。

表 4-17 显示了该电子垃圾拆解园区及其周围地区 $PM_{2.5}$ 中 tri-OPEs 和 di-OPEs 浓度的时间变化趋势,可以得出,2017 年、2018 年和 2019 年电子垃圾拆解园区 $PM_{2.5}$ 上对应的 tri-OPEs 浓度为 4.39×10^7 pg/m³、4.06×10^6 pg/m³ 和 4.16×10^6 pg/m³;周围地区 $PM_{2.5}$ 上对应的 tri-OPEs 中位值浓度为 3.35×10^5 pg/m³、2.26×10^4 pg/m³ 和 6.91×10^3 pg/m³。2017 年、2018 年和 2019 年电子垃圾拆解园区 $PM_{2.5}$ 上对应的 di-OPEs 浓度为 1.33×10^2 pg/m³、5.15 pg/m³ 和 11.0 pg/m³,周围地区 $PM_{2.5}$ 上对应的 di-OPEs 中位值浓度为 1.19 pg/m³、4.51 pg/m³ 和 5.78 pg/m³。从时间趋势看,电子垃圾拆解园区大气中 tri-OPEs 的浓度 2017 年最高,2018 年明显下降,2019 年呈现上升趋势;然而对于 di-OPEs 的浓度,2017 年和 2018 年的污染水平相似,而 2019 年则明显下降。上述结果表明,电子垃圾拆解活动可能是 tri-OPEs 和 di-OPEs 排放到周围大气环境中的重要来源之一,电子垃圾拆解园区内 tri-OPEs 或 di-OPEs 的浓度都是波动的,但总体趋势是下降的。电子垃圾拆解园区 tri-OPEs 和 di-OPEs 污染水平主要取决于采样日拆解活动的强度,而周边地区 tri-OPEs 和 di-OPEs 的浓度是逐年下降的,表明加强管理后,周围大气环境得到了改善。

2. 灰尘中的 OPFRs

有研究表明,当室内空气中的 OPFRs 浓度升高时,其附近灰尘中的 OPFRs 浓度也会升高,表明空气中的 OPFRs 会被灰尘吸附[141]。不同电子垃圾拆解排放造成的 OPFRs 的污染特征也有较大的差异。Wang 等[142]研究发现我国北方某个电子垃圾拆解区灰尘中 TCIPP、TCEP、TPhP 和 TMPP、TBEOP 是主要的同系物,而在我国广东清远电子垃圾拆解地灰尘中磷酸三甲苯酯(TCP)是主要的同系物(40.7%),而在城市地区(普通家庭室内和宿舍)灰尘中 TCEP 贡献最大(64.4%)[143]。除 tri-OPEs 外,di-OPEs 在灰尘中也有检出。Wang 等[124]对中国大陆地区室内和室外灰尘进行了采集,同时测量了 19 种 tri-OPEs 及 11 种 di-OPEs,TCIPP 和 DPHP 分别是灰尘中 tri-OPEs 和 di-OPEs 的主要化合物,对于 tri-OPEs,室内灰尘浓度(2380 ng/g dw)比室外灰尘浓度高一个数量级(446 ng/g dw)。Du 等[133]分析了华南地区某大型电子垃圾拆解园区车间和住宅中收集的室内灰尘中的 tri-OPEs 和 di-OPEs,其中,车间灰尘中 tri-OPEs 主要由 TCIPP、TMPP 和 TPhP 组成,而室内灰尘中 tri-OPEs 主要由 TCEP、TCIPP 和 TPhP 组成,DPhP 是室内外灰尘中 di-OPEs 的主要化合物。Stubbings 等[144]对加拿大电子电器废弃物设施拆解过程中收集的灰尘中的 OPFRs 进行了报道,结果发现 TPhP 含量最为丰富(中值浓度为 42 000 ng/g),占 Σ_{13}tri-OPEs 的 43%,其次是 TDCIPP,TCEP 和 TCIPP,分别占 Σ_{13}tri-OPEs 的 19%、12%和 8%。

表 4-16 电子垃圾拆解园区及其周边地区 TSP 样品中 tri-OPEs 和 di-OPEs 的浓度 (pg/m³)[140]

目标物	电子垃圾拆解园区									周边地区					
	2017年			2018年			2019年			2017年 (n=18)			2018年 (n=18)			2019年 (n=18)		
										中位值	范围	DF(%)	中位值	范围	DF(%)	中位值	范围	DF(%)

由于表格极宽，下面按目标物逐行列出：

目标物	园区2017年	园区2018年	园区2019年	周边2017中位	周边2017范围	DF(%)	周边2018中位	周边2018范围	DF(%)	周边2019中位	周边2019范围	DF(%)
TPrP	$3.77×10^1$	$1.10×10^1$	$5.30×10^1$	6.63	$1.44～4.59×10^1$	100	0.73	$0.30～2.38$	100	0.87	$0.42～4.08$	100
TEP	$4.55×10^3$	$3.61×10^4$	$7.11×10^4$	$4.32×10^3$	$5.25×10^2～2.52×10^4$	100	$5.29×10^2$	$1.41×10^2～1.78×10^3$	100	$5.39×10^2$	$3.19×10^2～1.40×10^3$	100
TBP	$8.37×10^3$	$12.0×10^3$	$5.28×10^3$	$7.33×10^1$	$3.37×10^1～1.47×10^2$	100	$1.77×10^1$	$7.61～8.72×10^1$	100	5.13	$1.19～4.98×10^1$	100
TCEP	$2.47×10^5$	$3.78×10^4$	$1.41×10^4$	$9.71×10^3$	$3.59×10^2～5.49×10^3$	100	$1.40×10^2$	$4.51×10^1～4.06×10^2$	100	$1.79×10^2$	$4.03×10^1～7.78×10^2$	100
TCPP	$2.26×10^7$	$1.30×10^6$	$1.01×10^7$	$1.86×10^4$	$2.80×10^3～2.09×10^5$	100	$1.12×10^4$	$2.83×10^2～1.37×10^5$	100	$6.98×10^2$	$2.59×10^2～4.09×10^3$	100
TCIPP	$8.42×10^6$	$3.84×10^5$	$2.90×10^6$	$6.79×10^3$	$8.38×10^2～1.80×10^5$	100	$4.40×10^3$	$5.83×10^1～4.14×10^4$	100	$1.63×10^2$	$6.30×10^1～4.14×10^4$	100
TDCIPP	$1.77×10^5$	$8.99×10^3$	$6.58×10^4$	$1.51×10^2$	$3.56×10^1～4.55×10^2$	100	$8.41×10^1$	$2.78～1.82×10^3$	100	$2.93×10^2$	$1.59×10^2～2.79×10^2$	100
TPhP	$9.57×10^7$	$2.77×10^6$	$2.63×10^6$	$2.79×10^4$	$2.03×10^3～3.13×10^5$	100	$4.57×10^3$	$2.18×10^2～1.00×10^5$	100	$2.77×10^2$	$1.43×10^2～5.10×10^4$	100
EDP	$6.11×10^3$	$1.42×10^3$	$5.24×10^5$	$9.49×10^1$	$4.81×10^1～7.74×10^2$	100	$4.18×10^2$	$0.68～3.87×10^3$	100	$5.51×10^1$	$1.52×10^2～2.33×10^1$	100
p-TCP	$1.86×10^6$	$4.75×10^4$	$3.15×10^5$	$2.22×10^2$	n.d.～$2.30×10^3$	78	$1.21×10^1$	n.d.～$1.44×10^2$	68	n.d.	n.d.～$2.08×10^2$	50
m-TCP	$1.13×10^5$	$3.61×10^4$	$3.19×10^4$	2.31	n.d.～$1.08×10^2$	61	n.d.	n.d.～$2.15×10^1$	45	0.27	n.d.～$2.04×10^1$	50
o-TCP	$6.92×10^4$	$2.21×10^4$	$1.56×10^5$	$7.63×10^1$	n.d.～$7.81×10^2$	89	$1.30×10^1$	n.d.～$2.69×10^2$	95	1.54	n.d.～$9.93×10^1$	75
TiPPP	$8.90×10^4$	$2.29×10^3$	$4.86×10^1$	$2.14×10^1$	n.d.～$1.36×10^2$	100	2.39	n.d.～6.35	84	0.68	n.d.～$2.53×10^1$	94
Σtri-OPEs	$1.30×10^8$	$4.60×10^6$	$4.01×10^7$	$7.40×10^4$	$9.09×10^3～7.09×10^5$		$2.37×10^4$	$1.44×10^3～1.91×10^5$	83.3	$3.68×10^3$	$1.29×10^2～1.35×10^5$	88
BCEP	1.07	n.d.	0.71	0.44	n.d.～2.52	72	0.46	n.d.～1.47	100	0.23	n.d.～1.10	100
BCIPP	$3.75×10^2$	$1.84×10^2$	$1.49×10^2$	3.50	n.d.～$1.34×10^1$	94	0.64	$0.11～3.66$	94	0.79	$0.28～3.27×10^1$	100
DPhP	$7.06×10^2$	$8.46×10^1$	$1.82×10^2$	$2.31×10^1$	$2.59～1.26×10^2$	100	2.78	$0.77～9.30$	100	2.56	$0.71～1.43×10^1$	100
DBP	7.05	4.75	2.33	2.04	$0.84～4.43$	100	0.32	$0.14～0.77$	100	0.40	$0.17～2.71$	100
BDCIPP	5.72	1.50	1.24	0.97	$0.43～2.06$	100	0.25	n.d.～1.09	94	0.29	n.d.～0.59	94
mDCP	$2.39×10^1$	$3.08×10^1$	4.73	0.25	$0.04～2.25$	100	0.04	n.d.～0.15	89	0.02	n.d.～0.20	81
BBOEP	1.95	1.00	0.94	0.16	n.d.～1.11	83	0.09	n.d.～0.69	89	0.06	n.d.～0.46	88
Σdi-OPEs	$1.14×10^3$	$1.10×10^3$	$3.46×10^2$	$3.16×10^1$	$8.27～1.41×10^2$		4.51	$1.36～1.43×10^1$		5.23	$2.13～4.85×10^1$	

a. DF: 检出频率; b. n.d.: 未检出

表 4-17 电子垃圾拆解园区及其周边地区 $PM_{2.5}$ 样品中 tri-OPEs 和 di-OPEs 的浓度 (pg/m³)

目标物	电子垃圾拆解园区						周边地区					
	2017年	2018年	2019年	2017年(n=18)			2018年(n=18)			2019年(n=18)		
				中位值	范围	DF(%)	中位值	范围	DF(%)	中位值	范围	DF(%)
TPrP	$3.35×10^1$	$2.26×10^1$	$1.67×10^1$	7.28	$0.36~1.94×10^1$	100	3.74	$1.24~2.26×10^1$	100	0.61	$0.34~1.67×10^1$	100
TEP	n.d.	$3.00×10^4$	$1.49×10^4$	$1.94×10^4$	$1.29×10^3~1.08×10^5$	100	$3.30×10^3$	$6.24~6.65×10^4$	100	$3.53×10^3$	$3.54×10^2~5.68×10^3$	100
TBP	$3.69×10^3$	$7.49×10^2$	$8.11×10^2$	$1.11×10^2$	$3.53×10^1~1.86×10^2$	100	$7.54×10^1$	$3.73×10^1~7.49×10^2$	100	$1.15×10^1$	$4.82~1.07×10^2$	100
TCEP	$9.66×10^4$	$2.08×10^4$	$2.66×10^4$	$2.13×10^3$	$3.51×10^1~1.05×10^6$	100	$4.54×10^2$	$1.72×10^2~2.08×10^4$	100	$2.54×10^2$	$9.14~2.66×10^4$	100
TCPP	$1.04×10^7$	$7.75×10^5$	$2.17×10^6$	$5.90×10^4$	$4.05×10^3~3.90×10^5$	100	$4.35×10^3$	$7.62×10^2~7.75×10^5$	100	$3.39×10^2$	$9.42×10^1~3.18×10^4$	100
TCIPP	$3.40×10^6$	$2.28×10^5$	$6.16×10^5$	$2.34×10^4$	$1.21×10^2~1.66×10^5$	100	$1.39×10^3$	$2.08×10^2~2.28×10^5$	100	$7.44×10^2$	$1.68×10^1~9.61×10^4$	100
TDCIPP	$6.68×10^4$	$5.92×10^3$	$1.71×10^4$	$4.62×10^2$	$3.77×10^1~1.69×10^3$	100	$5.69×10^1$	$2.06~5.92×10^3$	100	$8.04×10^1$	$1.89~1.71×10^2$	100
TPhP	$2.97×10^7$	$2.92×10^6$	$1.29×10^6$	$2.04×10^5$	$1.97×10^3~9.54×10^5$	100	$1.39×10^4$	$1.30×10^3~2.92×10^6$	100	$2.00×10^2$	$9.77×10^1~1.29×10^4$	100
EDP	$1.40×10^3$	$1.05×10^3$	$2.22×10^2$	$1.05×10^2$	$4.57×10^1~9.10×10^2$	100	$1.21×10^2$	$5.39×10^1~1.05×10^3$	100	$3.40×10^1$	$7.54~2.22×10^2$	100
p-TCP	$1.19×10^4$	$5.09×10^4$	$1.59×10^4$	$1.85×10^3$	n.d.^{b}~$5.54×10^3$	64	$1.18×10^2$	n.d.~$5.09×10^4$	79	1.55	n.d.~$1.59×10^2$	79
m-TCP	$3.28×10^4$	$3.61×10^3$	$1.51×10^3$	$2.46×10^1$	n.d.~$3.49×10^2$	64	0.44	n.d.~$3.61×10^2$	58	0.28	n.d.~$1.51×10^1$	53
o-TCP	$1.85×10^5$	$1.96×10^4$	$8.06×10^3$	$4.47×10^2$	n.d.~$1.28×10^3$	85	$6.09×10^1$	n.d.~$1.96×10^2$	79	1.11	n.d.~$8.06×10^1$	79
TiPPP	$2.15×10^4$	$2.19×10^4$	$1.61×10^3$	$4.54×10^1$	n.d.~$1.29×10^2$	93	9.52	n.d.~$2.19×10^3$	84	2.05	n.d.~$1.61×10^1$	
Σtri-OPES	$4.39×10^7$	$4.06×10^6$	$4.16×10^6$	$3.53×10^5$	$1.59×10^4~2.27×10^6$		$2.26×10^4$	$4.29×10^3~4.06×10^5$		$6.91×10^3$	$4.26×10^3~4.16×10^5$	
BCEP	n.d.	n.d.	n.d.	0.16	n.d.~3.27	89	0.46	n.d.~1.47	83.3	0.31	n.d.~1.46	79
BCIPP	n.d.	1.33	2.69	1.33	$0.06~1.01×10^2$	100	0.64	0.11~3.66	100	0.94	n.d.~$4.49×10^1$	95
DPhP	$1.28×10^2$	2.00	5.52	6.86	$2.00~5.98×10^2$	100	2.78	0.77~9.30	100	3.14	$0.72~1.47×10^2$	100
DBP	2.00	1.22	0.76	1.15	0.41~4.20	100	0.32	0.14~0.77	100	0.41	0.20~1.21	100
BDCIPP	n.d.	0.60	0.59	0.60	n.d.~3.12	95	0.25	n.d.~1.09	94.4	0.25	n.d.~0.94	89
mDCP	1.43	n.d.	0.08	0.11	$0.03~2.49×10^1$	100	0.04	n.d.~0.15	88.9	0.07	n.d.~2.83	81
BBOEP	0.42	n.d.	0.23	0.18	n.d.~2.10	95	0.09	n.d.~0.69	88.9	0.11	n.d.~0.78	88
Σdi-OPEs	$1.33×10^2$	5.15	$1.00×10^1$	1.19	$5.34~7.56×10^2$		4.51	$1.36~1.43×10^1$		5.78	$1.97~2.00×10^2$	

a. DF%: 检出频率; b. n.d.: 未检出

3. 土壤中的 OPFRs

土壤也是重要的环境介质之一，OPFRs 可以沉降到土壤中，由于与土壤有机质等的强结合，它们的半衰期增加到 50~60 天[130]。Ge 等[49]对中国南方电子垃圾拆解区及其周围地区的土壤中的 tri-OPEs 进行了研究，其中 TPhP 是最主要的同系物组成，电子垃圾拆解园区内土壤中 tri-OPEs 浓度(1.07×10^4 ng/g)高于周围地区(115 ng/g)，研究结果表明电子垃圾拆解是 tri-OPEs 的主要来源，而且对周围地区的土壤造成了一定的影响。Wang 等[142]研究发现我国北方某个电子垃圾拆解区土壤中 TCIPP 是最主要的同系物，其次是 TPhP 和 TBOEP，而 TPhP(11~3300 ng/g)是越南北部某电子垃圾回收区土壤中的主要 OPFRs[145]。

4.2.6 邻苯二甲酸酯

邻苯二甲酸酯(phthalates，PAEs)是一类工业化学品，自 1950 年以来作为添加剂和增塑剂广泛使用于各类产品中[146]。根据烷基侧链的碳原子数的大小，PAEs 可分为低分子量和高分子量的 PAEs，二者在工业生产中有不同的应用。低分子量的 PAEs，如邻苯二甲酸二甲酯(DMP)和邻苯二甲酸二乙酯(DEP)，常用作个人护理产品和化妆品的溶剂或载体，广泛添加入身体乳、护发产品、指甲油和香水等产品中；而高分子量 PAEs，如邻苯二甲酸二丁基苄基酯(BBzP)和邻苯二甲酸二(2-乙基)己酯(DEHP)，主要用作各种建筑材料、家庭家具和塑料产品的增塑剂，如 PVC 地板、塑料玩具等，添加含量比例可达 10%~60%[147]。其他常见的邻苯二甲酸二异丁酯(DiBP)和邻苯二甲酸二正丁酯(DnBP)则作为溶剂和增塑剂，在以上几类产品中都有应用。从 2007 年到 2017 年，全球 PAEs 的年产量从 270 万吨增加到 600 万吨，其中，发展中国家消耗了大部分的 PAEs[148]。在过去的几十年里，由于我国快速的城市化和现代化进程，增塑剂消费量快速增加。2017 年中国增塑剂消费量占全球的 42%，成为世界上最大的增塑剂市场，其中最主要的产品是 DEHP 和 DnBP[149]。

PAEs 的沸点在 250~450℃，是一种典型的半挥发性有机化合物(SVOCs)[150]，常见邻苯二甲酸酯的理化性质见表 4-18。由于 PAEs 的沸点较高而饱和蒸气压较低，并且在结构上与材料不是通过共价键结合，所以在产品制造、使用或处置过程中很容易通过浸出、挥发、迁移、磨损等途径释放到环境中，成为直接或间接的暴露源[151]。PAEs 在环境中的分布又因其物理化学性质的差异有所不同。低分子量的 PAEs(DMP，DEP，DnBP 和 DiBP 等)挥发性和水溶性更强，主要存在于大气气相中；而高分子量的 PAEs(DEHP 和 BBzP 等)的挥发性和水溶性相对较差，容易被气溶胶颗粒或水体悬浮颗粒物吸附，易在灰尘及沉积物中蓄积。在一项对法国室内空气中半挥发性有机化合物的研究中,发现 DEP 是气相中浓度最高的 PAEs,而 DEHP 是颗粒相中浓度最高的 PAEs[152,153]。由于被广泛地应用,PAEs 在各种环境基质中都能检测到,有研究发现我国几个城市的室内环境中 PAEs 浓度高于其他国家报道的浓度[154,155]。

表 4-18 邻苯二甲酸酯的物理化学性质

物质名称	英文全称	缩写	分子式	分子量	结构式	log K_{ow}	log K_{oa}
邻苯二甲酸二甲酯	dimethyl phthalate	DMP	$C_{10}H_{10}O_4$	194.2		1.61	7.01
邻苯二甲酸二乙酯	diethyl phthalate	DEP	$C_{12}H_{14}O_4$	222.2		2.54	7.55
邻苯二甲酸二烯丙酯	diallyl phthalate	DAP	$C_{14}H_{14}O_4$	246.2		3.11	7.87
邻苯二甲酸二正丁酯	di-n-butyl phthalate	DBP	$C_{16}H_{22}O_4$	278.4		4.27	8.54
邻苯二甲酸二异丁酯	diisobutyl phthalate	DiBP	$C_{16}H_{22}O_4$	278.4		4.27	8.54
邻苯二甲酸二(2-甲氧基)乙酯	bis(2-methoxy ethyl) phthalate	DMEP	$C_{14}H_{18}O_6$	282.3		1.11	9.77
邻苯二甲酸二戊酯	di-n-pentyl phthalate	DnPP	$C_{18}H_{26}O_4$	306.4		5.12	9.03
邻苯二甲酸二(2-乙氧基)乙酯	bis(2-ethoxyethyl) phthalate	DEEP	$C_{16}H_{22}O_6$	310.3		2.10	10.5
邻苯二甲酸二丁基苄基酯	butyl benzyl phthalate	BBP	$C_{19}H_{20}O_4$	312.4		4.70	8.78

续表

物质名称	英文全称	缩写	分子式	分子量	结构式	log K_{ow}	log K_{oa}
邻苯二甲酸二苯酯	diphenyl phthalate	DPHP	$C_{20}H_{14}O_4$	318.3		4.10	10.0
邻苯二甲酸二环己酯	dicyclohexyl phthalate	DCHP	$C_{20}H_{26}O_4$	330.4		6.20	11.6
邻苯二甲酸二己酯	dihexyl phthalate	DHXP	$C_{20}H_{30}O_4$	334.4		6.00	9.80
邻苯二甲酸二(4-甲基-2-戊基)乙酯	bis(4-methyl-2-pentyl) phthalate	BMPP	$C_{20}H_{30}O_4$	334.5		6.28	10.1
邻苯二甲酸二(2-丁氧基)乙酯	bis(2-n-utoxyethyl) phthalate	DBEP	$C_{20}H_{30}O_6$	366.5		4.06	12.0
邻苯二甲酸二(2-乙基)己酯	bis(2-ethylhexyl) phthalate	DEHP	$C_{24}H_{38}O_4$	390.6		7.73	10.5

由于 PAEs 的广泛应用以及频繁检出,一些 PAEs 产生的健康危害及控制问题已经成为全球关注的焦点,多种 PAEs,如 DMP、DEP、DnBP、BBzP、DEHP 和 DnOP 等,已被美国、欧盟和中国列为优先关注污染物[151]。美国消费者产品安全委员会(CPSC)在其 2008 年和 2017 年更新的《消费者产品安全改进法案》中提出禁止在儿童玩具和儿童护理用品中使用 DiBP、DnBP、BBzP、DEHP 和 DiNP[156]。2016 年,加拿大颁布法规,将儿童护理用品中 DEHP、DnBP 和 BBzP 的使用量限制在 0.1%以下[157]。另外,欧盟修订后的《电子电气设备中限制使用某些有害物质指令》(RoHS 指令,EU 2015/863)也新增 DEHP、BBP、DBP、DiBP 等 4 类 PAEs 列入限制物质清单,其在儿童玩具、化妆品和电子电器等产品中的限量都是均匀材质中的质量含量不能大于 0.1%[157]。2014 年,中国发布修订后的《国家玩具安全技术规范》(GB 6675—2014)增加了玩具中 6 种 PAEs 的限量要求,规定儿童玩具中 DBP、BBP、DEHP 的总含量不得超过 0.1%,可放在嘴里的儿童玩具中邻苯二甲酸二异壬酯(DINP)、邻苯二甲酸二异癸酯(DIDP)、邻苯二甲酸二正辛酯

(DNOP)的总含量不得超过 0.1%[158]。我国对饮用水中 DEHP 的浓度限值也有国标进行了规定(小于 8 μg/L)(GB 9685—2008)[159]。

1. 大气和灰尘中的 PAEs

Gu 等[160]在浙江省某电子垃圾拆解区域以及附近居民区采集了大气 $PM_{2.5}$ 样品，检测到 4 种 PAEs，其中 DEHP 占主导地位，并且电子垃圾拆解区域的浓度(216.41 ng/m^3)显著高于居民区(106.16 ng/m^3)，表明电子垃圾拆解活动会造成大量 PAEs 进入大气中。Tang 等[161]在广东省某电子垃圾拆解区域附近的室内灰尘中检测到 6 种 PAEs，2013 年灰尘样品中的总浓度范围为 54 000~734 900 ng/g(中值浓度 167 200 ng/g)，2017 年灰尘样品中的总浓度范围为 29 400~167 700 ng/g(中值浓度 95 000 ng/g)。其中 DEHP、DiNP、DiDP 和 DnBP 是灰尘中主要的 PAEs 同系物，占总浓度的 85%以上。从 2013 年到 2017 年，灰尘中 PAEs 的浓度显著降低，表明国家和地方禁止粗放式电子垃圾拆解活动的规定对缓解这些化学品的污染状况是有效的(图 4-25)。

在广东某电子垃圾拆解车间以及附近室内灰尘的研究中，检测到 9 种 PAEs，在电子垃圾拆解车间灰尘的总浓度范围为 170~5300 μg/g(中值浓度 1400 μg/g)，显著高于当地农村(中值浓度 300 μg/g)和城市(中值浓度 680 μg/g)室内灰尘浓度，表明电子垃圾拆解处理活动加剧了工作环境中 PAEs 的排放。在检测到的 9 种 PAEs 中，高分子量 DiNP 在电子垃圾拆解车间灰尘中占主导地位，中值浓度高达 860 μg/g，占 PAEs 总浓度的 66%。另一种高分子量 DEHP 的浓度比 DiNP 低两倍，并且电子垃圾拆解车间灰尘中的 DEHP 浓度(中位数：390 μg/g)显著高于农村(中位数：170 μg/g)，但与城市室内灰尘中的 DEHP 浓度相当(中位数：410 μg/g)。农村和城市室内灰尘 PAEs 的组成以 DEHP 为主，这与以往相关报道一致，而电子垃圾拆解车间灰尘 PAEs 的组成以 DiNP 为主，表明 DiNP 广泛应用于电子电器产品，而 DEHP 作为一种常用的增塑剂还广泛应用于其他家庭用品中。在电子垃圾拆解车间灰尘中检测到的 6 种低分子量 PAEs 中，DnBP 和 DiBP 的浓度与 DnOP 相当，其余低分子量 PAEs 浓度均低于 μg/g 水平。值得注意的是，电子垃圾拆解车间灰尘中 4 种低分子量 PAEs(DMP、DEP、DnBP 和 DiBP)的浓度甚至低于家庭室内灰尘，这可能由于低分子量 PAEs 被广泛应用于家庭日常用品中[162]。

一项研究调查了泰国某电子垃圾拆解区域和附近社区的室内地板和室外道路灰尘中 PAEs 的水平[163]，其中电子垃圾拆解区域和附近社区的室内灰尘中 PAEs 总浓度范围分别为 86 000~790 000 ng/g 和 44 000~2 700 000 ng/g，对应室外道路灰尘中 PAEs 总浓度范围分别为 40 000~670 000 ng/g 和 27 000~6 500 000 ng/g，在所有灰尘样品中，DEHP 都占主导地位，但在室外灰尘的占比小于室内灰尘，室内家具可能是 DEHP 的主要来源。

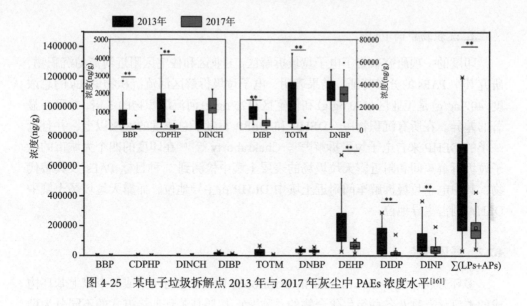

图 4-25 某电子垃圾拆解点 2013 年与 2017 年灰尘中 PAEs 浓度水平[161]

2. 土壤中的 PAEs

Zhang 等[164]对广东某电子垃圾拆解区域附近的居民区和农业区土壤中 PAEs 的水平进行了研究,在所有土壤样品中检测到 14 种 PAEs,总浓度范围为 2950.22~67 154.18 ng/g,中值浓度为 8163.19 ng/g。其中,居民区 A 的 PAEs 浓度较高,中值浓度为 21 609.37 ng/g,分别比居民区 B 和农业区高约 1.97 倍和 3.26 倍。根据《中国土壤环境质量标准》(GB 15618—2008),居民区 B 和农业区土壤中 PAEs 浓度仍属于背景浓度,而居民区 A 土壤则属于轻微污染土壤。大部分电子垃圾拆解车间位于居民区 A 中,可能导致居住 A 土壤中 PAEs 浓度高于居民区 B。此外,在三个取样区域发现了类似的组成特征(以 DBEP、DCHP 和 BMPP 为主),表明 PAEs 的来源相似。另外,土壤 PAEs 浓度与居民区到农业区土壤的距离上有明显的负相关,可以推测农业区 PAEs 可能主要来自居民区 A 中 PAEs 的长距离大气迁移。Liu 等[165]研究了广东省某电子垃圾拆解区域土壤以及附近菜地和稻田等耕地土壤中 6 种 PAEs 的分布情况,发现菜地、稻田和电子垃圾拆解区域土壤中 6 种 PAEs 的总浓度分别为 1.149~7.323 mg/kg(平均值 2.155 mg/kg)、1.233~1.543 mg/kg(平均值 1.392 mg/kg)和 11.77~17.87 mg/kg(平均值为 14.49 mg/kg),且所有土壤样品中 PAEs 均以 DEHP 和 DnBP 为主[166]。电子垃圾拆解区域土壤中 PAEs 的总浓度显著高于菜地和稻田,表明电子垃圾拆解是 PAEs 进入环境的重要原因之一。浙江省某电子垃圾拆解区域土壤中也有 PAEs 检出的报道,在采集的土壤样品中检测到 5 种 PAEs,总浓度在 12.566~46.669 mg/kg 之间,其中 DEHP、DBP 和 DEP 是主要的 PAEs,占总 PAEs 的 94%以上。

3. 沉积物中的 PAEs

印度的一项研究采集了电子垃圾拆解区、工业区和住宅区附近河流的沉积物，研究其中 PAEs 的分布特征。结果表明，电子垃圾拆解区河流沉积物 PAEs 的总浓度(407 ng/g)是工业区(150 ng/g)和住宅区(125 ng/g)河流沉积物的 3 倍，并且有显著的差异。在所有沉积物中，DEHP 和 DnBP 贡献了总浓度的 80%以上，并且近一半的 DEHP 来自电子垃圾拆解[167]。Chakraborty 等[168]在印度的四个大城市的电子垃圾拆解车间和附近露天垃圾场的表层土壤中检测到 7 种目标 PAEs 及其替代物，其中电子垃圾拆解车间附近土壤中 DEHP 占主导地位，而露天垃圾场土壤中 DEHA 则占主导地位。

4.2.7 多环芳烃

多环芳烃(polycyclic aromatic hydrocarbon, PAHs)是由两个或两个以上苯环构成的碳氢化合物及各种衍生化合物的总称[169]。按照其苯环连接方式的不同分为稠环和非稠环型两大类，其中稠环型是指相邻的苯环至少共享两个碳原子，即通常所说的 PAHs，如萘、菲、蒽等；非稠环型是指相邻苯环之间通过单键连接，如联苯、三联苯等。常温下，PAHs 是有颜色的结晶固体，其物理性质随着分子量和结构的不同而变化。一般而言，PAHs 的蒸气压和水溶性随着分子量的增加而降低；PAHs 具有高亲脂性，易溶于有机溶剂，难溶于水。此外，PAHs 还具有光敏性、耐热性、导电性、应激性、抗腐蚀等特性；PAHs 的毒性随着其分子量的增加而增加，而急性毒性反而降低[170]。不同的 PAHs 所引起的健康效应不同，鉴于其严重的致癌、致畸、致突变等毒性，美国环保署于 20 世纪 70 年代划定了 16 种优先控制的 PAHs[171]，并被全世界所采纳沿用至今，其化学结构如图 4-26 所示，其理化性质如表 4-19 所示。

1. 大气中 PAHs 的污染

化石燃料、汽车尾气和生物质燃烧是中国城市地区大气中 PAHs 的主要来源[172]。进入大气环境的 PAHs 在通过大气沉降作用沉积到土壤、植物和水体之前可进行长距离的传输，其在大气中的行为受物理化学反应、其他污染物的相互作用、光化学转化和干湿沉降等诸多因素影响；并且根据大气状态(温度、相对湿度等)、气溶胶的种类以及 PAHs 本身的特性分布在气相或颗粒相中[173]。通常二环等低环 PAHs 比较容易挥发，主要存在于气相；三至四环的 PAHs 则在气相和固相中均有分布，五环以上的 PAHs 主要吸附在颗粒物上。由于环数越多，PAHs 的同分异构体就越多。此外，湿度和悬浮颗粒物的类型，如炭黑、灰尘、飞灰、花粉等均会影响 PAHs 在颗粒相的吸附作用[174,175]。

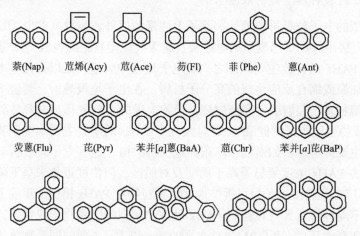

图 4-26　美国环保署优先控制 16 种 PAHs 的化学结构

表 4-19　美国环保署优先控制 16 种 PAHs 的理化性质

化合物名称	分子式	分子量	熔点(℃)	沸点(℃)	CAS 号
萘	$C_{10}H_8$	128.18	80-82	218	91-20-3
苊烯	$C_{12}H_8$	152.20	92-93	265-280	208-96-8
苊	$C_{12}H_{10}$	154.20	90-96	278-279	83-32-9
芴	$C_{13}H_{10}$	166.23	116-118	293-295	86-73-7
蒽	$C_{14}H_{10}$	178.24	216-219	340	120-12-7
菲	$C_{14}H_{10}$	178.24	96-101	339-340	85-01-8
荧蒽	$C_{16}H_{10}$	202.26	107-111	375-393	206-44-0
芘	$C_{16}H_{10}$	202.26	150-156	360-404	129-00-0
苯并[a]蒽	$C_{18}H_{12}$	228.30	157-167	435	56-55-3
䓛	$C_{18}H_{12}$	228.30	252-256	441-448	218-01-9
苯并[b]荧蒽	$C_{20}H_{12}$	252.32	167-168	481	205-99-2
苯并[k]荧蒽	$C_{20}H_{12}$	252.32	198-217	471-480	207-08-9
苯并[a]芘	$C_{20}H_{12}$	252.32	177-179	493-496	50-32-8
苯并[g,h,i]芘	$C_{22}H_{12}$	276.34	275-278	525	191-24-2
茚并[1,2,3-cd]芘	$C_{22}H_{12}$	276.34	162-163	-	193-39-5
二苯并[a,h]蒽	$C_{20}H_{14}$	278.35	266-270	524	53-70-3

1) 电子垃圾拆解场地的浓度水平

在典型的电子垃圾拆解地区，除了上述来源以外，粗放式的电子垃圾拆解手段，包括直接用煤炉烘烤废印刷电路板及露天焚烧塑料绝缘电线[176]，是电子垃圾拆解地区 PAHs 的排放主要来源之一。电子垃圾中废塑料是以石化基本单体为原料，经过加聚或缩合反应生成的高分子材料。在电子垃圾热解、焚烧等处理过程中，这些塑料发生裂解或不完全燃烧，导致大量的 PAHs 形成并释放到周围环境中，严重威胁当地居民的健康[1]。针对大气中 PAHs 的研究主要集中于我国早期电子垃圾的主要拆解点，如浙江、广东等地(表 4-20)，这些电子垃圾拆解场地大气颗粒物中的 PAHs 浓度要显著高于周围及对照区，但浓度近年来呈下降趋势，其中 2016 年广东某典型电子垃圾拆解地大气颗粒物中 PAHs 相比十年前下降了一个数量级；不同拆解点大气颗粒中 PAHs 呈现出相似的同系物组成特征，其中高环数 PAHs(四至六环)贡献了 PAHs 总水平的 80%以上，主要的同系物单体包括苯并[k]荧蒽(B[k]F)、苯并[b]荧蒽(B[b]F)、茚并[1,2,3-cd]芘(IcdP)和苯并[g,h,i]芘(B[ghi]P)等。此外，电子垃圾拆解场地大气颗粒物中分子量大于 300 的 PAHs 的浓度水平(中位数为 3.05 ng/m³)也显著高于城区(中位数为 1.42 ng/m³)($p<0.001$)，表明粗放式电子垃圾拆解活动可能是这些强致癌性高环 PAHs 的特定来源之一。大部分研究报道了美国环保署优先控制的 16 种 PAHs 的污染特征，只有少部分研究关于 16 种以外的 PAHs 同系物，如 1,3,5-三苯基苯(1,3,5-TPB)(图 4-27)，这类化合物被证实来自于聚乙烯塑料等的燃烧，在电子垃圾拆解场地大气颗粒物中被频繁检出，其中我国华南某典型电子垃圾拆解区(清远)大气颗粒物中 1,3,5-TPB 的浓度约为周边对照区的 3~10 倍，高于广州(0.27 ng/m³)20 多倍[177]。大气颗粒物中 1,3,5-TPB 随季节变化波动不大(冬季为 0.36~51.6 ng/m³，夏季为 0.16~49.6 ng/m³)[178]，这与常规 PAHs 的季节变化趋势不一致，可能来自电子垃圾拆解活动的持续排放，即 1,3,5-TPB 可作为电子垃圾拆解排放 PAHs 的特征指示物[179]。

表 4-20　不同电子垃圾拆解点大气中 PAHs 浓度水平和同系物组成(ng/m³)[179]

样品信息	电子垃圾拆解场地		周边位置		文献
	中位值(范围)	组成分布	中位值(范围)	组成分布	
广东 PM$_{2.5}$ (2004 年)	102±63.6 (22.7~263)[a]	B[bk]F>IcdP> B[ghi]P>B[a]P> Chr>Pyr[b]	—	—	[20]
浙江 PM$_{2.5}$ (2005~2006 年)	248.5±25.36[a]	B[bk]F>Chr>B[a]A >B[a]P≈ IcdP≈B[ghi]P	129.4±24.94[a]	B[bk]F>Chr> B[ghi]P≈IcdP≈ B[a]P≈B[a]A	[160]
广东气态和 TSP (2007 年)	(313~1041)	Phe>Flu>Pyr	50.1	Flu>Phe>Pyr	[35]

续表

样品信息	电子垃圾拆解场地		周边位置		文献
	中位值(范围)	组成分布	中位值(范围)	组成分布	
广东 TSP (2009~2010 年)	(7.38~87.9)	B[*ghi*]P>IcdP> B[*b*]F>B[*a*]P>Fluo >B[*k*]F	(16.7~113)[c]	B[*ghi*]P>IcdP> B[*b*]F>B[*a*]P> B[*k*]F>Fluo	[178]
广东 TSP (2010~2011 年)	18.3 (4.4~89.7)	B[*b*]F>B[*ghi*]P> IcdP>B[*a*]P>Fluo >B[*k*]F	14.3 (3.95~61.4)[c]	B[*b*]F>B[*ghi*]P> IcdP>Fluo>B[*a*]P >B[*k*]F	[19]
广东 PM (2012 年)	(15.1~17.7)[d]	B[*ghi*]P>Phe>IcdP >B[*b*]F>B[*k*] F≈Chr≈Fluo	(5.8~6.7)[c]	B[*ghi*]P>IcdP> Phe≈Fluo>B[*b*]F ≈B[*k*]F≈Chr	[1]
广东 $PM_{2.5}$ (2015~2016 年)	5.7 (2.1~45.0)	B[*b*]F>B[*k*]F>IcdP >B[*ghi*]P	5.7 (14.0~45.7)	B[*b*]F>B[*k*]F>IcdP >B[*ghi*]P	[17]
广东气态 (2015~2016 年)	62.1±9.4[a] (37.7~109)	Phe>Pyr	(12.3~36.4)	Phe>Pyr	[17]

a. 均值；b. B[*b*]F 和 B[*k*]F 浓度之和；c. 城市位置；d. 平均值

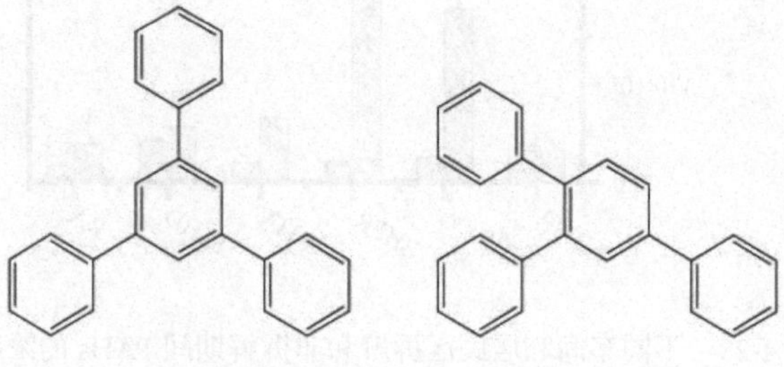

图 4-27　1,3,5-三苯基苯和 1,2,4-三苯基苯的化学结构

2) 电子垃圾处理车间的排放特征

早期的研究主要利用反应炉模拟高温焚烧或热解过程中，各种聚合物塑料产生 PAHs 的污染特征及排放因子。从已有的研究来看，不同的塑料种类对 PAHs 的排放因子影响最大，聚氯乙烯(PVC)焚烧或热解过程产生的 PAHs 相比其他的塑料多；此外，燃烧温度、反应气种类均影响 PAHs 的排放。在较高的炉温(例如>850℃)，PAHs 的排放因子会降低，且高环的 PAHs 易分解形成低环 PAHs 或其他挥发性有机物，而电子垃圾焚烧(温度在 400~800℃之间)是 PAHs 排放的主要原因，低温烘烤电路板(<400℃)排放因子相对较小。这些研究均证实了电子垃圾的热解和焚烧处理是造成 PAHs 大量排放的主要原因。

Liu 等[180]研究了广东某循环经济产业园中 5 个电子垃圾处理车间大气中 PAHs

的排放特征，包括电吹风处理手机车间（EBMP）、电烤锡炉处理电视机车间（EHFTV）、电烤锡炉处理路由器车间（EHFR）、旋转灰化炉处理电视机车间（RITV）和旋转灰化炉处理硬盘车间（RIHD）中 16 种优先控制 PAHs 在拆解期间和非拆解期间的浓度如图 4-28 所示，其中拆解过程中Σ_{16}PAHs 的浓度范围为 $1.53\times10^4 \sim 2.02\times10^5$ pg/m³。五个拆解车间内 PAHs 的平均浓度顺序依次为：EHFR（2.02×10^5 pg/m³）＞EHFTV（5.31×10^4 pg/m³）＞RITV（1.95×10^4 pg/m³）≈ RIHD（1.67×10^4 pg/m³）≈ EBMP（1.53×10^4 pg/m³）。不同拆解工艺过程释放的 PAHs 浓度均比居民区浓度高（1.16×10^4 pg/m³），这表明废电路板拆解回收过程可能是造成当地大气环境中 PAHs 污染的一个主要排放源。

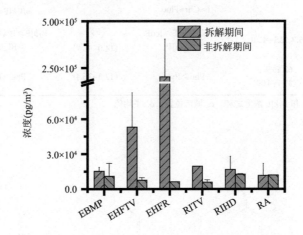

图 4-28　不同车间和居民区拆解和非拆解期间 PAHs 的浓度

拆解车间内的 PAHs 主要来源于各种电子垃圾材料的不完全燃烧。电烤锡炉拆解工艺过程排放的 PAHs 浓度高于其他拆解工艺过程的释放，可能由两方面的原因造成：一是电烤锡炉工艺较高的熔融温度和较大的加热面积有利于 PAHs 的释放；二是其拆解体系属于敞开体系。虽然拆解炉上方安装有集气罩，但是大量的有机废气仍随着拆解过程释放到周围的大气环境中，进而附着在颗粒物上。虽然旋转灰化炉的温度更高，有机物燃烧分解释放出大量的 PAHs，但是旋转灰化炉大部分时间处于闭合状态，并且焚烧期间产生的大量废气被上方的集气罩直接收集，因此仅有少量的 PAHs 进入大气中。

EHFR 车间内大气 TSP 颗粒相中 PAHs 的浓度与垃圾焚烧车间内浓度（434.2 ng/m³）相当，比采用煤炉拆解线路板车间（57.8 ng/m³）和塑料回收车间（50.5 ng/m³）内排放的 PAHs 高一个数量级[35]，低于 Chen 等[181]报道的煤炉拆解电视机线路板回收车间内 PAHs 的浓度（1100 ng/m³）。通过与其他职业环境内 PAHs 释放浓度比较发现，PAHs 的浓度比纽扣生产厂（2300 ng/m³）[182]、炭黑生产厂（1777

ng/m³)[183]和道路交叉口(4771 ng/m³)[184]等环境中 PAHs 浓度的低 1 到 2 个数量级，这可能由于不同的污染源释放强度以及不同有机质燃烧造成的。

不同拆解车间内非拆解期间 PAHs 的污染浓度与拆解车间呈现不同的顺序，其 RIHD 车间内 PAHs 浓度最高，为 1.27×10^4 pg/m³，其次为 EBMP(1.10×10^4 pg/m³)、EHFTV(7.59×10^3 pg/m³)、EHFR(6.26×10^3 pg/m³)和 RITV(5.93×10^3 pg/m³)。拆解车间内拆解期间 PAHs 的污染要高于非拆解期间，说明电子垃圾拆解造成了室内一定的 PAHs 污染。尤其是 EHFR 车间，其拆解期间内 PAHs 的浓度是非拆解期间的 30 倍多。在居民区，其白天主要来源于机动车尾气贡献，虽然在夜间人为排放减少，但是较低的大气混合高度对 PAHs 有很强的富集作用，抵消了较少的人为排放源。此外，非拆解期间大量的非法露天焚烧致使大气中高浓度的 PAHs 通过传输、沉降等污染周围的环境。

不同拆解工艺单体 PAHs 所占的百分比如图 4-29 所示。从图可以看出，PAHs 同系物百分比随着拆解工艺和线路板类型而变化。总体来说，在拆解过程中四环和五环 PAHs 占主导，其中 EBMP 车间内 FLU 和 BghiP 贡献最大，分别贡献了 Σ_{16}PAHs 的 13.8%和 12.7%；EHFTV 拆解车间内以 FLU 和 PYR 为主，其百分含量分别为 35.2%和 32.4%；EHFR 车间内 BbF 和 CHR 百分含量最高，分别为 14.7%和 12.3%；而在回收温度较高的 RITV 和 RIHD 车间内，FLU、PYR、BaA 和 CHR 含量百分比最大。而在回收温度较高的 RITV 和 RIHD 车间内，FLU、PYR、BaA 和 CHR 含量百分比最大。虽然低环(2~3 环)PAHs 挥发性强，很容易随着高温的拆解活动释放，但是其主要分布在气相中[185]，而四环及以上 PAHs 主要是通过高温燃烧产生的，主要存在于颗粒相中。与拆解期间不同的是，不同车间非拆解期间内 PAHs 同系物百分含量比较一致。

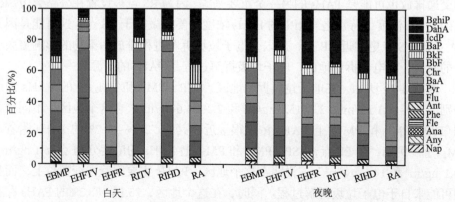

图 4-29 不同车间和居民区拆解期间和非拆解期间 PAHs 的组分含量

在过去的几十年间，早期典型的电子垃圾区(如中国华南地区等)的大气中

PAHs 水平急剧下降,这得益于政府对非法拆解活动的严厉管控。这些早期拆解点土壤中的 PAHs 浓度也低于印度、巴基斯坦和加纳等国家新兴涌现的拆解点,表明电子垃圾污染在全球呈扩散趋势。从实验室模拟到现场观测等多方面数据都表明,PAHs 的排放与电子垃圾的种类以及拆解工艺有关,废电路板的热解和燃烧过程是 PAHs 排放的主要来源。然而,当前大部分研究限定于美国环保署优先控制16 种 PAHs,事实上,一些特定的 PAHs 及其衍生物,如三苯基苯、卤代和含氧PAHs,被证实可作为指示电子垃圾拆解 PAHs 排放的特征标志物,应该被纳入今后的监测范围,其潜在的形成机制及潜在的健康效应还有待进一步阐明,从而更加全面地评估粗放式电子垃圾拆解活动造成的 PAHs 污染及其健康风险。

3)时空分布及源解析

对我国华南某电子垃圾拆解工业园区及其周边地区的大气样品进行采样分析[18],包括 $PM_{2.5}$、PM_{10} 和 TSP 以及对应的气相样品,剖析 PAHs 的时空变化趋势及潜在来源。采样地点共有五个,分别是电子垃圾拆解工业园区(EP)和周边地区包括公路旁边(RS)、乡村(VS)、居民区对照点(RA),以及代表典型城市生活的对照点——距离电子垃圾拆解地区大约 400 km 的广州市(GZ)(采样点位置见图 4-7)。如图 4-30 所示,在电子垃圾拆解地区,TSP 加气相中总 PAHs(除萘外)的浓度为 $25.9\sim88.7$ ng/m^3。其中,EP 点的 PAHs 浓度最高,中位数为 88.7 ng/m^3,其次是 RS 点、RA 点、GZ 点和 VS 点,分别是 58.5 ng/m^3、39.9 ng/m^3、30.1 ng/m^3 和 25.9 ng/m^3。EP 点和 RS 点的 PAHs 浓度明显高于对照点 GZ(Mann-Whitney U test,$P<0.01$),但 RA 点与对照点 GZ 的 PAHs 浓度不存在显著差异($P=0.114$)。这说明在 EP 点,电子垃圾拆解过程可能是 PAHs 的重要来源,而在 RS 点,其他污染源如交通排放也可能是 PAHs 的另一个重要来源,因为 RS 点被设置在公路旁边。RA 点的 PAHs 浓度,特别是气相中的 PAHs 浓度,要稍微高于 VS 点,这可能是因为 RA 点比 VS 点更靠近 EP 点,所以受电子垃圾拆解过程排放污染物的影响更大。总的来说,在电子拆解地区,电子垃圾拆解过程是 PAHs 的来源之一。

如图 4-30 所示,在电子垃圾拆解地区,TSP、PM_{10} 和 $PM_{2.5}$ 上的 PAHs 浓度分别为 $2.9\sim16.1$ ng/m^3、$2.7\sim15.1$ ng/m^3 和 $2.2\sim14.6$ ng/m^3。在这个地区,不同颗粒相(TSP、PM_{10} 和 $PM_{2.5}$)上 PAHs 浓度很接近,其中 72.7%~90.7%的 PAHs 附着在 $PM_{2.5}$ 上。例如,在 EP 点,TSP、PM_{10} 和 $PM_{2.5}$ 上 PAHs 的浓度分别是 16.1 ng/m^3、15.1 ng/m^3 和 14.6 ng/m^3。这说明在这个地区,PAHs 主要附着在细颗粒上,而且很可能来自于电子垃圾拆解过程。同时,在这个地区,75.1%~88.2%的 PAHs 存在于气相中,例如,在 EP 点,气相中的 PAHs 浓度为 (62.1 ± 9.4) ng/m^3。对照点 GZ 也有类似的结果,85.1%的 PAHs 存在于气相中,而颗粒相中 81.6%的 PAHs 附着在 $PM_{2.5}$ 上。在文献中,也有类似的报道,珠三角地区大气中 78.9%的 PAHs 存在

于气相中[186]。这些结果说明空气中绝大部分的 PAHs 存在于气相和 $PM_{2.5}$ 上。在电子垃圾拆解地区 EP 点的 PAHs 浓度最高，其中，$PM_{2.5}$ 上 PAHs 浓度中位数为 14.6 ng/m^3。相比于文献中报道电子垃圾拆解地区的结果，EP 点的 PAHs 浓度比 3 个电子垃圾拆解地区的浓度(分别是 27.6 ng/m^3、129 ng/m^3 和 102 ng/m^3)[19,20,160]都低。由于之前报道的电子垃圾拆解地区尚未建立专门的工业园区，这说明电子垃圾拆解工业园区的建设，对减少 PAHs 排放起到了积极的作用。

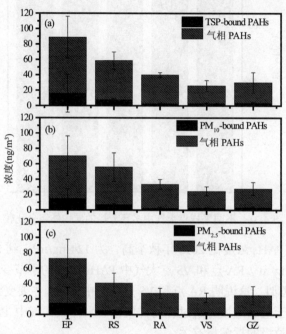

图 4-30　不同采样点的颗粒相(TSP(a)、PM_{10}(b)和 $PM_{2.5}$(c))和气相中多环芳烃浓度空间分布

PAHs 异构体在气固相中分布方面，二环和三环的 PAHs 如 Phe，大部分存在于气相中，四环 PAHs 如 FluA 和 Pyr，在气相和颗粒相中均有分布，而五环和六环 PAHs 则大部分存在于颗粒相中。如图 4-31 所示，在气相中，Phe 和 Pyr 浓度最高，占了总 PAHs 的 64.6%；在颗粒相中，BbF、BkF、IcdP 和 BghiP 的浓度最高，占了总 PAHs 的 65.9%。在中国珠三角地区的农村采样点，也有类似的结果[186]。此外，我国上海和韩国也有类似的报道[187,188]。PAHs 由多个苯环构成，其辛醇-空气分配系数(K_{OA})随着苯环数量增加而增加[35]。因此，低环 PAHs 倾向于分布在气相中，而高环 PAHs 倾向于分布在颗粒相中，特别是在 $PM_{2.5}$ 上。

在 EP 点，2015 年四季中春季的大气中 PAHs 浓度(TSP 加气相)最高，浓度中位数为 147 ng/m^3，其次是冬季和秋季，分别为 49.2 ng/m^3 和 42.1 ng/m^3，最低是夏季，为 32.1 ng/m^3，如图 4-32 所示。2015 年至 2017 年的秋季中，EP 点的 2016

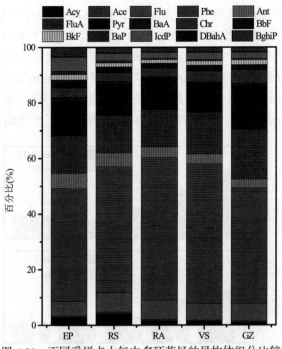

图 4-31 不同采样点大气中多环芳烃的异构体组分比较

年秋季的大气中 PAHs 浓度比 2015 年秋季高，为 124 ng/m³，到 2017 年秋季时有所下降，为 58.6 ng/m³。RA 点和 VS 点大气中 PAHs 浓度的季节变化趋势和年变化趋势和 EP 点的类似，这说明 RA 点和 VS 点大气中 PAHs 主要受到 EP 点的影响，可能大部分来自于电子垃圾拆解过程的排放，而采样点大气中 PAHs 浓度的时间变化可能与电子垃圾的拆解量有关。

如图 4-33 所示，RS 点大气中 PAHs 的年变化趋势与 EP 点、RA 点和 VS 点相同，不过，其季节变化趋势与这三个采样点不同。从 2015 年的春季到冬季，RS 点大气中 PAHs 浓度逐渐下降，范围为 39.9~77.9 ng/m³。除了电子垃圾拆解排放的影响，RS 点的 PAHs 浓度四季变化可能与交通车辆经过的数量有关，因为 RS 点被设置在电子垃圾拆解地区的一条公路旁边。这意味着 RS 点的 PAHs 主要受到电子垃圾拆解排放和交通排放的控制。至于对照点 GZ，其大气中 PAHs 的年变化趋势也与其他四个采样点相同，但季节变化趋势与 RS 点更相近。受到多种污染来源的影响，GZ 点在春季时 PAHs 浓度最高，其次是夏季、冬季和秋季，分别为 43.1 ng/m³、24.3 ng/m³、22.8 ng/m³ 和 21.7 ng/m³。GZ 点大气中 PAHs 浓度的季节变化也可能与交通排放有关，因为机动车排放是城市大气中 PAHs 的主要来源之一[189]。

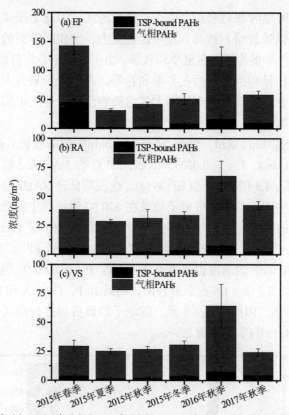

图 4-32 EP 点(a)、RA 点(b)和 VS 点(c)的 TSP 和气相中多环芳烃浓度的时间变化

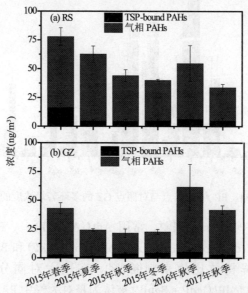

图 4-33 RS 点(a)和 GZ(b)点的 TSP 和气相中 PAHs 浓度的时间变化

电子垃圾拆解地区和对照点都位于亚热带地区，因此冬季煤炭燃烧供暖导致的 PAHs 排放可以被忽略。然而，在一定程度上，这些采样点的大气污染物浓度受到亚热带季风气候的影响。在夏季和秋季，由于受到来自海洋的夏季风的影响，大气污染物会被稀释和扩散；而在冬季和春季，由于受到来自大陆的冬季风的影响，中国北方冬春季燃煤供暖产生的大气污染物会被转运到中国南方地区，大气污染物会有所增加[41]。

EP 点的 Acy、Flu、Ant、Phe、FluA 和 BghiP 的浓度显著高于 GZ 对照点 (Mann-Whitney U test, P = 0~0.033)。例如，EP 点的 Acy、Ant 和 Phe 的浓度分别是 GZ 点的 7.5 倍、4.8 倍和 2.7 倍 (图 4-34)。这说明低环 PAHs 是电子垃圾拆解过程排放的特征异构体，与文献中电子垃圾在 320 ℃ 热解产生的低分子量 PAHs 异构体占主要成分的报道一致[190]。因此，这个地区的低分子量 PAHs 异构体，例如三至四环 PAHs，可能来自于电子垃圾拆解过程排放。和 EP 点不同的是，RS 点的 BbF、DBahA 和 IcdP 的浓度也显著高于 GZ 点 (P = 0~0.007)。例如，RS 点的 IcdP 的浓度是 GZ 点的 2.3 倍。RS 点靠近公路，而且 BbF、DBahA 和 IcdP 又是汽车排放的主要污染物[191]。因此，在 RS 点，除电子垃圾拆解过程排放 PAHs 外，汽车交通排放也是 PAHs 的重要来源之一。

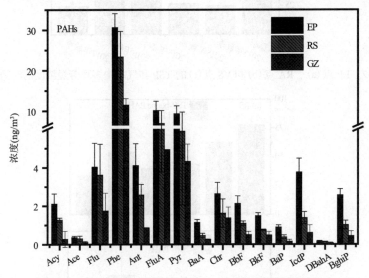

图 4-34　EP 点、RS 点与对照点 GZ 的多环芳烃浓度的对比

为了研究大气中 PAHs 的来源，常用 Ant/(Ant + Phe)、FluA/(FluA + Pyr)、BaA/(BaA + Chr)、IcdP/(IcdP + BghiP)、BbF/(BbF + BkF) 和 BaP/(BaP + BghiP) 等这些 PAHs 同分异构体比值来判断[190]。在这六对同分异构体比值中，FluA/(FluA + Pyr) 和 IcdP/(IcdP + BghiP) 被认为是对大气中 PAHs 源解析最保险的

诊断参数比值。根据文献报道[192]，FluA/(FluA+Pyr)＞0.5 或 IcdP/(IcdP+BghiP)＞0.5，说明 PAHs 主要由生物质和煤炭燃烧产生；而 $0.4 \leq$ FluA/(FluA+Pyr) ≤ 0.5 或 $0.2 \leq$ IcdP/(IcdP+BghiP) ≤ 0.5，说明 PAHs 主要由石油燃烧产生。如图 4-35 所示，EP 点的 FluA/(FluA + Pyr)和 IcdP/(IcdP + BghiP)比值都超过 0.5，说明 EP 点大气中的 PAHs 主要来源于"煤炭和生物质燃烧"。RS 点、VS 点、RA 点和 GZ 点都有类似的结果。

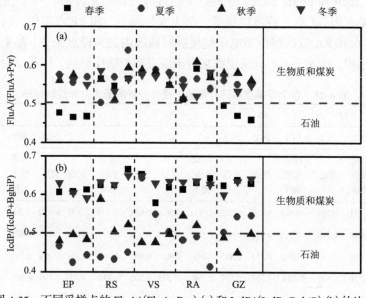

图 4-35　不同采样点的 FluA/(FluA+Pyr)(a)和 IcdP/(IcdP+BghiP)(b)的比值

然而，煤炭、生物质和塑料不完全燃烧形成 PAHs 的化学机理是类似的[193-195]，煤炭、生物质和塑料的不完全燃烧或热解会产生许多挥发性物质，包括一氧化碳、氢气、二氧化碳和烃类如乙烯和异丁烯等[196,197]，然后这些挥发性物质经过环戊二烯基自由基再结合机理或其他机理形成 PAHs[195]。在之前的报道中，传统的 PAHs 诊断参数比值法是不考虑电子垃圾拆解过程排放的[190]。在电子垃圾拆解地区，虽然电子垃圾拆解工业是当地最主要的产业，但 PAHs 的来源却被识别为"煤炭和生物质燃烧"。特别是在 EP 点，煤炭燃烧可以被忽略，电子垃圾拆解所用到的电路板原料，其材料组成大多为有机塑料，而这与生物质的元素构成和结构是类似的(塑料和生物质都是由 C、H、O、N 元素构成的有机聚合物)。因此，虽然 EP 点的 PAHs 来源用诊断参数法识别为"煤炭和生物质燃烧"，但其实来源于电子垃圾拆解过程排放。RS 点、VS 点和 RA 点的 PAHs 主要来源与 EP 点一样，都是"煤炭和生物质燃烧"，其实来源于电子垃圾拆解过程排放，这是因为这四个点都在电子垃圾拆解地区，PAHs 由 EP 点经大气扩散到 RS 点、VS 点和 RA 点。

采用主成分分析法对这个地区大气中 PAHs 的来源进一步分析。在 EP 点，得到了两个主成分因子(表 4-21)，因子 1 解释了 76.5%的总方差，主要由三至四环 PAHs 构成，包括 Phe、FluA 和 Pyr 等，这些化合物主要来源于煤炭和生物质燃烧[198,199]。然而，上文的讨论指出在电子垃圾拆解地区，PAHs 来源于煤炭和生物质燃烧，其实是来源于电子垃圾拆解过程。因此，在这个地区，因子 1 主要来自于电子垃圾拆解过程排放。因子 2 解释了 20.0%的总方差，主要由高环 PAHs 包括 BkF、IcdP 和 BghiP 构成，这些是交通排放的特征化合物[200,201]。至于 RS 点、VS 点、RA 点和 GZ 点，各自得到了三个主成分因子。和 EP 点类似，因子 1 对于 RS 点、VS 点和 RA 点这些位于电子垃圾拆解地区内的采样点来说，都来自于电子垃圾拆解过程，而对于 GZ 点来说则来自于煤炭和生物质燃烧。因子 2 对于所有采

表 4-21　各个采样点多环芳烃的主成分分析的最大方差旋转矩阵

	EP		RS			VS			RA			GZ		
	PC1	PC2	PC1	PC2	PC3	PC1	PC2	PC3	PC1	PC2	PC3	PC1	PC2	PC3
	煤炭和生物质	交通	煤炭和生物质	交通	烹饪	煤炭和生物质	交通	烹饪	煤炭和生物质	交通	烹饪	煤炭和生物质	交通	烹饪
Acy	**0.791**	0.582	−0.009	0.571	**0.788**	−0.151	0.580	**0.661**	−0.135	0.197	**0.872**	−0.231	0.313	**0.907**
Ace	**0.840**	0.344	−0.040	0.378	**0.915**	0.103	−0.053	**0.956**	0.273	0.292	**0.806**	0.683	0.441	0.088
Flu	**0.806**	0.512	−0.062	0.313	**0.935**	0.141	0.007	**0.897**	0.091	0.108	**0.904**	0.191	0.434	**0.858**
Phe	**0.730**	0.656	**0.934**	0.001	0.197	**0.790**	−0.528	0.205	**0.930**	−0.015	0.199	**0.909**	−0.155	−0.161
Ant	**0.729**	0.672	**0.764**	0.475	0.350	**0.657**	0.224	0.530	0.266	−0.194	0.749	**0.856**	−0.101	0.237
FluA	**0.803**	0.582	**0.955**	0.251	−0.110	**0.978**	−0.060	0.110	**0.965**	0.187	0.113	**0.931**	−0.152	−0.089
Pyr	**0.770**	0.634	**0.952**	0.050	−0.146	**0.976**	0.005	0.164	**0.943**	0.090	0.144	**0.971**	−0.139	−0.031
BaA	0.594	**0.801**	0.396	**0.859**	0.303	0.395	**0.858**	0.133	0.222	**0.905**	0.120	**0.835**	−0.066	0.036
Chr	0.633	**0.750**	**0.864**	0.408	−0.188	**0.963**	0.022	−0.234	**0.756**	0.605	−0.118	**0.816**	−0.332	−0.301
BbF	0.548	**0.829**	0.144	**0.883**	0.385	0.051	**0.937**	−0.252	0.216	**0.966**	0.048	−0.226	**0.876**	0.195
BkF	0.583	**0.804**	0.159	**0.886**	0.374	0.128	**0.878**	0.164	0.176	**0.939**	0.024	−0.161	**0.942**	0.106
BaP	0.375	**0.919**	0.142	**0.851**	0.473	−0.328	**0.875**	0.170	−0.078	**0.932**	0.255	0.024	**0.641**	0.414
IcdP	0.552	**0.825**	0.066	**0.891**	0.424	−0.247	**0.947**	0.064	−0.095	**0.963**	0.222	−0.014	**0.819**	0.549
DBahA	0.931	0.311	0.194	0.947	0.078	−0.100	0.605	0.054	0.078	0.958	−0.135	−0.092	**0.838**	0.453
BghiP	0.523	**0.846**	0.393	**0.862**	0.227	0.032	**0.965**	−0.009	0.238	**0.940**	0.158	−0.198	**0.966**	0.090
方差(%)	76.5	20.0	43.2	29.6	22.7	40.1	28.7	18.1	37.4	31.5	26.3	47.7	33.6	16.6
累计方差(%)	76.5	96.5	43.2	72.8	95.5	40.1	68.8	86.9	37.4	68.9	95.2	36.1	69.7	86.3

PC：主成分

样点来说都是交通排放。因子3来自于煤炭焦化,因为它主要由Acy和Flu组成,这与煤炭焦化以二至三环PAHs为主的排放特征相符[202]。

采用多元线性回归法对各个采样点的各个主成分因子对总PAHs的贡献率进行计算。如图4-36所示,在EP点,因子1(电子垃圾拆解过程)和因子2(交通排放)对总PAHs的贡献率分别为82.4%和17.6%。RA点和VS点有类似的结果,因子1、因子2和因子3(煤炭焦化)的贡献率分别是69.4%、19.5%和11.1%,以及69.8%、14.6%和15.6%,这是EP点的大气扩散污染和当地其他污染源混合的结果。RA点和VS点的PAHs来源主要受到电子垃圾拆解过程排放的影响,因为它们都在电子垃圾拆解地区且靠近EP点,具有相似的污染源贡献分布。RS点的结果与GZ点相近,因子1、因子2和因子3的贡献率分别是55.4%、28.7%和15.9%,以及52.6%、34.1%和13.3%。虽然RS点的PAHs主要来源于电子垃圾拆解过程,但交通排放仍然是重要来源。不过,GZ点的PAHs主要来自于煤炭和生物质燃烧,以及交通排放。这些结果与上文中诊断参数法的结果一致,两种方法都论证了这个地区大气中PAHs的主要来源是电子垃圾拆解过程。

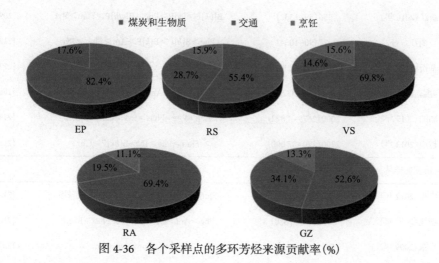

图4-36 各个采样点的多环芳烃来源贡献率(%)

2. 土壤和沉积物中的PAHs

PAHs也在拆解点附近的沉积物和土壤中被广泛检出,其总浓度在25~37 000 ng/g dw之间(表4-22),最高浓度是在越南北部某个典型的电子垃圾拆解车间收集的土壤样品中检测到,表明可能存在局部点源污染。我国早期典型的电子垃圾拆解场地已经运营多年,粗放式的拆解活动(如露天焚烧等)已经被禁止,因此,我国拆解点土壤中近期报道的PAHs中值水平普遍低于印度、巴基斯坦和加纳等近期兴起的电子垃圾拆解地区。

表 4-22 不同电子垃圾拆解点土壤和沉积物中 PAHs 的浓度和同系物组成（ng/g dw）

样品地	中位值（范围）	组成分布	文献
土壤样品			
广东（2004 年）	389（45~3206）	Phe＞Nap＞Chr＞Pyr＞B[bk]F＞B[ghi]P	[203]
浙江（2006 年）	640（330~20 000）	Fluo＞B[b]F≈BaA＞Pyr≈Chr	[204]
浙江（2008 年）	329（39~708）a	Fluo＞B[b]F＞Pyr＞Phe＞Ant＞Chr	[205]
香港（2008 年）	1008（107~2300）	B[b]F＞Fluo＞Pyr＞BaP≈IcdP≈B[ghi]P	[206]
广东（2009 年）	364（174~510）a	Phe＞Fluo	[207]
广东（2011 年）	514（25~4300）a	Nap＞Phe＞Fluo＞Pyr＞Chr＞B[b]F	[208]
浙江（2011 年）	480（191~1922）	IcdP＞B[ghi]P＞Phe＞B[bk]F＞Fluo＞D[a,h]A	[209]
广东（2012 年）	740（295~10 084）	Phe＞Fluo＞Pyr＞Nap＞Chr＞B[a]A	[210]
广东（2012 年）	（72~2506）	Nap＞Phe＞Fluo≈Pyr≈B[b]F≈B[a]A	[211]
浙江（2016 年）	356（59~1332）	B[bk]F＞IcdP＞Phe＞B[ghi]P＞Fluo＞Pyr	[209]
浙江（一）	327（90~1054）	Phe＞B[b]F＞B[k]F＞Ant＞Fluo＞Pyr	[212]
巴基斯坦（2012 年）	2350（2180~2940）	—	[213]
印度（2014 年）	866（255~3328）	Nap＞Phe＞Ant＞Pyr＞Fluo＞B[a]A	[168]
加纳（2015 年）	3000（2679~4 822）	Nap≈Phe≈Fluo＞Pyr＞Chr＞B[a]A	[214]
越南（2012 年）	2200（720~37 000）	Fluo＞Phe＞Pyr≈B[bk]F＞Chr	[215]
沉积物样品			
广东（2012 年）	479（67~4766）	Fluo＞B[a]A≈Chr＞B[b]F＞Flu＞B[k]F	[210]
浙江（2007 年）	3340（809~7880）	Phe＞Ant＞Pyr＞B[b]F＞Fluo＞B[k]F	[216]
广东（2009 年）	985（181~3034）a	Phe＞Fluo≈Pyr＞Chr	[207]
广东（2012 年）	（802~4902）	Nap＞Phe＞Fluo＞Pyr	[217]
越南（2012 年）	650（340~2100）	Pyr＞B[bk]F＞Fluo＞Chr＞Phe	[215]

a. 均值浓度

此外，很多研究对不同电子垃圾拆解工艺造成的 PAHs 污染特征进行了探讨，最常见报道的有 3 种工艺，包括露天焚烧点（OBS）、手工拆解点（EDS），以及储存或倾倒点（ESS）等。不同拆解点土壤中 PAHs 浓度由高到低依次为：OBS＞EDS＞ESS（图 4-37）。电子元器件外壳和电线/电缆等的露天焚烧可能是拆解地土壤中

PAHs 高污染的主要原因。此外，不同的拆解工艺点排放的 PAHs 同系物组成特征也有较大差异，如图 4-37(b) 所示，与 EDS 和 ESS 相比，在高污染的 OBS 土壤中发现了高浓度的低环 PAHs。

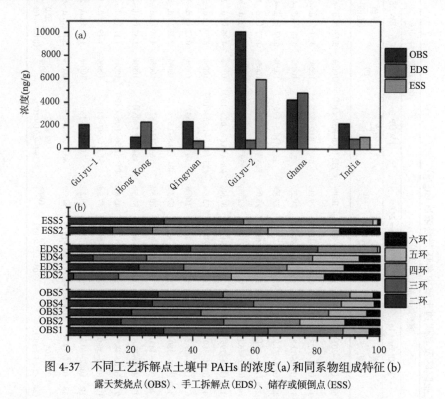

图 4-37 不同工艺拆解点土壤中 PAHs 的浓度(a)和同系物组成特征(b)
露天焚烧点(OBS)、手工拆解点(EDS)、储存或倾倒点(ESS)

Ren 等[218]对比研究了石化园区(PIP)、溴系阻燃剂生产园区(BFRP)和电子垃圾拆解园区(EWDP)及其周边(PIS、BFRS 和 EWDS)土壤中 PAHs 的污染水平、组成特征、异源空间分布特征及其主要影响因素。结果显示电子垃圾拆解园区土壤中 PAHs 总浓度为 394~2.01×10^4 ng/g，高于石化(340~2.43×10^3 ng/g)及阻燃剂生产园区(26.2~2.63×10^3 ng/g)，也高于周边 1~2 个数量级。工业园区内外土壤中 PAHs 浓度如表 4-23 所示，PIS、BFRP 和 EWDP 三个工业园区内 PAHs 总浓度分别为 340~2.43×10^3 ng/g、26.2~2.63×10^3 ng/g 和 394~2.01×10^4 ng/g，分别高于其周边区域 1~2 个数量级。工业园区内释放的 PAHs，可以通过干湿沉降过程沉积到园区及周围土壤中，因此，工业生产会引起场地污染并影响周边环境。园区内中位数浓度：EWDP＞PIP＞BFRP，园区外中位数浓度：PIS＞BFRP＞EWDP。电子垃圾拆解园区中废电路板烘烤燃烧等工业活动能够释放大量的 PAHs，其浓度是石化园区和溴系阻燃剂生产园区浓度的 2.80~3.66 倍。而电子垃圾拆解园区周边存在大范围农业区域，因为没有其他 PAHs 来源，所以该区域浓度最低。

表 4-23 工业园区及其周边土壤中多环芳烃浓度 (ng/g)

化学物质	石化园区 PIP (n=15) 中值(范围)	DF	PIS (n=26) 中值(范围)	DF	溴系阻燃剂生产园区 BFRP (n=16) 中值(范围)	DF	BFRS (n=24) 中值(范围)	DF	电子垃圾拆解园区 EWDP (n=24) 中值(范围)	DF	EWDS (n=83) 中值(范围)	DF
Nap	606 (36.3~1.96×10³)	100%	14.8 (n.d.~1.04×10³)	76.9%	68.0 (n.d.~201)	93.8%	53.5 (0.24~426)	100%	732 (27.6~6.19×10³)	100%	5.33 (n.d.~65.2)	77.1%
Acy	66.8 (25.3~126)	100%	14.6 (n.d.~120)	96.2%	3.34 (n.d.~7.23)	81.3%	0.20 (n.d.~16.4)	79.2%	67.9 (8.02~573)	100%	0.51 (n.d.~61.6)	83.1%
Ace	4.22 (1.90~6.40)	100%	0.42 (n.d.~10.5)	84.6%	3.99 (n.d.~12.5)	93.8%	n.d. (n.d.~7.07)	29.2%	65.3 (23.2~209)	100%	0.18 (n.d.~4.21)	63.9%
Flu	11.3 (4.58~16.0)	100%	5.09 (n.d.~20.6)	92.3%	22.3 (n.d.~71.2)	87.5%	n.d. (n.d.~7.22)	37.5%	43.3 (18.5~196)	100%	0.22 (n.d.~4.94)	61.5%
Phe	84.9 (34.7~186)	100%	35.5 (2.50~303)	100%	289 (2.61~791)	100%	1.47 (n.d.~63.0)	75.0%	805 (119~3.97×10³)	100%	1.72 (n.d.~148)	60.2%
Ant	3.10 (0.59~9.30)	100%	1.26 (n.d.~21.6)	69.2%	14.6 (0.32~46.1)	100%	0.51 (n.d.~10.9)	62.5%	39.1 (4.05~197)	100%	n.d. (n.d.~26.6)	27.7%
Fluo	52.3 (30.0~125)	100%	13.2 (0.05~223)	100%	133 (1.34~295)	100%	1.38 (n.d.~119)	87.5%	406 (36.6~2.02×10³)	100%	4.96 (n.d.~298)	84.3%
Pyr	33.3 (23.7~94.3)	100%	9.91 (n.d.~157)	96.2%	117 (0.64~360)	100%	1.23 (n.d.~92.8)	95.8%	425 (31.1~1.88×10³)	100%	3.79 (n.d.~371)	84.3%
BaA	33.7 (12.9~103)	100%	6.66 (n.d.~121)	96.2%	6.84 (n.d.~39.8)	87.5%	3.74 (0.47~120)	100%	143 (10.9~647)	100%	0.72 (n.d.~179)	81.9%
Chr	93.0 (33.6~380)	100%	29.3 (0.15~332)	100%	120 (0.86~380)	100%	1.24 (n.d.~98.6)	87.5%	481 (38.8~2.55×10³)	100%	9.51 (n.d.~562)	89.2%
BbF	32.9 (3.06~319)	100%	13.0 (n.d.~155)	96.2%	167 (n.d.~443)	93.8%	1.14 (n.d.~410)	70.8%	184 (13.6~821)	100%	4.09 (n.d.~615)	91.6%
BkF	15.1 (3.13~64.6)	100%	9.09 (n.d.~172)	100%	22.0 (0.50~75.5)	100%	1.78 (n.d.~88.3)	87.5%	154 (22.1~954)	100%	1.52 (n.d.~272)	90.4%
BaP	4.44 (0.25~18.7)	100%	n.d. (n.d.~24.4)	42.3%	9.03 (n.d.~38.9)	93.8%	n.d. (n.d.~13.6)	45.8%	108 (8.56~1.08×10³)	100%	1.11 (n.d.~328)	89.2%
IcdP	4.72 (1.11~168)	100%	5.59 (n.d.~47.2)	84.6%	10.1 (0.15~38.0)	100%	0.46 (n.d.~66.3)	87.5%	51.2 (7.97~248)	100%	1.49 (n.d.~336)	92.8%
DBahA	4.41 (1.06~106)	100%	2.57 (n.d.~33.1)	76.9%	10.5 (n.d.~40.5)	93.8%	0.39 (n.d.~28.6)	79.2%	33.0 (6.22~183)	100%	1.56 (n.d.~231)	92.8%
BghiP	13.0 (4.28~449)	100%	20.4 (n.d.~121)	96.2%	38.3 (0.54~143)	100%	2.27 (n.d.~149)	91.7%	86.3 (16.2~560)	100%	3.12 (n.d.~652)	92.3%
ΣPAHs	1.40×10³ (340~2.43×10³)	100%	268 (24.8~1.72×10³)	100%	1.07×10³ (26.2~2.63×10³)	100%	136 (9.03~1.31×10³)	100%	3.92×10³ (394~2.01×10⁴)	100%	50.0 (0.02~4.09×10³)	100%

DF: 检出频率; n.d.: 未检出

三个工业园区及其周边土壤中 PAHs 组成特征如图 4-38 所示。石化园区(73.0%)和阻燃剂生产园区周边(80.3%)以低环 PAHs(二至三环)为主,而其他区域以高环 PAHs(四至六环)为主。就单个化合物而言,Nap 在石化园区和溴系阻燃剂生产园区周边占主导地位,而 Phe 在石化园区周边、溴系阻燃剂园区和电子垃圾拆解园区占主导地位。说明不同工业园区及周边 PAHs 具有不同来源及其影响因素。石化园区和溴系阻燃剂生产园区周边低环 PAHs 主要来源于石油生产过程中石油产品的泄漏。根据前体理论可知,电子垃圾在烘烤或燃烧过程中,低环 PAHs 比高环 PAHs 优先合成,而且随着温度的升高,高环 PAHs 也会分解生成低环 PAHs,所以在电子垃圾拆解园区,虽然高环 PAHs 占比较高(54.2%),但 Phe(21.1%)占主导地位,其次是 Nap(19.1%)。而溴系阻燃剂生产园区、石化园区周边和电子垃圾拆解厂周边区域的高环 PAHs 主要来源于汽车燃料燃烧和煤燃烧。例如,溴系阻燃剂生产园区中煤或天然气不完全燃烧可能是高环 PAHs 的关键来源,另外在石化园区及电子垃圾拆解园区周边的商业区和住宅区有很多交通排放,这也可能是

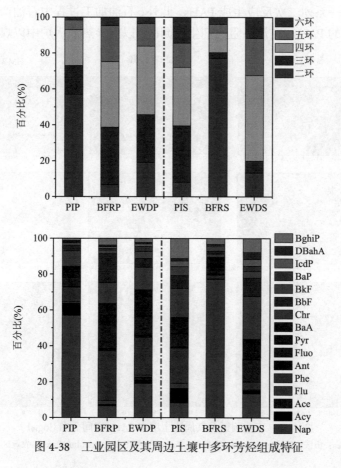

图 4-38 工业园区及其周边土壤中多环芳烃组成特征

高环 PAHs 的重要来源。在石化园区，Nap 的浓度最高（中位数浓度：606 ng/g），比其他物质高出 1~3 个数量级，且显著高于其周边土壤（$p<0.01$），所以 Nap 可作为石化园区的特征污染物。Phe 在溴系阻燃剂生产园区和电子垃圾拆解园区中位数浓度（中位数浓度：289 ng/g 和 805 ng/g）也显著高于其周边土壤（$p<0.01$），所以 Phe 可作为溴系阻燃剂生产园区和电子垃圾拆解园区的特征污染物。

三个工业园区及周边土壤中 PAHs、低环 PAHs（LMW PAHs）和高环 PAHs（HMW PAHs）的空间分布如图 4-39 所示。其中，蓝色区域表示污染程度最低，主要是农业区域或者山区，这些区域远离工业园区，附近没有其他排放源且具有极好的大气扩散条件。红色区域表示污染最严重。相比之下，绿色区域通常在工业园区周围或者是受其他污染源影响的一些区域。对于红色区域而言，高浓度 PAHs 主要来源于工业园区内化石燃料燃烧和石油产品的泄漏。例如，石化园区是中国西北最大的石化产业集聚区，包括乙烯厂、橡胶厂、石化厂、火电厂、化肥厂和污水处理厂等。在这些工业生产活动中，化石燃料在氧气不足的情况下燃烧会通过一系列二次反应生成 PAHs。此外，石油加工过程中石油产品的泄漏也会产生大量的 PAHs。在溴系阻燃剂生产园区，以前导热油锅炉中以煤作为燃料用

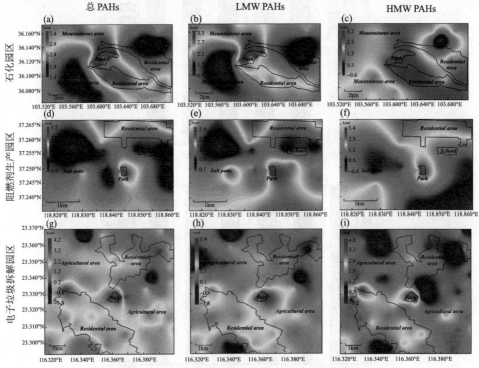

图 4-39 工业园区及其周边土壤中 PAHs 空间分布（$\log_{10}$）

Mountainous area：山区；Residential area：居民区；Park：工业园区；Agricultural area：农业区；School：学校

于加热合成多种阻燃剂,现在偶尔以天然气为燃料。在电子垃圾拆解园区,电子垃圾拆解过程中烘烤或焚烧也会产生大量的 PAHs。

而绿色区域多为工业园区周边的居民区,工业排放产生的 PAHs 可以吸附在颗粒上,后经干湿沉降过程蓄积在周边土壤中。由于兰州地处山谷,具有典型的河谷地形,夹在南北山丘之间,该区域盛行东风,促进了空气污染物(尤其是低环 PAHs),向下风向迁移[图 4-39(a)和(b)]。同时,山谷夹层结构阻止了污染物向两边山谷迁移。而溴系阻燃剂生产园区和电子垃圾拆解园区地处平原,导致污染物较大范围扩散。因此,土壤中 PAHs 的空间分布特征主要受工业活动的影响,同时受风向和地形的影响。另一方面,交通排放和家庭燃煤也会导致居民区土壤中 PAHs 污染,尤其是高环 PAHs[图 4-39(c)和(f)]。同时,在兰州和潍坊有 5 个月的冬季,以往燃煤锅炉系统是最常见的家庭供暖方式,在一些平房和旧住宅楼中,还存在家用燃煤炉灶,未经处理的燃煤废气不可避免地会导致严重的 PAHs 污染。此外,附近的一些工业制造厂也是不可忽视的 PAHs 污染源。

为了研究土壤中 PAHs 的来源,常用 Ant/(Ant + Phe)、Fluo/(Fluo + Pyr)、IcdP/(IcdP + BghiP)和 BaA/(BaA + Chr) 等 PAHs 同分异构体比值来判断。Ant/(Ant + Phe)和 Fluo/(Fluo + Pyr)分别位于临界值 0.1 和 0.5 的两侧。由 Ant/(Ant + Phe)可知 PAHs 主要来源于原油或石油加工过程中排放(石油源),由 Fluo/(Fluo + Pyr)可知 PAHs 主要来源于有机物不完全燃烧(燃烧源)。而 Ant/(Ant + Phe)被认为是相对保险的诊断参数比值,所以石油源是 PAHs 的主要来源[图 4-40(a)]。大多数 IcdP/(IcdP + BghiP)和 BaA/(BaA+Chr)处于 0.1~0.5 和 0~0.35 之间,表明 PAHs 具有混合来源。所以根据比值法可知,PAHs 主要来源于石油源和化石燃料燃烧源。

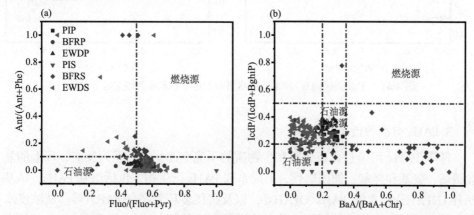

图 4-40 工业园区内外 Ant/(Ant + Phe)和 Fluo/(Fluo + Pyr)比值(a)以及 IcdP/(IcdP + BghiP)和 BaA/(BaA + Chr)比值(b)

为了更深入地了解 PAHs 来源,采用主成分分析法和多元线性回归法对各主成分因子贡献率进行研究。在三个工业园区[图 4-41(a)],因子 1 和因子 2 分别解释了 45.7%和 43.6%的总方差。因子 1 主要由 IcdP、DBahA 和 BghiP 等构成,这些化合物主要来源于化石燃料燃烧和交通排放,贡献了 42.8%的 PAHs。因子 2 主要由 Nap、Acy、Ant 和 Phe 构成,这些化合物主要来自石油产品挥发或电子垃圾烘烤焚烧,贡献了 57.2%的 PAHs。因此,工业园区土壤中 PAHs 主要来自石油源和燃烧源。在工业园区周边[图 4-41(b)],因子 1(62.6%)主要包括 Fluo、Pyr、Chr、BaA、BbF、BkF、IcdP、BghiP 和 DBahA,这些化合物通常与交通排放和燃煤有关,贡献了 67.8%的 PAHs。PC2(21.7%)主要与 Nap、Ace、Acy 和 Flu 相关,主要来源于工业排放,占 PAHs 总量的 32.2%。因此,工业园区周边土壤中 PAHs 来源于园区工业排放、历史燃煤及交通排放。与园区周边相比,工业园区内的各变量相对集中,尤其是三至五环 PAHs,说明在工业生产过程中可能存在 PAHs 的相互转化。

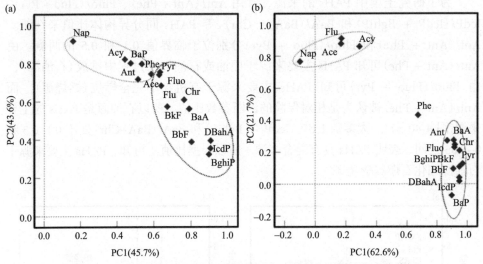

图 4-41　工业园区内(a)和工业园区外(b)土壤中多环芳烃主成分分析

3. PAHs 衍生物污染

母体 PAHs 可通过化石燃料和生物质的不完全燃烧等一系列反应,环上的氢被硝基、羰基和羟基官能团取代,形成系列 PAHs 衍生物,包括烷基 PAHs,硝基 PAHs(NPAHs)和含氧 PAHs(OPAHs),以及卤代多环芳烃(HPAHs)等。从浓度水平来看,采自非洲加纳某拆解点的土壤样品中甲基萘(1-MeNap 和 2-MeNap)的水平比母体萘还要高,在该电子垃圾拆解点附近采集的母乳样品中,甲基萘的浓度水平与萘相当，明显高于其他 PAHs 同系物；表明烷基化 PAHs 也是粗放式电子垃圾拆解活动排放的不容忽视的一类污染物。此外，一些拆解点土壤和沉积物中还发现了甲基菲(MePhe)和甲基苯并(a)蒽(MeBaA)等，应该在今后的研究中加以关注。

目前仅有少量文献报道了电子垃圾拆解地区环境介质中的 NPAHs 和 OPAHs，尽管部分 NPAHs 和 OPAHs 展现出比母体 PAHs 更强的毒性。对我国广东清远某电子垃圾拆解点大气颗粒物的研究显示，OPAHs 的浓度高于 NPAHs 1 个数量级，但是仍然低于母体化合物 1~2 个数量级，硝基取代的荧蒽、蒽及苯并(a)蒽是主要的 NPAHs 同系物，贡献了 NPAHs 总浓度的 65.1%~88.8%；三种主要 OPAH 单体(benzanthrone、9,10-anthraquinone 和 6H-benzo[cd]pyrene-6-one)贡献了 OPAHs 总浓度的 84.2%~95.3%。在模拟热解废电路板的实验中，硝基取代的苊和菲是主要的同系物单体；而在模拟热解电器塑料外壳的过程中，硝基蒽则是最主要排放的特征污染物，表明 NPAHs 和 OPAHs 的形成与电子垃圾的种类及拆解工艺有关。

Huang 等[219]测量了两种电子垃圾热解过程中 PAHs 衍生物(12 种 NPAHs 和 4 种 OPAHs)的排放因子，研究了成分分布、粒度分布、气体-颗粒分配、与前体物质的相关性以及热解温度的影响。图 4-42 显示了所研究的电子垃圾热解所排放的 NPAHs 和 OPAHs 的排放因子统计结果(结果以平均值±标准差表示)。对于 NPAHs，所有测试的线路板在气相、颗粒相和总计中的平均排放因子值分别为 (3.8 ± 13.2) ng/g、(81.8 ± 93.5) ng/g 和 (85.7 ± 92.4) ng/g[图 4-42(a)]。在塑料套管中，气相、颗粒相的平均 EF 值分别为 (3.9 ± 14.4) ng/g、(79.4 ± 66.4) ng/g 和 (83.3 ± 69.7) ng/g[图 4-42(c)]，表明与气相相比，颗粒相占压倒性的比例，而 NPAH 的相分布和排放因子总量在线路板和塑料外壳之间差异最小。考虑到 OPAH，线路板在气相、颗粒相和总量中的平均排放因子分别为 (0.33 ± 0.42) μg/g、(3.04 ± 3.80) μg/g 和 (3.37 ± 4.10) μg/g[图 4-42(b)]，塑料外壳在气相、颗粒相和总量中的平均排放因子分别为 (0.23 ± 0.12) μg/g、(32.32 ± 18.06) μg/g 和 (32.56 ± 18.09) μg/g [图 4-42(d)]，表明与气相相比，颗粒相中的主要贡献相似，并且在颗粒相中存在的线路板和塑料外壳之间存在很大差异。此外，NPAHs 的排放因子比其 PPAHs 低约 2~3 个数量级[190]，而 OPAHs 的排放因子与 PPAH 相近；其中，线路板和塑料外壳的平均排放因子分别为 (2.77 ± 1.41) μg/g 和 (23.65 ± 14.52) μg/g[190]。因此，所研究的电子垃圾热解产生的 PAH 衍生物主要是 OPAHs。然而，在另一项研究中，蜂窝煤和木柴燃烧产生的 NPAH 和 OPAH 对应的排放因子没有明显差异，均在 μg/g 的数量级上[220]，可能是由于不同的材料和燃烧条件。

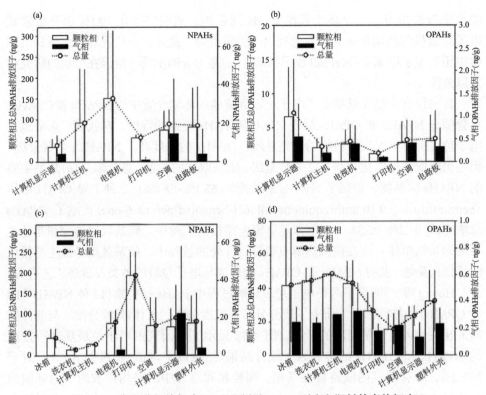

图 4-42 废电路板热解中 NPAHs(a) 和 OPAHs(b) 和塑料外壳热解中 NPAHs(c) 和 OPAHs(d) 的排放因子

Ma 等[221]采集了我国华南某电子垃圾拆解工业园区及其周边地区的大气样品，包括 $PM_{2.5}$、PM_{10} 和 TSP 以及对应的气相样品，剖析 NPAHs 和 OH-PAHs 的时空变化趋势，探讨 NPAHs、OH-PAHs 与电子垃圾拆解的关系。采样地点共有五个，分别是电子垃圾拆解工业园区(EP)和周边地区包括公路旁边(RS)、乡村(VS)、居民区对照点(RA)，以及代表典型城市生活的对照点广州市(GZ)(采样点位置见图 4-7)。如图 4-43 所示，在电子垃圾拆解地区，TSP 加气相的 NPAHs 和羟基多环芳烃(OH-PAHs)的浓度分别为 46.6~142 pg/m^3 和 75.9~291 pg/m^3。NPAHs 和 OH-PAHs 的浓度比原型 PAHs(25.9~88.7 ng/m^3)低 2~3 个数量级，NPAHs 在 RS 点的浓度最高，浓度中位数为 142 pg/m^3，其次是 EP 点、RA 点、GZ 点和 VS 点，分别是 131 pg/m^3、65.6 pg/m^3、57.4 pg/m^3 和 46.6 pg/m^3。和 NPAHs 的空间分布不同，OH-PAHs 在 EP 点的浓度最高，其次是 RS 点、VS 点、RA 点和 GZ 点，分别是 291 pg/m^3、170 pg/m^3、133 pg/m^3、92.7 pg/m^3 和 75.9 pg/m^3。EP 点和 RS 点的 NPAHs 和 OH-PAHs 浓度明显高于对照点 GZ(Mann-Whitney U test, $P<0.05$)，而 RA 点与对照点 GZ 的 NPAHs 和 OH-PAHs 浓度不存在显著差异($P>0.05$)。这说

明在 EP 点，电子垃圾拆解过程可能是 NPAHs 和 OH-PAHs 的重要来源，而在 RS 点，交通排放也可能是 NPAHs 和 OH-PAHs 的一个重要来源。

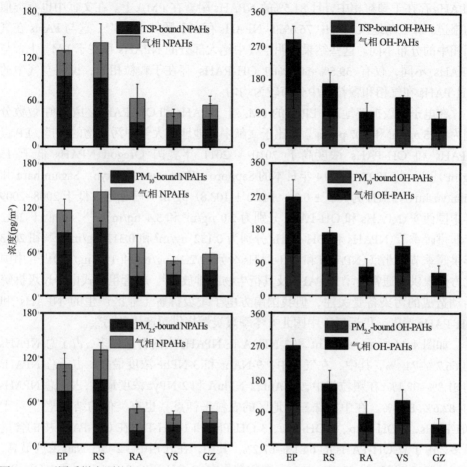

图 4-43　不同采样点颗粒相（TSP、PM_{10} 和 $PM_{2.5}$）和气相中硝基多环芳烃（NPAHs）和气相中羟基多环芳烃（OH-PAHs）浓度空间分布

如图 4-43 所示，在电子垃圾拆解地区，TSP、PM_{10} 和 $PM_{2.5}$ 上的 NPAHs 浓度分别为 29.9~95.6 pg/m³、26.4~92.8 pg/m³ 和 20.1~88.8 pg/m³，而 OH-PAHs 浓度分别为 41.6~145.4 pg/m³、38.9~113.8 pg/m³ 和 37.1~106.6 pg/m³。在这个地区，不同颗粒相（TSP、PM_{10} 和 $PM_{2.5}$）上 NPAHs 和 OH-PAHs 浓度很接近，其中 76.9%~95.1% 的 NPAHs 和 73.3%~91.6% 的 OH-PAHs 附着在 $PM_{2.5}$ 上，这和 PAHs 在颗粒相上的分布很类似。例如，在 EP 点，TSP、PM_{10} 和 $PM_{2.5}$ 上 NPAHs 浓度中位数分别是 95.6 pg/m³、86.1 pg/m³ 和 73.5 pg/m³，OH-PAHs 浓度中位数分别是 145 pg/m³、114 pg/m³ 和 107 pg/m³。这说明在这个地区，NPAHs 和 OH-PAHs 主要附着在细颗粒

上，而且很可能与电子垃圾拆解过程有关。同时，在这个地区(除 RA 点外)，60.1%~73.5%的 NPAHs 存在于颗粒相中。对照点 GZ 也有类似的结果，65.4%的 NPAHs 存在于颗粒相中，且 81.5%的 NPAHs 附着在 $PM_{2.5}$ 上。在文献中也有类似的报道，珠三角地区大气中 76.6%的 NPAHs 存在于颗粒相中[186]，这与 PAHs 在气固相中的分布不同。这些结果说明空气中绝大部分的 NPAHs 存在于 $PM_{2.5}$ 上。与 NPAHs 不同，仅有 39.5%~50.0% 的 OH-PAHs 存在于颗粒相中，说明空气中的 OH-PAHs 在气相和颗粒相中分布较为均匀。

在电子垃圾拆解地区，EP 点的 $PM_{2.5}$ 上 NPAHs 和 OH-PAHs 的浓度中位数分别为 73.5 pg/m^3 和 107 pg/m^3。相比于文献中其他地区大气中污染物的报道，EP 点 NPAHs 和 OH-PAHs 浓度高于 2010 ~ 2011 年美国 Oregon(NPAHs 低于 15 pg/m^3)[222]，与 1997 ~ 2014 年日本的 Sapporo、Kanazawa、Tokyo、Sagamihara 和 Kitakyushu(NPAHs 为 (0.5 ± 0.2 ~ 141.0 ± 108.8) pg/m^3)[223]相近，低于 2008~2009 年中国西安(NPAHs 和 OH-PAHs 分别为 2.0 ng/m^3 和 3.4 ng/m^3)[224]，也低于 2012 年非供暖季节(NPAHs 和 OH-PAHs 分别为 0.132 ng/m^3 和 0.312 ng/m^3)[192]和 2013 年供暖季节的北京(NPAHs 和 OH-PAHs 分别为 2.15 ng/m^3 和 18.3 ng/m^3)[192]。中国北方燃煤供暖通常会增加 PAHs 及其衍生物的排放[225]。尽管粗放式电子垃圾拆解活动造成的污染备受关注，但我国南方电子垃圾拆解工业区产生的 NPAHs 和 OH-PAHs 污染，仍远低于中国北方冬季煤炭燃烧供暖产生的污染。

如图 4-44 所示，9-NAnt 和 2-NFluA 在 NPAHs 中的含量最多，占了总 NPAHs 的 66.7%~74.4%。其中，在气相中，9-NAnt 和 3-NPhe 浓度最高，占了总 NPAHs 的 81.8%~92.1%；在颗粒相中，9-NAnt、2-NFluA 和 2-NPyr 浓度最高，占了总 NPAHs 的 82.6%~89.5%。在中国珠三角地区的农村采样点，也有类似的结果[186]。对于 OH-PAHs，2-OH-Nap、2-OH-Phe、3-OH-Phe 和 1-OH-Pyr 在 OH-PAHs 中的含量最多，占了总 OH-PAHs 的 64.1%~67.2%。其中，在气相中，2-OH-Nap 浓度最高，占了总 OH-PAHs 的 58.7%~69.2%；在颗粒相中，2-OH-Phe、3-OH-Phe 和 1-OH-Pyr 浓度最高，占了总 OH-PAHs 的 65.1%~86.5%。在气固相分布方面，三环 NPAHs 在气相和颗粒相中均有分布，四至五环 NPAHs 主要分布在颗粒相中。OH-PAHs 与 NPAHs 有类似的结果，三环及以上 OH-PAHs 更倾向于附着在颗粒相中。NPAHs 和 OH-PAHs 由于取代基的作用，分子量和极性变大，比原型 PAHs 更倾向于附着在颗粒相中。NPAHs 和 OH-PAHs 由多个苯环和取代基构成，其辛醇-空气分配系数(K_{OA})随着苯环和取代基数量增加而增加[186]。因此，二环 NPAHs 和 OH-PAHs 倾向于分布在气相中，而三环及以上 NPAHs 和 OH-PAHs 倾向于分布在颗粒相中，特别是在 $PM_{2.5}$ 上。

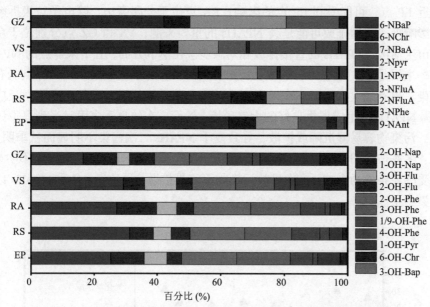

图 4-44　不同采样点大气中 NPAHs 和 OH-PAHs 的异构体组分比较

NPAHs 和 OH-PAHs 除了煤炭、生物质燃烧和机动车排放等一次排放外[226-228]，大气光化学反应也是重要的排放源[229,230]。NPAHs 的光化学途径有两种，一种是 PAHs 与·OH 自由基反应，生成 OH-PAHs，再与 NO_2 反应生成 NPAHs，主要在白天进行；另一种是 PAHs 与·NO_3 自由基反应，再与 NO_2 反应生成 NPAHs，主要在夜晚进行。其中，2-NFluA 被认为是 FluA 在气相中与·OH 和·NO_3 自由基反应生成，而 2-NPyr 是 Pyr 在气相中与·OH 自由基反应生成的[230,231]。1-NPyr 主要由一次排放产生，通过 2-NFluA 与 1-NPyr、2-NPyr 的比值可以判断 NPAHs 主要来源于一次污染还是二次污染。2-NFluA/1-NPyr＜5，说明 NPAHs 主要来源于一次排放；2-NFluA/2-NPyr 接近 10，说明·OH 自由基是二次排放 NPAHs 的主要生成途径；2-NFluA/2-NPyr 接近 100，说明·NO_3 自由基是二次排放 NPAHs 的主要生成途径[232,233]。

如图 4-45 所示，EP 点的 2-NFluA/1-NPyr 大部分低于 5（除秋季外），说明 EP 点的 NPAHs 来源主要是一次排放；EP 点的 2-NFluA/2-NPyr 都在 10 附近分布，说明 EP 点的 NPAHs 的二次排放主要是通过·OH 自由基途径生成。这可能与样品采集都在白天进行有关。RS 和 RA 点的 2-NFluA/1-NPyr 和 2-NFluA/2-NPyr 分布与 EP 点类似，其 NPAHs 来源主要是一次排放，而二次排放主要是通过·OH 自由基途径生成。VS 和 GZ 点的 2-NFluA/1-NPyr 分布与 EP 点不同，这两个采样点的 NPAHs 来源主要是二次排放，且主要是通过·OH 自由基途径生成。EP 点位于电子垃圾拆解工业区内，电子垃圾拆解和机动车运输货物都可能会排放 NPAHs；RA

点靠近 EP 点，可能受到 EP 点的影响；RA 点位于公路旁边，除电子垃圾拆解排放的影响外，交通排放可能是其 NPAHs 的重要来源；VS 和 GZ 点分别位于乡村和都市生活区，NPAHs 的直接排放相对较少。在季节方面，五个采样点的 NPAHs 在秋季主要是二次污染，而冬季主要是一次污染，春、夏季则不同采样点既有一次污染，也有二次污染。广东地区夏、秋季的紫外线较强，大气中光化学反应强烈，为 NPAHs 的生成提供良好的大气环境，然而广东的夏季多雨且受到夏季风的影响，污染物容易湿沉降或扩散，导致夏季 NPAHs 二次污染不明显而秋季明显。在春、冬季节，阳光辐射相对减弱，光化学反应减弱，NPAHs 二次污染不明显，同时受气象条件的影响，一次排放污染物不容易扩散，导致春、冬季 NPAHs 以一次污染为主。

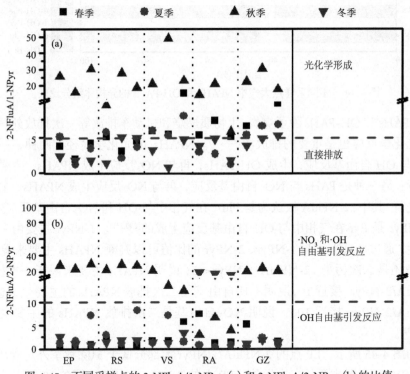

图 4-45　不同采样点的 2-NFluA/1-NPyr (a) 和 2-NFluA/2-NPyr (b) 的比值

采用主成分分析法对这个地区大气中 NPAHs 的来源进一步分析。在 EP 点，得到了三个主成分因子（表 4-24），因子 1 解释了 47.1%的总方差，主要由 9-NAnt、3-NPhe 和 3-NFluA 构成，煤炭燃烧会产生较多的 3-NPhe[227]，生物质燃烧会产生较多的 9-NAnt 和 3-NFluA[198, 233]，这些化合物主要来源于煤炭和生物质燃烧。在电子垃圾拆解地区，PAHs 来源于煤炭和生物质燃烧，其实是来源于电子垃圾拆解过程，NPAHs 也一样。因此，在这个地区，因子 1 主要来自于电子垃圾拆解过程排放。因子 2 解释了 21.6%的总方差，主要由 2-NFluA 构成，这是 PAHs 光化学反应的特征污染物[234,235]。因子 3 解释了 11.4%的总方差，因子 3 来自于交通排放，因为它主要由 6-NChr 组成，这与交通排放会产生较多的 7-NBaA 和 6-NChr[236]的排放特征相符。至于 RS 点、VS 点和 GZ 点，也得到了三个主成分因子，RA 点得到了两个主成分因子。和 EP 点类似，对于 RS 点、VS 点和 RA 点这些位于电子垃圾拆解地区内的采样点来说，NPAHs 主要来源于"煤炭和生物质燃烧"(因子 1)，其实是来自于电子垃圾拆解过程，而对于 GZ 点来说则主要来自于煤炭和生物质燃烧。其他两个因子(因子 2 和 3)对于所有采样点来说都分别是光化学反应和交通排放。

采用多元线性回归法对各个采样点的各个主成分因子对总 NPAHs 的贡献率进行计算。如图 4-46 所示，在 EP 点，因子 1(电子垃圾拆解过程)、因子 2(光化学反应)和因子 3(交通排放)对总 NPAHs 的贡献率分别为 78.2%、20.3%和 1.5%。在 VS 点，电子垃圾拆解过程、光化学反应和交通排放对总 NPAHs 的贡献率分别为 38.7%、49.6%和 11.7%。在 RA 点，电子垃圾拆解过程和光化学反应对总 NPAHs 的贡献率分别为 56.1%和 43.9%。这是 EP 点的大气扩散污染、大气光化学反应和当地其他污染源混合的结果。RA 点和 VS 点的 NPAHs 来源都受到电子垃圾拆解过程排放的影响，因为它们都在电子垃圾拆解地区且靠近 EP 点。RS 点的结果与 GZ 点相近，三个因子的贡献率分别是 36.0%、12.6%和 51.4%，和 15.1%、26.4%和 58.6%。虽然 RS 点的 NPAHs 也受到电子垃圾拆解过程的影响，但交通排放才是其最重要的来源。GZ 点的 NPAHs 也主要来自于交通排放，以及大气光化学反应。这些结果与上文中诊断参数法的结果基本一致，两种方法都论证了这个地区大气中 NPAHs 的主要来源是电子垃圾拆解过程、光化学反应和交通排放。

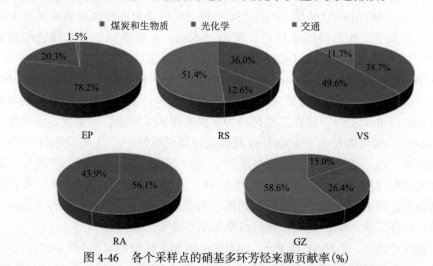

图 4-46 各个采样点的硝基多环芳烃来源贡献率(%)

表 4-24　各个采样点硝基多环芳烃(NPAHs)的主成分分析的最大方差旋转矩阵

	EP			RS			VS			RA			GZ		
	PC1	PC2	PC3	PC1	PC2	PC3	PC1	PC2	PC3	PC1	PC2	PC3	PC1	PC2	PC3
	煤炭和生物质	光化学	交通	煤炭和生物质	交通	光化学	光化学	煤炭和生物质	交通	光化学	煤炭和生物质	交通	光化学	交通	煤炭和生物质
9-NAnt	**0.948**	0.183	0.041	**0.967**	−0.071	−0.005	0.759	0.524	−0.353	0.582	0.789	0.154	**0.968**	−0.048	
3-NPhe	**0.804**	0.497	−0.044	**0.871**	0.346	0.096	**0.826**	0.270	0.075	**0.852**	0.445	−0.004	**0.929**	0.112	
2-NFluA	0.240	**0.871**	−0.035	−0.177	−0.106	0.698	**0.897**	−0.077	0.314	**0.853**	0.406	**0.809**	0.375	−0.087	
3-NFluA	**0.942**	−0.225	0.133	**0.887**	−0.100	−0.136	0.043	**0.919**	−0.320	−0.120	**0.971**	0.661	0.312	0.451	
1-NPyr	0.450	−0.475	−0.199	**0.884**	0.134	0.056	0.336	**0.682**	0.239	0.474	**0.699**	−0.124	0.190	0.931	
2-Npyr	0.811	0.007	−0.067	0.102	−0.040	**0.981**	0.333	0.847	−0.054	**0.808**	0.539	**0.704**	0.189	0.547	
7-NBaA	0.912	−0.042	0.137	−0.013	−0.006	0.991	−0.066	0.393	0.052	0.253	**0.955**	0.973	−0.080	−0.056	
6-NChr	0.084	−0.044	**0.978**	−0.327	**0.821**	0.030	0.223	−0.030	**0.879**	0.988	0.059	0.293	**0.684**	−0.473	
6-NBaP	−0.064	0.805	−0.072	0.518	0.771	−0.052	−0.667	0.083	0.671	0.974	−0.022	0.120	0.107	0.814	
方差(%)	47.1	21.6	11.4	56.8	15.8	11.7	41.5	29.5	16.1	70.3	21.9	39.9	27.1	16.9	
累计方差(%)	47.1	68.7	80.1	56.8	72.6	84.3	41.5	71.0	87.1	70.3	92.2	39.9	67.0	83.9	

PC：主成分

通过 2-NFluA 与 1-NPyr、2-NPyr 的比值可以判断各个采样点不同时期的 NPAHs 主要来源于一次污染还是二次污染。NPAHs 和 OH-PAHs 化合物中，只有 1-NPyr 的浓度在不同 NPAHs 污染来源（一次污染和二次污染）的采样点存在显著差异（Mann–Whitney U 检验，$P = 0.034$），且一次污染为主要来源的采样点 1-NPyr 浓度明显高于二次污染的采样点，这与 1-NPyr 主要来源于一次排放的报道一致[232,233]。1-NPyr 与 Pyr、1-OHPyr 的相关性分析显示（图 4-47），1-NPyr 与 Pyr 呈现中等相关（$P = 0.041$，$R=0.483$），1-NPyr 与 1-OHPyr 呈现强相关（$P = 0.000$，$R=0.728$）。在一次污染为主要来源的采样点，1-NPyr 与 1-OHPyr 依然呈现强相关（$P = 0.005$，$R=0.839$）；而在二次污染为主要来源的采样点，两者则显示线性无关（$P>0.05$）。这可能是由于 1-OHPyr 既能通过煤炭燃烧[226]、生物质燃烧[237]等一次排放产生，也可通过光化学反应生成[238]；而 1-NPyr 主要来源于一次排放，其光化学非均相反应产生量极少[239]，虽然 1-NPyr 光降解产生 1-OHPyr[240]，但产生量也是极少。因而，在电子垃圾拆解地区，两者在一次污染严重的采样点线性关系显著，而在二次污染严重的采样点线性关系不显著。

在这个地区，2-NPyr 与 Pyr、1-OHPyr 的相关性分析显示（图 4-47），2-NPyr 与 Pyr 呈现中等相关（$P = 0.007$，$R=0.572$），与 1-OHPyr 呈现强相关（$P = 0.000$，

$R=0.906$)。而且,无论是一次污染为主要来源的采样点,还是二次污染为主要来源的采样点,2-NPyr 与 1-OHPyr 都呈现强相关($P<0.01$)。这与烟雾箱模拟 PAHs 转化为 NPAHs 实验中,2-NPyr 主要是通过气相 1-OHPyr 途径生成的报道一致[234,235]。2-FluA 与 FluA 也是显著线性相关($P = 0.000$, $R=0.730$)[图 4-47(d)],类似 2-NPyr,无论是一次污染为主要来源的采样点,还是二次污染为主要来源的采样点,2-FluA 与 FluA 都呈现强相关($P<0.01$)。这可能是因为 2-FluA 主要是通过气相光化学反应途径生成[234,235]。

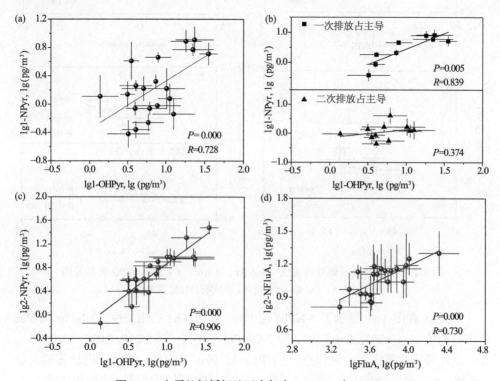

图 4-47 电子垃圾拆解地区大气中 1-OHPyr 与 1-NPyr

(a)总的;(b)一次污染和二次污染分别占主导的情况;1-OHPyr 与 2-NPyr(c)、2-NFluA 与 FluA(d)的相关性

对其他 NPAHs,包括 9-NAnt、3-NPhe、7-NBaA、6-NChr 和 6-NBaP,既能通过一次排放产生[198, 233, 236, 241],也能通过二次排放产生[230, 242]。9-NAnt 与 Ant 的相关性分析显示(图 4-48),两者呈现强相关($P = 0.001$, $R=0.669$)。在一次污染为主要来源的采样点,9-NAnt 与 Ant 依然呈现强相关($P = 0.012$, $R=0.789$);在二次污染为主要来源的采样点,两者的相关性有所下降($P = 0.049$, $R=0.598$),说明这个地区大气环境中 9-NAnt 更多地来自于一次排放,Lin 等[192]也有类似的报道。

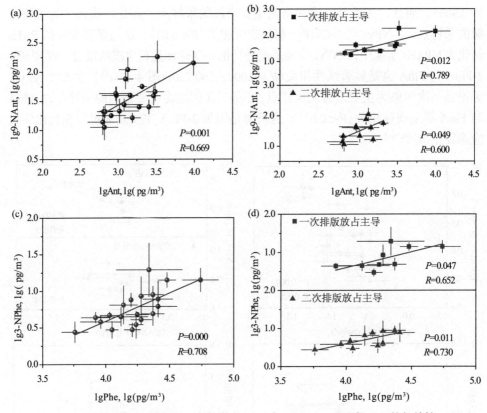

图 4-48　电子垃圾拆解地区大气中 9-NAnt 与 Ant、3-NPhe 与 Phe 的相关性
(a, c) 总的；(b, d) 一次污染和二次污染分别占主导的情况

图 4-48 和图 4-49 显示了 3-NPhe 与 Phe、2-OHPhe 的相关性，3-NPhe 与 Phe 呈现强相关（$P = 0.000$，$R=0.708$），与 2-OHPhe 呈现中等相关（$P=0.024$，$R=0.501$）。在一次污染为主要来源的采样点，3-NPhe 与 Phe、OHPhe 依然呈现显著相关（$P<0.05$，$R=0.425 \sim 0.705$），说明 3-NPhe、2-OHPhe 均能通过一次排放生成，与 Shen[233]和 Avagyan[237]的报道相符。在二次污染为主要来源的采样点，3-NPhe 与 Phe 线性相关性比一次污染为主要来源的采样点更强（$P = 0.011$，$R=0.730$），说明这个地区大气环境中的 3-NPhe 更多地来自于二次排放，Lin 等[192]也有类似的报道；而 3-NPhe 与 OHPhe 线性无关，这可能是因为大气环境中 3-NPhe 的二次污染主要通过 Phe 与·NO_3 自由基的暗反应生成，而不是通过 Phe 与·OH 自由基光化学反应[230,243]。

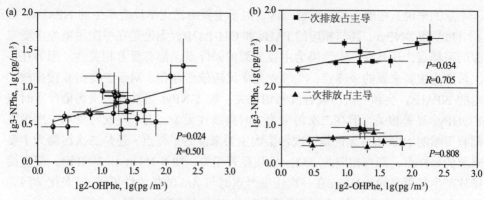

图 4-49 电子垃圾拆解地区大气中 3-NPhe 与 2-OHPhe 的相关性
(a)总的；(b)一次污染和二次污染分别占主导的情况

这个地区大气环境中 7-NBaA 与 BaA、6-NChr 与 Chr、6-NBaP 与 BaP 之间也存在相关性（$P<0.05$）（表 4-25），但相关性不强（$R=0.500\sim0.543$）。这可能是由于 BaA、Chr 和 BaP 主要分布在颗粒相中，非均相光化学反应产生的硝基衍生物较少[229-231]。另外，7-NBaA、6-NChr 和 6-NBaP 也可通过交通排放、生物质燃烧和煤炭燃烧[199, 227, 244-246]产生，不同的污染源的排放量不同。以上原因可能导致总体上 7-NBaA 与 BaA，6-NChr 与 Chr，6-NBaP 与 BaP 线性相关性不强。

表 4-25　lg NPAHs 与 lg PAHs、lg OH-PAHs 的相关系数和 P 值

	Source	lg 7-NBaA	lg 6-NChr	lg 6-NBaP
lg BaA		0.500*		
	1	0.521*		
	2	0.423*		
lg Chr			0.516*	
	1		0.476*	
	2		0.531*	
lg BaP				0.543*
	1			0.513*
	2			0.521*
lg 6-OHChr				
	1			
	2			
lg 3-OHBaP				
	1			−
	2			−

*：$P<0.05$，−：$P>0.05$；1：一次排放占主导地位；2：二次排放占主导地位

总的来说，在电子垃圾拆解地区，对于主要由光化学反应产生的 NPAHs，如 2-NFluA 和 2-NPyr，其与相应的 PAHs 和 OH-PAHs，无论是在一次污染为主要来源的采样点，还是在二次污染为主要来源的采样点，都有显著相关性。因为在一次污染为主要来源的采样点，大气光化学反应依然存在。对于主要由直接排放产生的 NPAHs，会和一次污染程度密切相关，如 1-NPyr 虽然在一次污染严重时与 1-OHPyr 显著相关，但在二次污染为主时则线性无关。对于一次污染和二次污染都有贡献的 NPAHs，无论是一次污染为主要来源的采样点，还是二次污染为主要来源的采样点，都与相应的 PAHs 呈现显著相关，如 9-NAnt 和 3-NPhe，但直接排放产生量更多的 9-NAnt 在一次污染严重时与 Ant 的相关性更强，而光化学污染产生量更多的 3-NPhe 在二次污染严重时与 Phe 的相关性更强。

4. 卤代 PAHs 污染

卤代多环芳烃(HPAHs)是 PAHs 母体化合物上带有一个或多个卤素取代基的一类有机物，根据卤原子数量和取代位的差异，H-PAHs 存在多种同分异构体。由于电子垃圾中有大量的溴氯源，在电子垃圾热解过程中产生的 HPAHs 一直以来也备受关注。总体而言，环境介质中发现的含氯 PAHs(Cl-PAHs)的浓度水平要高于含溴 PAHs(Br-PAHs)，这可能与碳氯键相比碳溴键更加稳定有关。表 4-26 总结了不同来源环境样品中 HPAHs 同系物组成特征，不同地区 HPAHs 的同系物差异很大，这可能与不同文献测定的 HPAHs 同系物种类不一致有关系。尽管如此，这些研究表明 HPAHs 同系物组成比其母体 PAHs 更为复杂。

表 4-26　不同来源卤代 PAHs 的组成

来源	组成分布	文献
电子垃圾拆解相关源的 Cl-PAHs 和 Br-PAHs		
破碎机废料	1-Cl-Pyr(24%)≈6-Cl-BaP(23%)≈8-Cl-Flu(22%)>7-Cl-BaA(17%)>3,9,10-Cl3-Phe	[247]
回收场地地面灰尘	6-Cl-BaP(43%)>7-Cl-BaA(19%)>1-Cl-Pyr(16%)>8-Cl-Flu(10%)>3,9,10-Cl3-Phe	[247]
回收场地土壤	6-Cl-BaP(49%)>1-Cl-Pyr(19%)>3,9,10-Cl3-Phe(13%)>7-Cl-BaA(8%)≈8-Cl-Flu	[247]
树叶样品	1-Cl-Pyr(32%)>9-Cl-Phe(20%)>3-Cl-Flu(10%)≈7-Cl-BaA(10%)	[247]
大气颗粒物	6-Cl-BaP>9-Cl-Ant>2-Cl-Ant>1-Cl-Pyr>8-Cl-Fluo; 6-Br-BaP>1-Br-Pyr	[19]
加纳露天焚烧土壤	Cl-Phe/Ant>Cl-Flu/Pyr>Cl$_2$-Phe/Ant>Cl$_2$-Flu/Pyr>Cl$_3$-Phe/Ant;Br$_2$-Phe/Ant>Br-Phe/Ant>1-Br-Pyr	[248]

续表

来源	组成分布	文献
电子垃圾拆解相关源的 Cl-PAHs 和 Br-PAHs		
越南露天焚烧土壤	9-Cl-Phe＞3,9-Cl$_2$-Phe≈9,10-Cl$_2$-Ant＞1-Cl-Pyr＞ 3-Cl-Flu;1,5-Br$_2$-Ant＞9,10-Br$_2$-Ant＞2,6-Br$_2$-Ant＞1-Br-Ant＞ 9-Br-Phe	[249]
越南、菲律宾和加纳等地露天焚烧土壤	9,10-Cl$_2$-Ant＞1-Cl-Pyr＞7-Cl-BaA＞9-Cl-Ant＞9-Cl-Phe＞ 3,9-Cl$_2$-Phe＞6-Cl-BaP;1,5-Br$_2$-Ant＞9,10-Br$_2$-Ant＞9-Br-Phe＞ 1-Br-Ant＞1-Br-Pyr＞7- Br-BaA	[250]
电子垃圾灰	2-Cl-Ant＞2-Cl-Phe＞9-Cl-Ant＞9,10-Cl$_2$-Phe＞1,5-Cl$_2$-Ant＞ 1-Cl-Pyr;1-Br-Pyr＞1,6-Br$_2$-Pyr＞7-Br-BaA＞4-Br-Pyr＞1,2-Br$_2$-Acy ＞9-Br-Phe	[251]
电子垃圾大气颗粒物	9,10-Cl$_2$-Phe＞1,5-Cl$_2$-Ant＞1-Cl-Pyr＞7-Cl-BaA;1,8-Br$_2$-Ant＞ 9,10-Br$_2$-Phe＞3-Br-Flu	[180]
其他源的 Cl-PAHs 和 Br-PAHs		
标准灰尘(SRM 2585)	9-Cl-Phe＞1-Cl-Pyr＞9-Cl-Ant＞1,5-Cl$_2$-Ant＞9,10-Cl$_2$-Phe＞ 2-Cl-Phe;5-Br-Ace＞9-Br-Phe	[251]
再生铜冶炼气体	9-Cl-Phe/2-Cl-Phe＞1,5-Cl$_2$-Ant/9,10-Cl$_2$-Ant＞3-Cl-Flu＞ 1-Cl-Pyr;9-Br-Phe(38%)＞3-Br-Flu	[252]
市政垃圾焚烧气体	6-Cl-BaP(31.2%)＞1-Cl-Pyr(12.8%)＞ 3-Cl-Flu(8.4%)≈7-Cl-BaA(8.1%);1-Br-Pyr＞6-Br-BaP＞7-Br-BaA	[253]
氯化工土壤	6-Cl-BaP(84%)＞1-Cl-Pyr(7%)＞7-Cl- BaA(4%)＞ 3,9,10-Cl$_3$-Phe(2%)	[247]

Tang 等[251]对电子垃圾拆解园区不同工艺车间灰尘中的 34 种 HPAHs 进行了定量，如表 4-27 列出了室内灰尘与标准灰尘(SRM 2585)样品中 Cl-PAHs 和 Br-PAHs 的分析结果。Σ_{16}Cl-PAHs 和 Σ_{18}Br-PAHs 在室内灰尘中的含量范围分别为 7.91~137 ng/g dw 和 8.80~399 ng/g dw。氯取代的同系物的浓度比溴取代的同系物的浓度略低，这可能是由于在电子垃圾拆解过程中大量的溴源(如 BFRs)析出，从而促使形成 Br-PAHs 更容易。在电子垃圾拆解手工车间，原料破碎车间和炼铜高炉车间中也观察到 Cl-PAHs 和 Br-PAHs 的浓度明显高于办公区。这进一步证实了电子垃圾处置的过程是影响 Cl-PAHs 和 Br-PAHs 形成的主要因素。在电子垃圾处理的过程中，煤粉等燃料充当碳源，形成 PAHs 前体。电子垃圾材料(例如印刷电路板、电子零件和组件以及电器外壳)中包含的有机杂质，例如聚氯乙烯和阻燃剂，可以为 Cl-/Br-PAH 的形成提供卤原子。而来自电子垃圾处置过程中产生的某些金属可能充当 Cl-/Br-PAH 形成的催化剂。各种 Cl-/Br-PAH 可能会从电子垃圾拆解处置过程中释放出来，在大气颗粒物中富集后最终沉降在灰尘中。各个车间的处理工艺、通风和光照等条件存在较大差异，这导致了灰尘中 Cl-PAHs 和 Br-PAHs 及其同系物的分布特征也不尽相同。

表 4-27　典型电子垃圾拆解园区不同车间灰尘样品中 X-PAHs 的平均浓度 (ng/g dw)

化学物质	EW	RW	BW	OA	RA	SRM
Cl-PAHs						
9-Cl-Fle	n.d.	n.d.	n.d.	n.d.[a]	n.d.	n.d.
9-Cl-Phe	1.81	2.67	6.34	0.97	0.52	37.9
2-Cl-Phe	38.4	4.50	2.64	0.38	28.5	6.02
1-Cl-Ant	n.d.	n.d.	n.d.	n.d.	28.0	5.74
2-Cl-Ant	46.7	5.14	3.83	0.81	30.4	4.47
9-Cl-Ant	15.6	2.46	1.48	n.d.	17.3	7.39
2,7-Cl$_2$-Fle	n.d.	0.80	n.d.	n.d.	n.d.	0.01
1,5-Cl$_2$-Ant	7.00	1.56	4.31	<LOQ[c]	9.87	7.40
9,10-Cl$_2$-Ant	0.49	<LOQ	0.83	<LOQ	0.72	0.49
9,10-Cl$_2$-Phe	21.5	6.83	14.6	0.33	17.2	6.40
3-Cl-Flu	1.92	1.95	4.34	0.59	0.50	3.02
1-Cl-Pyr	2.45	3.85	9.68	1.92	2.55	32.1
3,8-Cl$_2$-Flu	n.d.	n.d.	n.d.	n.d.	n.d.	<LOQ
7-Cl-BaA	1.51	1.67	2.51	0.34	1.86	2.30
1,5,9,10-Cl$_4$-Ant	n.d.	n.d.	n.d.	n.d.	n.d.	n.d.
6-Cl-BaP	n.d.	3.67	3.25	2.56	n.d.	n.d.
Σ_{16}Cl-PAHs	137	35.1	53.8	7.91	137	113
Br-PAHs						
5-Br-Ana	n.d.	n.d.	n.d.	<LOQ	n.d.	6.28
2-Br-Fle	n.d.	n.d.	n.d.	n.d.	n.d.	<LOQ
1,2-Br$_2$-Any	4.98	11.4	31.7	10.0	1.93	0.42
3-Br-Phe	0.54	0.85	1.16	<LOQ	n.d.	n.d.
9-Br-Phe	6.66	3.28	5.08	<LOQ	8.63	1.26
2-Br-Phe[b]	n.d.	9.50	n.d.	n.d.	n.d.	n.d.
1-Br-Ant[b]	n.d.	9.50	n.d.	n.d.	n.d.	n.d.
9-Br-Ant	3.10	8.63	1.27	0.43	0.94	0.35
2,7-Br$_2$-Fle	6.70	6.15	n.d.	n.d.	5.10	0.08
3-Br-Flu	3.72	12.2	14.4	1.28	0.27	0.10
1,8-Br$_2$-Ant	5.30	2.06	1.47	<LOQ	n.d.	n.d.
1,5-Br$_2$-Ant	n.d.	n.d.	n.d.	n.d.	n.d.	0.11
9,10-Br$_2$-Ant	5.58	1.59	<LOQ	n.d.	0.39	n.d.
4-Br-Pyr	3.34	1.57	28.8	7.35	n.d.	<LOQ
9,10-Br$_2$-Phe	n.d.	0.74	1.61	n.d.	0.30	n.d.
1-Br-Pyr	1.21	268	42.3	4.88	0.29	<LOQ
1,6-Br$_2$-Pyr	n.d.	44.4	22.7	3.02	n.d.	0.20
7-Br-BaA	18.8	19.3	1.18	<LOQ	7.84	<LOQ
Σ_{18}Br-PAHs	60.0	399	152	27.0	25.7	8.80

a. 未检出；b. 同分异构体的共洗脱；c. X-PAHs 的浓度低于检出限

EW：电子垃圾拆解车间，$n = 5$；RW：原料破碎车间，$n = 3$；BW：二级铜高炉车间，$n = 3$；OA：办公区，$n = 2$；RA：居民区，$n = 2$；SRM：标准灰尘样品 SRM 2585

如图 4-50 所示，氯取代的低环 Cl-PAHs(Cl-Phe/-Ant 和 Cl$_2$-Phe)，以及溴取代的高环 Br-PAHs(Br-Pyr 和 Br-BaA)是主要的同系物。此结果可能是由于低环 Cl-PAH 相对于高环同系物具有更高的光稳定性和大气反应惰性。此外，Cl-PAHs 的主要来源可能与 Br-PAHs 不同。电子垃圾中的 PVC 燃烧产生的 HCl 可以转化为 Cl$_2$ 并参与形成 Cl-PAHs。对于 Br-PAHs，它们可能是由电子垃圾燃烧过程飞灰介导的热化学反应产生的副产物，或者可能是通过高溴化的阻燃剂(例如十溴二苯乙烷)的前体形成的。Cl-PAHs 的主要异构体是 2-氯蒽(2-Cl-Ant)，仅在居民区的灰尘和标准灰尘 SRM 2585 中检测到了 1-氯蒽(1-Cl-Ant)，表明它们可能有相似的来源。Br-PAHs 的主要异构体是 1-溴芘(1-Br-Pyr)。在电子垃圾拆解园内的原料破碎车间和炼铜高炉车间的室内灰尘样品中发现的高环 Cl-PAHs(1-氯芘 Cl-Pyr，7-氯苯并[a]蒽 7-Cl-BaA 和 6-氯苯并[a]芘 6-Cl-BaP)和高环的 Br-PAHs(1-溴芘 1-Br-Pyr 和 1,6-二溴芘 1,6-Br$_2$-Pyr)占主导地位，这可能由于这两个车间中使用不同电子垃圾处理工艺有关。

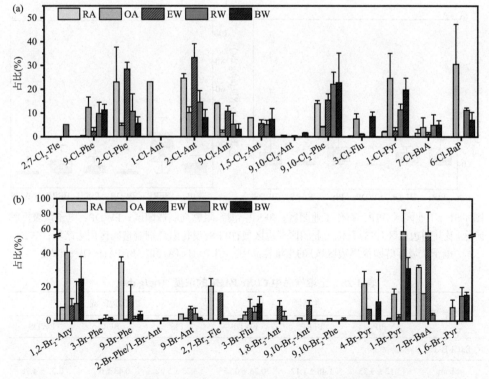

图 4-50 室内粉尘样品中 Cl-PAHs 和 Br-PAHs 的同系物分布

RA：居住区，$n=2$；OA：办公区，$n=2$；EW：电子垃圾拆解车间，$n=5$；RW：原料破碎车间，$n=3$；BW：炼铜高炉车间，$n=3$

(a) Cl-PAHs 的同族体分布；(b) Br-PAHs 的同族体分布

Wang 等[254]分析了代表我国典型石油化工(PIP)、溴系阻燃剂生产(BFRP)和电子垃圾拆解(EWDP)工业园区及其周边地区表层土壤中 HPAHs(Cl/Br-PAHs)的污染水平和组成情况,并对不同工业活动中 Cl/Br-PAHs 的产生、排放进行了分析。三个工业园区及其周边地区表层土壤中 Cl/Br-PAHs 的检出水平如图 4-51 和表 4-28 所示。溴系阻燃剂生产园区中 Cl/Br-PAHs 的检出水平最高,平均浓度达到 23.0 ng/g,并且其Σ_{17}Br-PAHs 的平均浓度达到 21.6 ng/g。石化园区和电子垃圾拆解园区 Cl/Br-PAHs 的平均检出浓度分别是 10.6 ng/g 和 6.94 ng/g。石油化工园区及其周边区域均检出较高浓度的 Cl-PAHs,并且其周边地区 Cl-PAHs 的检出水平略高于园区内。溴系阻燃剂生产园区内 Cl/Br-PAHs 的检出均显著高于其周边地区,这是因为在 BFRs 生产过程中大量使用含溴卤水和起氧化作用的氯气,这些卤素为 Cl/Br-PAHs 的形成提供了适宜条件。电子垃圾拆解园区中 Br-PAHs 的检出明显高于其周边地区,这可能是电子垃圾中含有的 BFRs 在处理过程中排放并转化成 Br-PAHs。

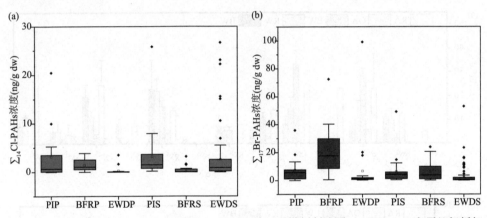

图 4-51 工业园区(PIP:石化工业园区;BFRP:溴系阻燃剂制造园区;EWDP:电子垃圾拆解园区)及其周边地区(PIS:石化工业园区周边区域;BFRS:溴化阻燃剂制造园区周边区域;EWDS:电子垃圾拆解园区周边区域)的土壤样品中Σ_{14}Cl-PAHs (A) 和Σ_{17}Br-PAHs (B) 的浓度

表 4-28 土壤样品中 Cl/Br-PAHs 的浓度 (ng/g·dw)

化学物质	工业园区			周边区域		
	PIP	BFRP	EWDP	PIS	BFRS	EWDS
Σ_{14}Cl-PAHs						
Mean	3.12 ± 5.33	1.48 ± 1.12	0.26 ± 0.75	3.23 ± 5.02	0.48 ± 0.76	2.57 ± 4.98
Median	0.69	1.05	0.05	1.48	0.09	0.88
Range	n.d.~20.5	n.d.~3.87	n.d.~3.52	0.19~25.8	n.d.~3.18	n.d.~26.7
DF	86.7%	100%	70.8%	100%	70.8%	89.2%

续表

化学物质	工业园区			周边区域		
	PIP	BFRP	EWDP	PIS	BFRS	EWDS
Σ_{17}Br-PAHs						
Mean	7.45 ± 9.03	21.6 ± 17.8	6.68 ± 19.9	6.27 ± 9.31	5.60 ± 6.18	2.99 ± 7.00
Median	5.64	17.61	1.09	4.31	3.14	0.61
Range	n.d.~35.9	0.52~72.4	0.25~99.1	0.31~49.1	0.11~23.8	0.01~53.0
DF	93.3%	100%	100%	100%	100%	100%
Total						
Mean	10.6 ± 11.6	23.0 ± 18.5	6.94 ± 20.1	9.49 ± 11.3	6.08 ± 6.77	5.56 ± 9.31
Median	7.39	19.7	1.13	6.43	3.16	1.97
Range	n.d.~39.4	0.52~75.3	0.30~99.3	1.26~57.1	0.11~27.0	0.08~56.7
DF	93.3%	100%	100%	100%	100%	100%

DF: 检出频率; PIP: 石化工业园区; BFRP: 溴化阻燃剂制造园区; EWDP: 电子垃圾拆解园区; PIS: 石化工业园区周边区域; BFRS: 溴系阻燃剂制造园区周边区域; EWDS: 电子垃圾拆解园区周边区域

为了追踪 Cl/Br-PAHs 的来源并阐明其对环境的影响,分析了其空间分布趋势。如图 4-52(a) 和 (b) 所示,在石油化工园区中均测得高浓度的 Cl/Br-PAHs,且主要分布在具有高温锅炉和原油裂解生产线附近,此生产区的高温锅炉为 Cl/Br-PAHs 的形成提供了温度和相关条件。此外,在居民区也分别检测到高浓度的 Cl/Br-PAHs,这表明该地区可能还有其他 Cl/Br-PAHs 排放源。溴系阻燃剂生产园区中 Cl/Br-PAHs 的空间分布与石油化工园区相似,如图 4-52(c) 和 (d) 所示,在园区内检测到较高水平的 Cl/Br-PAHs,并逐渐向北扩散。但 Br-PAHs 的检出水平显著高于石油化工园区,且高浓度检出点主要分布在生产区域,这可能是生产过程中挥发的溴和氯为 Cl/Br-PAHs 的形成提供了必要元素。

电子垃圾拆解园区的空间分布与另外两个园区有所不同,如图 4-52(e) 和 (f) 所示,在电子垃圾拆解园区的周边区域均能检出较高水平的 Cl/Br-PAHs。这是由于此地早期的电子垃圾处理模式主要为小作坊式拆解,采用酸浸、露天燃烧等落后且不规范的拆解方法所造成的污染。根据相关研究,园区南部曾大量使用汽油和煤块,这与高浓度 Cl/Br-PAHs 主要分布在园区南部的结果相呼应,其他研究也呈现出相似的空间分布。因园区采用了火法冶炼工艺回收贵金属,因此在园区东部的冶炼区测得较高浓度的 Br-PAHs,这可能是由于添加在印刷线路板中的 BFRs 在燃烧过程中释放并且转化成 Br-PAHs 或其前驱体。

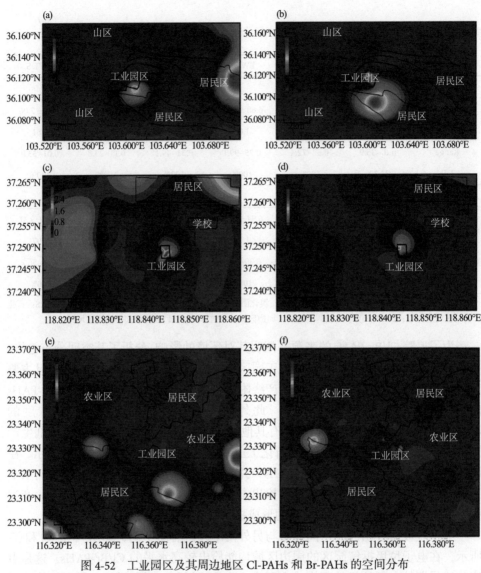

图 4-52 工业园区及其周边地区 Cl-PAHs 和 Br-PAHs 的空间分布
(a, b): 石化园区; (c, d): 溴系阻燃生产园区; (e, f): 电子垃圾拆解园区

为了进一步了解 Cl/Br-PAHs 的来源和转化,对其进行了主成分分析。园区及其周边地区 Cl-PAHs 的组成如图 4-53(a)所示, 1-/2-Cl-Ant 在石油化工园区的 Cl-PAHs 中占比较多,达到 47.6%,其次是 6-Cl-BaP、1,5-Cl_2-Ant 和 9,10-Cl_2-Ant,分别占 22.8%、9.7%和 9.3%。在园区周边,9,10-Cl_2-Ant、1,5-Cl_2-Ant 和 6-Cl-BaP 的组成占比分别为 25.5%、23.8%和 21.0%。据报道,石油原油及其副产品中的 PAHs 主要为含 2~3 个苯环的低分子量化合物。在溴系阻燃剂生产园区中,1-Cl-Pyr 和

3-Cl-Flu 分别占 29.6%和 17.9%。园区周边为 1-Cl-Pyr 所占比例最高，其次是 6-Cl-BaP 和 1,5,9,10-Cl$_4$-Ant，所占比例分别为 35.1%、16.8%和 15.7%。高环 Cl-PAHs 在溴系阻燃剂生产园区中占优势，园区内外四至五环 Cl-PAHs 的比例在 60%以上。在电子垃圾拆解园区中，含 2 个 Cl 的三环 Cl/Br-PAHs 占主导地位，尤其是氯化 Ant。园区内检出 1-/2-Cl-Ant 占比为 24.9%，其次是 9,10-Cl$_2$-Phe(24.1%)、1,5-Cl$_2$-Ant(15.0%)和 9,10-Cl$_2$-Ant(14.0%)。在其周边区域，1,5-Cl$_2$-Ant、9,10-Cl$_2$-Ant 和 7-Cl-BaA 分别占 35.7%、34.5%和 11.8%。

如图 4-53(b)所示，三个园区周边地区检出的主要 Br-PAHs 成分相似，但比例略有不同。1,8-Br$_2$-Ant(68.2%)和 4-Br-Pyr(21.5%)在石油化工园区内的 Br-PAHs

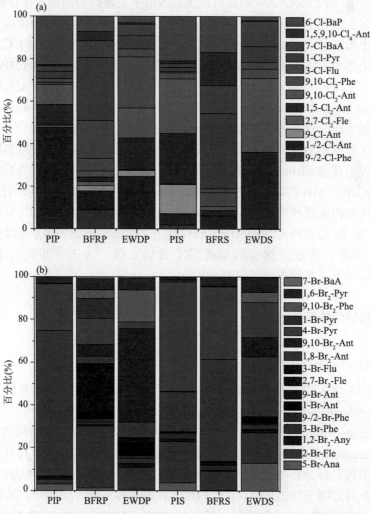

图 4-53 工业园区及其周边地区 Cl-PAHs (a) 和 Br-PAHs (b) 的组成成分

占主导地位，在园区周边中这两个化合物分别占 18.5%和 51.0%。此外，2-Br-Fle 占石化工业园区外 Br-PAHs 的 19.3%，而有研究认为 2-Br-Fle 可能是汽车行业的典型标志化合物。因此，1,8-Br_2-Ant 可能是石油化工行业生产活动中主要产生的 Cl/Br-PAHs，而园区外测得的 2-Br-Fle 可能来自汽车尾气排放。在溴系阻燃剂生产园区中，2-Br-Fle 和 2,7-Br_2-Fle 分别占 28.8%和 18.3%。园区外则以 1,8-Br_2-Ant(47.4%)和 4-BrPyr(34.0%)为主。园区内高浓度 Br-PAHs 的检出点均分布在生产线附近，这表明在溴系阻燃剂生产过程中可能产生 Br-PAHs，且主要产生排放低环 Br-PAHs。在电子垃圾拆解园区内，9,10-Br_2-Ant 和 9,10-Br_2-Phe 分别占组成成分的 43.7%和 15.0%，1,8-Br_2-Ant 和 4-Br-Pyr 分别占 28.1%和 16.3%。因此，含 2 个 Br 的 3 环 Cl/Br-PAHs 可认为是由电子垃圾拆解活动产生排放。

通过对石油化工园区、溴系阻燃剂生产园区和电子垃圾拆解园区内及其周边地区表层土壤中的 Cl/Br-PAHs 进行定量分析，石油化工园区内外的 Cl/Br-PAHs 检出平均浓度为 10.6 ng/g 和 9.49 ng/g，溴系阻燃剂生产园区内外的 HPAHs 检出平均浓度为 23.0 ng/g 和 6.08 ng/g，电子垃圾拆解园区内外的 Cl/Br-PAHs 检出平均浓度为 6.94 ng/g 和 5.56 ng/g。三个园区内检出的 Br-PAHs 都不同程度地高于 Cl-PAHs 的检出，这与两者的光稳定性和形成途径有关。通过空间分布分析，石油化工园区和溴系阻燃剂生产园区呈现点源污染模式，而电子垃圾拆解园区在周边地区发现了其他 Cl/Br-PAHs 高浓度点。从主成分分析和特征同系物探究中发现三个园区 Cl/Br-PAHs 的组成分布有所不同，石油化工园区和溴系阻燃剂生产园区分别以低环和四至五环的 Cl/Br-PAHs 为主，而电子垃圾拆解园区的特征同系物为含 2 个 Cl 或 Br 的三环 Cl/Br-PAHs。研究提出了特征污染物并建立指纹图谱以追踪工业排放源，为建立排放清单提供了数据支撑，为进一步研究工业排放 Cl/Br-PAHs 的形成机制提供了依据。

4.2.8 其他持久性有机污染物

1. PCDD/Fs 和 PBDD/Fs

PCDD/Fs 根据氯取代基的数量和位置的不同，分别有 75 种多氯代二苯并二噁英(polychlorinated dibenzo-p-dioxins，PCDDs)和 135 种多氯代二苯并呋喃(polychlorinated dibenzo furans，PCDFs)同分异构体。多溴代二苯并二噁英/呋喃(polybrominated dibenzo-p-dioxins/furans，PBDD/Fs)是与 PCDD/Fs 具有相同结构的共平面溴代三环芳香类化合物，也包括 135 种多溴二苯并呋喃(PBDFs)和 75 种多溴二苯并对二噁英(PBDDs)。它们的结构式如图 4-54 所示。此外，氯/溴混合卤代二噁英/呋喃(PHDD/Fs)种类复杂繁多(1550 种 PHDDs 和 3050 种 PHDFs)，由于缺乏参考标样,目前对此类化合物的研究较少[255]。

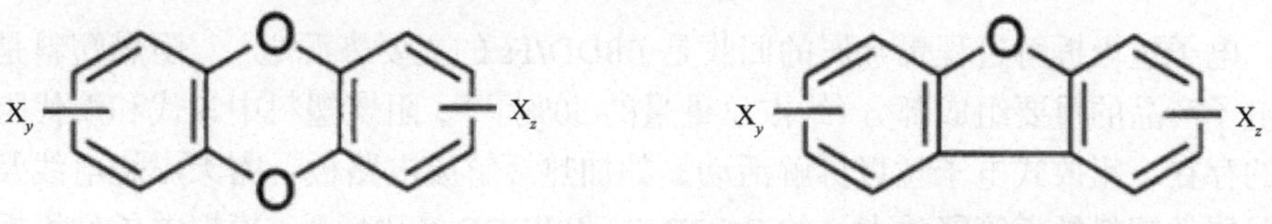

X=Cl,y+z=1~8(PCDDs)
X=Br,y+z=1~8(PBDDs)

X=Cl,y+z=1~8(PCDFs)
X=Br,y+z=1~8(PBDFs)

图 4-54 PCDD/Fs 和 PBDD/Fs 的结构示意图

PCDD/Fs 和 PBDD/Fs 性质相似,其平面型结构决定了其稳定的化学性质,在环境中难降解。与 PCDD/Fs 相比,PBDD/Fs 具有更高的分子量、更高的熔点、更低的蒸气压、更低的水溶性。它们通常溶于脂肪、油脂和有机溶剂。含溴取代 PBDD/Fs 的正辛醇-水分配系数(K_{ow})理论值比相应氯代化合物更高,因而表现出更强的脂溶性[256]。PCDD/Fs 除了科研用途而少量合成外,没有自然来源,主要的人为活动来源有以下几种:固废和化石燃料等焚烧、冶金、含氯农药生产使用、造纸业、水泥生产等[257]。

目前国内外关于 PBDD/Fs 的研究还处于初始阶段,已有研究主要集中在污染源/生成机制[258,259]、环境水平[260,261]和毒性方面[262,263]。与 PCDD/Fs 相似,目前不存在有意的 PBDD/Fs 工业生产活动,PBDD/Fs 主要源自含溴化合物各种过程的副产品:光化学反应、前体转化反应或从头合成反应等。研究表明,PBDD/Fs 存在以下四种主要来源[264]:①溴代阻燃剂(BFRs)以及含 BFRs 材料的生产、使用、回收及处理处置;②工业热过程,如生活垃圾、工业垃圾等各类废弃物焚烧、电子垃圾拆解回收和金属冶炼等;③不可控的热过程,如意外住宅火灾、森林火灾、露天焚烧垃圾等;④海洋生物合成 PBDD/Fs。其中,广泛使用的 BFRs 的生产是 PBDD/Fs 的主要来源之一。文献表明,BFRs 如 PBDEs、十溴联苯(DBB)、1,2-双(三溴苯氧)乙烷、四溴双酚 A 和其他阻燃剂中均存在 PBDD/Fs 杂质[265]。添加了 PBDEs 的树脂中 PBDD/Fs 含量通常高达 μg/g 级别[255],1987~1995 年间日本制造的电视机 PE 塑料、电路板中的 PBDD/Fs 含量高达 3000~130 000 ng/g[266]。商业品 PBDEs(BDE-71、-79 和-83)中 4~8 溴代 PBDFs 含量在 257~49 600 ng/g 之间[265]。同时,十溴联苯醚(deca-BDE)产品中也发现了相当水平的 PBDD/Fs(3400~13 600 ng/g),根据 2001 年全球 deca-BDE 产量估算伴随产生的 PBDD/Fs 量约在 0.21~0.78 吨之间[267]。由此可知,BFRs 产品的生产与使用可导致 PBDD/Fs 直接排放到环境中。有限的研究表明,PBDD/Fs 在大气[260,261]、室内灰尘[268,269]、土壤[270]、沉积物[271,272]、废水污泥[273]等环境介质和生物体[274,275]、人体样品[276,277]中被广泛检出,表明 PCDD/Fs 和 PBDD/Fs 的普遍污染。

2. PCDD/Fs 和 PBDD/Fs 的污染特征

电子垃圾拆解过程贵金属的回收是 PBDD/Fs 的主要来源之一。阻燃塑料是现代电子产品的重要组成部分(约占总重量的 30%)[278]，阻燃塑料中氯代和溴代阻燃剂的存在，粗放式电子垃圾拆解活动，如加热拆解废电路板、露天燃烧电线及电子元器件塑料外壳等释放大量的 PCDD/Fs 和 PBDD/Fs[269, 279]。模拟电子垃圾拆解的露天燃烧实验发现，电子垃圾本身的卤素成分影响排放物的组成特征。对两种溴和氯含量不同的电子垃圾废料进行模拟燃烧实验，即电路板(Br、Cl 含量分别为 17 mg/g 和 1.9 mg/g)和绝缘电线(几乎检测不到溴，但 Cl 含量高达 88.4 mg/g)，分析其废气排放含量。结果表明，电路板燃烧产生废气中的 PCDD/Fs 平均浓度为 11 ng/g，比绝缘电线(649 ng/g)低 1 个数量级，但其 1~5 溴代 PBDD/Fs 的排放量却很高。电路板燃烧产生废气中溴代同系物的含量是相应氯代同系物的 50~500 倍，可能是由于电路板中的溴/氯质量比高(Br/Cl>300)。相比之下，绝缘电线样品中 PBDD/Fs 含量极低，与绝缘电线中的微量溴一致。对于电路板燃烧，还观察到 PBDFs 排放量比 PBDDs 高几倍[280]。Duan 等[281]在管式炉中于 250~650 ℃ 焚烧废电路板组件的研究也呈现类似结果，焚烧后的气态、液态和固态残留物样品中 4~8 溴代 PBDD/Fs 均以 PBDFs 为主，含量远远高于 PBDDs。同时发现在 250~400 ℃ 温度下具有最大的 PBDD/Fs 和 PCDD/Fs 形成速率，在 325℃ 时，12 种 2,3,7,8-PBDD/Fs 总产量最大，固体、冷凝水和气态分别为时的 19 ng TEQ/g、160 ng TEQ/g 和 0.057 ng TEQ/g[281]。

电子垃圾不正规拆解场地主要分布在亚非拉等发展中国家，由于技术不规范和环保监督不到位，研究表明这些电子垃圾拆解场地是周边环境中 PBDD/Fs 的重要污染源[282]。Li 等[283]测量了我国华南地区某典型电子垃圾拆解区及其邻近的陈店镇环境大气中的 PBDD/Fs 和 PCDD/Fs，结果表明，电子垃圾拆解地环境空气的 PCDD/Fs 水平 (64.9~2365 pg/m^3，0.909~48.9 pg WHO-TEQ/m^3) 和 4~6 溴代 2,3,7,8-PBDD/Fs 浓度(8.124~61 pg/m^3，1.6~2104 pg I-TEQ/m^3)均处于全球高污染水平，比邻近的陈店测得的 2,3,7,8-PBDD/Fs 浓度高约 12~18 倍，比广州高 37~133 倍。PBDD/Fs 和 PCDD/Fs 的组成研究结果显示电子垃圾拆解地和陈店相似，推断陈店的高污染水平来源于电子垃圾拆解地的大气迁移[283]。电子垃圾回收过程中的加热拆解和露天燃烧导致大气中相对较高水平的 PBDD/Fs，而酸浸会直接污染土壤和水[270]。非洲加纳首都阿克拉电子垃圾露天焚烧区的土壤中检出了高浓度的 PCDD/Fs 和 PBDD/Fs，浓度范围分别为 18~520 ng/g 和 83~3800 ng/g，PBDD/Fs 浓度高于全球其他区域报道的浓度水平[279]。

由目前报道的电子垃圾拆解活动造成 PCDD/Fs 和 PBDD/Fs 的污染的相关研究可以总结出以下几条规律：①焚烧源相关(垃圾焚烧，电子垃圾拆解回收等)的

PBDD/Fs 排放过程，PBDFs 往往比 PBDDs 高出数倍，这与 PBDEs 脱溴形成 PBDFs 相关[279,284,285]；②电子垃圾中的溴氯来源影响最终焚烧产物中 PCDD/Fs 和 PBDD/Fs 的组成比例；③与 BFRs 相关的生产和处理过程及粗放式电子垃圾拆解回收均是 PBDD/Fs 的主要来源。

（马盛韬　余应新　郭　杰　安太成）

参 考 文 献

[1] Luo P, Bao L J, Li S M, et al. Size-dependent distribution and inhalation cancer risk of particle-bound polycyclic aromatic hydrocarbons at a typical e-waste recycling and an urban site [J]. Environmental Pollution, 2015, 200: 10-15.

[2] Rissler J, Gudmundsson A, Nicklasson H, et al. Deposition efficiency of inhaled particles (15-5000 nm) related to breathing pattern and lung function: an experimental study in healthy children and adults [J]. Particle and Fibre Toxicology, 2017, 14: 10.

[3] Makkonen U, Hellén H, Anttila P, et al. Size distribution and chemical composition of airborne particles in south-eastern Finland during different seasons and wildfire episodes in 2006 [J]. Science of The Total Environment, 2010, 408(3): 644-651.

[4] Wiseman C L, Zereini F. Airborne particulate matter, platinum group elements and human health: a review of recent evidence [J]. Science of The Total Environment, 2009, 407(8): 2493-2500.

[5] Nti A A A, Arko-Mensah J, Botwe P K, et al. Effect of particulate matter exposure on respiratory health of e-waste workers at agbogbloshie, Accra, Ghana [J]. International Journal of Environmental Research and Public Health, 2020, 17(9): 3042.

[6] Zeng Z J, Xu X J, Wang Q H, et al. Maternal exposure to atmospheric $PM_{2.5}$ and fetal brain development: Associations with BAI1 methylation and thyroid hormones [J]. Environmental Pollution, 2022, 308: 119665.

[7] Issah I, Arko-Mensah J, Rozek L S, et al. Association between global DNA methylation (LINE-1) and occupational particulate matter exposure among informal electronic-waste recyclers in Ghana [J]. International Journal of Environmental Health Research, 2022, 32(11): 2406-2424.

[8] Yang F X, Jin S W, Xu Y, et al. Comparisons of IL-8, ROS and p53 responses in human lung epithelial cells exposed to two extracts of PM2.5 collected from an e-waste recycling area, China [J]. Environmental Research Letters, 2011, 6(2): 024013.

[9] Nti A A A, Robins T G, Mensah J A, et al. Personal exposure to particulate matter and heart rate variability among informal electronic waste workers at Agbogbloshie: A longitudinal study [J]. Bmc Public Health, 2021, 21(1): 2161.

[10] Chen L Y, Cai C Y, Yu S Y, et al. Emission factors of particulate matter, CO and CO_2 in the pyrolytic processing of typical electronic wastes [J]. Journal of Environmental Sciences, 2019, 81: 93-101.

[11] Bungadaeng S, Prueksasit T, Siriwong W. Inhalation exposure to respirable particulate matter among workers in relation to their e-waste open burning activities in Buriram Province, Thailand [J]. Sustainable Environment Research, 2019, 29(1): 26.

[12] An T, Zhang D, Li G, et al. On-site and off-site atmospheric PBDEs in an electronic dismantling workshop in south China: Gas-particle partitioning and human exposure assessment [J]. Environmental Pollution, 2011, 159(12): 3529-3535.

[13] Guo J, Lin K, Deng J, et al. Polybrominated diphenyl ethers in indoor air during waste TV recycling process

[J]. Journal of Hazardous Materials, 2015, 283: 439-446.

[14] Choi J-K, Heo J-B, Ban S-J, et al. Chemical characteristics of $PM_{2.5}$ aerosol in Incheon, Korea [J]. Atmospheric Environment, 2012, 60: 583-592.

[15] 李静姝. 农村固定源燃烧产生的多环芳烃、含氧及硝基多环芳烃的排放因子的研究 [D]. 济南: 山东大学, 2020.

[16] 任照芳, 黄渤, 刘明, 等. 典型线路板回收过程排放颗粒物的主要成分和特征 [J]. 中国环境科学, 2012, 32(8): 1447-1451.

[17] Chen H J, Ma S T, Yu Y X, et al. Seasonal profiles of atmospheric PAHs in an e-waste dismantling area and their associated health risk considering bioaccessible PAHs in the human lung [J]. Science of the Total Environment, 2019, 683: 371-379.

[18] 陈浩佳. 电子垃圾拆解区大气典型SVOCs污染特征及健康风险评估 [D]. 广州: 广东工业大学, 2019.

[19] Chen S J, Wang J, Wang T, et al. Seasonal variations and source apportionment of complex polycyclic aromatic hydrocarbon mixtures in particulate matter in an electronic waste and urban area in south China [J]. Science of the Total Environment, 2016, 573: 115-122.

[20] Deng W J, Louie P K K, Liu W K, et al. Atmospheric levels and cytotoxicity of PAHs and heavy metals in TSP and $PM_{2.5}$ at an electronic waste recycling site in southeast China [J]. Atmospheric Environment, 2006, 40(36): 6945-6955.

[21] Guo J, Ji A, Xu Z M. On-site characteristics of airborne particles at a formal electronic waste recycling plant: size distribution and lung deposited surface area [J]. Journal of Material Cycles and Waste Management, 2023, 25: 346-358.

[22] Guo J, Ji A, Wang J, et al. Emission characteristics and exposure assessment of particulate matter and polybrominated diphenyl ethers (PBDEs) from waste printed circuit boards de-soldering [J]. Science of the Total Environment, 2019, 662: 530-536.

[23] Reche C, Viana M, Brines M, et al. Determinants of aerosol lung-deposited surface area variation in an urban environment [J]. Science of the Total Environment, 2015, 517: 38-47.

[24] Gomes J F P, Bordado J C M, Albuquerque P C S. On the assessment of exposure to airborne ultrafine particles in urban environments [J]. Journal of Toxicology and Environmental Health-Part a-Current Issues, 2012, 75(22-23): 1316-1329.

[25] Martin M, Lam P K, Richardson B J. An Asian quandary: Where have all of the PBDEs gone? [J]. Mar Pollut Bull, 2004, 49(5-6): 375-382.

[26] Mackay D, Shiu W-Y, Shiu W-Y, et al. Handbook of Physical-Chemical Properties and Environmental Fate for Organic Chemicals [M]. 2nd Edition. CRC Press, 2006.

[27] Hites R A. Polybrominated diphenyl ethers in the environment and in people: A meta-analysis of concentrations [J]. Environmental Science & Technology, 2004, 38(4): 945-956.

[28] U.S. Environmental Protection Agency (EPA). An exposure assessment of polybrominated diphenyl ethers [R]. National Center for Environmental Assessment, Washington, DC: National Technical Information Service, Springfield, VA, 2010.

[29] Lu C Y, Zhang L, Zhong Y G, et al. An overview of e-waste management in China [J]. Journal of Material Cycles and Waste Management, 2015, 17(1): 1-12.

[30] Pariatamby A, Victor D. Policy trends of e-waste management in Asia [J]. Journal of Material Cycles and Waste Management, 2013, 15(4): 411-419.

[31] Bernardeau F, Perrin D, Caro-Bretelle A S, et al. Development of a recycling solution for waste thermoset material: Waste source study, comminution scheme and filler characterization [J]. Journal of Material Cycles and Waste Management, 2018, 20(2): 1320-1336.

[32] Betts K S. Rapidly rising PBDE levels in North America [J]. Environmental Science & Technology, 2002, 36(3): 50A-52A.

[33] UNEP. Technical review of the implications of recycling commercial pentabromodiphenyl ether and commercial octabromodiphenyl ether [Z]. 2010: UNEP/POPS/POPRC.6/2.

[34] Abbasi G, Buser A M, Soehl A, et al. Stocks and flows of PBDEs in products from use to saste in the US and Canada from 1970 to 2020 [J]. Environmental Science & Technology, 2015, 49(3): 1521-1528.

[35] Zhang D L, An T C, Qiao M, et al. Source identification and health risk of polycyclic aromatic hydrocarbons associated with electronic dismantling in Guiyu town, South China [J]. Journal of Hazardous Materials, 2011, 192(1): 1-7.

[36] Liu R R, Ma S T, Li G Y, et al. Comparing pollution patterns and human exposure to atmospheric PBDEs and PCBs emitted from different e-waste dismantling processes [J]. Journal of Hazardous Materials, 2019, 369: 142-149.

[37] Liu R, Ma S, Li G, et al. Comparing pollution patterns and human exposure to atmospheric PBDEs and PCBs emitted from different e-waste dismantling processes [J]. Journal of Hazardous Materials, 2019, 369: 142-149.

[38] Chen D, Bi X, Liu M, et al. Phase partitioning, concentration variation and risk assessment of polybrominated diphenyl ethers (PBDEs) in the atmosphere of an e-waste recycling site [J]. Chemosphere, 2011, 82(9): 1246-1252.

[39] Han W L, Feng J L, Gu Z P, et al. Polybrominated diphenyl ethers in the atmosphere of Taizhou, a major e-waste dismantling area in China [J]. Bulletin of Environmental Contamination and Toxicology, 2009, 83(6): 783-788.

[40] Wang Y, Hou M, Zhao H, et al. Factors influencing the diurnal atmospheric concentrations and soil-air exchange of PBDEs at an e-waste recycling site in China [J]. Atmospheric Pollution Research, 2018, 9(1): 166-171.

[41] Jiang Y, Lin T, Wu Z, et al. Seasonal atmospheric deposition and air-sea gas exchange of polycyclic aromatic hydrocarbons over the Yangtze River Estuary, East China Sea: Implications for source–sink processes [J]. Atmospheric Environment, 2018, 178: 31-40.

[42] Lu S Y, Kang L, Liao S C, et al. Phthalates in $PM_{2.5}$ from Shenzhen, China and human exposure assessment factored their bioaccessibility in lung [J]. Chemosphere, 2018, 202: 726-732.

[43] Die Q Q, Nie Z Q, Huang Q F, et al. Concentrations and occupational exposure assessment of polybrominated diphenyl ethers in modern Chinese e-waste dismantling workshops [J]. Chemosphere, 2019, 214: 379-388.

[44] Chen Z, Luo X, Zeng Y, et al. Polybrominated diphenyl ethers in indoor air from two typical E-waste recycling workshops in Southern China: Emission, size-distribution, gas-particle partitioning, and exposure assessment [J]. Journal of Hazardous Materials, 2021, 402: 123667.

[45] Wang J, Ma Y-J, Chen S-J, et al. Brominated flame retardants in house dust from e-waste recycling and urban areas in South China: Implications on human exposure [J]. Environment International, 2010, 36(6): 535-541.

[46] Qiao L, Zheng X B, Zheng J, et al. Legacy and currently used organic contaminants in human hair and hand wipes of female e-waste dismantling workers and workplace dust in South China [J]. Environmental Science & Technology, 2019, 53(5): 2820-2829.

[47] Lin C M, Zeng Z J, Xu R B, et al. Risk assessment of PBDEs and PCBs in dust from an e-waste recycling area of China [J]. Science of the Total Environment, 2022, 803: 150016.

[48] Xu P J, Tao B, Zhou Z G, et al. Occurrence, composition, source, and regional distribution of halogenated flame retardants and polybrominated dibenzo-p-dioxin/dibenzofuran in the soils of Guiyu, China [J]. Environmental Pollution, 2017, 228: 61-71.

[49] Ge X, Ma S T, Zhang X L, et al. Halogenated and organophosphorous flame retardants in surface soils from an e-waste dismantling park and its surrounding area: Distributions, sources, and human health risks [J]. Environment International, 2020, 139: 105741.

[50] Wu Z, Han W, Xie M, et al. Occurrence and distribution of polybrominated diphenyl ethers in soils from an e-waste recycling area in northern China [J]. Ecotoxicology and Environmental Safety, 2019, 167: 467-475.

[51] McGrath T J, Morrison P D, Ball A S, et al. Spatial distribution of novel and legacy brominated flame retardants in soils surrounding two Australian electronic waste recycling facilities [J]. Environmental Science & Technology, 2018, 52(15): 8194-8204.

[52] Wang P, Zhang H D, Fu J J, et al. Temporal trends of PCBs, PCDD/Fs and PBDEs in soils from an E-waste dismantling area in East China [J]. Environmental Science-Processes & Impacts, 2013, 15(10): 1897-1903.

[53] Leung A O W, Luksemburg W J, Wong A S, et al. Spatial distribution of polybrominated diphenyl ethers and polychlorinated dibenzo-p-dioxins and dibenzofurans in soil and combusted residue at Guiyu, an electronic waste recycling site in southeast China [J]. Environmental Science & Technology, 2007, 41(8): 2730-2737.

[54] Ohajinwa C M, Van Bodegom P M, Xie Q, et al. Hydrophobic organic pollutants in soils and dusts at electronic waste recycling sites: occurrence and possible impacts of polybrominated diphenyl ethers [J]. International Journal of Environmental Research and Public Health, 2019, 16(3): 360.

[55] Luo Q, Wong M H, Wang Z J, et al. Polybrominated diphenyl ethers in combusted residues and soils from an open burning site of electronic wastes [J]. Environmental Earth Sciences, 2013, 69(8): 2633-2641.

[56] Zhang S H, Xu X J, Wu Y S, et al. Polybrominated diphenyl ethers in residential and agricultural soils from an electronic waste polluted region in South China: Distribution, compositional profile, and sources [J]. Chemosphere, 2014, 102: 55-60.

[57] Li N, Chen X W, Deng W J, et al. PBDEs and dechlorane plus in the environment of Guiyu, Southeast China: A historical location for e-waste recycling (2004,2014) [J]. Chemosphere, 2018, 199: 603-611.

[58] Wu Z N, Han W, Xie M M, et al. Occurrence and distribution of polybrominated diphenyl ethers in soils from an e-waste recycling area in northern China [J]. Ecotoxicology and Environmental Safety, 2019, 167: 467-475.

[59] Matsukami H, Suzuki G, Someya M, et al. Concentrations of polybrominated diphenyl ethers and alternative flame retardants in surface soils and river sediments from an electronic waste-processing area in northern Vietnam, 2012-2014 [J]. Chemosphere, 2017, 167: 291-299.

[60] Wang N, Kong D Y, Cai D J, et al. Levels of polychlorinated biphenyls in human adipose tissue samples from southeast China [J]. Environmental Science & Technology, 2010, 44(11): 4334-4340.

[61] Sun S J, Zhao J H, Leng J H, et al. Levels of dioxins and polybrominated diphenyl ethers in human milk from three regions of northern China and potential dietary risk factors [J]. Chemosphere, 2010, 80(10): 1151-1159.

[62] Deng B, Zhang J Q, Zhang L S, et al. Levels and profiles of PCDD/Fs, PCBs in mothers' milk in Shenzhen of China: Estimation of breast-fed infants' intakes [J]. Environment International, 2012, 42: 47-52.

[63] Zhao X R, Qin Z F, Yang Z Z, et al. Dual body burdens of polychlorinated biphenyls and polybrominated diphenyl ethers among local residents in an e-waste recycling region in southeast China [J]. Chemosphere, 2010, 78(6): 659-666.

[64] Zhao G F, Xu Y, Li W, et al. Prenatal exposures to persistent organic pollutants as measured in cord blood and meconium from three localities of Zhejiang, China [J]. Science of the Total Environment, 2007, 377(2-3): 179-191.

[65] Zhao G F, Wang Z J, Zhou H D, et al. Burdens of PBBs, PBDEs, and PCBs in tissues of the cancer patients in the e-waste disassembly sites in Zhejiang, China [J]. Science of the Total Environment, 2009, 407(17): 4831-4837.

[66] Zhao G F, Wang Z J, Dong M H, et al. PBBs, PBDEs, and PCBs levels in hair of residents around e-waste disassembly sites in Zhejiang Province, China, and their potential sources [J]. Science of the Total Environment, 2008, 397(1-3): 46-57.

[67] Zhang J Q, Jiang Y S, Zhou J, et al. Elevated body burdens of PBDEs, Dioxins, and PCBs on thyroid hormone homeostasis at an electronic waste recycling site in China [J]. Environmental Science & Technology, 2010, 44(10): 3956-3962.

[68] Xing G H, Wu S C, Wong M H. Dietary exposure to PCBs based on food consumption survey and food basket analysis at Taizhou, China - The World's major site for recycling transformers [J]. Chemosphere, 2010, 81(10):

1239-1244.

[69] Wen S, Yang F X, Gong Y, et al. Elevated levels of urinary 8-hydroxy-2′-deoxyguanosine in male electrical and electronic equipment dismantling workers exposed to high concentrations of polychlorinated dibenzo-p-dioxins and dibenzofurans, polybrominated diphenyl ethers, and polychlorinated biphenyls [J]. Environmental Science & Technology, 2008, 42(11): 4202-4207.

[70] Breivik K, Sweetman A, Pacyna J M, et al. Towards a global historical emission inventory for selected PCB congeners—A mass balance approach 3. An update [J]. Science of the Total Environment, 2007, 377(2-3): 296-307.

[71] Breivik K, Gioia R, Chakraborty P, et al. Are reductions in industrial organic contaminants emissions in rich countries achieved partly by export of toxic wastes? [J]. Environmental Science & Technology, 2011, 45(21): 9154-9160.

[72] Chen S J, Tian M, Zheng J, et al. Elevated levels of polychlorinated biphenyls in plants, air, and soils at an e-waste site in southern China and enantioselective biotransformation of chiral PCBs in plants [J]. Environmental Science & Technology, 2014, 48(7): 3847-3855.

[73] Li Q L, Xu Y, Li J, et al. Levels and spatial distribution of gaseous polychlorinated biphenyls and polychlorinated naphthalenes in the air over the northern South China Sea [J]. Atmospheric Environment, 2012, 56: 228-235.

[74] Wang Y, Luo C L, Wang S R, et al. The abandoned e-waste recycling site continued to act as a significant source of polychlorinated biphenyls: An *in situ* assessment using fugacity samplers [J]. Environmental Science & Technology, 2016, 50(16): 8623-8630.

[75] Han W L, Feng J L, Gu Z P, et al. Polychlorinated biphenyls in the atmosphere of Taizhou, a major e-waste dismantling area in China [J]. Journal of Environmental Sciences, 2010, 22(4): 589-597.

[76] Zhang Z, Liu L, Li Y F, et al. Analysis of polychlorinated biphenyls in concurrently sampled Chinese air and surface soil [J]. Environmental Science & Technology, 2008, 42(17): 6514-6518.

[77] 孟庆昱, 毕新慧, 储少岗, 等. 污染区大气中多氯联苯的表征与分布研究初探 [J]. 环境化学, 2000, (6): 501-506.

[78] Li Y M, Jiang G B, Wang Y W, et al. Concentrations, profiles and gas-particle partitioning of PCDD/Fs, PCBs and PBDEs in the ambient air of an E-waste dismantling area, southeast China [J]. Chinese Science Bulletin, 2008, 53(4): 521-528.

[79] Yu S-X, Jiang S X, Li J U. Environmental pollution by dioxin-like PCBs around a disassembly of obsolete solid waste [J]. Journal of Environment & Health, 2007, 24(5): 304-307.

[80] Xing G H, Liang Y, Chen L X, et al. Exposure to PCBs, through inhalation, dermal contact and dust ingestion at Taizhou, China—A major site for recycling transformers [J]. Chemosphere, 2011, 83(4): 605-611.

[81] Wang Y, Hu J, Lin W, et al. Health risk assessment of migrant workers' exposure to polychlorinated biphenyls in air and dust in an e-waste recycling area in China: Indication for a new wealth gap in environmental rights [J]. Environment International, 2016, 87: 33-41.

[82] Tue N M, Takahashi S, Suzuki G, et al. Contamination of indoor dust and air by polychlorinated biphenyls and brominated flame retardants and relevance of non-dietary exposure in Vietnamese informal e-waste recycling sites [J]. Environment International, 2013, 51: 160-167.

[83] Jakobsson K, Thuresson K, Rylander L, et al. Exposure to polybrominated diphenyl ethers and tetrabromobisphenol A among computer technicians [J]. Chemosphere, 2002, 46(5): 709-716.

[84] Environment Canada. Guidance manual for the categorization of organic and inorganic substances on Canada's Domestic Substances List: Determining persistence, bioaccumulation potential, and inherent toxicity to non-human organisms. Existing Substances Program (CDROM) [R]. Ottawa: Environment Canada, 2004.

[85] Choo G, Lee I S, Oh J E. Species and habitat-dependent accumulation and biomagnification of brominated flame retardants and PBDE metabolites [J]. Journal of Hazardous Materials, 2019, 371: 175-182.

[86] Guo J H, Romanak K, Westenbroek S, et al. Current-use flame retardants in the water of lake michigan tributaries [J]. Environmental Science & Technology, 2017, 51 (17): 9960-9969.

[87] Drage D S, Heffernan A L, Cunningham T K, et al. Serum measures of hexabromocyclododecane (HBCDD) and polybrominated diphenyl ethers (PBDEs) in reproductive-aged women in the United Kingdom [J]. Environmental Research, 2019, 177: 108631.

[88] Jaksic K, Saric M M, Culin J. Knowledge and attitudes regarding exposure to brominated flame retardants: A survey of Croatian health care providers [J]. Environmental Science and Pollution Research, 2020, 27 (7): 7683-7692.

[89] Kacew S, Hayes A W. Absence of neurotoxicity and lack of neurobehavioral consequences due to exposure to tetrabromobisphenol A (TBBPA) exposure in humans, animals and zebrafish [J]. Archives of Toxicology, 2020, 94 (1): 59-66.

[90] S.E. M M, R.A. M M, T.B. K J, et al. Respiratory and dermal exposure to organophosphorus flame retardants and tetrabromobisphenol A at five work environments [J]. Environmental Science & Technology, 2009, 43 (3): 941-947.

[91] 王爽, 路珍, 李斐, 等. 典型溴系阻燃剂四溴双酚 A 和十溴二苯乙烷的污染现状及毒理学研究进展 [J]. 生态毒理学报, 2020, 15 (6): 24-42.

[92] Liu K, Li J, Yan S, et al. A review of status of tetrabromobisphenol A (TBBPA) in China [J]. Chemosphere, 2016, 148 (Apr.): 8-20.

[93] 肖潇, 陈德翼, 梅俊, 等. 贵屿某电子垃圾拆解点附近大气颗粒物中氯代/溴代二噁英、四溴双酚 A 污染水平研究 [J]. 环境科学学报, 2012, 32 (5): 1142-1148.

[94] Carlsson S, Sjolin T, Ostman B. Flame retardants in indoor air at an electronics recycling plant and at other work environments [J]. Environ Science & Technology, 2001, 35 (3): 448-454.

[95] Barghi M, Shin E-S, Kim J-C, et al. Human exposure to HBCD and TBBPA via indoor dust in Korea: Estimation of external exposure and body burden [J]. Science of The Total Environment, 2017, 593–594: 593-594.

[96] Liu J, Ma S T, Lin M Q, et al. New mixed bromine/chlorine transformation products of tetrabromobisphenol A: Synthesis and identification in dust samples from an e-waste dismantling site [J]. Environmental Science & Technology, 2020, 54 (19): 12235-12244.

[97] Wang J X, Liu L L, Wang J F, et al. Distribution of metals and brominated flame retardants (BFRs) in sediments, soils and plants from an informal e-waste dismantling site, South China [J]. Environmental Science and Pollution Research, 2015, 22 (2): 1020-1033.

[98] Xiang G, Smab C, Yh D, et al. Mixed bromine/chlorine transformation products of tetrabromobisphenol A: Potential specific molecular markers in e-waste dismantling areas [J]. Journal of Hazardous Materials, 423: 127-126.

[99] Matsukami H, Tue N M, Suzuki G, et al. Flame retardant emission from e-waste recycling operation in northern Vietnam: Environmental occurrence of emerging organophosphorus esters used as alternatives for PBDEs [J]. Science of The Total Environment, 2015, 514: 492-499.

[100] Yang S W, Wang S R, Wu F C, et al. Tetrabromobisphenol A: Tissue distribution in fish, and seasonal variation in water and sediment of Lake Chaohu, China [J]. Environmental Science and Pollution Research, 2012, 19 (9): 4090-4096.

[101] Pierre L, Khawla T, Fabrice A, et al. Development of analytical procedures for trace-level determination of polybrominated diphenyl ethers and tetrabromobisphenol A in river water and sediment [J]. Analytical & Bioanalytical Chemistry, 2010, 396 (2): 865-875.

[102] 江田田, 朱丽岩, 韩萃, 等. 渤、黄海浮游动物对四溴双酚 A 生物富集的研究 [J]. 中国海洋大学学报 (自然科学版), 2018, 48 (5): 51-58.

[103] Liu H H, Hu Y J, Luo P, et al. Occurrence of halogenated flame retardants in sediment off an urbanized

coastal zone: Association with urbanization and industrialization [J]. Environmental Science & Technology, 2014, 48(15): 8465-8473.

[104] Oberg K, Warman K, Oberg T. Distribution and levels of brominated flame retardants in sewage sludge [J]. Chemosphere, 2002, 48(8): 805-809.

[105] Ashizuka Y, Nakagawa R, Hori T, et al. Determination of brominated flame retardants and brominated dioxins in fish collected from three regions of Japan [J]. Molecular Nutrition & Food Research, 2010, 52(2): 273-283.

[106] Barghi M, Shin E S, Choi S D, et al. HBCD and TBBPA in human scalp hair: Evidence of internal exposure [J]. Chemosphere, 2018, 207: 70-77.

[107] Tong F, Gu X Y, Gu C, et al. Insights into tetrabromobisphenol A adsorption onto soils: Effects of soil components and environmental factors [J]. Science of the Total Environment, 2015, 536: 582-588.

[108] Morf L S, Tremp J, Gloor R, et al. Brominated flame retardants in waste electrical and electronic equipment: substance flows in a recycling plant [J]. Environmental Science & Technology, 2005, 39(22): 8691-8699.

[109] Zhu Z C, Chen S J, Zheng J, et al. Occurrence of brominated flame retardants (BFRs), organochlorine pesticides (OCPs), and polychlorinated biphenyls (PCBs) in agricultural soils in a BFR-manufacturing region of north China [J]. Science of the Total Environment, 2014, 481: 47-54.

[110] Huang D, Zhao H, Liu C P, et al. Characteristics, sources, and transport of tetrabromobisphenol A and bisphenol A in soils from a typical e-waste recycling area in South China [J]. Environmental Science & Pollution Research, 2014, 21(9): 5818-5826.

[111] Xu T, Wang J, Liu S Z, et al. A highly sensitive and selective immunoassay for the detection of tetrabromobisphenol A in soil and sediment [J]. Analytica Chimica Acta, 2012, 751: 119-127.

[112] Ma Y L, Stubbings W A, Cline-Cole R, et al. Human exposure to halogenated and organophosphate flame retardants through informal e-waste handling activities — A critical review [J]. Environmental Pollution, 2021, 268: 115727.

[113] Chen S J, Tian M, Wang J, et al. Dechlorane Plus (DP) in air and plants at an electronic waste (e-waste) site in south China [J]. Environmental Pollution, 2011, 159(5): 1290-1296.

[114] Hong W J, Jia H L, Ding Y S, et al. Polychlorinated biphenyls (PCBs) and halogenated flame retardants (HFRs) in multi-matrices from an electronic waste (e-waste) recycling site in Northern China [J]. Journal of Material Cycles and Waste Management, 2018, 20(1): 80-90.

[115] Iqbal M, Syed J H, Breivik K, et al. E-waste driven pollution in Pakistan: The first evidence of environmental and human exposure to flame retardants (FRs) in Karachi City [J]. Environmental Science & Technology, 2017, 51(23): 13895-13905.

[116] Tian M, Chen S J, Wang J, et al. Atmospheric deposition of halogenated flame retardants at urban, e-waste, and rural locations in southern China [J]. Environmental Science & Technology, 2011, 45(11): 4696-4701.

[117] Wang J, Tian M, Chen S-J, et al. Dechlorane plus in house dust from e-waste recycling and urban areas in South China: Sources, degradation, and human exposure [J]. Environmental Toxicology and Chemistry, 2011, 30(9): 1965-1972.

[118] Zheng J, Wang J, Luo X J, et al. Dechlorane plus in human hair from an e-waste recycling area in south China: Comparison with dust [J]. Environmental Science & Technology, 2010, 44(24): 9298-9303.

[119] Yu Z Q, Lu S Y, Gao S T, et al. Levels and isomer profiles of dechlorane plus in the surface soils from e-waste recycling areas and industrial areas in south China [J]. Environmental Pollution, 2010, 158(9): 2920-2925.

[120] Xiao K, Wang P, Zhang H D, et al. Levels and profiles of dechlorane plus in a major e-waste dismantling area in China [J]. Environmental Geochemistry and Health, 2013, 35(5): 625-631.

[121] Tao W Q, Zhou Z G, Shen L, et al. Determination of dechlorane flame retardants in soil and fish at Guiyu, an electronic waste recycling site in south China [J]. Environmental Pollution, 2015, 206: 361-368.

[122] van den Eede N, Dirtu A C, Neels H, et al. Analytical developments and preliminary assessment of human

exposure to organophosphate flame retardants from indoor dust [J]. Environment International, 2011, 37(2): 454-461.

[123] Li H R, La Guardia M J, Liu H H, et al. Brominated and organophosphate flame retardants along a sediment transect encompassing the Guiyu, China e-waste recycling zone [J]. Science of the Total Environment, 2019, 646: 58-67.

[124] Wang Y, Yao Y M, Han X X, et al. Organophosphate di- and tri-esters in indoor and outdoor dust from China and its implications for human exposure [J]. Science of the Total Environment, 2020, 700: 134502.

[125] Wang Y, Kannan P, Halden R U, et al. A nationwide survey of 31 organophosphate esters in sewage sludge from the United States [J]. Science of the Total Environment, 2019, 655: 446-453.

[126] Ma Y X, Xie Z Y, Lohmann R, et al. Organophosphate ester flame retardants and plasticizers in ocean sediments from the North Pacific to the Arctic Ocean [J]. Environmental Science & Technology, 2017, 51(7): 3809-3815.

[127] Kim U J, Oh J K, Kannan K. Occurrence, removal, and environmental emission of organophosphate flame retardants/plasticizers in a wastewater treatment plant in New York State [J]. Environmental Science & Technology, 2017, 51(14): 7872-7880.

[128] Lee J S, Morita Y, Kawai Y K, et al. Developmental circulatory failure caused by metabolites of organophosphorus flame retardants in zebrafish, Danio rerio [J]. Chemosphere, 2020, 246: 125738.

[129] Chen X M, An H, Ao L, et al. The combined toxicity of dibutyl phthalate and benzo(a)pyrene on the reproductive system of male Sprague Dawley rats in vivo [J]. Journal of Hazardous Materials, 2011, 186(1): 835-841.

[130] Veen I V d, Boer J d. Phosphorus flame retardants: Properties, production, environmental occurrence, toxicity and analysis [J]. Chemosphere, 2012, 88(10): 1119-1153.

[131] Li W H, Wang Y, Kannan K. Occurrence, distribution and human exposure to 20 organophosphate esters in air, soil, pine needles, river water, and dust samples collected around an airport in New York state, United States [J]. Environment International, 2019, 131: 105054.

[132] He M J, Lu J F, Ma J Y, et al. Organophosphate esters and phthalate esters in human hair from rural and urban areas, Chongqing, China: Concentrations, composition profiles and sources in comparison to street dust [J]. Environmental Pollution, 2018, 237: 143-153.

[133] Du B B, Shen M J, Chen H, et al. Beyond traditional organophosphate triesters: Prevalence of emerging organophosphate triesters and organophosphate diesters in indoor dust from a Mega e-waste recycling industrial park in south China [J]. Environmental Science & Technology, 2020, 54(19): 12001-12012.

[134] Wu Y, Venier M, Salamova A. Spatioseasonal variations and partitioning behavior of organophosphate esters in the Great Lakes atmosphere [J]. Environmental Science & Technology, 2020, 54(9): 5400-5408.

[135] Bi X, Simoneit B, Wang Z, et al. The major components of particles emitted during recycling of waste printed circuit boards in a typical e-waste workshop of South China [J]. Atmospheric Environment, 2010, 44(35): 4440-4445.

[136] Wang T, Ding N, Wang T, et al. Organophosphorus esters (OPEs) in $PM_{2.5}$ in urban and e-waste recycling regions in southern China: Concentrations, sources, and emissions [J]. Environmental Research, 2018, 167: 437-444.

[137] Nguyen L V, Diamond M L, Venier M, et al. Exposure of Canadian electronic waste dismantlers to flame retardants [J]. Environment International, 2019, 129: 95-104.

[138] Liu D, Lin T, Shen K J, et al. Occurrence and concentrations of halogenated flame retardants in the atmospheric fine particles in Chinese cities [J]. Environmental Science & Technology, 2016, 50(18): 9846-9854.

[139] Yue C C, Ma S T, Liu R R, et al. Pollution profiles and human health risk assessment of atmospheric organophosphorus esters in an e-waste dismantling park and its surrounding area [J]. Science of the Total

Environment, 2022, 806: 10.

[140] 岳聪聪. 电子垃圾拆解区域典型大气 SVOCs 时空污染特征及其健康风险评估 [D]. 广州: 广东工业大学, 2021.

[141] 高小中, 许宜平, 王子健. 有机磷酸酯阻燃剂的环境暴露与迁移转化研究进展 [J]. 生态毒理学报, 2015, 10(2): 56-68.

[142] Wang Y, Sun H W, Zhu H K, et al. Occurrence and distribution of organophosphate flame retardants (OPFRs) in soil and outdoor settled dust from a multi-waste recycling area in China [J]. Science of the Total Environment, 2018, 625: 1056-1064.

[143] He C T, Zheng J, Qiao L, et al. Occurrence of organophosphorus flame retardants in indoor dust in multiple microenvironments of southern China and implications for human exposure [J]. Chemosphere, 2015, 133: 47-52.

[144] Stubbings W A, Nguyen L V, Romanak K, et al. Flame retardants and plasticizers in a Canadian waste electrical and electronic equipment (WEEE) dismantling facility [J]. Science of the Total Environment, 2019, 675: 594-603.

[145] Luna-Acosta A, Budzinski H, Le Menach K, et al. Persistent organic pollutants in a marine bivalve on the Marennes-Oleron Bay and the Gironde Estuary (French Atlantic Coast)-Part 1: Bioaccumulation [J]. Science of the Total Environment, 2015, 514: 500-510.

[146] Net S, Sempere R, Delmont A, et al. Occurrence, fate, behavior and ecotoxicological state of phthalates in different environmental matrices [J]. Environmental Science & Technology, 2015, 49(7): 4019-4035.

[147] Benning J L, Liu Z, Tiwari A, et al. Characterizing gas-particle interactions of phthalate plasticizer emitted from vinyl flooring [J]. Environmental Science & Technology, 2013, 47(6): 2696-2703.

[148] Gao D W, Li Z, Wang H, et al. An overview of phthalate acid ester pollution in China over the last decade: Environmental occurrence and human exposure [J]. Science of the Total Environment, 2018, 645: 1400-1409.

[149] Bu S B, Wang Y L, Wang H Y, et al. Analysis of global commonly-used phthalates and non-dietary exposure assessment in indoor environment [J]. Building and Environment, 2020, 177: 106853.

[150] Yang T, Wang H, Zhang X, et al. Characterization of phthalates in sink and source materials: Measurement methods and the impact on exposure assessment [J]. Journal of Hazardous Materials, 2020, 396: 122689.

[151] Tang Z, Chai M, Wang Y, et al. Phthalates in preschool children's clothing manufactured in seven Asian countries: Occurrence, profiles and potential health risks [J]. Journal of Hazardous Materials, 2020, 387: 121681.

[152] Weschler C J, Salthammer T, Fromme H. Partitioning of phthalates among the gas phase, airborne particles and settled dust in indoor environments [J]. Atmospheric Environment, 2008, 42(7): 1449-1460.

[153] Giovanoulis G, Alves A, Papadopoulou E, et al. Evaluation of exposure to phthalate esters and DINCH in urine and nails from a Norwegian study population [J]. Environmental Research, 2016, 151: 80-90.

[154] Bu Z, Zhang Y, Mmereki D, et al. Indoor phthalate concentration in residential apartments in Chongqing, China: Implications for preschool children's exposure and risk assessment [J]. Atmospheric Environment, 2016, 127: 34-45.

[155] Wang X, Tao W, Xu Y, et al. Indoor phthalate concentration and exposure in residential and office buildings in Xi'an, China [J]. Atmospheric Environment, 2014, 87: 146-152.

[156] Hammel S C, Levasseur J L, Hoffman K, et al. Children's exposure to phthalates and non-phthalate plasticizers in the home: The TESIE study [J]. Environment International, 2019, 132: 105061.

[157] Yang C Q, Harris S A, Jantunen L M, et al. Phthalates: Relationships between air, dust, electronic devices, and hands with implications for exposure [J]. Environmental Science & Technology, 2020, 54(13): 8186-8197.

[158] Gao C J, Wang F, Shen H M, et al. Feminine hygiene products—A neglected source of phthalate exposure in women [J]. Environmental Science & Technology, 2020, 54(2): 930-937.

[159] Lu S, Yang D, Ge X, et al. The internal exposure of phthalate metabolites and bisphenols in waste incineration plant workers and the associated health risks [J]. Environment International, 2020, 145: 106101.

[160] Gu Z P, Feng J L, Han W L, et al. Characteristics of organic matter in PM$_{2.5}$ from an e-waste dismantling area in Taizhou, China [J]. Chemosphere, 2010, 80(7): 800-806.

[161] Tang B, Christia C, Luo X J, et al. Changes in levels of legacy and emerging organophosphorus flame retardants and plasticizers in indoor dust from a former e-waste recycling area in south China: 2013—2017 [J]. Environmental Science and Pollution Research, 2022, 29(22): 33295-33304.

[162] Deng M, Han X, Ge J, et al. Prevalence of phthalate alternatives and monoesters alongside traditional phthalates in indoor dust from a typical e-waste recycling area: Source elucidation and co-exposure risk [J]. Journal of Hazardous Materials, 2021, 413: 125322.

[163] Muenhor D, Moon H B, Lee S, et al. Organophosphorus flame retardants (PFRs) and phthalates in floor and road dust from a manual e-waste dismantling facility and adjacent communities in Thailand [J]. Journal of Environmental Science and Health Part a-Toxic/Hazardous Substances & Environmental Engineering, 2018, 53(1): 79-90.

[164] Zhang S H, Guo A J, Fan T T, et al. Phthalates in residential and agricultural soils from an electronic waste-polluted region in south China: distribution, compositional profile and sources [J]. Environmental Science and Pollution Research, 2019, 26(12): 12227-12236.

[165] Liu W L, Shen C F, Zhang Z, et al. Distribution of phthalate esters in soil of e-waste recycling sites from Taizhou city in China [J]. Bulletin of Environmental Contamination and Toxicology, 2009, 82(6): 665-667.

[166] Liu S S, Peng Y F, Lin Q T, et al. Di-(2-Ethylhexyl) phthalate as a chemical indicator for phthalic acid esters: An investigation into phthalic acid esters in cultivated fields and e-waste Dismantling Sites [J]. Environmental Toxicology and Chemistry, 2019, 38(5): 1132-1141.

[167] Mukhopadhyay M, Sampath S, Munoz-Arnanz J, et al. Plasticizers and bisphenol A in Adyar and Cooum riverine sediments, India: occurrences, sources and risk assessment [J]. Environmental Geochemistry and Health, 2020, 42(9): 2789-2802.

[168] Chakraborty P, Sampath S, Mukhopadhyay M, et al. Baseline investigation on plasticizers, bisphenol A, polycyclic aromatic hydrocarbons and heavy metals in the surface soil of the informal electronic waste recycling workshops and nearby open dumpsites in Indian metropolitan cities [J]. Environmental Pollution, 2019, 248: 1036-1045.

[169] Laxman K, Kumar A, Ravikanth M. Polycyclic aromatic hydrocarbon-/heterocycle-embedded porphyrinoids [J]. Asian Journal of Organic Chemistry, 2020, 9(2): 162-180.

[170] Akyuz M, Cabuk H. Gas-particle partitioning and seasonal variation of polycyclic aromatic hydrocarbons in the atmosphere of Zonguldak, Turkey [J]. Science of the Total Environment, 2010, 408(22): 5550-5558.

[171] Keith L H. The source of U.S. EPA's sixteen PAH priority pollutants [J]. Polycyclic Aromatic Compounds, 2015, 35(2): 147-160.

[172] Mandalakis M, Gustafsson O, Reddy C M, et al. Radiocarbon apportionment of fossil versus biofuel combustion sources of polycyclic aromatic hydrocarbons in the Stockholm metropolitan area [J]. Environmental Science & Technology, 2004, 38(20): 5344-5349.

[173] Tan L, Wang N, Dong Y, et al. Characterization of 56 airborne persistent organic pollutants (POPs) in gas-phase and particle-phase [J]. Environmental Forensics, 2022, 23(1-2): 208-220.

[174] Zhang Y X, Tao S. Global atmospheric emission inventory of polycyclic aromatic hydrocarbons (PAHs) for 2004 [J]. Atmospheric Environment, 2009, 43(4): 812-819.

[175] Li X, Li P F, Yan L L, et al. Characterization of polycyclic aromatic hydrocarbons in fog-rain events [J]. Journal of Environmental Monitoring, 2011, 13(11): 2988-2993.

[176] Font R, Moltó J, Egea S, et al. Thermogravimetric kinetic analysis and pollutant evolution during the pyrolysis and combustion of mobile phone case [J]. Chemosphere, 2011, 85(3): 516-524.

[177] Bi X H, Sheng G Y, Peng P A, et al. Extractable organic matter in PM$_{10}$ from LiWan district of Guangzhou city, PR China [J]. Science of the Total Environment, 2002, 300(1-3): 213-228.

[178] Wei S L, Huang B, Liu M, et al. Characterization of PM$_{2.5}$-bound nitrated and oxygenated PAHs in two industrial sites of South China [J]. Atmospheric Research, 2012, 109: 76-83.

[179] Ma S T, Lin M Q, Tang J, et al. Occurrence and fate of polycyclic aromatic hydrocarbons from electronic waste dismantling activities: A critical review from environmental pollution to human health [J]. Journal of Hazardous Materials, 2022, 424: 127683.

[180] Liu R R, Ma S T, Yu Y Y, et al. Field study of PAHs with their derivatives emitted from e-waste dismantling processes and their comprehensive human exposure implications [J]. Environment International, 2020, 144: 106059.

[181] Chen J Y, Zhang D L, Li G Y, et al. The health risk attenuation by simultaneous elimination of atmospheric VOCs and POPs from an e-waste dismantling workshop by an integrated de-dusting with decontamination technique [J]. Chemical Engineering Journal, 2016, 301: 299-305.

[182] Goriaux M, Jourdain B, Temime B, et al. Field comparison of particulate PAH measurements using a low-flow denuder device and conventional sampling systems [J]. Environmental Science & Technology, 2006, 40(20): 6398-6404.

[183] Tsai P J, Shieh H Y, Lee W J, et al. Health-risk assessment for workers exposed to polycyclic aromatic hydrocarbons (PAHs) in a carbon black manufacturing industry [J]. Science of the Total Environment, 2001, 278(1-3): 137-150.

[184] Tsai P J, Shih T S, Chen H L, et al. Assessing and predicting the exposures of polycyclic aromatic hydrocarbons (PAHs) and their carcinogenic potencies from vehicle engine exhausts to highway toll station workers [J]. Atmospheric Environment, 2004, 38(2): 333-343.

[185] Maskaoui K, Hu Z. Contamination and ecotoxicology risks of polycyclic aromatic hydrocarbons in Shantou coastal waters, China [J]. Bulletin of Environmental Contamination and Toxicology, 2009, 82(2): 172-178.

[186] Huang B, Liu M, Bi X, et al. Phase distribution, sources and risk assessment of PAHs, NPAHs and OPAHs in a rural site of Pearl River Delta region, China [J]. Atmospheric Pollution Research, 2014, 5(2): 210-218.

[187] Liu Y, Wang S, Lohmann R, et al. Source apportionment of gaseous and particulate PAHs from traffic emission using tunnel measurements in Shanghai, China [J]. Atmospheric Environment, 2015, 107: 129-136.

[188] Nguyen T N T, Jung K-S, Son J M, et al. Seasonal variation, phase distribution, and source identification of atmospheric polycyclic aromatic hydrocarbons at a semi-rural site in Ulsan, South Korea [J]. Environmental Pollution, 2018, 236: 529-539.

[189] Liu J, Man R, Ma S, et al. Atmospheric levels and health risk of polycyclic aromatic hydrocarbons (PAHs) bound to PM$_{2.5}$ in Guangzhou, China [J]. Marine Pollution Bulletin, 2015, 100(1): 134-143.

[190] Cai C Y, Yu S Y, Li X Y, et al. Emission characteristics of polycyclic aromatic hydrocarbons from pyrolytic processing during dismantling of electronic wastes [J]. Journal of Hazardous Materials, 2018, 351: 270-276.

[191] Keyte I J, Albinet A, Harrison R M. On-road traffic emissions of polycyclic aromatic hydrocarbons and their oxy- and nitro-derivative compounds measured in road tunnel environments [J]. Science of the Total Environment, 2016, 566: 1131-1142.

[192] Lin Y, Ma Y Q, Qiu X H, et al. Sources, transformation, and health implications of PAHs and their nitrated, hydroxylated, and oxygenated derivatives in PM$_{2.5}$ in Beijing [J]. Journal of Geophysical Research-Atmospheres, 2015, 120(14): 7219-7228.

[193] Ross A B, Bartle K D, Hall S, et al. Formation and emission of polycyclic aromatic hydrocarbon soot precursors during coal combustion [J]. Journal of the Energy Institute, 2011, 84(4): 220-226.

[194] Williams A, Jones J M, Ma L, et al. Pollutants from the combustion of solid biomass fuels [J]. Progress in Energy and Combustion Science, 2012, 38(2): 113-137.

[195] Vejerano E P, Holder A L, Marr L C. Emissions of polycyclic aromatic hydrocarbons, polychlorinated

dibenzo-*p*-dioxins, and dibenzofurans from incineration of nanomaterials [J]. Environmental Science & Technology, 2013, 47(9): 4866-4874.

[196] Iniguez M E, Conesa J A, Fullana A. Effect of sodium chloride and thiourea on pollutant formation during combustion of plastics [J]. Energies, 2018, 11(8): 2014.

[197] Wijayanta A T, Alam M S, Nakaso K, et al. Optimized combustion of biomass volatiles by varying O_2 and CO_2 levels: A numerical simulation using a highly detailed soot formation reaction mechanism [J]. Bioresource Technology, 2012, 110: 645-651.

[198] Shen G F, Tao S, Wei S Y, et al. Field measurement of emission factors of PM, EC, OC, Parent, Nitro-, and Oxy- Polycyclic aromatic hydrocarbons for residential briquette, coal cake, and wood in rural Shanxi, China [J]. Environmental Science & Technology, 2013, 47(6): 2998-3005.

[199] Yang X Y, Igarashi K, Tang N, et al. Indirect- and direct-acting mutagenicity of diesel, coal and wood burning-derived particulates and contribution of polycyclic aromatic hydrocarbons and nitropolycyclic aromatic hydrocarbons [J]. Mutation Research—Genetic Toxicology and Environmental Mutagenesis, 2010, 695(1-2): 29-34.

[200] Li X L, Zheng Y, Guan C, et al. Effect of biodiesel on PAH, OPAH, and NPAH emissions from a direct injection diesel engine [J]. Environmental Science and Pollution Research, 2018, 25(34): 34131-34138.

[201] Zhou S, Zhou J X, Zhu Y Q. Chemical composition and size distribution of particulate matters from marine diesel engines with different fuel oils [J]. Fuel, 2019, 235: 972-983.

[202] Mu L, Peng L, Liu X, et al. Characteristics of polycyclic aromatic hydrocarbons and their gas/particle partitioning from fugitive emissions in coke plants [J]. Atmospheric Environment, 2014, 83: 202-210.

[203] Yu X Z, Gao Y, Wu S C, et al. Distribution of polycyclic aromatic hydrocarbons in soils at Guiyu area of China, affected by recycling of electronic waste using primitive technologies [J]. Chemosphere, 2006, 65(9): 1500-1509.

[204] Shen C F, Chen Y X, Huang S B, et al. Dioxin-like compounds in agricultural soils near e-waste recycling sites from Taizhou area, China: Chemical and bioanalytical characterization [J]. Environment International, 2009, 35(1): 50-55.

[205] Tang X J, Shen C F, Chen L, et al. Inorganic and organic pollution in agricultural soil from an emerging e-waste recycling town in Taizhou area, China [J]. Journal of Soils and Sediments, 2010, 10(5): 895-906.

[206] Lopez B N, Man Y B, Zhao Y G, et al. Major pollutants in soils of abandoned agricultural land contaminated by e-waste activities in Hong Kong [J]. Archives of Environmental Contamination and Toxicology, 2011, 61(1): 101-114.

[207] Xu P J, Tao B, Ye Z Q, et al. Polycyclic aromatic hydrocarbon concentrations, compositions, sources, and associated carcinogenic risks to humans in farmland soils and riverine sediments from Guiyu, China [J]. Journal of Environmental Sciences, 2016, 48: 102-111.

[208] Huang D Y, Liu C P, Li F B, et al. Profiles, sources, and transport of polycyclic aromatic hydrocarbons in soils affected by electronic waste recycling in Longtang, south China [J]. Environmental Monitoring and Assessment, 2014, 186(6): 3351-3364.

[209] He M J, Yang S Y, Zhao J, et al. Reduction in the exposure risk of farmer from e-waste recycling site following environmental policy adjustment: A regional scale view of PAHs in paddy fields [J]. Environment International, 2019, 133: 105136.

[210] Gao Y Y, Wang Y Y, Zhou Q X. Distribution and temporal variation of PCBs and PAHs in soils and sediments from an e-waste dismantling site in China [J]. Environmental Earth Sciences, 2015, 74(4): 2925-2935.

[211] Luo J, Qi S H, Xie X M, et al. The assessment of source attribution of soil pollution in a typical e-waste recycling town and its surrounding regions using the combined organic and inorganic dataset [J]. Environmental Science and Pollution Research, 2017, 24(3): 3131-3141.

[212] Gu W H, Bai J F, Yuan W Y, et al. Pollution analysis of soil polycyclic aromatic hydrocarbons from informal electronic waste dismantling areas in Xinqiao, China [J]. Waste Management & Research, 2019, 37(4): 394-401.

[213] Jiang L F, Cheng Z N, Zhang D Y, et al. The influence of e-waste recycling on the molecular ecological network of soil microbial communities in Pakistan and China [J]. Environmental Pollution, 2017, 231: 173-181.

[214] Daso A P, Akortia E, Okonkwo J O. Concentration profiles, source apportionment and risk assessment of polycyclic aromatic hydrocarbons (PAHs) in dumpsite soils from Agbogbloshie e-waste dismantling site, Accra, Ghana [J]. Environmental Science and Pollution Research, 2016, 23(11): 10883-10894.

[215] Hoa N T Q, Anh H Q, Tue N M, et al. Soil and sediment contamination by unsubstituted and methylated polycyclic aromatic hydrocarbons in an informal e-waste recycling area, northern Vietnam: Occurrence, source apportionment, and risk assessment [J]. Science of the Total Environment, 2020, 709: 135852.

[216] Chen L, Yu C N, Shen C F, et al. Study on adverse impact of e-waste disassembly on surface sediment in East China by chemical analysis and bioassays [J]. Journal of Soils and Sediments, 2010, 10(3): 359-367.

[217] Liu J, Chen X, Shu H Y, et al. Microbial community structure and function in sediments from e-waste contaminated rivers at Guiyu area of China [J]. Environmental Pollution, 2018, 235: 171-179.

[218] Ren H, Su P, Kang W, et al. Heterologous spatial distribution of soil polycyclic aromatic hydrocarbons and the primary influencing factors in three industrial parks [J]. Environmental Pollution, 2022, 310: 119912.

[219] Huang H, Cai C, Yu S, et al. Emission behaviors of nitro- and oxy-polycyclic aromatic hydrocarbons during pyrolytic disposal of electronic wastes [J]. Chemosphere, 2019, 222: 267-274.

[220] Yang C, Su Y, Chen Y, et al. Emission factors of particulate matter and polycyclic aromatic hydrocarbons for residential solid fuels [J]. Asian Journal of Ecotoxicology, 2014, 9: 545-555.

[221] Ma S T, Chen H J, Yue C C, et al. Atmospheric occurrences of nitrated and hydroxylated polycyclic aromatic hydrocarbons from typical e-waste dismantling sites [J]. Environmental Pollution, 2022, 308: 119713.

[222] Lafontaine S, Schrlau J, Butler J, et al. Relative influence of trans-pacific and regional atmospheric transport of PAHs in the Pacific Northwest, US [J]. Environmental Science & Technology, 2015, 49(23): 13807-13816.

[223] Hayakawa K, Tang N, Nagato E G, et al. Long term trends in atmospheric concentrations of polycyclic aromatic hydrocarbons and nitropolycyclic aromatic hydrocarbons: A study of Japanese cities from 1997 to 2014 [J]. Environmental Pollution, 2018, 233: 474-482.

[224] Bandowe B A M, Meusel H, Huang R-j, et al. $PM_{2.5}$-bound oxygenated PAHs, nitro-PAHs and parent-PAHs from the atmosphere of a Chinese megacity: Seasonal variation, sources and cancer risk assessment [J]. Science of The Total Environment, 2014, 473-474: 77-87.

[225] Yu Q Q, Yang W Q, Zhu M, et al. Ambient $PM_{2.5}$-bound polycyclic aromatic hydrocarbons (PAHs) in rural Beijing: Unabated with enhanced temporary emission control during the 2014 APEC summit and largely aggravated after the start of wintertime heating [J]. Environmental Pollution, 2018, 238: 532-542.

[226] Simoneit B R T, Bi X H, Oros D R, et al. Phenols and Hydroxy-PAHs (Arylphenols) as tracers for coal smoke particulate matter: Source tests and ambient aerosol assessments [J]. Environmental Science & Technology, 2007, 41(21): 7294-7302.

[227] Huang W, Huang B, Bi X H, et al. Emission of PAHs, NPAHs and OPAHs from residential honeycomb coal briquette combustion [J]. Energy & Fuels, 2014, 28(1): 636-642.

[228] Cao X Y, Hao X W, Shen X B, et al. Emission characteristics of polycyclic aromatic hydrocarbons and nitro-polycyclic aromatic hydrocarbons from diesel trucks based on on-road measurements [J]. Atmospheric Environment, 2017, 148: 190-196.

[229] Jariyasopit N, McIntosh M, Zimmermann K, et al. Novel nitro-PAH formation from heterogeneous reactions of PAHs with NO_2, NO_3/N_2O_5, and OH radicals: Prediction, laboratory studies, and mutagenicity [J]. Environmental Science & Technology, 2014, 48(1): 412-419.

[230] Jariyasopit N, Zimmermann K, Schrlau J, et al. Heterogeneous reactions of particulate matter-bound PAHs and NPAHs with NO_3/N_2O_5, OH radicals, and O_3 under simulated long-range atmospheric transport conditions: Reactivity and mutagenicity [J]. Environmental Science & Technology, 2014, 48(17): 10155-10164.

[231] Zimmermann K, Jariyasopit N, Simonich S L M, et al. Formation of nitro-PAHs from the heterogeneous reaction of ambient particle-bound PAHs with $N_2O_5/NO_3/NO_2$ [J]. Environmental Science & Technology, 2013, 47(15): 8434-8442.

[232] Wada M, Kido H, Kishikawa N, et al. Assessment of air pollution in Nagasaki city: Determination of polycyclic aromatic hydrocarbons and their nitrated derivatives, and some metals [J]. Environmental Pollution, 2001, 115(1): 139-147.

[233] Shen G F, Tao S, Wei S Y, et al. Emissions of parent, nitro, and oxygenated polycyclic aromatic hydrocarbons from residential wood combustion in rural China [J]. Environmental Science & Technology, 2012, 46(15): 8123-8130.

[234] Arey J, Zielinska B, Atkinson R, et al. Nitroarene products from the gas-phase reactions of volatile polycyclic aromatic hydrocarbons with the OH radical and N_2O_5 [J]. International Journal of Chemical Kinetics, 1989, 21(9): 775-799.

[235] Atkinson R, Arey J, Zielinska B, et al. Kinetics and nitro-products of the gas-phase OH and NO_3 radical-initiated reactions of naphthalene-d8, Fluoranthene-d10, and pyrene [J]. International Journal of Chemical Kinetics, 1990, 22(9): 999-1014.

[236] Alves C A, Vicente A M P, Gomes J, et al. Polycyclic aromatic hydrocarbons (PAHs) and their derivatives (oxygenated-PAHs, nitrated-PAHs and azaarenes) in size-fractionated particles emitted in an urban road tunnel [J]. Atmospheric Research, 2016, 180: 128-137.

[237] Avagyan R, Nystrom R, Lindgren R, et al. Particulate hydroxy-PAH emissions from a residential wood log stove using different fuels and burning conditions [J]. Atmospheric Environment, 2016, 140: 1-9.

[238] Miet K, Le Menach K, Flaud P M, et al. Heterogeneous reactions of ozone with pyrene, 1-hydroxypyrene and 1-nitropyrene adsorbed on particles [J]. Atmospheric Environment, 2009, 43(24): 3699-3707.

[239] Zimmermann K, Atkinson R, Arey J, et al. Isomer distributions of molecular weight 247 and 273 nitro-PAHs in ambient samples, NIST diesel SRM, and from radical-initiated chamber reactions [J]. Atmospheric Environment, 2012, 55: 431-439.

[240] Vandenbrakenvanleersum A M, Tintel C, Vantzelfde M, et al. Spectroscopic and photochemical properties of mononitropyrenes [J]. Recueil Des Travaux Chimiques Des Pays-Bas-Journal of the Royal Netherlands Chemical Society, 1987, 106(4): 120-128.

[241] Vicente E D, Vicente A M, Bandowe B A M, et al. Particulate phase emission of parent polycyclic aromatic hydrocarbons (PAHs) and their derivatives (alkyl-PAHs, oxygenated-PAHs, azaarenes and nitrated PAHs) from manually and automatically fired combustion appliances [J]. Air Quality Atmosphere and Health, 2016, 9(6): 653-668.

[242] Cochran R E, Jeong H, Haddadi S, et al. Identification of products formed during the heterogeneous nitration and ozonation of polycyclic aromatic hydrocarbons [J]. Atmospheric Environment, 2016, 128: 92-103.

[243] Lee J, Lane D A. Formation of oxidized products from the reaction of gaseous phenanthrene with the OH radical in a reaction chamber [J]. Atmospheric Environment, 2010, 44(20): 2469-2477.

[244] Zhao J B, Zhang J, Sun L N, et al. Characterization of $PM_{2.5}$-bound nitrated and oxygenated polycyclic aromatic hydrocarbons in ambient air of Langfang during periods with and without traffic restriction [J]. Atmospheric Research, 2018, 213: 302-308.

[245] de Jesus R M, Mosca A C, Guarieiro A L N, et al. *In vitro* evaluation of oxidative stress caused by fine particles ($PM_{2.5}$) exhausted from heavy-duty vehicles using diesel/biodiesel blends under real world

conditions [J]. Journal of the Brazilian Chemical Society, 2018, 29(6): 1268-1277.

[246] Yang X Y, Liu S J, Xu Y S, et al. Emission factors of polycyclic and nitro-polycyclic aromatic hydrocarbons from residential combustion of coal and crop residue pellets [J]. Environmental Pollution, 2017, 231: 1265-1273.

[247] Ma J, Horii Y, Cheng J P, et al. Chlorinated and parent polycyclic aromatic hydrocarbons in environmental samples from an electronic waste recycling facility and a chemical industrial complex in China [J]. Environmental Science & Technology, 2009, 43(3): 643-649.

[248] Tue N M, Goto A, Takahashi S, et al. Soil contamination by halogenated polycyclic aromatic hydrocarbons from open burning of e-waste in Agbogbloshie (Accra, Ghana) [J]. Journal of Material Cycles and Waste Management, 2017, 19(4): 1324-1332.

[249] Wang Q, Miyake Y, Amagai T, et al. Halogenated polycyclic aromatic hydrocarbons in soil and river sediment from e-waste recycling sites in Vietnam [J]. Journal of Water and Environment Technology, 2016, 14(3): 166-176.

[250] Nishimura C, Horii Y, Tanaka S, et al. Occurrence, profiles, and toxic equivalents of chlorinated and brominated polycyclic aromatic hydrocarbons in e-waste open burning soils [J]. Environmental Pollution, 2017, 225: 252-260.

[251] Tang J, Ma S T, Liu R R, et al. The pollution profiles and human exposure risks of chlorinated and brominated PAHs in indoor dusts from e-waste dismantling workshops: Comparison of GC-MS, GC-MS/MS and GC x GC-MS/MS determination methods [J]. Journal of Hazardous Materials, 2020, 394: 122573.

[252] Jin R, Liu G R, Zheng M H, et al. Secondary copper smelters as sources of chlorinated and brominated polycyclic aromatic hydrocarbons [J]. Environmental Science & Technology, 2017, 51(14): 7945-7953.

[253] Horii Y, Ok G, Ohura T, et al. Occurrence and profiles of chlorinated and brominated polycyclic aromatic hydrocarbons in waste incinerators [J]. Environmental Science & Technology, 2008, 42(6): 1904-1909.

[254] Wang Y, Su P, Ge X, et al. Identification of specific halogenated polycyclic aromatic hydrocarbons in surface soils of petrochemical, flame retardant, and electronic waste dismantling industrial parks [J]. Journal of Hazardous Materials, 2022, 436: 129160.

[255] WHO. Environmental Health Criteria 205: Polybrominated dibenzo-*p*-dioxins and dibenzofurans [Z]. Geneva; Environmental Health Criteria 205. 1998.

[256] Fiedler H. Sources of PCDD/PCDF and impact on the environment [J]. Chemosphere, 1996, 32(1): 55-64.

[257] WHO. Environmental Health Criteria 88: Polychlorinated dibenso-*para*-dioxins and dibenzofurans [Z]. Geneva. 1989

[258] Liang J H, Lu G N, Wang R, et al. The formation pathways of polybrominated dibenzo-*p*-dioxins and dibenzofurans (PBDD/Fs) from pyrolysis of polybrominated diphenyl ethers (PBDEs): Effects of bromination arrangement and level [J]. Journal of Hazardous Materials, 2020, 399: 123004.

[259] Shen X J, Yang Q T, Shen J, et al. Characterizing the emissions of polybrominated dibenzo-*p*-dioxins and dibenzofurans (PBDD/Fs) from electric arc furnaces during steel-making [J]. Ecotoxicology and Environmental Safety, 2021, 208: 111722.

[260] Zhou Z-G, Zhao B, Qi L, et al. Level and congener profiles of polybrominated dibenzo-p-dioxins and dibenzofurans in the atmosphere of Beijing, China [J]. Atmospheric Environment, 2014, 95: 225-230.

[261] Zhang T, Huang Y R, Chen S J, et al. PCDD/Fs, PBDD/Fs, and PBDEs in the air of an e-waste recycling area (Taizhou) in China: Current levels, composition profiles, and potential cancer risks [J]. Journal of Environmental Monitoring, 2012, 14(12): 3156-3163.

[262] Kimura E, Suzuki G, Uramaru N, et al. Behavioral impairments in infant and adult mouse offspring exposed to 2,3,7,8-tetrabromodibenzofuran *in utero* and *via* lactation [J]. Environment International, 2020, 142: 105833.

[263] Frawley R, DeVito M, Walker N J, et al. Relative potency for altered humoral immunity induced by polybrominated and polychlorinated dioxins/furans in female B6C3F1/N mice [J]. Toxicological Sciences, 2014, 139(2): 488-500.

[264] Yang L, Liu G, Shen J, et al. Environmental characteristics and formations of polybrominated dibenzo-p-dioxins and dibenzofurans [J]. Environ Int, 2021, 152: 106450.

[265] Hanari N, Kannan K, Miyake Y, et al. Occurrence of polybrominated biphenyls, polybrominated dibenzo-p-dioxins, and polybrominated dibenzofurans as impurities in commercial polybrominated diphenyl ether mixtures [J]. Environmental Science & Technology, 2006, 40(14): 4400-4405.

[266] Sakai S, Watanabe J, Honda Y, et al. Combustion of brominated flame retardants and behavior of its byproducts [J]. Chemosphere, 2001, 42(5-7): 519-531.

[267] Ren M, Peng P A, Cai Y, et al. PBDD/F impurities in some commercial deca-BDE [J]. Environmental Pollution, 2011, 159(5): 1375-1380.

[268] Ma J, Addink R, Yun S H, et al. Polybrominated dibenzo-p-dioxins/dibenzofurans and polybrominated diphenyl ethers in soil, vegetation, workshop-floor dust, and electronic shredder residue from an electronic waste recycling facility and in soils from a chemical industrial complex in eastern China [J]. Environmental Science & Technology, 2009, 43(19): 7350-7356.

[269] Tue N M, Suzuki G, Takahashi S, et al. Evaluation of dioxin-like activities in settled house dust from Vietnamese e-waste recycling sites: Relevance of polychlorinated/brominated dibenzo-p-dioxin/furans and dioxin-like PCBs [J]. Environmental Science & Technology, 2010, 44(23): 9195-9200.

[270] Xiao X, Hu J F, Peng P A, et al. Characterization of polybrominated dibenzo-p-dioxins and dibenzofurans (PBDDs/Fs) in environmental matrices from an intensive electronic waste recycling site, south China [J]. Environmental Pollution, 2016, 212: 464-471.

[271] Goto A, Tue N M, Someya M, et al. Spatio-temporal trends of polybrominated dibenzo-p-dioxins and dibenzofurans in archived sediments from Tokyo Bay, Japan [J]. Science of the Total Environment, 2017, 599: 340-347.

[272] Zhou L, Li H, Yu Z, et al. Chlorinated and brominated dibenzo-p-dioxins and dibenzofurans in surface sediment from Taihu Lake, China [J]. Journal of Environmental Monitoring, 2012, 14(7): 1935-1942.

[273] Venkatesan A K, Halden R U. Contribution of polybrominated dibenzo-p-dioxins and dibenzofurans (PBDD/Fs) to the toxic equivalency of dioxin-like compounds in archived biosolids from the U.S. EPA's 2001 national sewage sludge survey [J]. Environmental Science and Technology, 2014, 48(18): 10843-10849.

[274] Falandysz J, Smith F, Fernandes A R. Polybrominated dibenzo-p-dioxins (PBDDs) and - dibenzofurans (PBDFs) in cod (Gadus morhua) liver-derived products from 1972 to 2017 [J]. Science of the Total Environment, 2020, 722: 137840.

[275] Zacs D, Rjabova J, Viksna A, et al. Method development for the simultaneous determination of polybrominated, polychlorinated, mixed polybrominated/chlorinated dibenzo-p-dioxins and dibenzofurans, polychlorinated biphenyls and polybrominated diphenyl ethers in fish [J]. Chemosphere, 2015, 118: 72-80.

[276] Chen M W, Castillo B A A, Lin D Y, et al. Levels of PCDD/Fs, PBDEs, and PBDD/Fs in breast milk from southern Taiwan [J]. Bulletin of Environmental Contamination and Toxicology, 2018, 100(3): 369-375.

[277] Fromme H, Hilger B, Albrecht M, et al. Occurrence of chlorinated and brominated dioxins/furans, PCBs, and brominated flame retardants in blood of German adults [J]. International Journal of Hygiene and Environmental Health, 2016, 219(4-5): 380-388.

[278] Zhang M M, Buekens A, Li X D. Brominated flame retardants and the formation of dioxins and furans in fires and combustion [J]. Journal of Hazardous Materials, 2016, 304: 26-39.

[279] Tue N M, Goto A, Takahashi S, et al. Release of chlorinated, brominated and mixed halogenated dioxin-related compounds to soils from open burning of e-waste in Agbogbloshie (Accra, Ghana) [J]. Journal of Hazardous Materials, 2016, 302: 151-157.

[280] Gullett B K, Linak W P, Touati A, et al. Characterization of air emissions and residual ash from open burning of electronic wastes during simulated rudimentary recycling operations [J]. Journal of Material Cycles and Waste Management, 2007, 9(1): 69-79.

[281] Duan H B, Li J H, Liu Y C, et al. Characterization and inventory of PCDD/Fs and PBDD/Fs emissions from the incineration of waste printed circuit board [J]. Environmental Science & Technology, 2011, 45(15): 6322-6328.

[282] Sepúlveda A, Schluep M, Renaud F G, et al. A review of the environmental fate and effects of hazardous substances released from electrical and electronic equipments during recycling: Examples from China and India [J]. Environmental Impact Assessment Review, 2010, 30(1): 28-41.

[283] Li H R, Yu L P, Sheng G Y, et al. Severe PCDD/F and PBDD/F pollution in air around an electronic waste dismantling area in China [J]. Environmental Science & Technology, 2007, 41(16): 5641-5646.

[284] Wang L C, Chang-Chien G P. Characterizing the emissions of polybrominated dibenzo-*p*-dioxins and dibenzofurans from municipal and industrial waste incinerators [J]. Environmental Science & Technology, 2007, 41(4): 1159-1165.

[285] Wang M, Liu G R, Jiang X X, et al. Formation and emission of brominated dioxins and furans during secondary aluminum smelting processes [J]. Chemosphere, 2016, 146: 60-67.

第 5 章　电子垃圾污染的人体暴露特征及其风险

　　电子垃圾拆解处理过程中排放的污染物容易通过大气呼吸、膳食摄入和皮肤接触等途径进入人体内,从而对电子垃圾拆解从业人员和拆解区域附近的居民造成潜在的健康危害。人体生物监测是了解污染物在人体内浓度水平和分布特征,评估污染物对人体的暴露负荷及其潜在健康风险的常用手段。本章主要介绍电子垃圾拆解产生污染物经不同暴露途径对特定人群造成的健康风险,阐述了人体样品(血液、尿液、头发、指甲等)中持久性有机物检测分析的预处理方法,以及污染物进入人体后在不同组织中的赋存特征。

5.1　呼吸暴露及风险

5.1.1　车间工人的暴露

　　人们对污染物的暴露风险主要来自食物的消化吸收、呼吸空气或颗粒物以及皮肤接触污染。而对于在电子垃圾拆解车间环境内工作的人群,呼吸暴露是人体暴露的重要途径之一。对于车间工人的呼吸暴露,Cahill 等[1]的报告称,美国职业工人的多溴联苯醚(PBDEs)吸入暴露量(男性和女性分别为 38 ng/(kg·d) 和 40 ng/(kg·d))低于最低参考剂量(RfD),不会对工人构成威胁。Guo 等[2]指出,上海四个电子垃圾回收车间工人的 PBDEs 吸入暴露量从 2.3 ng/(kg·d) 到 782 ng/(kg·d) 不等,不戴口罩工人的 BDE-47 暴露量(118 ng/(kg·d))超过了参考剂量(100 ng/(kg·d))。Zhang 等[3]报告称,不同电子垃圾家庭作坊车间的工人 PBDEs 的暴露量为 4~26.3 ng/(kg·d),其中在印刷线路板处理车间中,BDE-47 和 BDE-99 对工人造成了主要的职业暴露风险,而在处理其他类型的作坊车间中的 BDE-209 是工人面临风险的主要因素。Die 等的研究[4]表明,中国正规电子垃圾回收设施的 PBDEs 暴露水平处于安全水平,比美国环保署参考剂量低一个数量级。在台州电子垃圾处理车间的工人吸入气相和颗粒相的总多氯联苯(PCBs)平均暴露量估计为 162 ng/d,显著高于相应的台州和临安的居民区的平均暴露量[5]。Liu 等[6]在研究中评估了电子垃圾拆解区的五个车间,包括电吹风处理手机车间(EBMP)、电烤锡炉处理电视机车间(EHFTV)和电烤锡炉处理路由器车间(EHFR),以及旋转灰化炉处理电视机车间(RITV)和旋转灰化炉处理硬盘车间(RIHD)中 PBDEs 和 PCBs 的非致癌危害

指数 (HI)。在 EHFTV 车间中 PBDEs 和 PCBs 总的吸入 HI 值为 23.7，其非致癌风险最高。这一水平分别是 EBMP(0.03)、EHFR(0.1)、RITV(1.32) 和 RIHD(0.22) 车间的 790、158、18 和 108 倍。在 EHFTV 和 RITV 车间中，只有 PBDEs 的吸入 HI 值超过非致癌风险警戒值 1.0，表明这些车间内的 PBDEs 对拆解工人构成了非癌症风险，其他三个车间的非癌风险均低于 1.0；然而，EHFR 和 RIHD 车间中 PBDEs 的 HI 均超过了潜在风险水平 (≥ 0.1)。BDE-47 和 BDE-99 是两种主要的危险污染物，在 EHFTV 车间的 HI 值分别为 10.1 和 5.9，在 RITV 车间的值分别为 0.55 和 0.36。此外，RITV(0.16) 和 EHFTV(0.29) 车间中的 PCBs 也构成潜在风险。结果表明，电视机回收车间是车间工人 PBDEs 和 PCBs 吸入暴露量最高的场所，车间工人在这些场所面临职业暴露的高风险。Liu 等同时也对车间工人吸入 PBDEs 和 PCBs 的终身致癌风险 (LCR) 进行了评估。BDE-209 通过吸入的 LCR 在所有车间中均低于可接受的风险水平 (1.0×10^{-6})。通过吸入的总 PCBs 导致的最高 LCR 是在 EHTTV 车间 (3.9×10^{-6})，这种风险分别是 RITV (2.1×10^{-6}) 车间和 RIHD (2.1×10^{-7}) 车间风险的大约 2 倍和 19 倍。处理电视机的车间通过吸入途径的总 PCBs 的 LCR 高于致癌物的可接受 LCR (1.0×10^{-6})。在 EHFTV 车间中，癌症风险的最主要贡献者是五氯联苯 (3.3×10^{-6})，占总 LCR 的 85%。此外，两种重要的危险 PCBs 是 PCB-82 和 PCB-118，LCR 分别为 1.2×10^{-6} 和 1.2×10^{-6}，表明这两种污染物可能对 EHFTV 车间的拆解工人构成癌症威胁。Liu 等[7]的研究中指出，PAHs 及其衍生物吸入途径的癌症风险占男性和女性工人总风险的 99%以上，表明空气吸入是电子垃圾回收工人接触污染物的主要途径。在 EHFR 车间观察到的吸入风险最高 (3.13×10^{-5})，其次是 EBMP (1.99×10^{-6})。在其他三个车间，EHFTV、RITV 和 RIHD 的风险值分别为 1.63×10^{-6}、1.60×10^{-6} 和 1.08×10^{-6}。五个车间的致癌水平均高于美国环保署的可接受范围 ($10^{-6} \sim 10^{-4}$)，表明在废弃印刷线路板拆解过程中需要更加关注 PAHs 及其衍生物的致癌风险。此外发现二苯并 (a, h) 蒽和苯并芘的致癌风险占到大于所有 PAHs 及其衍生物致癌风险的 85.2%。因此，为了降低职业暴露工人的癌症风险，应该强制引入一些暴露削减技术，例如通过除尘与净化设施的集成去除电子垃圾加热处理产生烟雾中的二苯并 (a, h) 蒽和苯并芘。

5.1.2 周边居民的暴露

对于居民，Chen 等[8]的研究报告称，广东某典型电子垃圾拆解地成人和儿童夏季通过空气吸入的 PBDEs 暴露量分别为 54 ng/(kg·d) 和 70.6 ng/(kg·d)，冬季分别为 225 ng/(kg·d) 和 295 ng/(kg·d)，这意味着潜在的更高暴露风险。Luo 等[9]的研究指出，广东清远居民通过空气吸入 PBDEs 的日摄入量为 4.6 ng/(kg·d)，高于美国 (1.11 ng/(kg·d))，但低于美国环保署综合建议的单个 BDE 同系物的 RfD 值 ($1 \times 10^2 \sim 7 \times 10^3$

ng/(kg·d))。根据 BDE-209 的癌症斜率因子为 7×10^{-4} (kg·d)/mg，对 BDE-209 的终生癌症风险增量进行评估得出，BDE-209 诱发的癌症风险为 1.36×10^{-10}，远低于由美国环保署确定的安全可接受范围（$1.0 \times 10^{-6} \sim 1.0 \times 10^{-4}$）。Niu 等[10]的报告中显示，广东某电子垃圾拆解地区成人居民空气吸入的污染物中氯化萘（PCNs）的 ILCR 最高（8.72×10^{-5}），其次是多氯联苯并二噁英/呋喃（PCDD/Fs）（4.17×10^{-5}），PCBs（1.40×10^{-6}）和 PAHs（2.52×10^{-9}），ILCR 最低的是 PBDEs（4.32×10^{-13}）。Chen 等[11]的研究中指出，电子垃圾拆解区婴儿的空气 PAHs 的估计日摄入量最高，是对照区成人的 2~3 倍，主要是因为婴儿的吸氧率与体重比率较大，从而吸入 PAHs 的潜在健康风险比成人高得多。

5.1.3 个体暴露监测

个人空气采样泵具有恒定的流量，这可能无法完全模拟人类呼吸的变化，但与固定空气采样相比，它们能更准确地评估个人吸入暴露。Papadopoulou 等[12]的研究中介绍，个人空气样品的采集是使用一台低流量 SKC 224-PCMTX4 泵（SKC Inc.，Eight Four，PA，USA）以 1 L/min 的流速采集个人空气样本的污染物。采样泵装在背包中，在整个 24 h 采样期间陪伴受试者。采样器固定在受试者的肩膀上，并建议受试者在整个 24 h 采样事件（包括睡眠时间）中将采样器靠近他们的脸。Xu 等[13]比较分析了个人空气采样和固定空气采样两种空气采样方式下，人体对有机磷阻燃剂（OPFRs）的呼吸暴露风险情况。研究受试者 24 h 携带便携式 SKC 224-PCMTX4 泵，其中泵与 ENV+ SPE 小柱（1 g, 25 mL, Biotage Uppsala, Sweden）相连，小柱固定在受试者胸部上方，距离受试者的鼻子约 30 cm 处，从而模拟个人 24 h 吸入的环境空气。

5.2 摄食暴露及风险

5.2.1 饮用水暴露

环境污染物会释放到水体等环境介质中，日常摄入饮用水是污染物进入人体内的重要暴露途径。Zhao 等[14]在浙江某电子垃圾拆解区采集了饮用水样品，估算出当地居民饮用水中 PBBs、PBDEs 和 PCBs 的日摄入量分别为 9.4 ng/d，4.3 ng/d 和 6.5 ng/d。一项研究表明[15]，在广东清远市电子垃圾拆解区的饮用水（一共 25 个饮用水样本，其中 2 个来自地下水样本）中 Cd、Pb、Cu 和 Zn 的几何平均值分别为 0.048 μg/L、1.91 μg/L、66.5 μg/L 和 80.7 μg/L。地下水中 Cd（几何平均值：0.32 μg/L）、Pb（8.84 μg/L）和 Zn（222 μg/L）的浓度显著高于自来水中的浓度。当从计算中排除地下水样品时，电子垃圾拆解区中 Cd、Pb、Cu 和 Zn 的几何平均值分别降低到小于定量限、1.63 μg/L、66.2 μg/L 和 66.9 μg/L；这些值与该研究中从对照区测得的值相当。虽然电子垃圾拆解区中存在电子垃圾拆解活动，但结果表明电子垃圾拆解区使用的自来水没有被回收的电子垃圾污染，因为当地政府对清远市所有地区进行了统一供水。然而，地下水被污染了，这应该是需要关注的一个问题，因为有些人仍然在电子垃圾拆解区中使用地下水作为饮用水。关于饮用水重金属暴露的另一项研究指出，地下水中 Cd、Pb、Zn、Cu 和 Ni 的几何平均值分别为 0.09 mg/L、1.40 mg/L、4.16 mg/L、1.34 mg/L 和 0.35 mg/L，分别是国家地下水标准值（Cd：0.005 mg/L；Pb：0.01 mg/L；Zn：1 mg/L；Cu：1 mg/L；Ni：0.02 mg/L）的 1.3~140 倍。该研究中电子垃圾区地下水的金属浓度比非电子垃圾区高几个数量级，表明电子垃圾回收活动显著影响该地区地下水。由于饮用水中这些金属的浓度非常低，该地区居民通过饮用水接触这些金属的风险可以忽略不计。相比之下，地下水的摄入将对居民构成潜在的重大风险。地下水中 Cd、Pb、Cu、Zn 和 Ni 的 HQ 对于成人分别为 3.0、13、4.0、0.2 和 0.5，对于儿童分别为 4.0、20、6.0、0.4 和 0.8。然而，由于只有少数（约 6%）的居民使用地下水作为饮用水，因此这一结果可能对高暴露的人群构成风险[16]。

5.2.2 食物暴露

食物消费是一般人群污染物暴露的主要途径，约占每日可耐受摄入量的 90%。食物摄入，尤其是污染地区的高脂肪食物，是 PBDEs 暴露的重要来源。当被污染的食物被摄入并进入胃肠道时，吸附物质发生脱附作用，导致 PBDEs 从食物中释放到胃肠液中并最终进入血液循环[17]。由于人体代谢能力有限，被污染的食物不断被摄入，甚至人体 PBDEs 的负担会超过健康风险水平。研究表明，在我国广东省食物 PBDEs 摄入量占 PBDEs 摄入总量的 79%[18]。食品消费被认为是普通人群日常接触 PBDEs 的主要来源。

电子垃圾回收地区周围的当地食品消费已被证明是人类接触 PBDEs 的主要途径。例如，浙江省台州市电子垃圾回收区域对新鲜鸭蛋中 PBDEs 的摄入量为 159~5100 ng/d[19]，而浙江和广东等电子垃圾拆解地膳食中 PBDEs 的总摄入量（包括淡水和海洋鱼类、贝类、肉类、家禽、鸡蛋、动物内脏和蔬菜）分别为 2700 ng/d 和 56000 ng/d[20]，均表明了饮食摄入是电子垃圾回收地区周围居民接触 PBDEs 的主要来源。据估计，广东和浙江等典型电子垃圾拆解地淡水鱼对总膳食暴露的贡献分别为 98%和 61%，当地河流淡水鱼的 PBDEs 浓度至少是当地市场淡水鱼的 30 倍[20]。如果居民只食用当地市场上出售的鱼，而这些鱼可能来自电子垃圾回收区域之外，饮食暴露可减少 61%~98%。然而，当从当地杂货店或居民家庭取样当地食品供应(包括水、蔬菜、豆类、大米、鸡蛋、鱼、猪肉和鸡肉)时,PBDEs 的总膳食摄入量估计为 196 g/d[14],这些研究表明,如果居民不食用当地生产的食品,电子垃圾回收地区居民接触 PBDEs 的情况并不比世界上一般人群更严重。

在浙江某电子垃圾拆解区,超过 60%的猪肉和鸡肉样本中检出 BDE-209,占 PBDEs 总摄入量的 8.4%。广东某电子垃圾拆解区居民 PBDEs 食物摄入量在目前所有研究中最高(931 ng/(kg·d))[21],其中摄入的 BDE-47(584 ng/(kg·d))超过了美国环保署的参考剂量(100 ng/(kg·d))。与广东相比,台州市居民的饮食摄入量相对较低(44.7 ng/(kg·d)),这一结果也与台州市其他研究结果存在差异,如 Zhao 等[22]报道的居民 PBDEs 饮食摄入量为 3.8 ng/(kg·d),Labunska 等[23]报道的台州市成人和儿童的 PBDEs 饮食摄入量分别为 109.8 ng/(kg·d)和 432.9 ng/(kg·d)。还有报道称,电子垃圾回收工厂区域的食物中普遍存在高浓度的 PBDEs,儿童比成年人有更高的 PBDEs 暴露风险[24]。有研究[25]表明,越南电子垃圾拆解区居民的 PBDEs 平均摄入量为 14.6 ng/(kg·d)。研究发现,清远市儿童每天从鸡蛋中摄入的 PBDEs(1365 ng/(kg·d))远高于成年人(317 ng/(kg·d))[26,27]。Huang 等[27]在 2013 年和 2016 年调查了鸡蛋中 PBDEs 的暴露情况(成人和儿童分别为 120 ng/(kg·d)和 517 ng/(kg·d)),与以往研究相比,PBDEs 摄入量显著降低[26],但仍高于台州另一个电子垃圾回收工厂的 PBDEs 的摄入量(成人和儿童分别为 101 ng/(kg·d) 和 435 ng/(kg·d))[23]。Labunska 等[19, 23]指出,鸭蛋(成人和儿童分别为 61.6 ng/(kg·d) 和 265 ng/(kg·d))中估计的日摄入量是鸡蛋(成人和儿童分别为 17.3 ng/(kg·d) 和 74.5 ng/(kg·d))的 3 倍。母乳是婴儿摄入 PBDEs 的重要途径。在台州 6 个月大婴儿(母乳喂养)的 PBDEs 日摄入量为 572 ng/(kg·d),高于参考值(10.1 ng/(kg·d))[28]。我国浙江温岭市的另一项研究表明,居住时间大于 20 年的母亲和居住时间大于 3 年的母亲的婴儿 PBDEs 摄入量中位数分别为 45.3 ng/(kg·d)和 11.4 ng/(kg·d)。研究表明,如果母亲在电子垃圾回收区域附近居住的时间越长,其婴儿接触 PBDEs 的风险越高[29]。

Zhao 等[14]的研究指出,在浙江台州电子垃圾拆解区居民通过饮用水和特定食物摄入 PCBs 的量高达 12 372.9 ng/d。PCBs 是电子垃圾拆解地区食品样品中的主要污染物,并且 PCB-138 是拆解地区膳食中 PCBs 摄入量中的主要贡献者(1088.8 ng/d)。PCBs 的最高摄入量以大米消费为主,占 PCBs 估计每日膳食摄入总量的 69.5%,其次 PCBs 摄入量来自鱼肉消费,摄入量为 2526.3 ng/d。在拆解区 ΣTEQ_{PCBs} 摄入量为 196.6 pg/d,摄入也主要通过大米和鱼肉的消耗。这里的 ΣTEQ_{PCBs} 水平指的是毒性当量的总和,国际上普遍采用毒性当量因子来对二噁英和类二噁英物质的毒性进行量化和评估。据 Niu 等[10]的估计显示,在广东某电子垃圾拆解地区的幼童、儿童、青少年和成人所有年龄组中,多氯萘(PCNs)饮食摄入对终生致癌风险(ILCR)的贡献比其他暴露途径的风险高 3~10 倍。

5.3 皮肤接触暴露风险

皮肤作为人体与环境最大的防御体系，同时也是人体最大的器官，无时无刻不受到环境污染物的直接暴露。研究表明，可以使用直接或间接方法对人体研究人群中的皮肤暴露进行估计。直接方法进一步细分为三类：去除(擦拭和洗手)、拦截(贴片采样器、腕带)和原位技术(直接测量，例如使用荧光示踪剂进行视频成像)。间接方法包括监测人体样本中的母体化合物或代谢物，以及调查皮肤接触发生前的过程。由于低成本和易用性，擦拭法已被广泛用于评估人类研究人群中各种化学物质的皮肤暴露[30]。皮肤暴露途径是不容忽视的重要的暴露途径，而擦手样品作为一种非侵入性介质经常被用作皮肤暴露途径样品的采集。擦拭法是一种有效的采集方法，可通过测量皮肤表面污染物的浓度来直接评估皮肤暴露情况[31]。目前环境污染物的皮肤暴露途径越来越受到国内外研究学者的广泛关注，并且可以用来指示皮肤暴露的擦手样品中环境污染物，例如阻燃剂的暴露特征[32-34]，从而评估环境污染物通过皮肤暴露途径进入人体后造成的潜在健康影响。

关于皮肤吸收对人体外暴露于卤代阻燃剂(HFRs)和有机磷阻燃剂(OPFRs)的贡献存在一些争论。虽然一些研究人员表示通过皮肤接触的暴露风险几乎可以忽略不计[35]，但 Wu 等[36]表明皮肤吸收对人类外暴露于 HFRs 和 OPFRs 的贡献被低估了，因为之前对皮肤暴露的评估只考虑了受污染的灰尘或土壤的无意接触的皮肤暴露途径，而忽略了通过空气对颗粒和气体污染物的皮肤吸收-皮肤转移，以及皮肤与电子垃圾物品的直接接触的暴露途径。

我国的 Qiao 等[37]研究了来自电子垃圾拆解地区工人的擦手样品中典型的 BFRs 和 OPFRs，发现 BDE-209 是 PBDEs 中含量最高的同系物，在擦手样品中占 90.0%，而 OPFRs 在所有分析的阻燃剂中浓度最高。之前的一项研究发现[36]，空气中半挥发性有机物的皮肤吸收是当地居民接触电子垃圾拆解活动产生的燃烧烟雾的主要途径，成年人通过空气介导传递的气态 BDE-47 和 BDE-99 的每日皮肤摄入量估计分别为 0.65 ng/(kg·d) 和 0.61 ng/(kg·d)，超过了通过吸入气态和颗粒结合的 BDE-47(0.55 ng/(kg·d))和 BDE-99(0.33 ng/(kg·d))的暴露量。Shen 等[38]也报道了类似的结果，发现清远电子垃圾回收工人、当地成年人和当地儿童通过皮肤吸收而不是吸入途径的新型溴代阻燃剂、四溴双酚 A 和六溴环十二烷的估计每日摄入量(EDI)更高。此外，迄今为止似乎没有考虑到通过直接接触电子垃圾物品的皮肤暴露，鉴于最近的数据表明直接接触经阻燃处理的织物皮肤接触的重要性[39]，这可能是研究的一个重大遗漏。总而言之，对于电子垃圾拆解和回收工人或居住在电子垃圾拆解和回收区的居民而言，皮肤吸收似乎是人类外暴露于 HFRs(可能还有 OPFRs)的潜在途径。最近有证据发现，电子垃圾拆解工人的职业皮肤接触是重要的半挥发性有机物暴露途径[37,40,41]。Estill 等[42]研究了电子垃圾拆解行业职业工人暴露于阻燃剂的情况,发现在擦手样品中 BDE-209 是 PBDEs 中检出浓度较高的同系物(135.3 ng/sample)。Beaucham 等[40]的研究中,调查了美国的一家电子垃圾回收工厂的员工分别采用斜纹纱布和薄纱纱布进行 3 次手部皮肤擦拭过程污染物的变化。对于每位员工,两种擦拭材料的第一次擦拭去除了更高比例的阻燃剂总量。然而,第一次斜纹擦拭布通常比第一次薄纱擦拭布去除了更多的阻燃剂。所有 19 种阻燃剂在第一次擦拭样品上的总和分别为薄纱纱布中 97 000 ng/sample,斜纹纱布中 170 000 ng/sample。

据估计,清远电子垃圾回收地区居民的颗粒结合卤代阻燃剂的皮肤暴露量为 0.33 ng/d[9]。Wu 等[36]估计,夏季成人的 BDE-47 和 BDE-99 的 EDI 分别为 0.65 ng/(kg·d) 和 0.61 ng/(kg·d)。夏季儿童皮肤接触 BDE-47(1.1 ng/(kg·d)) 和 BDE-99(1.03 ng/(kg·d)) 的 EDI 高于成人。夏季居民皮肤接触的 BDE-47 是冬季的 3 倍。Abafe 和 Martincigh[43]发现,南非电子垃圾回收工人的 PBDEs 皮肤接触量为 1.59 ng/(kg·d),低于华东地区的工人(7.9 ng/(kg·d) 和 11.4 ng/(kg·d))[44],但高于中国清远居民的水平(0.07 ng/(kg·d))[9]。然而,PBDEs 的皮肤暴露量可能被低估了,因为没有考虑到其他皮肤途径如空气到皮肤的运输,尤其是那些较低溴化的同系物,由于其较低的 K_{OW},被认为更容易被人体皮肤吸收[36, 45]。Nguyen 等[46]报告称,加拿大电子垃圾拆解工人的 PBDEs 皮肤暴露量仅占 PBDEs 暴露总量的 2%,并且男女工人之间没有差异。但仍需进一步了解 PBDEs 的皮肤接触途径。Niu 等[10]的研究中显示,广东某电子垃圾拆解地区幼童皮肤接触的污染物中 ILCR 最高的是 PCDD/Fs(1.12×10^{-2}),其次是 PCNs(8.13×10^{-4}),PAHs(5.29×10^{-6}),PCBs(4.90×10^{-7}) 和 PBDEs(1.73×10^{-10})。Liu 等[6]研究了电子垃圾拆解车间工人的皮肤接触 PBDEs 和 PCBs 的暴露风险。对于 PBDEs,EBMP、EHFTV、EHFR、RITV 和 RIHD 车间工人的皮肤暴露 HI 值分别为 1.3×10^{-5}、8.9×10^{-3}、5.7×10^{-5}、5.0×10^{-4} 和 8.5×10^{-5}。在 EHFTV、RITV 和 RIHD 车间中,PCBs 对工人皮肤暴露的贡献分别为 1.1×10^{-4}、5.9×10^{-5} 和 5.9×10^{-6}。这表明 PBDEs 和 PCBs 的皮肤接触总 HI 均低于 1.0 的风险水平。这意味着皮肤接触的 PBDEs 和 PCBs 可能不会对拆解工人造成重大的职业威胁。Liu 等[7]在一项研究中指出,车间工人 PAHs 及其衍生物的皮肤接触致癌风险均低于可接受范围($10^{-6} \sim 10^{-4}$),男性和女性工人的最高风险仅为 6.29×10^{-8} 和 6.17×10^{-8},表面皮肤暴露的 PAHs 可能不会对车间工人产生致癌风险。

5.4　电子垃圾拆解人群的人体生物监测方法

近年来,分析化学的进步使得测量生物组织中多种环境污染物的微量水平(即生物监测)成为可能。人体样品基质复杂,在分析过程中目标化合物难以避免会与样品中的脂质以及其他非目标化合物一起萃取出来。这些杂质在仪器分析的时候往往会对目标化合物的检出造成干扰。为了保证目标化合物定性定量的准确性和可靠性,开发灵敏度高、重现性好以及回收率优的样品前处理和仪器分析方法至关重要。

5.4.1 血液样品

血液样品成分复杂,除了蛋白等生物大分子以外,还有脂肪以及很多日常代谢产物,对前处理要求比较高。目前,电子垃圾拆解人群血清样品常用的检测方法主要参考 Hovander 等建立的方法[47],主要分为脂肪提取(图 5-1)、中性组分和极性组分分离(图 5-2)以及除脂净化(图 5-3 和图 5-4)等步骤。简述如下:量取 3~5 mL 解冻后并混匀的血清样品于 50 mL 聚四氟乙烯离心管,加入回收率指示物混匀后平衡 12 h。加入 1 mL 6 mol/L 的盐酸,振荡,然后迅速加入 6 mL 的异丙醇溶

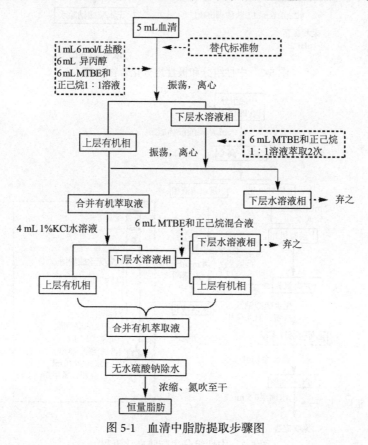

图 5-1 血清中脂肪提取步骤图

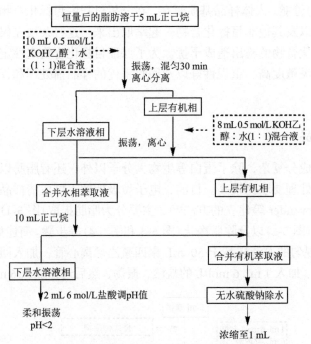

图 5-2　中性组分和极性组分化合物分离

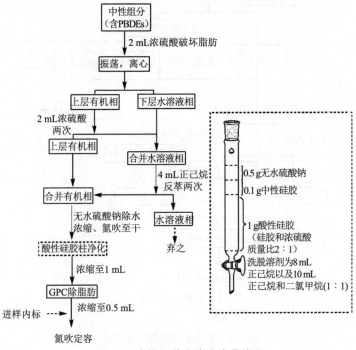

图 5-3　中性组分去脂肪净化流程

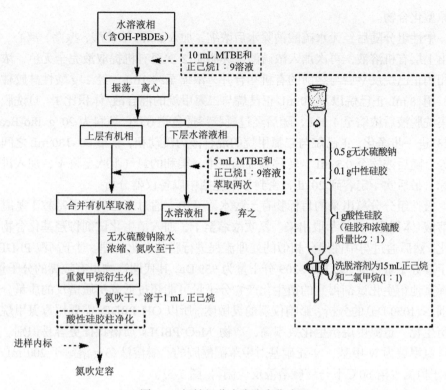

图 5-4 极性组分去脂肪净化流程

液,振荡,之后加入 6 mL 正己烷和甲基叔丁基醚(MTBE)(体积比 1∶1)的混合溶液,振荡,离心,分离出上层有机相。下层水相再分别用 6 mL 上述混合溶液萃取两次,合并三次所得的有机相,加入 4 mL 质量分数为 1%的 KCl 水溶液,混匀后离心分离出有机相后,下层 KCl 水溶液相再用 6 mL 正己烷和 MTBE 混合溶液萃取一次,水溶液相弃之,有机相合并后经无水硫酸钠除水,随后经旋转蒸发仪浓缩至 1 mL 转移至细胞瓶并氮吹至干,置于密闭干燥器中恒重,48 h 后称量脂肪的质量。

恒量后脂肪用 5 mL 正己烷溶解并转移到 50 mL 聚四氟乙烯离心管,加入 10 mL 0.5 mol/L KOH 的乙醇水混合碱液(体积比 1∶1),混匀,离心分离,取出无机相。极性化合物(如 OH-PBCs、OH-PBDEs 以及其他酚类化合物与碱形成的钠盐)在强碱溶液的作用下转变为其酚类盐溶液进入乙醇水混合液相中,中性组分(如含 PCBs、PBDEs 和 MeO-PBDEs)则保留在正己烷相中,正己烷相再用 8 mL 上述醇碱溶液萃取一次,合并两次所得的醇水相混合液,并用 10 mL 正己烷反萃取一次,两次所得正己烷相合并。至此,中性组分和极性组分分离。分离出来的极性组分再加入 2 mL 6 mol/L 的盐酸溶液混匀,调节 pH 值小于 2,酚盐类化合物还原成分子原型化合物。

中性组分随后经无水硫酸钠除水后浓缩，加入 2 mL 浓硫酸，振荡，离心，分离出上层有机溶液，再次加入浓硫酸破坏脂肪 2 次至有机提取液完全无色。浓硫酸相用正己烷反萃后与之前的有机相合并，浓缩至约 1mL，经 1 g 酸性硅胶柱净化，用 8 mL 正己烷以及 10 mL 正己烷与二氯甲烷的混合液(体积比 1:1)洗脱，收集洗脱液后浓缩至 1 mL，随后经过凝胶渗透色谱(GPC 填料为 20 g Bio-Beads S-X3)进一步净化，正己烷与二氯甲烷的混合液为流动相，收集 60~180 mL 之间的馏分，随后浓缩至 0.5 mL，转移到细胞瓶中，柔和的氮气下吹至将干，加入进样内标，最后吹干定容到 20 μL，密封于细胞瓶中以备仪器分析。

极性组分分离出来的目标物存于醇水溶液中，通过正己烷和甲基叔丁基醚混合溶液(体积比 9:1)萃取出来。氮吹浓缩至干，加入衍生化试剂将羟基化合物衍生化。然后通过与中性组分相似的处理流程进行脂肪净化处理。对于高溴 PBDEs，由于其分子量很高，如 BDE-209 分子量为 959 Da，其代谢产物也有较高的分子量，选择其他衍生化试剂得到的衍生化产物分子量可能超过低分辨质谱仪的质量分辨范围(<1050 Da)的分析，影响仪器的灵敏度。所以 OH-PBDEs 需要用重氮甲烷进行衍生化。重氮甲烷衍生化效率高，产物 MeO-PBDEs 质谱特征更容易识别。所用重氮甲烷为 N-甲基，N-亚硝基对甲苯磺酰胺的乙醚溶液(5 g 溶解于 200 mL 乙醚)于恒温水浴 60℃下与醇碱溶液反应制得(图 5-5)。

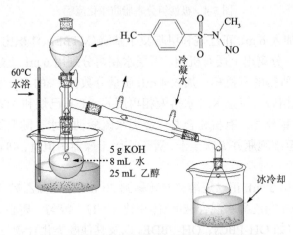

图 5-5 重氮甲烷衍生化试剂的制备装备图

对于监测电子垃圾拆解区人群暴露的溴代阻燃剂，如 PBDEs，中性组分及衍生化后的极性组分的分析方法均采用气相色谱串联质谱法，色谱条件如下：DB-5HT(15 m × 0.25 mm, i.d.; 0.1 μm; J&W Scientific, USA)高温毛细管柱；载气为 He，采用恒流模式，柱流量为 1.2 mL/min，脉冲不分流进样，进样量为 1 μL；

进样口温度 280℃。气相色谱的升温程序为：柱始温 110℃(5 min)，20℃/min 升至 200℃(4.5 min)，7.5℃/min 升至 300℃ 并保持 16 min。另一根 Rtx®-1614 色谱柱(30 m × 0.25 mm，i.d.；0.1 μm；Restek，USA)用来确证分析结果，升温程序如下：100℃(3 min)，20℃/min 升至 200℃，5℃/min 升至 320℃保持 15 min。

质谱采用电子捕获负化学电离(electron capture negative ion，ECNI)离子源：2 mL/min 甲烷为反应气；离子源温度 250℃；四极杆温度 150℃；传输线温度 300℃；选择性离子分段扫描(selected ion monitoring，SIM)，低溴化合物的扫描离子为 m/z 79 和 81，nona-BDEs 和 BDE-209 等化合物的扫描离子为 m/z 79、81、486.7 和 488.7；^{13}C-BDE-209 监测离子为 494.7 和 496.7；其他化合物监测离子如下：^{13}C-PCB-208 m/z 475.7 和 473.7；^{13}C-PCB-141 m/z 372 和 374；PCB-209 m/z 498 和 500；^{13}C$_{12}$-4-MeO-CB187 监测 m/z 421 和 423；6-MeO-BDE-196 和 6-MeO-BDE-199 监测 m/z 406.7 和 408.7；6'-MeO-BDE-206 监测 m/z 486.7 和 488.7，此外，高溴 MeO-PBDEs 化合物监测加入 m/z 438.6 和 440.6 进行确证。

用上述方法测定电子垃圾拆解人群血清中的 PBDEs 和 OH-PBDEs，各加标指示物回收率分别为 ^{13}C-PCB-141：83% ± 12%(mean ± SE)；PCB-209：79% ± 14%；^{13}C-4-OH-CB187：71% ± 18%；^{13}C-BDE-209：78% ± 23%；定量限(limits of quantification，LOQ)定义为 10 倍的信噪比，通过进样(1 μL)标准曲线最低浓度点来确定，tri-到 nona-BDEs 为 0.05~0.12 pg，BDE-209 为 1.2 pg，高溴 MeO-PBDEs 为 0.6 pg。方法经过反复的小牛血清加标回收实验进行确证，稳定可靠[48]。

5.4.2 尿液样品

有机污染物进入人体后，除了以原型化合物形式储存在各个组织器官，还会在人体内发生一系列复杂的代谢转化反应。一般而言，有机污染物在人体中的代谢分别由两个阶段的酶完成[49]。阶段Ⅰ，通过细胞色素 P450 酶的氧化，生成极性较强、毒性更高的单羟基或多羟基代谢产物；阶段Ⅱ是阶段Ⅰ代谢物的解毒过程，主要与谷胱甘肽、葡萄糖苷酸或磺酸结合，进一步生成水溶性较强的代谢物，通过尿液排出，从而达到解毒的目的。尿液中结合态代谢物在水解酶的作用下极易还原成羟基化合物。在电子垃圾拆解区，测定尿液中的 OH-PAHs 是最常用的评估人体 PAHs 暴露水平的方法，而相关卤代阻燃剂代谢物的监测比较少。

Lin 等[50]建立了尿液中多种酚类化合物同时监测的样品前处理和仪器分析方法。尿液样品自然解冻后充分混匀，取 2 mL 于 50 mL Teflon 离心管中，加入同位素内标化合物，充分混匀后加入 1 mL 的乙酸-乙酸钠缓冲液(1 mol/L)调节尿液 pH 为 5.5 和 10 μL β-葡糖苷酸-芳基硫酸酯酶，充分混匀后，盖上盖子将样品放入恒温振荡器中，在 37℃下酶解 12 h。酶解完成待样品冷却至室温，加入 1.7 mL

HCl(0.5 mol/L)调节尿液 pH 为 2 左右。然后用 Oasis® HLB 固相萃取小柱对尿液中的目标化合物进行萃取，流程如下：依次用 6 mL 体积比为 1∶1 的甲醇和二氯甲烷混合溶剂、6 mL 甲醇、6 mL 水和 8 mL 磷酸二氢钾水溶液(25 mmol/L)活化小柱；然后把尿液样品缓慢加载至小柱上，随后依次用 3 mL 磷酸二氢钾水溶液(25 mmol/L)和 3 mL 超纯水淋洗小柱；最后用真空泵把小柱中的水抽干。为了优化洗脱效率，分别用 8 mL 甲醇和 8 mL 体积比为 1∶1 的甲醇和二氯甲烷混合溶剂对目标化合物进行洗脱，比较两种溶剂的洗脱效果。该固相萃取过程保持流速为 1 mL/min 左右。洗脱液浓缩至 1 mL 左右，转移至细胞瓶中，在柔和氮气下吹至近干，用甲醇定容到 200 μL，密封，储存于 4℃冰箱，用 LC-MS/MS 分析。

尿液中的 OH-PAHs、BRPs、OH-PBDEs、TCS 和 TBBPA 分析采用 HPLC-MS/MS(Agilent 1260-6470)，在负模式下的喷射流电喷雾电离源(AJS ESI)MRM 模式下完成。干燥气温度为 300℃，气体流速为 5 L/min，雾化器压力为 45 psi[①]，鞘气温度为 350℃，鞘气流速为 12 L/min，喷嘴电压为 500 V，毛细管电压为 3500 V。目标化合物用色谱柱 Poroshell 120 EC-C_{18}(100 mm × 4.6 mm, 2.7 μm particle diameter, Agilent)进行分离，流动相(A)水相中添加 5 mmol/L 乙酸铵，(B)有机相为乙腈。流动相流速为 0.4 mL/min，洗脱程序如下：0~2 min, 5%~15% B；2~2.1 min, 15%~50% B；2.1~10 min, 50% B；10~15 min, 50%~80% B；15~18 min, 80%~95% B；18~19 min, 95% B；19~20 min, 95%~99% B；20~22 min, 99% B。该方法测定的所有目标分析物的低浓度、中浓度和高浓度加标回收率分别为 65%~143%、56%~135% 和 53%~123%，仪器检出限和方法检出限，分别为 0.2~15.0 pg 和 0.008~0.161 ng/mL。

5.4.3 头发和指甲样品

近年来，头发和指甲被逐渐用于评估人类受环境或职业有机污染物暴露的无创生物检材。Zheng 等[51]研究表明，头发中的 PCBs 有 64%来源于大气，27%来源于血液输送，9%来源于室内灰尘。头发与指甲性质相似，因此，长期以来，头发和指甲分析的困难主要在于内源污染物和外源污染物的区分。目前大部分研究主要通过清洗，将头发或指甲外源污染物去除，从而确保所分析的目标化合物来源于头发内源。清洗方式多种多样，主要表现为清洗剂的差异，不同清洗剂对头发和指甲外源污染物的清洗效果不同。Kucharska 等[52]对比了水、甲醇、体积比为 1∶1 的正己烷和二氯甲烷混合溶液、丙酮和洗发水对头发外源 PBDEs 的去除效率时发现，水对 PBDEs 的去除几乎没有作用，去除率低于 10%，其他溶剂对 PBDEs

---

① 1 psi=6.895 kPa

的去除效率差别不大。Zheng 等[53]通过扫描电镜观察不同溶剂清洗前后头发形态及其表面污染物时发现，被温水、正己烷、丙酮清洗后的头发表面均显得比较平滑，从感官上显示外源污染物能被有效去除；但是被二氯甲烷清洗后的头发，结构上受到一定程度的破坏，洗发水清洗后的头发表面也还残留一些污染物；因此，温水、正己烷以及丙酮是较合适的清洗溶剂。但是，与温水和正己烷相比，丙酮的极性较强，对脂溶性和水溶性的化合物均有较好的溶解效果，更有利于头发表面污染物的去除。Lu 等[54]研究丙酮清洗对头发表面 PCBs、PBDEs 和有机氯农药（OCPs）的去除效率时发现，清洗第 1 次、第 2 次和第 3 次头发外源污染物的累积去除率分别为 65%、85%和 95%。结合上述关于头发清洗的研究发现，无论是对污染物的去除还是对头发和指甲结构的保护，丙酮是头发清洗溶剂的首选。

为了开发电子垃圾拆解人群污染暴露的无创生物监测方法，Lin 等[55]开发了同时测定头发/指甲中原型化合物和羟基代谢物的分析方法（图 5-6）。称取约 1 g 头发/100 mg 指甲，加入 20 mL 丙酮，超声清洗 1 min，涡旋 1 min，收集清洗液至梨形瓶，重复清洗 2 次，合并 3 次丙酮清洗液，旋蒸至 1 mL，获得头发/指甲

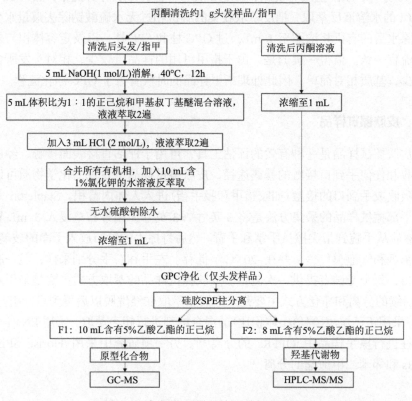

图 5-6　头发/指甲样品前处理方法流程

外源污染物。然后通过 GPC 净化柱，弃掉前 60 mL 含有脂肪等大分子干扰物的洗脱液，接收中间 60~160 mL 含有目标化合物的洗脱液，浓缩至 1 mL。并参考 Motorykinet 等[56]的方法，过 1 g/6 mL 硅胶固相萃取小柱分离原型化合物与羟基化合物，即把 1 mL 浓缩清洗液加载至先后经 6 mL 乙酸乙酯、6 mL 二氯甲烷和 12 mL 正己烷活化的固相萃取小柱上，用 10 mL 含有 5%乙酸乙酯的正己烷洗脱原型化合物(组分 1, F1)，随后用 8 mL 含有 50%乙酸乙酯的正己烷洗脱羟基化合物(组分 2, F2)。组分 1 洗脱液浓缩至 1 mL 左右，转移至细胞瓶中，在柔和氮气下吹至近干，用异辛烷定容到 30 μL，加入一定量的进样内标后密封，储存于 4℃冰箱，待仪器分析；组分 2 洗脱液浓缩至 1 mL 左右，转移至细胞瓶中，在柔和氮气下吹至近干，用甲醇定容到 200 μL，密封，储存于 4℃冰箱，待仪器分析。

头发/指甲内源化合物则将清洗干净的头发样品自然晾干后，用不锈钢剪刀剪碎(1~2 mm)，置于 Teflon 离心管，加入回收率指示物，加入 5 mL NaOH(1 mol/L)，盖上盖子，放入水浴锅中 40℃消解 12 h。然后加入 5 mL 体积比为 1∶1 的正己烷和甲基叔丁基醚混合溶液进行液液萃取，混合溶液涡旋混匀 1 分钟后于 4000 r/min 离心 10 分钟，取出上层有机相，重复 2 次。然后合并所有有机相，加入 10 mL 含 1% KCl 的水溶液反萃取，把反萃取收集的有机相过无水硫酸钠除去痕量水分。随后把除水后的有机相浓缩至 1 mL，过 GPC 柱和 SPE 柱，以及定容体积与外源污染物流程一致。值得一提的是，由于指甲可用的样品量较少，相对头发所含的脂肪较少，基质相对简单，因此处理指甲样品的时候省去了 GPC 净化流程。

5.4.4 皮肤擦拭样品

皮肤擦拭样品是一种有效的评估工具，可用于评估直接表面接触、经皮肤途径吸收和任何手到口接触的暴露途径。皮肤擦拭样品中发现的化学物质可以通过皮肤吸收或手到口的接触(如咬指甲和吸手指)进入人体内蓄积。Stapleton 等[57]报道的手部擦拭样品的采集方法是将 3 英寸①×3 英寸无菌纱布垫浸入 3 mL 异丙醇中，然后从手腕到指尖擦拭手掌和手背。然后将擦手的纱布放入干净的玻璃瓶中，用锡箔纸和气泡膜包裹，并在−20 ℃下储存。左手和右手分别采样，但一起提取和分析，每个受试者提供一次测量。皮肤擦拭样品的萃取方式主要是采用超声萃取，后续的分离和净化方式主要是采用固相萃取柱经洗脱以后得到目标组分[13,57]。Liu 等采用正己烷/丙酮(3/1 体积比)的混合溶剂进行超声萃取，采用 ENVI-Florisil 萃取柱进行擦手样品中 OPFRs 的分离[34]。另一项研究中采用 Florisil SPE 进行 OPFRs 和邻苯二甲酸酯的分离[32]。

---

① 1 英寸=2.54 厘米

5.5 拆解工人的暴露特征

5.5.1 阻燃剂暴露特征

1. PBDEs

PBDEs 是电子垃圾拆解工人典型的职业暴露污染物之一。国外对电子垃圾拆解工人 PBDEs 内暴露研究较早。瑞典斯德哥尔摩大学 Bergman 教授领导的研究小组早在 1999 年就对职业人群的 PBDEs 暴露进行了系统研究[58],发现电子垃圾拆解工人血液中 PBDEs 的含量(中值 37 pmol/g lipid)显著高于长期使用电脑的白领(中值 7.3 pmol/g lipid)。在同系物组成上,电子垃圾拆解工人体内富集了较高比例的六溴和七溴联苯醚,其中 BDE-183 和 BDE-153 是两个最主要的同系物,这与使用电脑办公的白领以 BDE-47 为主的同系物组成显著不同。此外,Thomsen 等[59]对挪威电子垃圾拆解工人,印刷电路板制造劳工和实验室员工血液中 PBDEs 含量进行了研究,发现电子垃圾拆解工人体内的 BDE-153 显著高于另外两组,且 BDE-183 只有在电子垃圾拆解工人体内才可以检出;而低溴代的 BDE-47 在所有人群体内都是主要的同系物,且在三组人群之间没有显著差异。从这些早期的职业暴露研究结果来看,以 BDE-47 为代表的低溴代联苯醚暴露主要来自背景暴露,比如饮食等,而 BDE-153 和 BDE-183 更多反映职业暴露的影响。

早期粗放式的电子垃圾拆解活动,对拆解区生态环境造成严重污染,对拆解工人造成污染暴露风险,损害了人民的健康安全。2007 年,Bi 等[60]首次报道了广东某电子垃圾拆解人群血清中 ΣPBDEs 浓度为 140~8500 ng/g lipid,BDE-209 为主要同系物,含量高达 3100 ng/g lipid,为其他国家职业暴露人群的 50~200 倍,也显著高于我国普通背景人群暴露水平。Qu 等[61]对我国广东地区电子垃圾拆解工人血清中 PBDEs 含量的研究中也发现,工人血清中 PBDEs 的含量比对照组高出 10~20 倍,最高含量达 3483 ng/g lipid。Yuan 等[62]测得广东电子垃圾拆解区工人血清中 PBDEs 含量为 77~8452 ng/g lipid。Zheng 等[63]测得广东电子垃圾拆解区男性工人血清中 PBDEs 含量为 105~1806 ng/g lipid,女性工人为 206~35902 ng/g lipid,BDE-209 为主要同系物,含量高达 34482 ng/g lipid,是迄今世界上报道的最高值。Zhao 等[64]在浙江台州路桥和温岭电子垃圾拆解区工人全血中测得 PBDEs 含量为 40~396 ng/g lipid,低于广东地区电子垃圾拆解工人血清中 PBDEs 的含量,BDE-209 均是其主要的同系物。而我国北方最大的电子垃圾拆解地职业暴露工人血清中 PBDEs 含量最低,为 10.9~34.8 ng/g lipid,BDE-209 仍是主要的同系物[65]。

随着我国"洋垃圾"禁令的出台,欧美地区大量的洋垃圾转而流入到环境管控相对宽松的东南亚和非洲国家,如印度、越南、泰国等,粗放式的电子垃圾拆解作坊在这些国家重蹈中国污染之覆辙。近年来,不少科学家对这些地区电子垃圾拆解工人 PBDEs 的内暴露进行了研究。Eguchi 等[66,67]和 Kuo 等[68]分别对印度、越南和泰国电子垃圾拆解工人血清中的 PBDEs 含量进行测定,结果分别为 22~2900 ng/g lipid、37~2300 ng/g lipid 和 22.51~2738 ng/g lipid,其浓度中值与我国广东地区电子垃圾拆解工人血清中 PBDEs 含量在同一数量级,但是,低溴代联苯醚(如 BDE-47、BDE-99、BDE-153)是其主要的同系物,与我国不同。这反映了各个地区进口的"洋垃圾"来自不同的地方。此外,Eguchi 等[67]在印度电子垃圾拆解工人血清中测得较高浓度的八溴 OH-PBDEs 和四溴酚(tetra-BPhs)。Yu 等[69]也在我国电子垃圾拆解工人血清中检测到 BDE-209 的潜在代谢产物八溴和九溴 OH-PBDEs。

Shen 等[70]报道了电子垃圾拆解区儿童血液中 PBDEs 含量为(32.1±17.5)ng/g lipid,显著高于对照区。Wu 等[71]研究表示广东某电子垃圾拆解区新生儿均暴露于高水平的 PBDEs,脐带血中 PBDEs 的含量为 1.14~504.97 ng/g lipid,浓度中值为 13.84 ng/g lipid,显著高于对照区,其中 BDE-209 是主要同系物,其次是 BDE-47、BDE-153 和 BDE-99,电子垃圾家庭式拆解作坊是新生儿 PBDEs 暴露的重要因素,而且 PBDEs 暴露水平与不良分娩结局(包括死产、低出生体重和早产等)具有明显关联。2010 年,Han 等[72]研究了电子垃圾拆解区儿童血液样本中 PBDEs 与促甲状腺激素(TSH)的关联,表明儿童体内 TSH 水平随着血清中污染物含量的显著增加而改变。Xu 等[73]研究表明广东某电子垃圾拆解区儿童血液中 TSH 与几乎所有 PBDEs 同系物呈正相关,与胰岛素样生长因子结合蛋白-3(IGFBP-3)呈负相关,而游离三碘甲状腺原氨酸(FT3)和游离甲状腺素(FT4)与 BDE154 呈负相关,说明电子垃圾拆解释放的 PBDEs 污染升高可能是儿童激素改变的重要危险因素。

头发作为无创生物检材,近年来也被逐渐用于评估电子垃圾拆解区暴露人群有机污染物暴露的生物检材。Wen 等[74]、Zhao 等[75]和 Ma 等[76]均测定了台州电子垃圾拆解工人头发中 PBDEs 的含量分别为 18~9400 ng/g dw、0.72~59.52 ng/g dw 和 21.5~1020 ng/g dw;而广东地区电子垃圾拆解工人头发中 PBDEs 含量分别由 Zheng 等[77]和 Qiao 等[37]测得,浓度分别为 12.4~845 ng/g dw 和 49.8~2104 ng/g dw;Tang 等[78]测得河北地区电子垃圾拆解工人头发中 PBDEs 含量为 2.14~861 ng/g dw;BDE-209 均为我国电子垃圾拆解工人头发中主要的 PBDEs 同系物,占 ΣPBDEs 浓度的 48.5%~92.8%。Zheng 等[79]和 Liang 等[80]分别对广东和台州地区电子垃圾拆解工人配对血清和头发中 PBDEs 进行分析,结果显示 BDE-209 均是血清和头发中的主要 PBDEs 同系物,且在两者中的占比相似。

与头发相似,指甲也是无创生物检材的新检材。指甲用于电子垃圾拆解工人 PBDEs 暴露的分析不多。Meng 等[81]测得的广东电子垃圾拆解工人指甲中 PBDEs 含量为 168~1280 ng/g dw,BDE-209 为主要同系物。

2. PCBs

台州电子垃圾拆解工人全血中 PCBs 的含量为 27~1044 ng/g lipid[64],血清中的含量为 24.7~3410 ng/g lipid[82],PCB-28、PCB-118、PCB-74、PCB-153、PCB-132 和 PCB-138 均为主要的同系物。广东电子垃圾拆解工人血清中,Bi 等[60]所研究地区的含量为 17~180 ng/g lipid,而 Zheng 等[63]所研究区域的含量较高,为 136~11846 ng/g lipid,均比同期所研究的 PBDEs 含量低,主要的同系物为 PCB-28、PCB-118、PCB-153、PCB-138,说明两个地区电子垃圾拆解工人受到的 PCBs 暴露量是有差异的,主要与拆解的电子垃圾的种类和组分有关。PCBs 曾被广泛用于变压器和电容器等产品,而台州存在大量废变压器的拆解活动,容易造成含 PCBs 变压器油的泄漏及污染。我国北方最大的电子垃圾拆解地工人血清中 PCBs 的含量最低,为 20~72.9 ng/g lipid[65]。印度、越南和加纳的电子垃圾拆解工人血液样本中 PCBs 的含量分别为 19~3200 pg/g(血清湿重)[66]、150~6400 ng/g lipid(血清)[67]和 0.03~14.7 ng/mL(血浆)[83],PCB-138、PCB-153 和 PCB-118 均是其主要的同系物。可见不同地区电子垃圾拆解工人血液中 PCBs 含量尽管有差异,但是低氯代 PCBs 均是其主要的同系物,这与 PBDEs 显著的地域差异不同。Shen 等[70]报道了电子垃圾拆解区儿童血液中 PCBs 含量为 (40.6 ± 7.01) ng/g lipid,显著高于对照区。Han 等[72]研究了电子垃圾拆解区儿童血液样本中 PCBs 与 TSH 的关联,表明儿童体内 TSH 水平随着血清中 PCBs 含量的显著增加而改变。Guo 等[84]的研究也表明,受电子垃圾拆解污染暴露儿童血清中的 PCBs 和 HFRs 与 TH 相关蛋白、基因表达之间存在剂量关系,而且,暴露浓度越高,碘甲状腺原氨酸去碘酶 I 的表达越高,促甲状腺激素的浓度降低。

台州电子垃圾拆解工人头发中 PCBs 的含量为 12.3~7200 ng/g dw[74, 85],广东电子垃圾拆解工人头发中 PCBs 的含量为 90.2~1657 ng/g dw[37, 66]。与血清一致,低氯代 PCBs 也是电子垃圾拆解工人头发 PCBs 主要的同系物。此外,Zheng 等[51]研究了电子垃圾拆解工人配对的头发和血清中的 PCBs 含量的男女差异,结果显示,男性和女性血清中 PCBs 浓度相似,但是男性头发中的浓度显著低于女性,这主要是由于女性较长的头发在外暴露的时间比男性长,这也间接说明头发中的 PCBs 主要来自大气环境。Meng 等[81]在电子垃圾拆解工人指甲中也检出了 PCBs,含量为 59~341 ng/g;与头发不同,指甲中的 PCBs 没有显著的性别差异,且与年龄之间也没有显著的相关性。

3. 新型阻燃剂

随着各类 PBDEs 和 PCBs 的逐步禁用，它们的替代产品有了迅速发展，如得克隆(DP)、十溴二苯乙烷(DBDPE)、1,2-双(2,4,6-三溴苯氧基)乙烷(BTBPE)和 OPFRs。已有少量报道关于电子垃圾拆解工人体内新型阻燃剂的暴露特征。Ren 等[86]报道我国电子垃圾拆解工人血清中 DP 含量为 7.8~465 ng/g lipid，且拆解工人血清 DP 浓度与 BDE-209 呈正相关关系，anti-DP 异构体的质量分数(f_{anti})为 0.58±0.11，部分样品中初步鉴定出脱氯产物，表明 DP 在人体中的积累主要以立体异构选择的方式进行，并进行脱氯代谢。而 Yan 等[87]测得的电子垃圾拆解工人血清中 DP 的含量为 22~2200 ng/g lipid，且发现女性血清中 DP 含量与年龄有关，男性则无关，而且女性体内 DP 的浓度以及 f_{anti} 值显著高于男性。北方最大的电子垃圾拆解地电子垃圾拆解工人血清中的 DP 含量最低，为 4.21~12.4 ng/g lipid[65]。Zheng 等[88]在电子垃圾拆解区人群头发中测定了 DP 及其脱氯产物(anti-Cl-11-DP)，DP 的浓度为 0.02~58.32 ng/g dw，anti-Cl-11-DP 浓度为 n.d.~0.23 ng/g dw，f_{anti} 值为 0.55±0.11，与灰尘样品相当，且头发和灰尘中的 anti-Cl-11-DP 与 anti-DP 均有显著正相关关系，表明人体内的 DP 主要来源于环境暴露的积累，而不是生物转化。Qiao 等[37]在电子垃圾拆解女工头发中测得的 DP 含量为 1.64~360 ng/g dw。Chen 等[89]测定了电子垃圾拆解工人配对头发和血清样本中的 DP 及其脱氯产物，DP 含量分别为 6.3~1100 ng/g dw 和 22~1400 ng/g lipid，anti-Cl-11-DP 分别为 0.02~1.8 ng/g dw 和 n.d.~7.9 ng/g lipid，且头发中 DP 的含量具有性别差异，女性头发中的 syn-DP、anti-DP 和 anti-Cl-11-DP 与血清均有显著的关联，但男性没有。

DBDPE 和 BTBPE 在广东电子垃圾拆解工人头发中研究较多。Zheng 等[77]测得的电子垃圾拆解工人头发 DBDPE 含量为 5.92~365 ng/g dw，BTBPE 为 0.15~29.2 ng/g dw。Qiao 等[37]测得的电子垃圾拆解女工头发中 DBDPE 含量为 15.8~980 ng/g dw，BTBPE 为 0.64~18.2。可见工人从电子垃圾拆解暴露的 DBDPE 比 BTBPE 至少高一个数量级。Liang 等[80]测定了电子垃圾拆解工人配对的血清和头发中的 DBDPE，含量分别为 27~440 ng/g lipid 和 21~239 ng/g dw，但与 BDE-209 相比，DBDPE 不是电子垃圾拆解工人暴露的溴代阻燃剂。

目前，电子垃圾拆解工人 OPFRs 内暴露主要通过头发中的原型化合物和尿液中的代谢物监测。Lu 等[90]测定了广东电子垃圾拆解工人尿液中含氯和不含氯 OPFRs 代谢物，浓度分别为 <LOQ~58 ng/mL 和 0.32~8.4 ng/mL，二(2-氯乙基)磷酸酯(BCEP)是主要的不含氯 OPFRs 代谢产物(mOPFRs)，且 BCEP 和 DPHP 浓度显著高于对照人群。Shi 等[91]测定了电子垃圾拆解区工人一天内不同时间点尿液中的 6 种 2 酯 OPFRs 和 mOPFRs，BCEP、DBP、BDCIPP 和 DPHP 在 50%以上尿液

样品中均有检出，浓度分别为 2.43~4.80 ng/mL、0.09~2.65 ng/mL、0.46~0.89 ng/mL 和 0.66~1.83 ng/mL，且夜尿浓度高于晨尿。Qiao 等[37]在测定电子垃圾拆解女工头发中其他阻燃剂的同时，对 OPFRs 也进行了分析，含量为 189~1558 ng/g dw。

5.5.2 塑料添加剂暴露特征

电子垃圾拆解过程中，其塑料外壳在焚烧、破碎过程中还会释放出许多塑料添加剂，如 BPA、PAE、PFAS 等。但是，目前对电子垃圾拆解工人塑料添加剂暴露的研究不如阻燃剂研究多，主要以尿液中的污染暴露为主。Zhang 等[82]测定了某区域 50%家庭为电子垃圾拆解作坊的人群尿液中的 BPAs 和 7 种其他双酚化合物，其中 BPA、BPS 和 BPF 检出率均高于 90%，平均浓度分别为 2.99（或 3.75）、0.361（或 0.469）和 0.349（或 0.435）ng/mL（或 mu g/g creatinine），其中 BPA 和 BPF 的浓度显著高于对照人群，表明电子垃圾拆解活动是人群暴露 BPA 和 BPF 的来源，且 BPA 和 BPF 均与 8-OHdG 有显著正相关关系。Zhang 等[92]对电子垃圾拆解区人群尿液中 11 种邻苯二甲酸酯代谢物（mPAEs）进行分析，浓度为 11.1~3380 ng/mL，显著高于对照区，且 mono-(2-isobutyl) phthalate 和 mono-n-butyl phthalate 是主要的同系物，有 8 种 mPAEs 与尿液中的 8-OHdG 有显著正相关，对电子垃圾拆解人群带来了潜在风险。Li 等[93]结果显示根据人群与电子垃圾拆解核心区距离，PAEs 具有空间分布上的差异，开展集中电子垃圾拆解的循环经济产业园中人群尿液的 mPAEs 浓度（389 ng/mL）显著高于其他两个地方（285 ng/mL 和 207 ng/mL）；此外，电子垃圾拆解的方式显著影响 PAEs 暴露水平，家庭车间工人尿液中 mPAEs 浓度（401 ng/mL）显著高于集中管理车间工人尿液中 mPAEs 浓度（298 ng/mL），即使尿液中 mPAEs 浓度在年龄、体重指数（BMI）、性别亚组间存在明显差异，PAEs 暴露水平与这些社会人口学指标之间没有显著的统计学关联。Zhang 等[94]对电子垃圾拆解区人群配对血液和尿液中的全氟辛酸（PFASs）以及健康指标进行分析，结果显示，在所有健康指标中，电子垃圾拆解地人群的空腹血糖（FBG）和泌尿临床指标均显著高于参考地区，电子垃圾拆解区老年人血清中 16 种 PFASs 浓度显著高于对照区，表明电子垃圾拆解活动是 PFASs 暴露的一个来源，而且 BMI 与血清 PFHpA、PFHpA、PFOA 和 PFNA 浓度呈显著正相关，FBG 与血清 PFUnDA 和 PFTeDA 呈负相关。

5.5.3 多环芳烃暴露

电子垃圾拆解过程中涉及的燃烧和热解过程还会产生 PAHs。关于电子垃圾拆解工人 PAHs 暴露的研究主要以尿液 OH-PAHs、头发和指甲 PAHs 及 OH-PAHs 为主。Lu 等[95]从电子垃圾拆解工人尿液样本中测定了 10 种 OH-PAHs 和 2 种氧化应激生物标志物 8-OHdG 和 MDA,其中拆解工人尿液中 OH-PAHs 浓度为 36.6 mu g/g creatinine,显著高于附近居民,OH-PAHs 与 8-OHdG 的升高显著相关,1-羟基芘(1-OH-Pyr)与 MDA 呈显著正相关,表明电子垃圾拆解产生的 PAHs 对暴露工人的 DNA 氧化损伤有一定影响。Feldt 等[96]对非洲最大的电子垃圾回收站工人尿液中的 PAHs 代谢产物(OH-PAHs)的研究结果表明,电子垃圾回收活动是工人 PAHs 暴露的最重要因素。Lin 等[97]对电子垃圾拆解工人头发中 PAHs 以及头发和尿液中 OH-PAHs 进行研究发现,Phe、Flua 和 Pyr 是头发中主要的 PAHs 同系物,2-OH-Nap 是头发和尿液中主要的 OH-PAHs 同系物,但头发与尿液中 OH-PAHs 没有显著差异。Ma 等[98]对电子垃圾拆解工人指甲中的 PAHs 和 OH-PAHs 进行分析得出,与头发相似,Phe、Pyr 和 Flua 也是主要的 PAHs 同系物,2-OH-Nap 是主要的 OH-PAHs 同系物。

电子垃圾拆解区儿童 PAHs 暴露也不容忽视。Xu 等[99]评估了 16 种 PAHs 同系物在电子垃圾拆解区儿童体内的负荷及潜在健康风险,结果发现 PAHs 与儿童身高和胸围均呈负相关关系,而且在男孩中相关性更显著。Zheng 等[100]研究了电子垃圾拆解区儿童 Pb 和 PAHs 共暴露与心血管内皮炎症之间的关系,结果表明血液 Pb、尿 2-OH-Nap 和 OH-Flu 水平升高与 IL-6、IL-12p70、IP-10、CD4(+) T 细胞百分比、中性粒细胞和单核细胞计数升高相关,表明电子垃圾拆解区儿童 Pb 和 PAHs 共暴露会加剧其血管内皮炎症。Dai 等[101]通过电子垃圾拆解区儿童尿液 OH-PAHs 暴露与低级别炎症和血小板参数的相关性研究也表明,表明电子垃圾拆解排放的 PAHs 对当地儿童健康具有潜在风险。Cheng 等[102]通过测定电子垃圾拆解区儿童尿液 OH-PAHs 发现,暴露组 OH-PAHs 浓度明显高于对照组,特别是 1-OH-Nap 和 2-OH-Nap,通过 OH-PAHs 暴露与 AhR、NLRP3 和细胞因子之间的关系表明电子垃圾拆解区儿童 PAHs 暴露与细胞因子风暴的关系可能是由 AhR 和 NLRP3 表达介导的。Wang 等[103,104]研究发现,与对照区相比,电子垃圾拆解区儿童尿液中 OH-PAHs 浓度较高,收缩压、脉压、血清 SOD、GPx 浓度较低,PAHs 暴露与儿童低血压有关,而且电子垃圾拆解区儿童血清甘油三酯浓度和血脂异常发生率较高,血清高密度脂蛋白(HDL)浓度较低。

5.5.4 其他持久性有机污染物暴露

除了上述阻燃剂、塑料添加剂和 PAHs 外,研究人员还对电子垃圾拆解工人其他持久性有机物的暴露进行了研究,如 OCP 和 PCDD/Fs 等。Bi 等[60]研究了广东某电子垃圾拆解工人血清中 OCPs 的水平,结果表明六氯苯、HCH 和 DDTs 是检出率最高的 OCPs,浓度中值分别为 12 ng/g lipid 和 610 ng/g lipid,但是显著低于以渔业为主的对照区,表明 OCPs 并不是电子垃圾拆解区的特征污染物。

我国对电子垃圾拆解工人 PCDD/Fs 的暴露主要以头发分析为主。Wen 等[74]测定了电子垃圾拆解暴露工人头发中 PCDD/Fs 的浓度为 (2.6 ± 0.6) ng/g dw，同系物组成表明高浓度 PCDD/Fs 主要来源于露天电子垃圾焚烧。Ma 等[76]分析了台州电子垃圾拆解工人头发中 PCDD/Fs，含量为 126~5820 pg/g dw，平均值 1670 pg/g dw 是对照人群的 18 倍，其中 TCDFs 是主要的同系物。而 Wittsiepe 等[105]分析了非洲最大的电子垃圾拆解地工人血液中的 PCDD/Fs，基于 WHO2005-TEQ，血液中 PCDD/F 浓度为 2.1~42.7 pg/g lipid，浓度中值为 6.18 pg/g lipid。

综上所述，电子垃圾拆解工人污染物的暴露以传统的阻燃剂 PBDEs 和 PCBs 为主，且含量比其他新型阻燃剂高，其在血液样品中的研究较多，近年来也不断开展了头发和指甲无创样品中的监测分析。而对于新型阻燃剂，DP 在拆解工人血清中的含量较高，与一些电子垃圾拆解场地中工人血清中的 PBDEs 和 PCBs 相当；其次是 DBDPE，但显著低于目前报道的普通居民的浓度，可能与 DBDPE 的使用尚未普及有关；血清中浓度最低的新型阻燃剂为 BTBPE；而 OPFRs 主要针对尿液中的代谢物进行监测，且对拆解工人的暴露研究较少。新型阻燃剂对于电子垃圾拆解工人较低的暴露水平主要可能是现今拆解的电子垃圾还是老一代的添加 PCBs 和 PBDEs 为主，而大部分添加新型阻燃剂的电子产品还在使用中。这给了我们一个警醒，研究人员日后对电子垃圾拆解工人暴露研究时，需要更加关注新型阻燃剂的暴露研究。

（林美卿　马盛韬　唐　僭　安太成）

参 考 文 献

[1] Cahill T M, Groskova D, Charles M J, et al. Atmospheric concentrations of polybrominated diphenyl ethers at near-source sites [J]. Environmental Science & Technology, 2007, 41(18): 6370-6377.

[2] Guo J, Lin K, Deng J, et al. Polybrominated diphenyl ethers in indoor air during waste TV recycling process [J]. Journal of Hazardous Materials, 2015, 283: 439-446.

[3] Zhang M, Shi J, Meng Y, et al. Occupational exposure characteristics and health risk of PBDEs at different domestic e-waste recycling workshops in China [J]. Ecotoxicology and Environmental Safety, 2019, 174: 532-539.

[4] Die Q, Nie Z, Huang Q, et al. Concentrations and occupational exposure assessment of polybrominated diphenyl ethers in modern Chinese e-waste dismantling workshops [J]. Chemosphere, 2019, 214: 379-388.

[5] Xing G H, Liang Y, Chen L X, et al. Exposure to PCBs, through inhalation, dermal contact and dust ingestion at Taizhou, China - A major site for recycling transformers [J]. Chemosphere, 2011, 83(4): 605-611.

[6] Liu R, Ma S, Li G, et al. Comparing pollution patterns and human exposure to atmospheric PBDEs and PCBs emitted from different e-waste dismantling processes [J]. Journal of Hazardous Materials, 2019, 369: 142-149.

[7] Liu R R, Ma S T, Yu Y Y, et al. Field study of PAHs with their derivatives emitted from e-waste dismantling processes and their comprehensive human exposure implications [J]. Environment International, 2020, 144: 106059.

[8] Chen D, Bi X, Liu M, et al. Phase partitioning, concentration variation and risk assessment of polybrominated diphenyl ethers (PBDEs) in the atmosphere of an e-waste recycling site [J]. Chemosphere, 2011, 82(9): 1246-1252.

[9] Luo P, Bao L J, Wu F C, et al. Health risk characterization for resident inhalation exposure to particle-bound halogenated flame retardants in a typical e-waste recycling zone [J]. Environmental Science & Technology, 2014, 48(15): 8815-8822.

[10] Niu S, Tao W Q, Chen R W, et al. Using polychlorinated naphthalene concentrations in the soil from a southeast China e-waste recycling area in a novel screening-level multipathway human cancer risk assessment [J]. Environmental Science & Technology, 2021, 55(10): 6773-6782.

[11] Chen H, Ma S, Yu Y, et al. Seasonal profiles of atmospheric PAHs in an e-waste dismantling area and their associated health risk considering bioaccessible PAHs in the human lung [J]. Science of the Total Environment, 2019, 683: 371-379.

[12] Papadopoulou E, Padilla-Sanchez J A, Collins C D, et al. Sampling strategy for estimating human exposure pathways to consumer chemicals [J]. Emerging Contaminants, 2016, 2(1): 26-36.

[13] Xu F C, Giovanoulis G, van Waes S, et al. Comprehensive study of human external exposure to organophosphate flame retardants *via* air, dust, and hand wipes: The importance of sampling and assessment strategy [J]. Environmental Science & Technology, 2016, 50(14): 7752-7760.

[14] Zhao G, Zhou H, Wang D, et al. PBBs, PBDEs, and PCBs in foods collected from e-waste disassembly sites and daily intake by local residents [J]. Science of the Total Environment, 2009, 407(8): 2565-2575.

[15] Zhang T, Ruan J, Zhang B, et al. Heavy metals in human urine, foods and drinking water from an e-waste dismantling area: Identification of exposure sources and metal-induced health risk [J]. Ecotoxicology and Environmental Safety, 2019, 169: 707-713.

[16] Zheng J, Chen K H, Yan X, et al. Heavy metals in food, house dust, and water from an e-waste recycling area in south China and the potential risk to human health [J]. Ecotoxicology and Environmental Safety, 2013, 96: 205-212.

[17] Collins C D, Craggs M, Garcia-Alcega S, et al. 'Towards a unified approach for the determination of the bioaccessibility of organic pollutants' [J]. Environment International, 2015, 78: 24-31.

[18] Meng X-Z, Zeng E Y, Yu L-P, et al. Assessment of human exposure to polybrominated diphenyl ethers in China *via* fish consumption and inhalation [J]. Environmental Science & Technology, 2007, 41(14): 4882-4887.

[19] Labunska I, Harrad S, Santillo D, et al. Domestic duck eggs: An important pathway of human exposure to PBDEs around e-waste and scrap metal processing areas in eastern China [J]. Environmental Science & Technology, 2013, 47(16): 9258-9266.

[20] Chan J K Y, Man Y B, Wu S C, et al. Dietary intake of PBDEs of residents at two major electronic waste recycling sites in China [J]. Science of the Total Environment, 2013, 463: 1138-1146.

[21] Chan J K Y, Wong M H. A review of environmental fate, body burdens, and human health risk assessment of PCDD/Fs at two typical electronic waste recycling sites in China [J]. Science of the Total Environment, 2013, 463: 1111-1123.

[22] Zhao G F, Wang Z J, Zhou H D, et al. Burdens of PBBs, PBDEs, and PCBs in tissues of the cancer patients in the e-waste disassembly sites in Zhejiang, China [J]. Science of the Total Environment, 2009, 407(17): 4831-4837.

[23] Labunska I, Harrad S, Wang M J, et al. Human dietary exposure to PBDEs around e-waste recycling sites in eastern China [J]. Environmental Science & Technology, 2014, 48(10): 5555-5564.

[24] Jiang H, Lin Z, Wu Y, et al. Daily intake of polybrominated diphenyl ethers *via* dust and diet from an e-waste recycling area in China [J]. Journal of Hazardous Materials, 2014, 276: 35-42.

[25] Anh H Q, Nam V D, Tri T M, et al. Polybrominated diphenyl ethers in plastic products, indoor dust, sediment

and fish from informal e-waste recycling sites in Vietnam: A comprehensive assessment of contamination, accumulation pattern, emissions, and human exposure [J]. Environmental Geochemistry and Health, 2017, 39(4): 935-954.

[26] Zheng X B, Wu J P, Luo X J, et al. Halogenated flame retardants in home-produced eggs from an electronic waste recycling region in South China: Levels, composition profiles, and human dietary exposure assessment [J]. Environment International, 2012, 45: 122-128.

[27] Huang C C, Zeng Y H, Luo X J, et al. Level changes and human dietary exposure assessment of halogenated flame retardant levels in free-range chicken eggs: A case study of a former e-waste recycling site, south China [J]. Science of the Total Environment, 2018, 634: 509-515.

[28] Leung A O W, Chan J K Y, Xing G H, et al. Body burdens of polybrominated diphenyl ethers in childbearing-aged women at an intensive electronic-waste recycling site in China [J]. Environmental Science and Pollution Research, 2010, 17(7): 1300-1313.

[29] Li X, Tian Y, Zhang Y, et al. Accumulation of polybrominated diphenyl ethers in breast milk of women from an e-waste recycling center in China [J]. Journal of Environmental Sciences, 2017, 52: 305-313.

[30] Tay J H, Sellstrom U, Papadopoulou E, et al. Assessment of dermal exposure to halogenated flame retardants: Comparison using direct measurements from hand wipes with an indirect estimation from settled dust concentrations [J]. Environment International, 2018, 115: 285-294.

[31] Liu X, Yu G, Cao Z, et al. Occurrence of organophosphorus flame retardants on skin wipes: Insight into human exposure from dermal absorption [J]. Environment International, 2017, 98: 113-119.

[32] Hammel S C, Hoffman K, Phillips A L, et al. Comparing the use of silicone wristbands, hand wipes, and dust to evaluate children's exposure to flame retardants and plasticizers [J]. Environmental Science & Technology, 2020, 54(7): 4484-4494.

[33] Larsson K, de Wit C A, Sellstrom U, et al. Brominated flame retardants and organophosphate esters in preschool dust and children's hand wipes [J]. Environmental Science & Technology, 2018, 52(8): 4878-4888.

[34] Liu X T, Cao Z G, Yu G, et al. Estimation of exposure to organic flame retardants *via* hand wipe, surface wipe, and dust: Comparability of different assessment strategies [J]. Environmental Science & Technology, 2018, 52(17): 9946-9953.

[35] Wu Y Y, Li Y Y, Kang D, et al. Tetrabromobisphenol A and heavy metal exposure *via* dust ingestion in an e-waste recycling region in southeast China [J]. Science of the Total Environment, 2016, 541: 356-364.

[36] Wu C C, Bao L J, Tao S, et al. Dermal uptake from airborne organics as an important route of human exposure to e-waste combustion fumes [J]. Environmental Science & Technology, 2016, 50(13): 6599-6605.

[37] Qiao L, Zheng X B, Zheng J, et al. Legacy and currently used organic contaminants in human hair and hand wipes of female e-waste dismantling workers and workplace dust in South China [J]. Environmental Science & Technology, 2019, 53(5): 2820-2829.

[38] Shen M J, Ge J L, Lam J C W, et al. Occurrence of two novel triazine-based flame retardants in an e-waste recycling area in south China: Implication for human exposure [J]. Science of the Total Environment, 2019, 683: 249-257.

[39] Abdallah M A-E, Harrad S. Dermal contact with furniture fabrics is a significant pathway of human exposure to brominated flame retardants [J]. Environment international, 2018, 118: 26-33.

[40] Beaucham C C, Ceballos D, Mueller C, et al. Field evaluation of sequential hand wipes for flame retardant exposure in an electronics recycling facility [J]. Chemosphere, 2019, 219: 472-481.

[41] Wang Y, Peris A, Rifat M R, et al. Measuring exposure of e-waste dismantlers in Dhaka Bangladesh to organophosphate esters and halogenated flame retardants using silicone wristbands and T-shirts [J]. Science of the Total Environment, 2020, 720: 137480.

[42] Estill C F, Slone J, Mayer A, et al. Worker exposure to flame retardants in manufacturing, construction and service industries [J]. Environment International, 2020, 135: 105349.

[43] Abafe O A, Martincigh B. An assessment of polybrominated diphenyl ethers and polychlorinated biphenyls in the indoor dust of e-waste recycling facilities in south Africa: Implications for occupational exposure [J]. Environmental Science and Pollution Research, 2015, 22(18): 14078-14086.

[44] Deng J, Guo J, Zhou X, et al. Hazardous substances in indoor dust emitted from waste TV recycling facility [J]. Environmental Science and Pollution Research, 2014, 21(12): 7656-7667.

[45] Abdallah M A E, Pawar G, Harrad S. Effect of bromine substitution on human dermal absorption of polybrominated diphenyl ethers [J]. Environmental Science & Technology, 2015, 49(18): 10976-10983.

[46] Nguyen L V, Diamond M L, Venier M, et al. Exposure of Canadian electronic waste dismantlers to flame retardants [J]. Environment International, 2019, 129: 95-104.

[47] Hovander L, Athanasiadou M, Asplund L, et al. Extraction and cleanup methods for analysis of phenolic and neutral organohalogens in plasma [J]. Journal of Analytical Toxicology, 2000, 24(8): 696-703.

[48] 马盛韬. 职业暴露及背景人群体内多溴联苯醚及其代谢产物的初步研究 [D]. 北京: 中国科学院研究生院, 2012.

[49] Ashrap P, Zheng G, Wan Y, et al. Discovery of a widespread metabolic pathway within and among phenolic xenobiotics [J]. Proceedings of the National Academy of Sciences of the United States of America, 2017, 114(23): 6062-6067.

[50] Lin M Q, Ma S T, Yu Y X, et al. Simultaneous determination of multiple classes of phenolic compounds in human urine: Insight into metabolic biomarkers of occupational exposure to e-waste [J]. Environmental Science & Technology Letters, 2020, 7(5): 323-329.

[51] Zheng J, Yu L H, Chen S J, et al. Polychlorinated biphenyls (PCBs) in human hair and serum from e-waste recycling workers in southern China: Concentrations, chiral signatures, correlations, and source identification [J]. Environmental Science & Technology, 2016, 50(3): 1579-1586.

[52] Kucharska A, Covaci A, Vanermen G, et al. Non-invasive biomonitoring for PFRs and PSDEs: New insights in analysis of human hair externally exposed to selected flame retardants [J]. Science of the Total Environment, 2015, 505: 1062-1071.

[53] Zheng J, Yan X, Chen S J, et al. Polychlorinated biphenyls in human hair at an e-waste site in China: Composition profiles and chiral signatures in comparison to dust [J]. Environment International, 2013, 54: 128-133.

[54] Lu D, Feng C, Lin Y, et al. Determination of organochlorines, polychlorinated biphenyls and polybrominated diphenyl ethers in human hair: Estimation of external and internal exposure [J]. Chemosphere, 2014, 114: 327-336.

[55] Lin M Q, Tang J, Ma S T, et al. Simultaneous determination of polybrominated diphenyl ethers, polycyclic aromatic hydrocarbons and their hydroxylated metabolites in human hair: A potential methodology to distinguish external from internal exposure [J]. Analyst, 2019, 144(24): 7227-7235.

[56] Motorykin O, Schrlau J, Jia Y L, et al. Determination of parent and hydroxy PAHs in personal $PM_{2.5}$ and urine samples collected during Native American fish smoking activities [J]. Science of the Total Environment, 2015, 505: 694-703.

[57] Stapleton H M, Kelly S M, Allen J G, et al. Measurement of polyhrominated diphenyl ethers on hand wipes: Estimating exposure from hand-to-mouth contact [J]. Environmental Science & Technology, 2008, 42(9): 3329-3334.

[58] Sjodin A, Hagmar L, Klasson-Wehler E, et al. Flame retardant exposure: Polybrominated diphenyl ethers in blood from Swedish workers [J]. Environmental Health Perspectives, 1999, 107(8): 643-648.

[59] Thomsen C, Lundanes E, Becher G. Brominated flame retardants in plasma samples from three different occupational groups in Norway [J]. Journal of Environmental Monitoring, 2001, 3(4): 366-370.

[60] Bi X H, Thomas G O, Jones K C, et al. Exposure of electronics dismantling workers to polybrominated diphenyl ethers, polychlorinated biphenyls, and organochlorine pesticides in south China [J]. Environmental

Science & Technology, 2007, 41(16): 5647-5653.
[61] Qu W, Bi X, Sheng G, et al. Exposure to polybrominated diphenyl ethers among workers at an electronic waste dismantling region in Guangdong, China [J]. Environment International, 2007, 33(8): 1029-1034.
[62] Yuan J, Chen L, Chen D, et al. Elevated serum polybrominated diphenyl ethers and thyroid-stimulating hormone associated with lymphocytic micronuclei in Chinese workers from an e-waste dismantling site [J]. Environmental Science & Technology, 2008, 42(6): 2195-2200.
[63] Zheng J, He C-T, Chen S-J, et al. Disruption of thyroid hormone (TH) levels and TH-regulated gene expression by polybrominated diphenyl ethers (PBDEs), polychlorinated biphenyls (PCBs), and hydroxylated PCBs in e-waste recycling workers [J]. Environment International, 2017, 102: 138-144.
[64] Zhao X R, Qin Z F, Yang Z Z, et al. Dual body burdens of polychlorinated biphenyls and polybrominated diphenyl ethers among local residents in an e-waste recycling region in southeast China [J]. Chemosphere, 2010, 78(6): 659-666.
[65] Yang Q Y, Qiu X H, Li R, et al. Exposure to typical persistent organic pollutants from an electronic waste recycling site in northern China [J]. Chemosphere, 2013, 91(2): 205-211.
[66] Eguchi A, Nomiyama K, Devanathan G, et al. Different profiles of anthropogenic and naturally produced organohalogen compounds in serum from residents living near a coastal area and e-waste recycling workers in India [J]. Environment International, 2012, 47: 8-16.
[67] Eguchi A, Nomiyama K, Tue N M, et al. Residue profiles of organohalogen compounds in human serum from e-waste recycling sites in north Vietnam: Association with thyroid hormone levels [J]. Environmental Research, 2015, 137: 440-449.
[68] Kuo L J, Cade S E, Cullinan V, et al. Polybrominated diphenyl ethers (PBDEs) in plasma from e-waste recyclers, outdoor and indoor workers in the puget sound, WA region [J]. Chemosphere, 2019, 219: 209-216.
[69] Yu Z Q, Zheng K W, Ren G F, et al. Identification of hydroxylated octa- and nona-bromodiphenyl ethers in human serum from electronic waste dismantling workers [J]. Environmental Science & Technology, 2010, 44(10): 3979-3985.
[70] Shen H, Ding G, Han G, et al. Distribution of PCDD/Fs, PCBs, PBDEs and organochlorine residues in children's blood from Zhejiang, China [J]. Chemosphere, 2010, 80(2): 170-175.
[71] Wu K S, Xu X J, Liu J X, et al. Polyhrominated diphenyl ethers in umbilical cord blood and relevant factors in neonates from Guiyu, China [J]. Environmental Science & Technology, 2010, 44(2): 813-819.
[72] Han G G, Ding G Q, Lou X M, et al. Correlations of PCBs, dioxin, and PBDE with TSH in children's blood in areas of computer e-waste recycling [J]. Biomedical and Environmental Sciences, 2011, 24(2): 112-116.
[73] Xu X J, Liu J X, Zeng X, et al. Elevated serum polybrominated diphenyl ethers and alteration of thyroid hormones in children from Guiyu, China [J]. Plos One, 2014, 9(11): e113699.
[74] Wen S, Yang F X, Gong Y, et al. Elevated levels of urinary 8-hydroxy-2'-deoxyguanosine in male electrical and electronic equipment dismantling workers exposed to high concentrations of polychlorinated dibenzo-p-dioxins and dibenzofurans, polybrominated diphenyl ethers, and polychlorinated biphenyls [J]. Environmental Science & Technology, 2008, 42(11): 4202-4207.
[75] Zhao G F, Wang Z J, Dong M H, et al. PBBs, PBDEs, and PCBs levels in hair of residents around e-waste disassembly sites in Zhejiang province, China, and their potential sources [J]. Science of the Total Environment, 2008, 397(1-3): 46-57.
[76] Ma J, Cheng J P, Wang W H, et al. Elevated concentrations of polychlorinated dibenzo-p-dioxins and polychlorinated dibenzofurans and polybrominated diphenyl ethers in hair from workers at an electronic waste recycling facility in Eastern China [J]. Journal of Hazardous Materials, 2011, 186(2-3): 1966-1971.
[77] Zheng J, Luo X J, Yuan J G, et al. Levels and sources of brominated flame retardants in human hair from urban, e-waste, and rural areas in South China [J]. Environmental Pollution, 2011, 159(12): 3706-3713.
[78] Tang Z W, Huang Q F, Cheng J L, et al. Polybrominated diphenyl ethers in soils, sediments, and human hair in

a plastic waste recycling area: A neglected heavily polluted area [J]. Environmental Science & Technology, 2014, 48(3): 1508-1516.

[79] Zheng J, Chen K H, Luo X J, et al. Polybrominated diphenyl ethers (PBDEs) in paired human hair and serum from e-waste recycling workers: Source apportionment of hair PBDEs and relationship between hair and serum [J]. Environmental Science & Technology, 2014, 48(1): 791-796.

[80] Liang S, Xu F, Tang W B, et al. Brominated flame retardants in the hair and serum samples from an e-waste recycling area in southeastern China: The possibility of using hair for biomonitoring [J]. Environmental Science and Pollution Research, 2016, 23(15): 14889-14897.

[81] Meng H J, Tang B, Zheng J, et al. Levels and sources of PBDEs and PCBs in human nails from e-waste, urban, and rural areas in south China [J]. Environmental Science-Processes & Impacts, 2020, 22(8): 1710-1717.

[82] Zhang T, Xue J C, Gao C Z, et al. Urinary concentrations of bisphenols and their association with biomarkers of oxidative stress in people living near e-waste recycling facilities in China [J]. Environmental Science & Technology, 2016, 50(7): 4045-4053.

[83] Kaifie A, Schettgen T, Bertram J, et al. Informal e-waste recycling and plasma levels of non-dioxin-like polychlorinated biphenyls (NDL-PCBs) - A cross-sectional study at Agbogbloshie, Ghana [J]. Science of the Total Environment, 2020, 723: 138073.

[84] Guo L C, Liu T, Yang Y, et al. Changes in thyroid hormone related proteins and gene expression induced by polychlorinated biphenyls and halogen flame retardants exposure of children in a Chinese e-waste recycling area [J]. Science of the Total Environment, 2020, 742: 140597.

[85] Zhao C, Li J F T, Li X H, et al. Measurement of polychlorinated biphenyls with hand wipes and matched serum collected from Chinese e-waste dismantling workers: Exposure estimates and implications [J]. Science of the Total Environment, 2021, 799: 149444.

[86] Ren G F, Yu Z Q, Ma S T, et al. Determination of dechlorane pus in serum from electronics dismantling workers in south China [J]. Environmental Science & Technology, 2009, 43(24): 9453-9457.

[87] Yan X, Zheng J, Chen K H, et al. Dechlorane plus in serum from e-waste recycling workers: Influence of gender and potential isomer-specific metabolism [J]. Environment International, 2012, 49: 31-37.

[88] Zheng J, Wang J, Luo X J, et al. Dechlorane plus in human hair from an e-waste recycling area in south China: Comparison with dust [J]. Environmental Science & Technology, 2010, 44(24): 9298-9303.

[89] Chen K H, Zheng J, Yan X, et al. Dechlorane plus in paired hair and serum samples from e-waste workers: Correlation and differences [J]. Chemosphere, 2015, 123: 43-47.

[90] Lu S Y, Li Y X, Zhang T, et al. Effect of e-waste recycling on urinary metabolites of organophosphate flame retardants and plasticizers and their association with oxidative stress [J]. Environmental Science & Technology, 2017, 51(4): 2427-2437.

[91] Shi Y, Zheng X, Yan X, et al. Short-term variability in levels of urinary phosphate flame retardant metabolites in adults and children from an e-waste recycling site [J]. Chemosphere, 2019, 234: 395-401.

[92] Zhang B, Zhang T, Duan Y, et al. Human exposure to phthalate esters associated with e-waste dismantling: Exposure levels, sources, and risk assessment [J]. Environment International, 2019, 124: 1-9.

[93] Li J, Xu Z M. Compound tribo-electrostatic separation for recycling mixed plastic waste [J]. Journal of Hazardous Materials, 2019, 367: 43-49.

[94] Zhang T, Zhang B, Bai X Y, et al. Health status of elderly people living near e-waste recycling sites: Association of e-waste dismantling activities with legacy perfluoroalkyl substances (PFASs) [J]. Environmental Science & Technology Letters, 2019, 6(3): 133-140.

[95] Lu S Y, Li Y X, Zhang J Q, et al. Associations between polycyclic aromatic hydrocarbon (PAH) exposure and oxidative stress in people living near e-waste recycling facilities in China [J]. Environment International, 2016, 94: 161-169.

[96] Feldt T, Fobil J N, Wittsiepe J, et al. High levels of PAH-metabolites in urine of e-waste recycling workers from

Agbogbloshie, Ghana [J]. Science of the Total Environment, 2014, 466: 369-376.

[97] Lin M Q, Tang J, Ma S T, et al. Insights into biomonitoring of human exposure to polycyclic aromatic hydrocarbons with hair analysis: A case study in e-waste recycling area [J]. Environment International, 2020, 136: 105432.

[98] Ma S T, Zeng Z H, Lin M Q, et al. PAHs and their hydroxylated metabolites in the human fingernails from e-waste dismantlers: Implications for human non-invasive biomonitoring and exposure [J]. Environmental Pollution, 2021, 283: 117059.

[99] Xu X, Liu J, Huang C, et al. Association of polycyclic aromatic hydrocarbons (PAHs) and lead co-exposure with child physical growth and development in an e-waste recycling town [J]. Chemosphere, 2015, 139: 295-302.

[100] Zheng X, Huo X, Zhang Y, et al. Cardiovascular endothelial inflammation by chronic coexposure to lead (Pb) and polycyclic aromatic hydrocarbons from preschool children in an e-waste recycling area [J]. Environmental Pollution, 2019, 246: 587-596.

[101] Dai Y F, Huo X, Cheng Z H, et al. Alterations in platelet indices link polycyclic aromatic hydrocarbons toxicity to low-grade inflammation in preschool children [J]. Environment International, 2019, 131: 105043.

[102] Cheng Z H, Huo X, Dai Y F, et al. Elevated expression of AhR and NLRP3 link polycyclic aromatic hydrocarbon exposure to cytokine storm in preschool children [J]. Environment International, 2020, 139: 105720.

[103] Wang Q H, Xu X J, Zeng Z J, et al. Antioxidant alterations link polycyclic aromatic hydrocarbons to blood pressure in children [J]. Science of the Total Environment, 2020, 732: 138944.

[104] Wang Q H, Xu X J, Zeng Z J, et al. PAH exposure is associated with enhanced risk for pediatric dyslipidemia through serum SOD reduction [J]. Environment International, 2020, 145: 106132.

[105] Wittsiepe J, Fobil J N, Till H, et al. Levels of polychlorinated dibenzo-p-dioxins, dibenzofurans (PCDD/Fs) and biphenyls (PCBs) in blood of informal e-waste recycling workers from Agbogbloshie, Ghana, and controls [J]. Environment International, 2015, 79: 65-73.

第6章 电子垃圾排放污染物的毒理学与健康危害

电子垃圾本身含有毒害物质,其不恰当的拆解回收、处理处置等过程均可能产生毒害组分,经无意排放进入到各种环境介质。因此生活在电子垃圾拆解场地附近居民和拆解作业工人不可避免地暴露于这些毒害物质,从而对暴露人群造成了一系列的副作用和健康负担,随之产生各种毒理学特征和健康风险。本章从毒性来源和效应为出发点,讨论了有毒重金属(如铅、镉、铬、汞、铜和镍)的毒性效应,如神经、遗传、肝肾、心血管系统、肠胃系统和呼吸系统毒性;还阐述了典型有机污染物(如多溴联苯醚、多氯联苯和双酚 A 等)引起的内分泌干扰、免疫毒性、生殖毒性等对人体健康的负面作用。同时,结合流行病学证据对电子垃圾拆解场地的暴露人群及与之引起的健康负担的相关证据进行了探讨。

6.1 毒性来源与效应概述

6.1.1 电子垃圾产生毒性效应的原因

电子垃圾的不当处理处置已经在发展中国家引起了不同程度的环境污染问题,并对人体健康构成了威胁。电子电器产品的制造材料内所含有的成分十分复杂:有重金属(如电脑、电视机显像管内的铅,电脑元件中含有的汞等有害物质,手机的原材料中的镉、铅等有毒物质)和各类有机物质(包括阻燃剂和助剂等)。电子垃圾对环境以及人体产生毒性效应的原因归根结底是其内部所含有的具有潜在危害的成分,进入环境介质及人体中进而产生的潜在风险与严重后果。对重金属来说,镉、铅和汞等导致的生物毒性比较显著,究其原因一是电子垃圾在填埋过程中处理不当,重金属能够渗入土壤、污染地下水,从而在水环境、沉积物、悬浮物质等多种环境介质中长期地累积并且存在一定的迁移,进而对植物、动物、微生物等造成严重的损伤,最终又能通过食物链进入人体引起一定的中毒甚至是导致病变。二是焚烧过程所排放出来的大量有害气体(如 PCBs、二噁英等致癌物质)通过生物吸入进而产生负面伤害。有机物质中以持久性毒性化学污染物含量居多的电子垃圾又会对人体产生不同于重金属物质的毒性。这类物质的持久性、生物富集性、半挥发性、高毒性,使得通过各种环境介质能够远距离迁移从而给人类和环境带来危害和风险,并且一些新的持久性有毒物质污染事件引起了各国科学家们的高度关注，各类持久性有毒化学污染物对环境造成的影响（包括对污染水平和污染特征的研究）以及在生物体内的累积及其毒理已成为目前迫切需要开展的工作。

6.1.2 重金属的毒性效应

电子垃圾中含有大量的有色金属和贵金属，将这些金属回收利用可以带来巨大的经济效益，但是如果将其随意丢弃或者采用不适当的利用方式或者使用不当的处理处置方法就会成为环境污染中重要的污染源之一，并通过各种暴露途径进入人体进而对人体产生一定的伤害。本小节对重金属铅、镉、铬、镍、汞、钡、铍金属或其化合物进行毒性效应的分析。

铅是一种被广泛应用于电子产品制造业中的重要金属材料，电脑显示器的荧幕（电脑中含有金属铅的主要部位）、线路板和其他元件的焊接点是电脑中含有金属铅的部分。铅在环境中累积，从而对动植物、微生物都有强烈且长久的影响。铅大多经由胃肠道或呼吸道进入人体，而铅的职业性暴露人群主要通过呼吸的方式吸入铅。铅损伤人体健康，对中枢神经、造血、心血管、肾脏和生殖等多系统产生毒性。铅的神经毒性作用将影响神经系统，尤其对胎儿和小于 6 岁儿童的大脑智力损伤更为显著。大量研究表明，当儿童血铅浓度大于 100 μg/L 时，智力、认知及行为能力明显低下[1]。此外，铅引起的致癌作用应该越来越受到人们的重视，尤其肾脏引起的肿瘤和癌症值得引起人们的关注。

镉常被用在电阻器、红外线发生器、半导体、阴极射线管等元器件中，也是塑料固化剂的重要组分。重金属镉的化合物对人也是非常有害的，会在人体中各个部位中积累。镉及其化合物主要与肺癌、肾癌和前列腺癌[2]、膀胱癌的发生有关。镉在人体内，主要累积在肾和肺器官，镉对肝肾的有害影响主要体现在镉进入人体后富集在体内，结合金属硫蛋白的毒性物质，在体内存留长达 10~30 年的时间。此外，对肺部的伤害主要表现为致使肺组织病变包括肺部纤维化和肺气肿，甚至可导致肺癌。镉也影响中枢神经系统，当镉含量超标时，直接影响儿童大脑发育障碍，镉还可能导致其他疾病的产生，包括高血压、动脉粥样硬化、骨质疏松和软骨症等疾病。此外，镉对生殖系统也有明显的影响，抑制了动物的胚胎的生长并停滞，导致出现畸胎、死胎的现象，更严重的是出现不孕不育的症状。

自然界中的铬通常以三价和六价的形式存在，人体通过食物和水等方式摄入铬，三价铬是人体必需微量元素之一，但是电子垃圾产生的六价铬具有强腐蚀和氧化性，会通过人的呼吸等途径进入人体而产生危害。它很容易穿过细胞膜而被吸收，而后对被污染了的细胞产生毒害影响。六价铬于 1990 年就被国际癌症研究机构(IARC)列为已知的人类致癌物。六价铬是一种存在于环境中的毒性极高的物质,会对 DNA 产生损伤作用,职业性铬暴露人群的鼻咽部恶性肿瘤肺癌的发病率明显增高。铬引起的对呼吸道的疾病有鼻炎、支气管炎、肺气肿等,作为一种极易致敏的有毒金属,还可引发哮喘、过敏性皮炎或湿疹等副作用,同时在消化道表现出呕吐、恶心、溃疡和腹泻等不良反应。只有在高暴露剂量条件下,铬会表现出致癌作用,且致癌作用只发生在肺和鼻腔这些特定部位[3],且 10%的被人体吸收的六价铬中在人体内停留可能长达五年。此外,实验研究表明,六价铬会影响生命体的繁殖功能。

镍主要存在于阴极射线管、印刷电路板和外壳等电子垃圾中。有研究表明,镍可导致 DNA 链断裂,接触镍血细胞会出现的 DNA 断裂频率显著高于对照组人群[4]。也有研究表明,其致癌性是与 RNA 结合的结果。职业性接触镍与心血管系统病变、皮肤变态反应和肺纤维化等有关[5]。国际癌症机构很早就将镍列为人类确证致癌物,其致癌机理是进入细胞后通过自由基作用、DNA-蛋白质交联、碱基甲基化、多基因变异等方式,其中镍的硫化物、氧化物会引发以肺癌为主的癌症。

广泛存在于电池、线路板、转换器和平面显示器中的汞主要以汞蒸气形式存在于空气中,经呼吸道进入人体。有机汞混合物则通过肺、消化道、皮肤等途径被人体吸收。汞具有脂溶性使得在其被人体吸收之后迅速经血液到达全身各器官中,肾脏中含量分布较多,在体内高达总汞量的 70%~80%。汞的毒性作用主要表现在神经毒性、肾毒性、生殖毒性和免疫毒性等,且具有一定的持久性。

钡主要存在于电脑显示器阴极射线管荧屏上,目前还没有关于长期暴露于钡对人体产生影响的资料,但是短期暴露于钡即可导致脑肿、肌肉无力,还会损伤心脏、肝脏和脾脏。

铍适于各种工业用途,如电脑这类电子产品,被广泛用于主板和键盘底片之中。铍被认为是与肺癌有关的致癌物,长期接触即使是很小剂量的铍的职业人群,也会容易导致一种肺病和皮肤病——"铍长期症"。

6.1.3 有机污染物的毒性效应

电子垃圾中包含的毒害有机物主要以持久性有机污染物为主,如多溴联苯醚(PBDEs)、多氯联苯(PCBs)、得克隆(DP)、邻苯二甲酸酯(PAEs)、多环芳烃(PAHs)、二噁英(Dioxin)等。

PCBs 被广泛应用于电力装置的制造行业之中,包括电容器和变压器设备的绝缘油、热交换剂、增塑剂、润滑剂、填料添加剂和浸渍剂等。PCBs 经由空气暴露、食物暴露、饮用水暴露以及皮肤暴露等各类暴露途径对人体产生一定的毒性效应。人体经由空气吸入 PCBs 会损害呼吸系统，使得呼吸道很容易被传染病感染；这种持久性有机物的脂溶性使得其在脂肪含量较高的动物类食品中大量存在，经食物的摄入，会在进入体内后很好地为人体所吸收并广泛分布于全身组织并产生相应的毒性效应。研究表明，PCBs 会集中存在于肝脏之中，即肝脏是 PCBs 中毒的主要靶器官。另有大鼠、小鼠的实验的研究表明 PCBs 具有致癌性，并且这种癌变作用于肝脏；职业性的 PCBs 接触又会造成工作人员体内脂肪中的 PCBs 的沉积。长期的职业暴露更会导致皮肤病多发现象的出现、肾功能和血尿素氮异常、发育毒性和神经毒性，诱发心血管系统疾病和恶性黑素瘤等癌症，甚至存在因长期暴露而致死的风险。另有动物实验研究结果显示，PCBs 的致突变性与胚胎的染色体畸变有关。此外，PCBs 中毒患者的免疫系统和淋巴系统受到抑制和损害作用；内分泌系统存在一定的毒性影响，表现为改变甲状腺激素的水平，使甲状腺机能衰退[6-8]。

在电子制造业中，DP 被广泛应用于电线、尼龙、电缆、电视、电子元件和计算机硬件连接器及外壳等高分子材料中[9]。DP 具有典型持久性污染物的特点，高亲脂性、不易被降解并且能够在生物体内富集。长期暴露于 DP 之中会对人体产生极大的危害，DP 的亚急性毒理学研究表明，长期的皮肤接触或者高剂量吸入会造成肺部、肝脏和生殖系统的病变。

以 PAEs 为例的增塑剂主要存在于电线中，有研究表明，PAEs 在人体和动物体内发挥着类似雌性激素的作用，可干扰内分泌，可导致男性生殖系统产生病变，此外，PAEs 还会提高女性的乳腺癌的发病率。

存在于废弃电子产品之中的 PAHs 属于典型的持久性有毒物质，具有致突变性、致癌性、毒性、遗传毒性，可能对肝脏、肾脏造成损害，对人体的呼吸系统、循环系统、神经系统损伤。含 PAHs 颗粒物的吸入可导致肺功能下降与肺部损伤；此外，PAHs 可导致如心脏自主神经功能紊乱、心血管疾病等循环系统的损伤；PAHs 主要对儿童的神经发育产生影响，对于拆解电子垃圾的职业人群可造成神经系统损伤；肾脏损害与接触 PAHs 有关。

电子垃圾的简易焚烧所产生排放的 Dioxin 属于一类剧毒物质，它无任何用途而是作为一种过程产生的副产物，在环境中具有热稳定性、脂溶性、高的稳定性和低挥发性，因其具有高的辛醇-水分配系数，很容易能通过食物链进行生物富集，又因其具有难降解性，会长久蓄积在体内可以对机体造成严重的损害。Dioxin 可以通过食物、空气和皮肤接触入侵人体，对生殖系统产生影响，对胎儿及婴儿的神经发育系统和免疫系统会造成一定的损害。此外，Dioxin 还会导致皮肤肿瘤和甲状腺肿瘤，还具有致死作用和"消瘦综合征"、胸腺萎缩、免疫毒性、肝脏毒性、氯痤疮、生殖毒性、发育毒性和致畸性、致癌性和环境内分泌干扰作用等多种毒性作用。

6.1.4 电子垃圾污染物的毒性分类

电子垃圾中毒害物质所产生的毒性效应大致可以分为神经毒性、遗传毒性、肝肾毒性、心血管系统毒性、肠胃系统毒性、呼吸系统毒性、内分泌干扰效应和免疫系统毒性等，表 6-1 列出了电子垃圾中典型毒害物质所具有的毒性效应。

表 6-1 电子垃圾中典型毒害物质的毒性效应

毒性效应	典型毒害物质及症状
神经毒性	锰和铅主要损伤人的脑细胞，造成痴呆、脑死亡等；汞主要破坏大脑的视力神经；铬会造成精神异常的不正常行为。有机污染物中重点关注 PBDEs、PCBs 和双酚 A 等污染物
遗传毒性	铬会引起 DNA 损坏
肝肾毒性	铜可导致肝硬化；锰导致肝脏的损伤；铅、汞和镉引起肾功能失调
心血管系统毒性	钒、锰会对心脏造成损伤；镉引起心脑血管疾病；钡、PCBs 等
肠胃系统	PCBs 等
呼吸系统	镉、汞、钒会损伤肺部；镍会导致哮喘和慢性支气管炎，并削弱肺部功能；铬会引发哮喘性支气管炎；铍、钒、PCBs 等
内分泌干扰效应	锰超量时会使人甲状腺功能亢进；铅、PBDEs、PAEs、双酚 A 以及全氟化物等
免疫系统毒性	双酚 A、PAEs 和 PCBs 等

6.2 重金属的毒性效应

电子垃圾中含有大量的重金属，如印刷线路板以及电子显示器中含有大量的铅、镉、铜以及汞等，而这些重金属通常会通过各种暴露途径进入人体，从而对人体系统产生毒性，造成一系列疾病甚至癌症。

6.2.1 神经毒性效应

神经毒性是指暴露在有毒物质中引起的神经生理变化，如认知、记忆障碍，以及情绪或情绪的改变或发作。一些重金属，如铅、汞、镉、砷和铝，被已知或怀疑具有发育性神经毒性。据一项研究表明[10]，由于在电子垃圾回收环境中可能接触到毒害物质，广东某电子垃圾拆解地新生儿脐带血和胎粪铅水平显著高于参照组，而新生儿行为神经评估(NBNA)评分低于参照组，且新生儿 NBNA 评分与脐带血血铅浓度呈负相关。电子垃圾拆解地儿童血液中铅、镉、锰水平与某些行为异常相关，如行为问题和反社会行为，血清 S100b 与血液重金属和某些行为异常相关。

接触镉会严重影响神经系统的功能[11,12]，症状包括头痛和眩晕，嗅觉障碍，帕金森等症状，血管运动功能减慢，周围神经病变，平衡能力下降，注意力下降和学习障碍[13,14]。Wang 等[15]以重金属镉为例研究其神经毒性效应，对镉的神经毒性效应的体内外证据进行论述。研究结果表明，镉的神经毒性效应与细胞生化改变和中枢神经系统功能改变有关，提示神经毒性效应可能在长期接触镉暴露的全身毒性效应中起一定作用。文中在大鼠的实验中，由于血脑屏障完整性的不同，镉对新生和幼龄大鼠的毒性比对成年大鼠更大。镉可增加大鼠血脑屏障的通透性，使其穿透并积聚在发育中和成年大鼠的大脑中[16,17]，导致脑细胞内积聚、细胞功能障碍和脑水肿。越来越多的证据表明，镉可能是阿尔茨海默病（AD）和帕金森病（PD）等神经退行性疾病的一个可能的病因[18]。另外，接触砷和锰与儿童的神经发育问题有关，而镉诱导的神经毒性可能是由于神经发生受损，导致神经元分化和轴突发生明显减少，导致神经细胞死亡[19]。

铅暴露可导致人类神经功能的显著下降。人体内的铅可通过多种机制对神经系统造成损害，包括对神经系统的形态学和药理学的两类直接影响[20]。形态效应改变神经系统的发育，特别是从产前到童年，包括在神经元迁移和分化过程中破坏关键分子[21]，减少神经元唾液酸产量干扰突触形成，以及胶质细胞过早分化。药理作用源于铅可充当中枢神经系统中药剂的作用。铅替代了钙，在较小程度上替代了锌，并不适当地触发了依赖钙调素的过程。铅还会干扰神经递质的释放，扰乱 γ-氨基丁酸（GABA）、多巴胺和胆碱能系统的功能，并在新生儿期抑制 N-甲基-D-天冬氨酸（NMDA）离子通道[22]。体外研究表明，铅激活毛细血管细胞中的蛋白激酶 C，抑制细胞膜上的 Na^+/K^+-ATP 酶，干扰能量代谢[23]。在细胞内，铅似乎干扰线粒体的钙释放，导致活性氧物种的形成，通过形成通透性转换孔加速线粒体的自我破坏，并启动程序性细胞死亡过程[24]。研究发现，急性和长期接触铅可以导致显著的神经生理和神经心理缺陷，这取决于暴露的水平。铅暴露会增加患多种疾病的风险，这些疾病可能会对神经系统功能产生不利影响，包括高血压、肾功能受损、甲状腺功能受损、维生素 D 缺乏和早产等。

6.2.2 遗传毒性效应

重金属可通过多种途径诱导遗传损伤，包括产生微核、染色体畸变（CA）以及其他异常和分子水平的 DNA 损伤。重金属能引起哪种形式的 DNA 损伤取决于它们的化学特性，因此作用机制也不相同。这些效应可以通过应用一系列遗传毒性检测来鉴定和定量，包括胞质分裂阻断微核检测、CA 检测、姐妹染色单体交换（SCE）检测和碱性彗星检测[25]。居住在电子垃圾拆解场地或从事电子垃圾相关工作的人在 DNA 上的损伤更为明显[26]，有研究[27]观察到广东某电子垃圾拆解地和对照区域中新生儿淋巴细胞 DNA 损伤存在显著差异。另一项研究探讨了新生儿 DNA 甲基化模式与母体妊娠期间暴露铅之间的关系，并强调了母体铅暴露[28]下胎儿的表观遗传修饰变化。这些研究表明了铅的细胞毒性和基因毒性作用及其潜在的诱变性，这可能是铅诱导的健康影响的分子机制。

镉在体内和体外都能诱导哺乳动物细胞发生各种表观遗传学变化，导致致病风险和各种类型癌症的发生。镉暴露可改变基因表达谱并改变表观遗传成分，主要有三个特征[29]：DNA 甲基化、组蛋白翻译后修饰和非编码小分子RNA(miRNAs)的调控。DNA 甲基化水平似乎与接触镉的时间有关。事实上，短期(24 h 至 1 周)镉暴露会导致低甲基化，而较长时间(8~10 周)镉暴露会导致高甲基化[30]。TRL1215 大鼠肝细胞体外镉暴露 1 周可抑制 DNA 甲基转移酶活性(最高达 40%)，最终导致 DNA 低甲基化，而长时间镉暴露(10 周)可引起 DNA 甲基转移酶活性升高，从而导致 DNA 高甲基化。Benbrahim-Tallaa 等[31]指出镉诱导的 DNA 高甲基化与人前列腺上皮细胞的恶性转化有关。镉在低剂量时也会对人类的男性和女性生殖产生不利影响，并影响怀孕或其结局[32]。这是因为胚胎中许多基因的表达发生了变化，导致它们在胎盘和胚胎中发生异常甲基化。镉诱导的表观遗传修饰模式与其容易与硫醇结合的能力有关，而甲基供体 S-腺苷蛋氨酸的耗尽导致了甲基组的改变，进而导致 DNA 甲基转移酶活性的改变。这可能导致胎盘和胎儿发育障碍[33]。

不同类型的细胞对铅的遗传毒性作用具有不同的敏感性。这种差异可能是由于蛋白质的存在，如红细胞中的金属硫蛋白，这种蛋白质将铅隔离成一种不可生物利用的形式，保护个体免受金属的毒性[34]。虽然铅的生化和分子作用机制仍不清楚，但据报道铅的遗传毒性可能是由间接机制引起的[30, 35-38]。有研究表明，电子垃圾拆解场地居民体内血铅和淋巴细胞双微核率水平升高呈现一致性的关系，在一定程度上反映了机体染色体损伤程度[39]。铅可以替代参与 DNA 加工和修复酶中的钙或锌，导致 DNA 修复受到抑制，当与其他 DNA 损伤剂(如烟草烟雾或太阳紫外线长波)结合时，其遗传毒性增强。此外，铅暴露引起的自由基水平升高所产生的氧化应激也可能是造成铅间接遗传毒性的原因之一。

铬诱导 DNA 损伤的主要机制是通过产生活性中间体来攻击 DNA、蛋白质和膜脂，从而破坏细胞的完整性和功能[40,41]。三价铬可诱导 DNA 链内交联、DNA 加合物和 DNA 单链断裂。暴露于三价铬中也会造成 DNA 双链断裂。因此，三价铬暴露不能直接与 DNA 相互作用，从而间接导致遗传毒性。此外，六价铬暴露可激活细胞周期检查点和凋亡细胞反应[42]。

一些研究表明汞具有遗传毒性，也有研究表明汞暴露与遗传毒性之间没有相关性[43]。在支持汞诱导遗传毒性的研究中，发现汞会增加活性氧的产生，可引起 DNA、有丝分裂纺锤体和染色体分离蛋白的构象变化[44]，从而破坏细胞中的DNA，并可能导致癌症[44,45]。

6.2.3 肝肾毒性效应

肾脏具有过滤、重吸收和浓缩二价离子的能力，是重金属中毒的靶器官。肾脏损伤的程度和表现取决于金属的种类、剂量和暴露时间。几乎所有的急性肾损害与慢性肾功能衰竭在其机制和结果的大小上都不同。血浆中的重金属以电离的形式存在，具有毒性，当金属与金属硫蛋白结合时，会导致急性毒性和结合的惰性形式，然后被输送到肝脏，在肝脏中会造成损伤和功能紊乱。肝脏损伤可以通过组织病理学发现来证实，并且通常伴随着血液酶水平的升高和蛋白质合成的减少[46-49]。重金属对肾脏的毒性作用表现为肾脏的结构损伤和排泄功能的变化[46,47,49,50]。

比如，在一项研究中[51]，通过分析267名住院患者血液中肝功能和肾功能指标的情况（其中158名来自广东某电子垃圾拆解地作为暴露组，109名来自贵州锦屏作为对照组），我们发现两组的血铅水平与血镉水平呈正相关。暴露组的血清 γ-谷氨酰基转移酶（GGT）活性中位数较高，是对照组的1.6倍。在血常规检查中，暴露组的红细胞比对照组高7.1%。暴露组的血铅中值水平高出71%。两组患者的血铅水平与丙氨酸转氨酶（ALT）、红细胞和血红蛋白活性之间存在显著正相关。患者的血镉水平也与ALT和天冬氨酸转氨酶（AST）相关。血铅和血镉均与ALT呈正相关。两组患者的血铅均与镉呈正相关，暴露组和对照组患者的血铅和血镉水平与血液学和肝脏参数升高相关。结果表明，有毒微量金属可能增加肝脏代谢负荷，导致肝功能异常。

本节主要介绍电子垃圾污染场地中几种重金属如镉、铅、汞、砷的肝肾毒性作用机制，表6-2列出了电子垃圾中几种典型重金属的毒性作用机制。

表6-2 重金属对肝肾的毒性作用机制

重金属	急性暴露	慢性暴露
镉	电离游离形式导致细胞毒性，减少磷酸盐和葡萄糖的转运，抑制线粒体呼吸，并导致肾单位近端肾小管细胞膜破裂[52]。中毒症状包括精神状态改变、呕吐、恶心和呼吸困难，并伴有低血压、休克和急性肾功能衰竭和肝功能衰竭	摄入或吸入后，镉通过金属硫蛋白运输到肝脏和肾脏，金属硫蛋白与镉结合。细胞凋亡和细胞因子途径激活的迹象在该综合征中是常见的。一种典型的慢性肾小管间质肾病是由这种金属在髓质和近端肾小管的S1段积聚而产生的。慢性接触会导致慢性肾脏病，肾损害表现为多尿、浓缩能力丧失、肾小管蛋白尿、肾性糖尿、氨基酸尿、高磷尿症和高钙尿症
铅	急性铅中毒可损害近端肾小管结构和组织学改变，如由铅-蛋白复合物组成的肾小管细胞中的嗜酸性核内包涵体和线粒体肿胀[53]	损伤可能表现为尿酸分泌增加、血管收缩，以及随之而来的肾小球硬化、高血压和间质纤维化[54,55]

续表

重金属	急性暴露	慢性暴露
汞	出现急性肾小管坏死，通常伴有少尿	汞储存在肾脏中，并导致近端小管的直肠部上皮损伤和坏死[56]。汞相关的肾损害可由尿白蛋白、转铁蛋白、视黄醇结合蛋白和β-半乳糖苷酶升高的肾小管功能障碍和膜性肾病综合征引起[56]
砷	急性肾损伤可能导致急性肾小管坏死、溶血所致的血红蛋白尿、血尿、少尿和蛋白尿	慢性中毒以周围神经病变和伴有认知功能障碍的脑病为主要表现

6.2.4 心血管系统毒性效应

心血管疾病(CVD)是一个日益严重的全球性健康问题。长期接触重金属(如砷、铅、镉、汞)与心血管疾病之间的潜在联系尚未明确。重金属增加心血管危险因素的作用机制仍然未知，尽管抗氧化剂代谢受损和氧化应激可能起作用。然而，重金属诱发心血管疾病的确切机制值得通过动物实验或分子和细胞研究进一步探讨。

以一项广东某电子垃圾拆解地的学龄前儿童研究[57]为例，暴露组儿童的血铅中位数显著高于对照组(7.2 mg/dL $vs.$ 3.9 mg/dL)。与参考组相比，暴露儿童的血清 S100A8/A9 水平(1067.5 ng/mL)显著高于参考儿童(843.8 ng/mL)，通过多介质模型表明血铅浓度与血清 S100A8/A9，经偏差校正的 95%可信区间表示显著的总间接效应。结果表明，较高水平的心血管炎性细胞、细胞因子/趋化因子和 S100A8/A9 可能与血铅水平浓度升高有关。

镉是一种强有力的心脏毒性环境重金属[58]，可通过肥料、电池等途径接触，镉中毒可导致心脏毒性和高血压[59,60]。研究表明，镉暴露会导致许多心血管疾病，如动脉粥样硬化、高血压和心肌梗死。镉增加了心肌细胞中活性氧(ROS)的体外形成，导致毒性级联和细胞膜退化，而镉中毒增加了心脏标志物(心肌肌酸激酶、乳酸脱氢酶、心肌肌钙蛋白)的水平、降低心脏组织中抗氧化剂(谷胱甘肽、谷胱甘肽过氧化物酶、超氧化物歧化酶和过氧化氢酶)的活性[61]。在镉处理的人乳头瘤病毒-巨细胞病毒中发现了细胞活力下降、凋亡增加、心肌肌节紊乱、高水平的活性氧以及电生理学和心律失常的改变[62]。RNA 序列分析揭示了镉处理的人类多能干细胞分化成心肌细胞(hPSC-CMs)中差异转录组谱和激活的 MAPK 信号通路以及抑制的 P38 MAPK 的存在。

铅与血压和高血压的研究中，显示血铅水平与血压或高血压呈正相关[63]。铅暴露导致收缩压变化，从 0.6 mmHg 到 1.25 mmHg 增加了 1 倍。铅暴露与临床心血管异常(如冠心病、中风和外周动脉疾病)以及其他心血管功能障碍(如左心室肥厚和心律变化)的发病率增加有关。当研究心室壁尺寸和功能参数之间的关系时，血铅水平升高与左心室肥厚、左心室重量增加、射血分数低和舒张功能恶化的患病率相关。长期暴露于低铅水平会导致遗传正常动物的永久性高血压[64,65]。各种模型系统、动物、培养的内皮和血管平滑肌细胞以及分离的组织已经证明了几种铅诱导的高血压机制，包括氧化应激、受损的一氧化氮系统、炎症、血管活性激素的不规则性以及改变细胞钙转运和细胞内钙分布[66]。先前的研究表明，铅暴露降低了一氧化氮的生物利用度，部分原因是增加了活性氧的产生[67,68]。据推测，活性氧的相应增加和细胞因子产生的变化可能支持动脉粥样硬化的发生。

铜作为一种强有力的促氧化剂，可能在病理动脉粥样硬化形成中起作用[69]。人们认为冠心病发病的关键因素是氧化，因为 LDL 胆固醇的形成和自由基的形成都是通过这些分子的氧化而发生的[70-72]。氧化型低密度脂蛋白是一种促动脉粥样硬化物质，在动脉粥样硬化的发生发展过程中起着重要作用。巨噬细胞泡沫细胞在动脉粥样硬化的发展过程中也起着重要作用，它们导致活性氧的产生和血管壁脂质形成的积累[73]。研究人员认为，铜可以影响动脉粥样硬化的发病机制[74]。总之，铜污染已对人体铜吸收及代谢机制产生影响，并威胁到人体健康[75]。

6.2.5 肠胃系统毒性效应

在一项实验[76]中，采取了浙江温岭电子垃圾拆解场的三种土壤，评估了它们对人类结肠上皮细胞 Caco-2 的毒性作用及其潜在机制。三种土壤中的总镉和镍超过了风险筛选值，分别为 3.8~8.8 mg/kg 和 42.4~155 mg/kg。此外，土壤-1、土壤-2 和土壤-3 中的可提取金属含量分别为 5.9 mg/kg、1.9 mg/kg 和 0.87 mg/kg 镉（20%~67%）和 4.6 mg/kg、6.4 mg/kg 和 12.4 mg/kg 镍（3.6%~29%）。所有三种提取物均引发细胞毒性，其中土壤-2 对细胞活力的抑制作用最强。在土壤-2 和土壤-3 中观察到更高的活性氧产生和更强的抗氧化酶 SOD1 和 CAT 抑制。观察到促炎介质（IL-1β、IL-8 和 TNF-α）和凋亡调节基因（GADD45α、半胱天冬酶-3 和半胱天蛋白酶-8）的上调。数据表明，土壤提取物在 Caco-2 细胞中诱导细胞毒性、氧化损伤、炎症反应和细胞凋亡，表明从电子垃圾拆解现场摄取土壤可能对人类健康产生不利影响。

根据文献资料，铅的肠道转运包括载体介导的成分和被动扩散。在各种体外研究[77,78]中，铅通过肠膜的扩散已被注意到。在这些研究中，铅进入浆膜间的速率和组织摄取在一定条件下随着铅阳离子浓度的增加而线性增加。成年人摄入食物中 10%~15%的铅，而儿童可能通过胃肠道吸收高达 50%的铅。在成年人中，无机铅不会穿透血脑屏障，而在儿童中，这一屏障较不发达。胃肠道的高摄取率和可渗透的血脑屏障使儿童特别容易受到铅暴露和随后的脑损伤的影响。有机铅化合物能穿透人体和细胞膜，四甲基铅和四乙基铅很容易渗透皮肤。这些化合物也可能穿过成年人的血脑屏障，因此成年人可能会患上与有机铅化合物急性中毒有关的铅脑病[79]。另外，急性铅暴露在血液浓度为 100~200 μg/dL 的成年人中会引起胃肠功能障碍，甚至在低至 40~60 μg/dL 的浓度下也观察到了影响[80]。

肠道对镉的吸收的特点是在肠黏膜内的高积累量和低转移率[81]。在一项以 Caco-2 细胞为模型的研究中，Jumarie 等[82]得出结论，镉的转运仅通过跨细胞途径发生，细胞内大容量结合位点的饱和是镉吸收的限速步骤。Oner 等[83]评估了大鼠口服高镉(饮用水中 $CdCl_2$ 15 μg/mL) 30 天的后果，阿尔新蓝(AB)染色显示胃黏膜屏障功能受损，黏蛋白含量降低($P<0.01$)。胃黏膜中黏蛋白含量与镉含量呈负相关。后来，Asar 等[84]结果证实，大鼠口服 15 μg/mL 镉 30 天后，胃黏膜黏蛋白含量显著降低($P<0.01$)，与未处理的动物相比，胃黏膜黏液厚度降低($P<0.01$)。镉对胃肠道远端区域(如结肠)黏液的影响也已被描述[85]。在大鼠和小鼠中，镉经口急性暴露>30 mg/kg dw 后，可观察到胃肠道上皮严重坏死、出血和溃疡。在人类中，一例致命病例的尸检显示胃、十二指肠和空肠出血性坏死[86]。

6.2.6 呼吸系统毒性效应

大量研究结果表明，暴露于电子垃圾中的重金属可能会对儿童的呼吸系统造成伤害或症状，如鼻子和喉咙刺激，随后出现咳嗽、喘息和呼吸困难，或更严重地导致哮喘，通常发生在暴露于某些重金属(如铅、镉、铬、锰、镍、砷、汞、钴或钒)水平增加后。与没有任何电子垃圾回收历史的同龄人相比，电子垃圾拆解地小学生的血液锰水平和血清镍水平都有所上升。使用用力肺活量(FVC)作为肺功能的标志，电子垃圾拆解地 8~9 岁的男孩锰和镍水平升高肺功能明显低于参考值[87]。Razi 等[88]研究表明，血清铅和汞水平的升高以及锌和硒水平的降低可能扰乱了反复喘息儿童的抗氧化系统。在人类[89-91]和啮齿动物模型[92,93]中，铅暴露及其血液浓度已被证明与免疫球蛋白 E(IgE)的产生直接相关。Smith 等[94]研究表明，哮喘儿童血铅水平升高的可能性是非哮喘儿童的 5 倍以上。

由于气道上皮细胞增殖、分化、肺泡形成和功能成熟，产前和产后早期是肺发育最脆弱的时期。环境毒物(如砷、镉等)可能会干扰细胞分化并改变肺结构，导致出生后肺功能下降和成年后对肺部疾病的高度易感性[95]。重金属对肺部疾病易感性的研究发现[95]，在子宫内暴露镉不影响胎儿存活率、体重和肺重，这与含有卵磷脂的饱和脂肪酸的减少有关，卵磷脂是胎鼠肺表面活性物质的主要成分，在 E21[96,97]进行检查时，减少的气道表面活性物质抑制了气道上皮细胞上有毒颗粒物和微生物的消除，还降低了气道的顺应性和肺泡的可塑性，从而增加了感染的风险，减少了气流和气体交换。目前的数据表明，早期暴露于镉可能会损害气流并增加肺部感染。

在子宫内，暴露于砷会导致 2 周大的 C57BL/6 小鼠的肺变小和肺机械受损，诱导气道黏液细胞化生和 CLCA3 蛋白表达增加[98]。此外，宫内暴露可导致 C57BL/6 小鼠黏液产生相关基因（*Clca3*、*MUC5b*、*Scgb3a1*）、先天免疫相关基因（*Reg3Gamma*、*Tff2*、*Dynlrb2*、*Lplunc1*）和肺形态发生相关基因（*Sox2*）表达上调[98]。这些数据表明，出生前暴露于砷会扰乱气道上皮细胞分化和正常实质组织的成熟，可能会改变出生后肺对刺激的抵抗力，并使后代容易患上肺部疾病，如哮喘和慢性阻塞性肺疾病（COPD）。

6.3　有机污染物的毒性效应

生物体是一个整体，体内的各个系统是相辅相成的，它们既是独立的个体又是相互关联的，它们通过激素或者其他信号构成一个平衡的网络，来保持机体正常的运行与调节。内分泌干扰物可以起到修饰靶细胞中受体水平的作用，并且每个细胞的受体数量对于确定靶细胞的信号响应至关重要，因此受体数量的任何改变（或多或少）都会改变激素作用。内分泌干扰物进入生物体内后，往往会因为相似的作用机制从而表现出复合效应。电子垃圾拆解过程中产生的典型有机污染物，例如阻燃剂、PAHs、PCBs、二噁英等，大多都是内分泌干扰物（EDCs）[99]，会干扰体内天然激素的合成、分泌、运输和结合[100]，使得内分泌系统发生紊乱，最后导致免疫、神经、生殖等系统不能保持正常运行，而免疫系统和神经系统被干扰后又会进一步使得内分泌系统不正常，产生连锁反应，从而导致生物的繁殖行为、感知行为以及种群行为失常。因此，电子垃圾拆解过程中的有机污染物通过不同途径进入人体后会引起一系列不良健康影响，尤其针对特殊的职业人群[101]。

6.3.1　内分泌干扰效应

电子垃圾拆解产生的内分泌干扰物所具有的效应机制主要是受体介导反应。内分泌干扰物能起到和激素相同的作用，首先内分泌干扰物与相对应的激素受体结合，形成受体-配体复合物，受体-配体结合物启动下游基因直到结合在 DNA 结合区的 DNA 反应元件上，诱导或抑制相对应基因的转录，启动一系列激素依赖性生理生化过程。主要会产生两种结果，一种是和雌激素一样的激活作用，一种是拮抗雌激素作用，如今报道的激素受体有很多种，包括肾上腺激素受体、甲状腺激素受体、雌激素受体、雄激素受体和细胞激素受体等。除此之外还有非受体介导反应机制，通过改变内分泌腺中激素的合成以及通过干扰结合酶的活性或竞争与载体蛋白的结合来改变生物利用度从而来发挥作用。内分泌干扰物与血清白蛋白和性激素结合蛋白有一定亲和力，影响正常体内的内分泌的作用。

电子垃圾拆解过程中产生的溴化阻燃剂、PCBs、PAEs、双酚 A 以及全氟化物对生物体内甲状腺激素影响很大[102,103]。在我们最新的研究结果中显示，在电子垃圾厂区有机磷阻燃剂职业性皮肤暴露干扰了甲状腺激素水平[104]。甲状腺激素对于正常的大脑发育和新陈代谢的调节至关重要，干扰甲状腺功能在所有年龄段都会产生多种后果，尤其是在发育过程中。甲状腺的主要产物是甲状腺素(四碘甲状腺原氨酸，T4)，T4 转变为三碘甲状腺原氨酸(T3)，存储了数周的 T3 和 T4 通常与甲状腺激素结合球蛋白、转甲状腺素蛋白或白蛋白一起被转运到靶组织，然后与细胞中的甲状腺激素受体(TR)结合发挥作用[105]。甲状腺激素是唯一由脊椎动物体内平衡产生和必需的复杂卤化(碘)分子，这使得生理学极易受到卤素基团(氯、溴)取代的内分泌干扰物的影响。而电子拆解过程中具有卤素取代的酚类部分可以作为一种拮抗剂或者类似物模拟天然甲状腺激素，从而与激素的产生、反馈、分布、进入细胞、甲状腺激素的细胞内代谢(脱碘、结合)以及受体水平的多个方面相互作用[106]。

PCBs 是一个联苯家族，其结构可被随机氯化产生多达 209 种氯化化学物，它们的生物活性各不相同，在结构上，至少有两类不同的 PCBs，共面(非邻位取代)和非共面(邻位取代)同系物。PCBs 的毒理学性质取决于氯取代的数量和位置，例如，PCBs 中 3,3,4,4,5-五氯联苯(共面)和 2,2,4,4,5,5-六氯联苯(非共面)对鸡甲状腺受限基因研究中，PCB-126 显著提高了 TPO 和 TG 基因的 mRNA 表达，降低了 NIS mRNA 表达，却未观察到 PCB-153 对这些基因表达的影响，并且 PCB-126 显著降低 T4 和 T3 的分泌。PCB-153 对这些激素分泌无显著影响。与非共面 PCB-153 相比共面 PCB-126 可直接影响甲状腺激素的合成和分泌，从而干扰蛋鸡甲状腺的内分泌功能[107]。除此之外，人们普遍认为接触 PCBs 会降低成年和新生啮齿动物的血清 T4 浓度，这种减少的机制之一可能是通过与甲状腺转运蛋白结合，之后减少 T4 的产生或者增加代谢，还有可能减少 T4 向 T3 的转化来干扰甲状腺的功能[108]。出生前后的大鼠在子宫内和通过牛奶暴露于 PCBs，会导致甲状腺结构的变化，血清中甲状腺激素水平降低[109]。一项对 105 对母婴的研究也表明，母乳中 PCBs 与妊娠后期和产后产妇甲状腺激素水平下降显著相关[110]。

人类研究和实验观察证明 PCBs 和二噁英通过干扰运输和新陈代谢对甲状腺功能产生负面影响。对于溴化阻燃剂、全氟化学品、邻苯二甲酸盐以及双酚 A 这些内分泌化合物能够干扰甲状腺激素代谢，破坏甲状腺稳态。但是需要进一步地研究对人类的影响[102]。

除了对甲状腺激素的影响，双酚 A、PCBs 和 PAEs 类也是一种外源性性激素。性激素又被叫做类固醇激素或甾体激素，是一类亲脂性的、低分子量物质，通过血液运输到其他器官，有调节、控制组织器官生理活动和代谢机能的作用，并对各种生理机能和代谢过程起着重要的协调作用。外源性激素影响激素受体家族，主要影响雌激素受体、雄激素受体和孕激素受体结合，或激活替代受体，然后通过下游信号传导，从而产生激动或拮抗作用，继而改变内分泌与生殖系统的正常功能，诱发生殖、发育及恶性肿瘤疾病。暴露于邻苯二甲酸单乙基己酯的雄性和雌性斑马鱼的排卵数量和肝脏卵黄蛋白原(VTG)mRNA 显著减少，且雄激素和雌激素水平显著增加，而在雌性斑马鱼中观察到生殖系统功能障碍[111]。高浓度暴露于邻苯二甲酸二丁酯下，女性的血浆中雌二醇水平显著降低，而男性的睾酮水平显著升高，下丘脑-垂体-性腺轴的基因表达被下调，表明具有抗雌激素活性[112]。双酚 A、PCBs 和 PAEs 是干扰正常雌激素/雄激素途径的主要毒物，通过多种途径导致两性不育，包括精子 DNA 损伤、甲基化模式改变、组蛋白修饰与 miRNA 表达[113]。

6.3.2 神经毒性效应

内分泌干扰物导致行为和神经发育改变的细胞和分子效应仍然不完全清楚，人们普遍认为内分泌干扰物能够影响神经系统发育和干扰神经内分泌功能。主要通过以下途径实现：作用于神经内分泌系统后，影响激素的释放及其在靶器官的效应，再通过反馈作用影响到神经系统；或者直接作用于神经系统，引起行为、精神等的改变。同时，甲状腺功能障碍也会导致神经发生严重受损，它们影响祖细胞的增殖、分化和迁移，尤其是在产前和产后早期等大脑发育的关键敏感时期，暴露会对啮齿动物和人类的认知功能产生长期影响。

随着神经回路的形成，发育中的大脑具有明显的可塑性，而这些回路的发展在很大程度上依赖于激素。由于这些原因，大脑在发育的关键时期非常容易受到干扰内分泌的化学品的影响。PCBs、PBDEs 和双酚 A 已知会破坏甲状腺功能并与神经行为缺陷有关，其主要的作用机制涉及甲状腺功能的破坏和神经递质水平、钙信号传导和神经毒性的改变[114]。这些机制中的每一个都会干扰正常的大脑回路发育。大脑中神经递质含量的改变可以解释 PCBs 引起的认知功能改变，大鼠在子宫内和哺乳期接触 PCB-47(非共面)会导致额叶皮层和尾状核中的多巴胺含量减少，直至成年。然而，PCB-77(类二噁英同系物)具有相反的效果[115]。PCBs 也能通过作用于兰尼碱受体(RyR)来增加细胞内钙浓度，进一步引起神经毒性效应[116]。PBDEs 的结构与 PCBs 近似，可以通过相似的作用机制起作用。而 BPA 在大脑中的作用机制仍未完全阐明，在可能的机制中，神经递质含量的改变可能是一个重要的解释。

6.3.3 免疫毒性效应

内分泌干扰物可以通过作用于不同水平的免疫调节网络来影响、调节免疫系统，包括免疫细胞的细胞和体液反应、存活、成熟和细胞因子合成。内分泌干扰物的调节作用分为降低免疫力、诱发过敏性疾病和炎症、诱发自身免疫性疾病等，从而造成免疫系统功能改变后的两种后果：一种是免疫力降低，导致免疫缺陷和对传染性微生物的易感性增加；另一种是增强免疫反应，导致诱发过敏或自身免疫性疾病[117]。

双酚A、PAEs[118]和PCBs能够对免疫系统造成影响。在18岁以下人群中，双酚A的水平与抗巨细胞病毒抗体滴度呈负相关，随着双酚A水平的升高，导致抗巨细胞病毒体滴度降低，从而降低其免疫力[119]。邻苯二甲酸盐在小鼠巨噬细胞中表现出促炎状态，其特征是IL-1、IL-6、TNF-alpha以及趋化因子CXCL1和ROS的表达和分泌增加[120]。349名接触PCBs的儿童的全血样本的研究表明，出生后接触PCBs会导致淋巴细胞亚群波动，这表明产后免疫系统受损[121]。免疫性疾病系统性红斑狼疮和类风湿性关节炎等，被认为是内源性雌激素在调节免疫反应中发挥作用，它们对内源性垂体-性腺轴的影响可能表明它们干预了自身免疫性疾病的发生和发展[122]。电子垃圾拆解过程中的有机污染物，尤其是内分泌干扰物引起的免疫毒性效应与内分泌干扰效应密不可分，应该进一步全面关注其对免疫系统的影响。

6.3.4 生殖系统毒性

内分泌干扰物会干扰不同的激素和代谢过程，并破坏器官和组织以及生殖系统的发育，但是对于内分泌干扰物影响生殖系统的机制尚未得到全面的阐明，可能的原因是通过细胞信号通路对生殖功能产生不利影响[112]。而在基因水平上，内分泌干扰物会导致减数分裂停滞，诱导减数分裂非整倍体和染色体畸变，并抑制减数分裂双链断裂的修复导致生殖系统的破坏[123]。双酚A和邻苯二甲酸盐在斑马鱼中观察到GnRH信号通路和丝裂原活化蛋白激酶（MAPK）信号通路的变化，从而扰乱生殖系统[124]。在小鼠的实验中观察到细胞凋亡的诱导，但是影响不同的凋亡因子[125,126]，表明不同的污染物质通过不同的途径影响生殖系统的毒性。

6.3.5 疾病

内分泌干扰物通过多种机制发挥作用，可以影响内分泌系统的各个方面，包括激素的合成、分泌、转运、结合、作用或消除。这反过来又可能影响发育、繁殖和代谢稳态。大量证据表明，内分泌干扰物与多种疾病有关，主要体现在生长发育与生殖疾病，比如多囊卵巢、影响精液质量、睾丸发育不良、不孕症、子宫内膜异位症、流产等[127]。此外，内分泌干扰物与甲状腺功能减退症、肥胖、糖尿病、神经系统疾病等也有关联，甚至与一些激素敏感性恶性肿瘤的发病率有关，如乳腺癌[128]、前列腺癌卵巢癌和子宫内膜癌[113, 129]。

流行病学和临床研究未能提供具体结果来证实内分泌受体对人类健康的影响。每项研究的差异和方案不同，暴露类型(职业、环境)、暴露时间范围、年龄组和其他因素也各不相同。污染物可能通过多种不同的机制发挥作用，但这些机制尚未在人体研究中完全阐明，需要进一步研究以确定其发挥病理生理作用的机制。因此，电子垃圾拆解所释放的内分泌干扰物质，对暴露人群，尤其是拆解工人的健康影响、流行病学数据等还需深入研究。

6.4 流行病学证据

中国是世界上面临电子垃圾污染和人类暴露影响最严重的国家之一，电子垃圾暴露区主要位于广东和浙江等典型电子垃圾拆解场地。在早期电子垃圾拆解回收过程中，存在于废弃元器件中的污染物随着不规范开放式的加热、酸洗和焚烧等过程而被排放到环境中，造成铅、镉、铬等重金属污染，以及 PBDEs、PAHs 和 PCBs 等持久性有机污染物含量在环境介质中明显较高。在早期电子垃圾拆解区域，人群通过膳食摄入、土壤/粉尘摄入、胎盘、脐带血、母乳、血液、头发、粪便和尿液等途径暴露在电子垃圾环境中，特别是通过食物摄入污染物(PBDEs、PCBs、PCDD/Fs、重金属)，导致污染物在人体组织样本中检测到较高水平。接触电子垃圾中污染物将对人体造成负担，并可能导致各种疾病。儿童和新生儿是人体影响最敏感的群体，污染物暴露将对儿童和新生儿的神经、智力、体格发育，以及新生儿出生结局、癌症、精神健康、甲状腺功能障碍和一般身体健康恶化(DNA 损伤和对基因表达的影响)[130]产生了一定的负面影响，因而可以通过流行病学研究揭示身体负担与疾病之间的相关性和关联，尤其要重视对新生儿和儿童的流行病学调查[131]。

6.4.1 重金属暴露引发健康疾病风险的研究案例

21 世纪初期，Huo 等[132]调查测定广东某典型电子垃圾拆解地儿童中血铅水平，以 226 名儿童为样本进行血铅水平研究。通过采集乡村幼儿园儿童的血样，测量血红蛋白(Hgb)和身体指标(身高、体重、头围和胸围)。统计分析表明，电子垃圾拆解地儿童的血铅水平 15.3 μg/dL 明显高于对照区 9.94 μg/dL，81.8%的暴露组儿童(165 人中 135 人)患有高血铅，还观察到电子垃圾拆解地区的血铅水平随着年龄的增加有显著的增加趋势，且儿童的血铅水平与电子垃圾拆解作坊的数量之间存在相关性。研究证实，暴露组儿童的血铅水平升高是由于接触了电子垃圾回收活动造成的铅污染，且铅污染已经对电子垃圾回收地区周围儿童的健康构成了严重威胁。

1. 单金属与相关疾病的研究

铅是一种高毒性的重金属，可影响接种疫苗儿童的免疫系统。通过对 2011~2012 年广东某电子垃圾拆解地(暴露组)和濠江(对照组)的 590 例儿童进行血铅水平和血乙型肝炎表面抗体滴度的二次探索性分析[133]，可以探讨铅长期暴露儿童血液中铅与乙型肝炎表面抗体水平的关系。与对照地区儿童相比，暴露地区儿童血铅水平较高，乙型肝炎表面抗体滴度较低。在每个阶段，广义线性混合模型显示乙型肝炎表面抗体滴度与儿童血铅水平显著负相关。研究发现，乙型肝炎疫苗免疫应答和免疫系统的降低可能对慢性铅暴露儿童有潜在的危害，并建议生活在慢性铅暴露条件下的儿童，需要采取不同的疫苗接种策略。

铅的心血管毒性主要表现为对血压的影响，并最终增加动脉粥样硬化和心血管事件的风险。现有的流行病学研究很少探讨细胞因子和血浆脂蛋白相关磷脂酶 A2(Lp-PLA2)在易感儿童血铅和心血管疾病危险因素之间关系中的促炎作用。在 2016 年 11~12 月，对共 590 名儿童(3~7 岁)，包括电子垃圾暴露组 337 名和对照区濠江 253 名进行了儿童血铅水平、收缩压和舒张压的测量，并检测血清生物标志物，包括脂质谱、炎症细胞因子和血浆脂蛋白相关磷脂酶(Lp-PLA2)，从而研究了铅暴露对儿童的血管炎症生物标志物和心血管的影响。结果表明：①血铅升高与低脂蛋白 pla2、白介素(IL)-6、甘油三酯(TG)和低高密度脂蛋白(HDL)有关。暴露组儿童的 Lp-PLA2 浓度高于濠江个体，表明暴露组儿童更容易患上血管炎症。Lp-PLA2 与血铅呈正相关，表明铅对促炎和促动脉粥样硬化标志物水平的增加有影响，类似于环境毒物的影响。电子垃圾暴露组的 IL-6、IL-8 和 TNF-α 水平较高，表明炎症升高，且 IL-6 水平的升高与所有结果变量相关。血管炎症参与了对环境重金属有毒物质的反应，并由此导致内皮功能障碍，而内皮功能障碍是动脉粥样硬化的起始和维持的主要因素，因此可以将血管炎症指标作为铅暴露心血管风险的早期标志物。②暴露于电子垃圾的儿童，其血铅水平较高，同时伴随有心血管生理异常，血管炎症和脂质紊乱的患病率增加，推测暴露组儿童收缩压和异常的主要原因之一可能是外周血管阻力的降低。尽管血压水平的微小波动在临床上并不显著，但引发适应或不适应反应的分子和细胞效应可能引发动脉粥样硬化斑块，形成恶性循环。巨噬细胞和内皮细胞可能因内皮细胞吞铅而产生泡沫细胞，导致内皮细胞受损、细胞凋亡。当内皮层被破坏时，铅可在平滑肌细胞内达到并积累，加重炎症和脂质过氧化，最终引发动脉粥样硬化。考虑到暴露组儿童铅暴露的患病率，即使是微小的生理变化也可能增加心血管疾病的风险[134]。

电子垃圾拆解地区学龄前儿童的口腔健康很容易受到铅影响，这增加了患龋齿和引起牙周炎和其他口腔疾病的风险。在 2017 年 11~12 月期间的一项研究中[135]，共招募了 574 名 2.5~6 岁学龄前儿童，包括 357 名广东某电子垃圾拆解地（暴露组）儿童和 217 名来自汕头市濠江，通过检测儿童血铅、唾液酸、血清白细胞介素-6(IL-6)和血清肿瘤坏死因子-α(TNF-α)水平，并调查乳牙龋齿患病率。研究发现暴露组儿童血铅、血清 IL-6、TNF-α 中位数均显著高于濠江儿童。同时，暴露组儿童的唾液酸低于濠江儿童，导致暴露组儿童乳牙龋病患病率明显高于濠江县。血铅水平与唾液中唾液酸呈负相关，IL-6 在血铅水平与唾液中唾液酸浓度的关系中起中介作用。

铅污染与儿童佝偻病发病率存在一定关系。一项关于儿童佝偻病（5082 例儿童佝偻病和 6054 例健康对照组）的研究发现，佝偻病儿童的体铅水平显著高于健康对照组，亚组分析显示一致的结果。中度至重度疾病活动病例的铅水平也明显高于轻至中度病例。这项分析表明，体内铅水平与儿童佝偻病之间存在关联，铅暴露可能是佝偻病的一个风险因素。在诊断儿童佝偻病时应考虑高铅水平，需要进一步研究以确定铅放电在抗佝偻病治疗中的作用，特别是对铅水平高的儿童[136]。

许多研究关注的是电子垃圾回收对工作或生活在电子垃圾回收附近的人的不利影响。然而，在回收工厂关闭后，接触电子垃圾对人类健康的持续影响研究较少。在 Xue 等[137]的研究中，在对电子垃圾回收设施关闭 2 年后的居民进行血液采集的研究中，所有血样中检测了 8 种重金属。结果显示，暴露组血铅、镍、钴、汞含量显著高于对照组，铜、锌、锡、镉含量无显著差异。转化生长因子-β(TGF-β)和 α-平滑肌肌动蛋白(α-SMA)是纤维化的重要指标，暴露组的这两项指标均显著高于对照组。8-异前列腺素(8-I)和丙二醛(MDA)作为氧化应激的生物标志物在暴露组升高。Spearman 相关及多元线性回归均显示钴与暴露组 TGF-β、α-SMA、8-I 呈正相关。因此，推测高浓度的钴溶解在血液中可能通过刺激肌成纤维细胞的活化而增加组织纤维化的风险，而氧化应激参与了这一过程，这可以为干预组织纤维化提供一些潜在的新线索。

研究电子垃圾拆解工人尿液中的汞含量和他们工作的电子废物商店中空气中汞含量[138]，评估尿液和空气中汞含量之间的关联，并评估电子垃圾拆解工人中汞接触相关健康影响的普遍程度。通过对居住在泰国那空西塔玛拉省 25 家电子垃圾商店的 79 名工人进行相关研究。通过问卷收集一般和职业特征、个人防护用品使用情况和个人卫生信息，采集尿液样本以测定汞水平，发现电子垃圾拆解工人尿汞水平为 11.60~5.23 μg/g creatinine，空气汞平均水平为 17.00~0.50 μg/m³。尿液和空气中的汞含量显著相关（$r=0.552$，$P<0.001$）。自我报告症状的发生率为：失眠 46.8%，肌肉萎缩 36.7%，无力 24.1%，头痛 20.3%。

2. 多金属与相关疾病的研究

在广东某非正规电子垃圾回收地区开展的以人群为基础的多种重金属相关疾病的研究发现，暴露于环境污染物在注意力缺陷多动障碍的病因学中起着重要作用。儿童急性或长期暴露于铅、镉、汞、铝或锰通常会留下永久性的神经后遗症，包括注意缺陷、情绪倾向和行为反应，即使是低水平的铅暴露也与注意力缺陷多动障碍（ADHD）有关。基于广东某典型电子垃圾拆解地区 240 名 3~7 岁幼儿园儿童的血清 S100 钙结合蛋白 β（S100β）、全血中铅、镉和锰含量等数据，得出以下研究结果：①儿童注意力缺陷多动障碍症状患病率为 18.6%，而根据家长或教师量表得分，怀疑有行为问题的儿童比例分别为 46.2%和 46.5%。儿童神经行为障碍患病率和行为问题检出率均高于其他地区的同类研究，可能是由于接触重金属造成的身体和神经行为健康损害的结果。研究观察到血液中金属含量的高低都与一系列行为结果的缺陷有关，如行为问题和反社会行为。特别有趣的是，发现低水平的镉和锰对行为问题、冲动/多动症或反社会行为有更大的影响，这是多项行为测量结果所表明的。在多动症患者中，暴露于一种以上的有毒金属，无论是高水平还是阈值水平，这些金属可能联合作用并增加总的毒性效应。因此，铅、镉和锰的组合可能具有协同毒性。②评估了 S100β 蛋白在环境病因介导儿童行为障碍，特别是多动症中的作用的研究，支持使用 S100β 蛋白作为脑损伤的一种新的生化标志物，这对注意力缺陷多动障碍的诊断模式具有重要意义。研究发现儿童血清中位 S100β 在正常人中接近正常上限（0.13 mg/L），还观察到儿童血清 S100β 水平的性别差异。在女孩中，血清 S100β 水平随血液镉浓度下降，表明多基因 S100 蛋白家族成员可能在电子垃圾回收领域的镉毒性反应中发挥调节作用，S100 蛋白可能为神经毒性的实验室评估提供信息[139]。

根据美国精神病学会制订的《诊断与统计手册：精神障碍》第四版标准，广东某典型电子垃圾拆解地 243 名父母接受了关于他们孩子（3~7 岁）多动症行为（ADHD）的调查[140]，并采集这些儿童的外周血标本，测定血铅和血镉水平。结果发现 12.8%的儿童符合 ADHD 标准，其中注意力不集中、多动/冲动性和合并亚型分别为 4.5%、5.3%和 2.9%。在所有儿童中，28.0%的血铅水平为 10 μg/dL，只有 1.2%的血镉水平为 2 μg/L，这一水平通常被认为较高。无论是单因素分析还是多变量分析，根据父母评定量表计算的注意力缺陷多动/冲动性和总分数与血铅水平呈显著正相关，而与血镉水平无显著正相关。高血铅水平儿童患

ADHD 的风险是低血铅水平儿童的 2.4 倍。此研究不同于其他研究，选择了调查孩子们的生活状况，而不是测量其他毒物的含量。在研究中，父亲的电子垃圾相关工作和家庭周围的电子垃圾拆解作坊都与 ADHD 得分相关，这可能反映了电子垃圾回收活动促成了儿童多动症的发生。从事电子垃圾回收工作的父母可能会携带电子垃圾从而污染家庭，而生活在电子垃圾车间附近的儿童可能会摄入更多电子垃圾中的有毒物质。

Xu 等[141]运用流行病学研究方法，将重金属暴露与易感人群听力损失联系起来，并将表观遗传标记的初步评估用于解释环境暴露导致听力发育缺陷的可能机制。研究检测了选定基因 Rb1、CASP8 和 MeCP2 启动子区域内的血液 DNA 甲基化水平，并评估了在典型的电子垃圾区域接触铅和镉后，儿童的甲基化与听力损失的关系。此研究通过可能的表观遗传机制，为了解早期生命的听力损失风险提供参考。实验研究揭示了化学暴露引起的耳毒性，以人群为基础的听力风险评估，旨在测量 116 名来自电子垃圾和参考区域的 3~7 岁学龄前儿童的铅和镉水平、基因 Rb1、CASP8 和 MeCP2 的血液 DNA 甲基化水平和听力状况，并评估暴露与表观遗传修饰可能影响听力损失之间的关系。研究结果表明，生活在电子垃圾区儿童的铅水平升高，儿童的左、右双耳听力阈值明显增强，铅可能破坏内皮细胞和边界细胞之间的紧密连接，增加血耳蜗屏障的渗透性，并允许其他毒物侵入内耳，引发耳蜗神经纤维中的凋亡细胞死亡，从而导致听力下降。铅是听力损失的危险因素，随着铅水平的升高，Rb1 中特定 CpGs 的甲基化水平发生改变，也干扰 MeCP2 和其他 DNA 甲基转移酶的蛋白质水平，可诱导特定基因启动子的表观遗传改变，导致儿童听力损失。总之研究结果显示，电子垃圾拆解地区儿童的铅水平升高，启动子 DNA 甲基化改变听力能力，表明特定基因的表观遗传变化涉及在早期接触环境化学物质时听觉系统的发展。

Zheng 等[87]选择广东某电子垃圾拆解地(暴露组 71 名)和良营(对照组 73 名)两所当地小学的 8~13 岁学童(共 144 名)为研究对象，评估了电子垃圾回收区铬、镍、锰暴露对学龄期儿童肺功能的损害，以及氧化应激的影响作用。这是第一次调查铬、锰和镍在电子垃圾回收领域的影响，并探索它们对肺功能潜在的负面影响。研究儿童在当地出生和生活，且没有呼吸系统和心血管疾病。每个儿童的基本信息，包括病史和住房特征，由儿童的父母或监护人填写问卷，没有儿童在日常生活中接触过电子垃圾。通过对儿童体内血清铬、镍、锰含量、抗氧化酶活性及脂质过氧化水平检测发现，暴露组血锰和血清镍浓度均显著高于对照组。暴露组学龄儿童血锰和血清镍升高可能与环境污染有关，这与当地地区不受控制的电子垃圾回收活动有关。由于暴露组环境的空气中有高浓度的过渡金属，儿童在成长和发育期间长期、持续地吸入含这些金属的空气，这种金属暴露可能会继续损害肺组织并产生较低的肺活量，其中暴露组 8~9 岁男孩的用力肺活量(FVC)

明显低于对照组，并可能随后通过氧化损伤导致 8~9 岁学龄儿童肺功能受损，尤其在年龄较小的儿童中。电子垃圾回收区的学生暴露在高水平的三种过渡金属中，其中 8~9 岁儿童的丙二醛水平和超氧化物歧化酶活性显著升高，但过氧化氢酶活性下降，表明肺对过渡金属颗粒的防御尚未完全进化。血锰和血清镍的积累可能是氧化损伤和肺功能下降的危险因素。相比 10~11 年和 12~13 年组的用力肺活量，发现两者之间没有显著差异，这可能与呼吸道上皮细胞对空气污染物的渗透性有关。此外，年龄较小的儿童呼吸道较窄，因此空气污染引起的刺激可能潜在地造成功能障碍和组织损伤。研究发现，长期暴露于环境铬镍颗粒可能对幼龄儿童的肺功能产生有害影响，这可以通过 FVC 值的下降来说明。然而，随着年龄的增长，功能可能会恢复，因为 12~15 岁的儿童在他们的生长发育阶段呼吸系统的生长速度最快，且女性和男性分别在 20 年和 25 年左右达到最大肺功能。随着年龄的增长，儿童的呼吸功能和发育更加成熟，从而增强了对废物的清除和对受损组织的修复。

除此之外，血液中的铅和镉水平也与儿童肺功能低下有关，铅、镉等重金属可通过影响血红素合成、铁代谢、促红细胞生成素的产生来干扰血红蛋白的合成，血红蛋白是氧气的主要运输载体，是氧气从肺运输到身体其他部位组织的必要条件。一项研究分别从广东某电子垃圾拆解地（暴露组 100 名）、夏山（对照组 54 名）和濠江（对照组 52 名）三所幼儿园招募 5~7 岁学龄前儿童 206 名，通过测量血液参数水平，铅、镉、血红蛋白、血小板计数，并进行肺活量测定，以获得 FVC 和 FEV1 肺功能水平。研究结果显示，与参考人群相比，生活在暴露区域的儿童血液铅和血小板计数水平较高，而血红蛋白、红细胞压积和肺功能（FVC 和 FEV1）水平较低。电子垃圾暴露地区儿童血铅平均水平为 5.53 mg/dL，高于对照区域（3.57 mg/dL）和更新的美国疾病控制与预防中心血液铅水平指南（5 mg/dL）。与血铅水平相似，暴露组儿童血镉平均浓度（0.58 mg/L）低于之前的年份（1.58 mg/L）。与女孩相比，男孩的肺功能更容易受居住地的影响，而且年龄较小的儿童对环境污染尤其敏感和脆弱，与年龄较大的儿童或成年人相比，更容易受到肺功能损害，原因有幼童通过其他途径接触毒素（母乳喂养，胎盘传播），有高危行为（手口接触和物体口接触），更活跃，呼吸能力与体重的比例更大，表面积与体重的比例更大，毒素清除率低。在研究中还指出，暴露组地区儿童血小板计数和血小板增多率的升高可能表明暴露组地区儿童感染风险较高，炎症反应较多。上述研究部分解释了暴露组儿童肺功能水平低于对照地区儿童的可能原因[142]。

铅和镉不仅对肺功能有损害，还与普通人群的听力损失有关，尤其是儿童，铅和镉可以通过氧化应激的生成和凋亡损伤耳蜗或前庭功能，导致听神经传导障碍，最终导致严重的听力损失。从重金属的耳毒性效应和对儿童听力损失流行病学研究有限的角度来看，特别是在电子垃圾污染地区，Liu 等[143]试图调查环境化学污染物暴露是否会影响学龄前儿童的听力,以找出血铅和尿镉暴露与儿童早期听力损伤的关系,对汕头市电子垃圾回收区及配套对照区 234 名 3~7 岁学龄前儿童进行血铅和尿镉检测。采用纯音空气传导(PTA)测试儿童在 0.25 kHz、0.5 kHz、1 kHz、2 kHz、4 kHz 和 8 kHz 频率下的听力阈值,PTA 25 dB 为听力损失,更高的平均血铅水平被发现在暴露组,仍有相当一部分儿童处于铅暴露的高危状态,长期积累可能会对健康产生不利影响。与对照组相比,暴露区域的儿童单耳或双耳的听力损失发生率高于参照组,暴露区域的儿童单耳或双耳的平均听力阈值(包括性别分层)也更高。这表明电子垃圾污染地区的儿童听力受损更严重。镉的生物学半衰期估计为 10 年至几十年,但即使在生命早期低水平接触镉也会导致不良的神经发育。儿童年龄、咬指甲习惯与铅呈显著正相关,父母教育程度、儿童饭前洗手与铅、镉暴露呈显著负相关。Logistic 回归分析显示,铅暴露对听力损失的调整为 1.24。研究数据表明,儿童早期接触铅可能是听力损失的一个重要风险因素,电子垃圾污染地区儿童的发育听觉系统可能受到影响。

最近的研究探究重金属暴露与免疫系统之间的变化。流行病学研究表明,铅或镉暴露与人免疫球蛋白 G(IgG)水平的变化有关,而 IgG 亚类的产生可由铅或镉暴露引起的 Th1 和 Th2 细胞因子的差异调节引起。然而,目前还没有研究关注铅和镉共同暴露通过调节 Th1/Th2 细胞因子对生活在电子垃圾拆解地区儿童 IgG 亚类产生的不利影响。在一项旨在分析儿童血液中铅、镉、血清中 Th1/Th2 细胞因子、IgG 亚类之间的关系的研究中[144],总共有 181 名 2~7 岁的健康儿童接受了检查,包括广东某电子垃圾拆解地 104 人(暴露组)和濠江 104 人(对照组),对全血铅、镉水平,血清中细胞因子、IgG 亚类等进行了检测分析。暴露儿童血铅、镉、血清 IgG1、IgG1 + IgG2、血清 Th1 细胞因子干扰素-γ(IFN-γ)水平升高,Th2 细胞因子白细胞介素(IL)-13 水平降低。血铅水平升高与血清 IFN-γ 水平呈正相关,与血清 IL-13 水平呈负相关。校正线性回归分析显示,血清 IL-13 水平与血清 IgG1 和 IgG1+IgG2 水平呈负相关。调节模型显示 IL-13 在血铅水平与血清 IgG1、血铅水平与血清 IgG1 + IgG2 的关系中具有显著的调节作用。血镉水平升高与血清 IgG1 水平呈正相关。研究结果表明,重金属(尤其是铅)暴露可通过调节暴露儿童的 Th1/Th2 细胞因子影响 IgG 亚类的产生,从而为体液免疫功能与环境暴露之间的关系提供了新的证据。

除了广东省,在我国浙江省,通过流行性病学研究评估了锡箔制造和电子垃圾回收领域中铅的潜在健康风险[145]。研究选取锡箔生产区(兰溪)、电子垃圾回收区(路桥)和对照区(淳安)的 11~12 岁 329 名儿童,测定儿童血铅水平,血尿素氮、血肌酐、血钙、δ-氨基乙酰丙酸和智商。调查结果发现,在暴露区域兰溪和路桥的血铅水平明显高于对照区淳安地区。血铅水平增加会对身体产生不良影响是因为铅通过抑制几种酶的活性来改变血液系统,特别是参与血红素生物合成的

δ-氨基乙酰丙酸脱水酶(δ-ALAD)，铅结合并抑制 δ-氨基乙酰丙酸脱水酶，从而防止 δ-氨基乙酰丙酸分解为前胆素原，导致血清和尿 δ-氨基乙酰丙酸水平升高。肾功能损害是铅引起的另一个重要的不良反应，在此研究中，兰溪患儿血清肌酐水平较淳安患儿明显升高，而路桥患儿则没有。肌酐与血铅水平之间也存在正相关关系。神经系统是铅毒性的主要目标，铅暴露与总体智商、表现智商、反应时间、视觉-运动整合、精细运动技能和注意力(包括执行功能、任务外行为和教师报告的退缩行为)等领域呈负相关。以血铅水平为生物标志物，将儿童血铅水平 10 μg/dL 归类为中国健康风险高血铅或铅中毒。研究发现，路桥地区(38.9%)和兰溪地区(35.1%)儿童血铅水平均在 10 μg/dL 以上，而淳安地区儿童血铅水平均在 10 μg/dL 以下。当血铅水平增加 1 μg/dL 时，智商值下降约 0.71。在此研究中，路桥和兰溪儿童与淳安儿童相比，智商没有明显下降，且路桥、兰溪和淳安的智商与血铅水平之间没有显著的关系。因此，生活在电子垃圾回收和锡箔制造地区周围的儿童可能面临铅污染的潜在健康风险。

通过流行病学调查发现，电子垃圾拆解处理可能导致多种疾病的发生，为预防这些危害仅仅改变政策是不够的，应尽快对参与回收电子垃圾的工人进行健康教育，因为多种化学品可能对健康产生重大影响。减少接触是公共卫生调查的最终目标。此外，所有国家采取实际行动是解决电子垃圾问题的最佳途径。未来，我们建议对电子垃圾回收区的风险进行全面检查，并加强监管。由于电子垃圾回收带来的污染已经成为一个全球性的环境问题，解决暴露于电子垃圾的影响必须成为国际社会关注的优先事项之一。

6.4.2 持久性有机污染物暴露引发健康疾病风险的研究案例

在我国北方处理电子垃圾的成年男性基因组不稳定性的流行病学的研究中，发现回收电子垃圾的工人血液中的 PCBs 浓度也明显高于对照组[146]。在分解电子垃圾的过程中，包括加热或燃烧，大量的 PCBs 被释放到人们工作和生活的环境中。PCBs 同属物的持久性和生物积累潜力使它们能够沉积在植物、鱼类和哺乳动物中，然后在人类消费者中积累。前期研究表明，累积的 PCBs 会对内分泌、免疫、代谢、心血管和神经系统产生毒性[147-149]。PCBs 对生育的不良结果还可能与铅的协同作用影响有关。虽然人们并不认为 PCBs 一定是引起有害影响的诱因，但它们确实是电子垃圾回收中产生的一个重要的污染物。非正规回收电子垃圾过程中释放了其他可能的污染物，也对接触者的健康产生了不利影响。

全氟辛酸(PFOA)在许多工业和消费品中都有应用，其在环境介质中的普遍存在也引起了人们的关注。在一项中国广东某电子垃圾回收区和中国潮南对照区母亲对 PFOA 的暴露程度和对新生儿的潜在危害的研究中[150]，对 167 名孕妇(广东某电子垃圾拆解地 108 例,潮南 59 例)进行问卷调查并收集孕妇血清样本,并对产妇血清样本、健康影响检查及相关因素进行分析。广东某电子垃圾拆解地孕妇血清中 PFOA 浓度高于潮南孕妇血清。暴露组的住宅、参与电子垃圾回收、丈夫参与电子垃圾处理以及将家庭住宅作为车间使用是导致 PFOA 暴露的重要因素。孕妇 PFOA 浓度在正常分娩和不良分娩结果(包括早产、足月出生体重和死产)之间存在显著差异。产妇高水平的 PFOA 水平与新生儿长度、体重、Apgar 评分和胎龄的改变有关,这表明 PFOA 水平可能会影响新生儿的健康和发育。没有发现脐带血清 PFOA 水平与母亲的年龄、教育程度或孩子的性别之间存在关联,调整了潜在的混杂因素,得到暴露组地区母亲 PFOA 暴露水平高于对照地区,产前暴露于 PFOA 与新生儿身体发育下降和不良出生结果相关的结论。

在浙江省台州市电子垃圾回收区域,调查从一组育龄妇女收集的母乳、胎盘和头发中的多氯二苯并对二噁英及呋喃(PCDD/Fs)水平,从每个位点采集五组样本(每组由母乳、胎盘和头发组成),胎盘中 PCDD/Fs 含量均显著高于对照点。虽然样本数量有限但是结果发现,电子垃圾回收操作导致人类和环境中的 PCDD/Fs 水平升高。由于化合物的半衰期较长,婴儿的免疫系统不成熟,在母乳喂养期间大量摄入超过毒理学极限的急性接触是令人担忧的。接触二噁英会影响婴儿和儿童的甲状腺激素系统和免疫功能。此外,婴儿和儿童由于生长速度快,吸收二噁英的速度比成年人快,这意味着他们会通过皮肤接触、吸入和产前暴露吸收更多的二噁英。在国际范围内,电子垃圾拆解场地人群体内二噁英的负担较高,这种副作用可能会对下一代的健康造成影响[151]。

6.4.3 阻燃剂暴露引发健康疾病风险的研究案例

通过对体内多种基质(母乳、胎盘和头发)中 PBDEs 的分析检测,结合在电子垃圾拆解场地从一组育龄妇女收集的数据,以确定 PBDEs 在不同人体基质中的分配情况、身体负担以及在电子垃圾拆解处理对婴儿可能造成的健康风险。从电子垃圾拆解场地(台州)和对照点(临安)采集 5 组母乳、胎盘和毛发样本[152],对有限数量的样本分析得出,电子垃圾拆解场地环境中的 PBDEs 导致当地育龄妇女体内 PBDEs 负担增加,其中膳食摄入和室内粉尘被怀疑是 PBDEs 的主要暴露途径。居民体内 PBDEs 含量高,引起了人们对 PBDEs 的生理影响和潜在健康风险的关注,特别是对儿童的影响。应采取控制措施,以尽量减少电子垃圾拆解作业对环境和人类造成的污染水平,应就电子垃圾回收作业对健康影响的流行病学研究进行深入和加大样本量的调查。

近年来,甲状腺癌(TC)的发病率在世界范围内迅速增加,而暴露在内分泌干扰物中会影响甲状腺激素,可能对人类有致癌作用。PBDEs 和一些重金属

(镉、铅、砷和汞)对 TC 风险的影响鲜有报道。因此,有研究探究 TC 风险与 PBDEs 和四种重金属暴露的关系[153]。这项病例-对照研究涉及 308 例 TC 病例和 308 名年龄和性别匹配的对照。测定血浆中 PBDEs 的浓度和检测尿液样本中的重金属浓度,采用条件 logistic 回归模型探讨 PBDEs 和 4 种重金属暴露与 TC 风险的关系。在 logistic 回归模型中加入联合效应交互项,评估 PBDEs-重金属对 TC 风险的倍增交互效应。部分 PBDEs 同系物(BDE-28、BDE-47、BDE-99、BDE-183、BDE-209)与 TC 风险呈正相关。砷和汞也与 TC 风险增加有关。与低暴露水平相比,高暴露水平砷和汞的参与者患 TC 的可能性分别是低暴露水平的 5.35 倍和 2.98 倍。BDE-209 和铅共同暴露对 TC 风险呈负交互作用,而 BDE-28、BDE-47、BDE-209 等同系物与汞具有显著的正交互作用,BDE-183 与汞联合暴露对 TC 风险呈负交互作用。PBDEs 与铅、汞联合暴露对 TC 风险的交互作用机制,还需要进一步的大样本前瞻性研究。

6.4.4 细颗粒物暴露引发健康疾病风险的研究案例

大气环境细颗粒物($PM_{2.5}$)是呼吸道疾病的危险因素。既往研究表明 $PM_{2.5}$ 暴露可下调气道抗菌蛋白和多肽(AMPs),从而加速气道病原体感染。然而,该方面的流行病学研究还很少。Zhang 等[154]开展了相对应的流行性病学研究,估计个体之间的 $PM_{2.5}$ 慢性每日摄入量(CDI)和气道 AMP 唾液凝集素的关联水平,以及外周白细胞计数和促炎细胞因子。研究人员在广东某电子垃圾拆解地(暴露组)和濠江两地招募了 581 名学龄前儿童,其中 222 名被纳入本研究,进行匹配设计(暴露组 110 名 vs. 濠江 112 名)。收集空气 $PM_{2.5}$ 污染数据,计算个体 $PM_{2.5}$ CDI。暴露组 $PM_{2.5}$ 平均浓度高于濠江,导致个体 $PM_{2.5}$ CDI 较高。暴露组儿童唾液 SAG 水平(5.05 ng/mL)低于濠江儿童(8.68 ng/mL),且与 CDI 呈负相关。暴露组患儿外周血白细胞计数、白细胞介素-8、肿瘤坏死因子-α 浓度均高于濠江患儿,且与 CDI 呈正相关。在中性粒细胞和单核细胞中也发现了类似的结果。据我们所知,这是第一次在电子垃圾拆解区域对 $PM_{2.5}$ 暴露与儿童气道固有抗菌活性关系的研究,表明 $PM_{2.5}$ 污染可能通过下调唾液 SAG 水平而削弱气道抗菌活性,从而加速儿童气道病原体感染。电子垃圾拆解地区的环境 $PM_{2.5}$ 污染仍然威胁着学龄前儿童的健康,为了保护儿童免受电子垃圾拆解处理所造成的 $PM_{2.5}$ 污染影响,相关政府职能部门未来应加强相应的管理。

6.4.5 多环芳烃暴露引发健康疾病风险的研究案例

暴露于 PAHs 与脂质代谢异常有关,但缺乏证据表明 PAHs 是血脂异常的危险因素,为了研究 PAHs 暴露和抗氧化剂摄入在儿童血脂异常风险中的作用机制,对 403 例儿童(其中电子垃圾拆解活动暴露地区 203 例,对照区濠江地区 200 例)的血清脂质、超氧化物歧化酶(SOD)和尿羟基多环芳烃(OH-PAHs)进行检测,生物相互作用采用可加模型计算。研究得出,暴露组儿童血清甘油三酯浓度和血脂异常发生率高于濠江儿童,高密度脂蛋白(HDL)浓度低于濠江儿童。OH-PAH 浓度升高伴随 SOD 降低都与较低的 HDL 浓度和较高的低 HDL 风险相关。萘、芴和菲暴露均与低 SOD 相关,且 SOD 与萘和芴的关联呈剂量依赖性。这一发现表明,不同水平的 PAHs 暴露会增加氧化应激和抗氧化剂消耗,即接触电子垃圾的儿童氧化应激增加,抗氧化能力适应性增强。PAHs 暴露和伴随的 SOD 降低均可导致儿童血脂异常风险,而 PAHs 暴露和 SOD 降低之间的相互作用使儿童血脂异常风险高于其单一风险之和。这些发现为 PAHs 暴露和氧化应激对儿童血脂异常风险的协同作用提供了深刻的见解,提示 PAHs 相关血脂异常风险评估应考虑抗氧化浓度。建议尿液 OH-PAHs 和血清 SOD 浓度作为预测儿童血脂异常风险的生物标志物,建议 SOD 补充剂作为对 PAHs 心血管毒性的干预[155]。

除此之外,有研究表明,2016~2019 年,连续监测中国南方一个电子垃圾回收区周边人群(包括 275 名儿童和 485 名成人)中氧化应激标志物和电子垃圾回收过程中排放的典型污染物(重金属、PAHs 和挥发性有机化合物)的尿液水平,发现儿童比成人遭受更高水平的氧化应激损伤,儿童更容易受到电子垃圾污染的影响。尽管对电子垃圾进行了四年的控制,但氧化应激损伤水平和尿液中的大多数重金属水平保持不变,这表明电子垃圾污染对人类健康的影响是持久而深刻的[156]。总之,电子垃圾拆解区污染物的复杂性和流行病学研究中的局限性仍然需要在未来的研究中进行充分考虑,未来需要大样本队列研究和临床试验来探索因果关系。其次,由于样本量有限,需要在之后的研究中继续完善方案和扩大样本容量。

(高艳蓬 王 梅 李桂英 安太成)

参 考 文 献

[1] Canfield R L, Henderson C R, Cory-Slechta D A, et al. Intellectual impairment in children with blood lead concentrations below 10 mu g per deciliter [J]. New England Journal of Medicine, 2003, 348 (16): 1517-1526.

[2] van der Gulden J W, Kolk J J, Verbeek A L. Work environment and prostate cancer risk [J]. Prostate, 1995, 27 (5): 250-257.

[3] De Flora S. Threshold mechanisms and site specificity in chromium (VI) carcinogenesis [J]. Carcinogenesis, 2000, 21 (4): 533-541.

[4] Werfel U, Langen V, Eickhoff I, et al. Elevated DNA single-strand breakage frequencies in lymphocytes of welders exposed to chromium and nickel [J]. Carcinogenesis, 1998, 19 (3): 413-418.

[5] Denkhaus E, Salnikow K. Nickel essentiality, toxicity, and carcinogenicity [J]. Critical Reviews in Oncology

Hematology, 2002, 42(1): 35-56.

[6] Emmett E A, Maroni M, Schmith J M, et al. Studies of transformer repair workers exposed to PCBs: I study design, PCB concentrations, questionnaire, and clinical examination results [J]. American Journal of Industrial Medicine, 1988, 13(4): 415-427.

[7] Guo Y L L, Yu M L, Hsu C C, et al. Chloracne, goiter, arthritis, and anemia after polychlorinated biphenyl poisoning: 14-year follow-up of the Taiwan Yucheng cohort [J]. Environmental Health Perspectives, 1999, 107(9): 715-719.

[8] Langer P, Tajtakova M, Fodor G, et al. Increased thyroid volume and prevalence of thyroid disorders in an area heavily polluted by polychlorinated biphenyls [J]. European Journal of Endocrinology, 1998, 139(4): 402-409.

[9] Ren N Q, Sverko E, Li Y F, et al. Levels and isomer profiles of dechlorane plus in Chinese air [J]. Environmental Science & Technology, 2008, 42(17): 6476-6480.

[10] Zeng X, Xu X, Boezen H M, et al. Children with health impairments by heavy metals in an e-waste recycling area [J]. Chemosphere, 2016, 148: 408-415.

[11] Lopez E, Figueroa S, Oset-Gasque M J, et al. Apoptosis and necrosis: Two distinct events induced by cadmium in cortical neurons in culture [J]. British Journal of Pharmacology, 2003, 138(5): 901-911.

[12] Cao Y, Chen A M, Radcliffe J, et al. Postnatal cadmium exposure, neurodevelopment, and blood pressure in children at 2, 5, and 7 years of age [J]. Environmental Health Perspectives, 2009, 117(10): 1580-1586.

[13] Pihl R O, Parkes M. Hair element content in learning disabled children [J]. Science, 1977, 198(4313): 204-206.

[14] Kim S D, Moon C K, Eun S Y, et al. Identification of ASK1, MKK4, JNK, c-Jun, and caspase-3 as a signaling cascade involved in cadmium-induced neuronal cell apoptosis [J]. Biochemical and Biophysical Research Communications, 2005, 328(1): 326-334.

[15] Wang B, Du Y L. Cadmium and its neurotoxic effects [J]. Oxidative Medicine and Cellular Longevity, 2013, 2013: 898034.

[16] Goncalves J F, Fiorenza A M, Spanevello R M, et al. N-acetylcysteine prevents memory deficits, the decrease in acetylcholinesterase activity and oxidative stress in rats exposed to cadmium [J]. Chemico-Biological Interactions, 2010, 186(1): 53-60.

[17] Mendez-Armenta M, Rios C. Cadmium neurotoxicity [J]. Environmental Toxicology and Pharmacology, 2007, 23(3): 350-358.

[18] Okuda B, Iwamoto Y, Tachibana H, et al. Parkinsonism after acute cadmium poisoning [J]. Clinical Neurology and Neurosurgery, 1997, 99(4): 263-265.

[19] Son J, Lee S E, Park B S, et al. Biomarker discovery and proteomic evaluation of cadmium toxicity on a collembolan species, Paronychiurus kimi (Lee) [J]. Proteomics, 2011, 11(11): 2294-2307.

[20] Goyer R A. Results of lead research: Prenatal exposure and neurological consequences [J]. Environmental Health Perspectives, 1996, 104(10): 1050-1054.

[21] Silbergeld E K. Mechanisms of lead neurotoxicity, or looking beyond the lamppost [J]. FASEB journal: official publication of the Federation of American Societies for Experimental Biology, 1992, 6(13): 3201-3206.

[22] Guilarte T R, Miceli R C, Altmann L, et al. Chronic prenatal and postnatal Pb^{2+} exposure increases [3H] MK801 binding sites in adult rat forebrain [J]. European Journal of Pharmacology, 1993, 248(3): 273-275.

[23] Markovac J, Goldstein G W. Lead activates protein kinase C in immature rat brain microvessels [J]. Toxicology and Applied Pharmacology, 1988, 96(1): 14-23.

[24] Brookes P S, Yoon Y S, Robotham J L, et al. Calcium, ATP, and ROS: a mitochondrial love-hate triangle [J]. American Journal of Physiology-Cell Physiology, 2004, 287(4): C817-C833.

[25] Kocadal K, Alkas F B, Battal D, et al. Cellular pathologies and genotoxic effects arising secondary to heavy

metal exposure: A review [J]. Human & Experimental Toxicology, 2020, 39(1): 3-13.

[26] Huang W L, Shi X L, Wu K S. Human body burden of heavy metals and health consequences of Pb exposure in Guiyu, an e-waste recycling town in China [J]. International Journal of Environmental Research and Public Health, 2021, 18(23): 12428.

[27] Li Y, Xu X J, Liu J X, et al. The hazard of chromium exposure to neonates in Guiyu of China [J]. Science of the Total Environment, 2008, 403(1-3): 99-104.

[28] Zeng Z J, Huo X, Zhang Y, et al. Differential DNA methylation in newborns with maternal exposure to heavy metals from an e-waste recycling area [J]. Environmental Research, 2019, 171: 536-545.

[29] Genchi G, Sinicropi M S, Lauria G, et al. The effects of cadmium toxicity [J]. International Journal of Environmental Research and Public Health, 2020, 17(11): 113782.

[30] Takiguchi M, Achanzar W E, Qu W, et al. Effects of cadmium on DNA-(cytosine-5) methyltransferase activity and DNA methylation status during cadmium-induced cellular transformation [J]. Experimental Cell Research, 2003, 286(2): 355-365.

[31] Benbrahim-Tallaa L, Waterlandz R A, Dill A L, et al. Tumor suppressor gene inactivation during cadmium-induced malignant transformation of human prostate cells correlates with overexpression of de novo DNA methyltransferase [J]. Environmental Health Perspectives, 2007, 115(10): 1454-1459.

[32] Kumar S, Sharma A. Cadmium toxicity: effects on human reproduction and fertility [J]. Reviews on Environmental Health, 2019, 34(4): 327-338.

[33] Geng H X, Wang L. Cadmium: Toxic effects on placental and embryonic development [J]. Environmental Toxicology and Pharmacology, 2019, 67: 102-107.

[34] de Restrepo H G, Sicard D, Torres M M. DNA damage and repair in cells of lead exposed people [J]. American Journal of Industrial Medicine, 2000, 38(3): 330-334.

[35] Hartwig A. Role of DNA repair inhibition in lead-induced and cadmium-induced genotoxicity: a review [J]. Environmental Health Perspectives, 1994, 102 Suppl 3: 45-50.

[36] Hartwig A, Schlepegrell R, Beyersmann D. Indirect mechanism of lead-induced genotoxicity in cultured mammalian cells [J]. Mutation Research, 1990, 241(1): 75-82.

[37] Landrigan P J, Boffetta P, Apostoli P. The reproductive toxicity and carcinogenicity of lead: A critical review [J]. American Journal of Industrial Medicine, 2000, 38(3): 231-243.

[38] Silbergeld E K. Facilitative mechanisms of lead as a carcinogen [J]. Mutation Research-Fundamental and Molecular Mechanisms of Mutagenesis, 2003, 533(1-2): 121-133.

[39] 陈兰, 徐国建, 张裕曾, 等. 电子垃圾拆解集散地居民铅、镉、铜内暴露水平与淋巴细胞双微核率的关系 [J]. 环境与职业医学, 2008, (5): 442-445.

[40] De Mattia G, Bravi M C, Laurenti O, et al. Impairment of cell and plasma redox state in subjects professionally exposed to chromium [J]. American Journal of Industrial Medicine, 2004, 46(2): 120-125.

[41] O'Brien T J, Ceryak S, Patierno S R. Complexities of chromium carcinogenesis: Role of cellular response, repair and recovery mechanisms [J]. Mutation Research-Fundamental and Molecular Mechanisms of Mutagenesis, 2003, 533(1-2): 3-36.

[42] Wakeman T P, Yang A M, Dalal N S, et al. DNA mismatch repair protein Mlh1 is required for tetravalent chromium intermediate-induced DNA damage [J]. Oncotarget, 2017, 8(48): 83975-83985.

[43] Valko M, Morris H, Cronin M T D. Metals, toxicity and oxidative stress [J]. Current Medicinal Chemistry, 2005, 12(10): 1161-1208.

[44] Valko M, Rhodes C J, Moncol J, et al. Free radicals, metals and antioxidants in oxidative stress-induced cancer [J]. Chemico-Biological Interactions, 2006, 160(1): 1-40.

[45] Ogura H, Takeuchi T, Morimoto K. A comparison of the 8-hydroxydeoxyguanosine, chromosome aberrations and micronucleus techniques for the assessment of the genotoxicity of mercury compounds in human blood lymphocytes [J]. Mutation Research, 1996, 340(2-3): 175-182.

[46] Cobbina S J, Chen Y, Zhou Z X, et al. Toxicity assessment due to sub-chronic exposure to individual and mixtures of four toxic heavy metals [J]. Journal of Hazardous Materials, 2015, 294: 109-120.

[47] El-Boshy M, Ashshi A, Gaith M, et al. Studies on the protective effect of the artichoke (cynara scolymus) leaf extract against cadmium toxicity-induced oxidative stress, hepatorenal damage, and immunosuppressive and hematological disorders in rats [J]. Environmental Science and Pollution Research, 2017, 24(13): 12372-12383.

[48] El-Sayed Y, El-Neweshy M. Influence of vitamin C supplementation on lead-induced histopathological alterations in male rats [J]. Toxicology Letters, 2010, 196: S299.

[49] Yuan G P, Dai S J, Yin Z Q, et al. Toxicological assessment of combined lead and cadmium: Acute and sub-chronic toxicity study in rats [J]. Food and Chemical Toxicology, 2014, 65: 260-268.

[50] Abdou H M, Hassan M A. Protective role of omega-3 polyunsaturated fatty acid against lead acetate-induced toxicity in liver and Kidney of female rats [J]. Biomed Research International, 2014, 2014: 435857.

[51] Chen Y, Xu X, Zeng Z, et al. Blood lead and cadmium levels associated with hematological and hepatic functions in patients from an e-waste-polluted area [J]. Chemosphere, 2019, 220: 531-538.

[52] Fowler B A. Mechanisms of kidney cell injury from metals [J]. Environmental Health Perspectives, 1993, 100: 57-63.

[53] Pai P, Thomas S, Hoenich N, et al. Treatment of a case of severe mercuric salt overdose with DMPS (dimercapo-1-propane sulphonate) and continuous haemofiltration [J]. Nephrology Dialysis Transplantation, 2000, 15(11): 1889-1890.

[54] M L-A. Renal effects of environmental and occupational lead exposure [J]. Environmental Health Perspectives, 1997, 105(9): 928-938.

[55] Patrick L. Lead toxicity, a review of the literature. Part 1: Exposure, evaluation, and treatment [J]. Alternative Medicine Review, 2006, 11(1): 2-22.

[56] Zalups R K. Molecular interactions with mercury in the kidney [J]. Pharmacological Reviews, 2000, 52(1): 113-143.

[57] Zheng X B, Huo X, Zhang Y, et al. Cardiovascular endothelial inflammation by chronic coexposure to lead (Pb) and polycyclic aromatic hydrocarbons from preschool children in an e-waste recycling area [J]. Environmental Pollution, 2019, 246: 587-596.

[58] Rafati Rahimzadeh M, Rafati Rahimzadeh M, Kazemi S, et al. Cadmium toxicity and treatment: An update [J]. Caspian journal of internal medicine, 2017, 8(3): 135-145.

[59] Meng J, Wang W X, Li L, et al. Cadmium effects on DNA and protein metabolism in oyster (Crassostrea gigas) revealed by proteomic analyses [J]. Scientific Reports, 2017, 7: 11716.

[60] Solenkova N V, Newman J D, Berger J S, et al. Metal pollutants and cardiovascular disease: Mechanisms and consequences of exposure [J]. American Heart Journal, 2014, 168(6): 812-822.

[61] Huang Y, Gang H S, Liang B T, et al. 3, 4-Dihydroxyphenylethanol attenuates cadmium-induced oxidative stress, inflammation and apoptosis in rat heart [J]. Tropical Journal of Pharmaceutical Research, 2019, 18(4): 713-719.

[62] Shen J X, Wang X C, Zhou D N, et al. Modelling cadmium-induced cardiotoxicity using human pluripotent stem cell-derived cardiomyocytes [J]. Journal of Cellular and Molecular Medicine, 2018, 22(9): 4221-4235.

[63] Obeng-Gyasi E, Armijos R X, Weigel M M, et al. Cardiovascular-related outcomes in US adults exposed to lead [J]. International Journal of Environmental Research and Public Health, 2018, 15(4): 040759.

[64] Vaziri N D, Sica D A. Lead-induced hypertension: Role of oxidative stress [J]. Current Hypertension Reports, 2004, 6(4): 314-320.

[65] Vaziri N D, Ding Y X. Effect of lead on nitric oxide synthase expression in coronary endothelial cells-role of superoxide [J]. Hypertension, 2001, 37(2): 223-226.

[66] Vaziri N D. Mechanisms of lead-induced hypertension and cardiovascular disease [J]. American Journal of

Physiology-Heart and Circulatory Physiology, 2008, 295(2): H454-H465.

[67] Bhatnagar A. Environmental cardiology - studying mechanistic links between pollution and heart disease [J]. Circulation Research, 2006, 99(7): 692-705.

[68] Peters J L, Kubzansky L D, Ikeda A, et al. Lead concentrations in relation to multiple biomarkers of cardiovascular disease: The normative aging study [J]. Environmental Health Perspectives, 2012, 120(3): 361-366.

[69] Sudhahar V, Das A, Horimatsu T, et al. Copper transporter ATP7A (copper-transporting p-type ATPase/Menkes ATPase) limits vascular inflammation and aortic aneurysm development role of microRNA-125b [J]. Arteriosclerosis Thrombosis and Vascular Biology, 2019, 39(11): 2320-2337.

[70] Klevay L M. Copper, coronary heart disease, and dehydroepiandrosterone [J]. Journal of the American College of Cardiology, 2015, 65(19): 2151-2152.

[71] Ford E. Serum copper concentration and coronary heart disease among US adults [J]. American Journal of Epidemiology, 1999, 149(11): S49.

[72] Linton M R F Y P G, Davies S S. The Role of Lipids and Lipoproteins in Atherosclerosis [Z]. EndotextSouth Dartmouth (MA). 2000.

[73] Lara-Guzman O J, Gil-Izquierdo A, Medina S, et al. Oxidized LDL triggers changes in oxidative stress and inflammatory biomarkers in human macrophages [J]. Redox Biology, 2018, 15: 1-11.

[74] Chowdhury R, Ramond A, O'Keeffe L M, et al. Environmental toxic metal contaminants and risk of cardiovascular disease: Systematic review and meta-analysis [J]. Bmj-British Medical Journal, 2018, 362: 13.

[75] 张裕曾, 陈兰, 居颖, 等. 电子垃圾处理环境中居民体内重金属水平及其影响因素的研究 [J]. 环境与健康杂志, 2007, (8): 563-566.

[76] Ma J Y, Bao X C, Tian W, et al. Effects of soil-extractable metals Cd and Ni from an e-waste dismantling site on human colonic epithelial cells Caco-2: Mechanisms and implications [J]. Chemosphere, 2022, 292: 133361.

[77] Blair J A, Coleman I P, Hilburn M E. The transport of the lead cation across the intestinal membrane [J]. The Journal of physiology, 1979, 286: 343-350.

[78] Aungst B J, Fung H L. Kinetic characterization of in vitro lead transport across the rat small intestine: Mechanism of intestinal lead transport [J]. Toxicology and applied pharmacology, 1981, 61(1): 39-47.

[79] Jarup L. Hazards of heavy metal contamination [J]. British Medical Bulletin, 2003, 68: 167-182.

[80] Abadin H A A, Stevens Y W. Toxicological profile for lead [Z]. Atlanta (GA). 2007.

[81] Elsenhans B, Strugala G J, Schafer S G. Small-intestinal absorption of cadmium and the significance of mucosal metallothionein [J]. Human & Experimental Toxicology, 1997, 16(8): 429-434.

[82] Jumarie C, Campbell P G C, Houde M, et al. Evidence for an intracellular barrier to cadmium transport through caco-2 cell monolayers [J]. Journal of Cellular Physiology, 1999, 180(2): 285-297.

[83] Oner G, Izgut-Uysal V N, Senturk U K. Role of lipid peroxidation in cadmium-induced impairment of the gastric mucosal barrier [J]. Food and chemical toxicology : an international journal published for the British Industrial Biological Research Association, 1994, 32(9): 799-804.

[84] Asar M, Kayisli U A, Izgut-Uysal V N, et al. Cadmium-induced changes in epithelial cells of the rat stomach [J]. Biological Trace Element Research, 2000, 77(1): 65-81.

[85] Liu Y, Li Y, Liu K, et al. Exposing to cadmium stress cause profound toxic effect on microbiota of the mice intestinal tract [J]. PLoS One, 2014, 9(2): e85323.

[86] Buckler H M, Smith W D, Rees W D. Self poisoning with oral cadmium chloride [J]. British medical journal (Clinical research ed), 1986, 292(6535): 1559-1560.

[87] Zheng G N, Xu X J, Li B, et al. Association between lung function in school children and exposure to three transition metals from an e-waste recycling area [J]. Journal of Exposure Science and Environmental Epidemiology, 2013, 23(1): 67-72.

[88] Razi C H, Akin O, Harmanci K, et al. Serum heavy metal and antioxidant element levels of children with recurrent wheezing [J]. Allergologia Et Immunopathologia, 2011, 39(2): 85-89.

[89] Lutz P M, Wilson T J, Ireland J, et al. Elevated immunoglobulin E (IgE) levels in children with exposure to environmental lead [J]. Toxicology, 1999, 134(1): 63-78.

[90] Sun L, Hu J, Zhao Z Y, et al. Influence of exposure to environmental lead on serum immunoglobulin in preschool children [J]. Environmental Research, 2003, 92(2): 124-128.

[91] Karmaus W, Brooks K R, Nebe T, et al. Immune function biomarkers in children exposed to lead and organochlorine compounds: a cross-sectional study [J]. Environmental health : a global access science source, 2005, 4(1): 5.

[92] Miller T E, Golemboski K A, Ha R S, et al. Developmental exposure to lead causes persistent immunotoxicity in Fischer 344 rats [J]. Toxicological Sciences, 1998, 42(2): 129-135.

[93] Snyder J E, Filipov N M, Parsons P J, et al. The efficiency of maternal transfer of lead and its influence on plasma IgE and splenic cellularity of mice [J]. Toxicological Sciences, 2000, 57(1): 87-94.

[94] Smith P P, Nriagu J O. Lead poisoning and asthma among low-income and African American children in Saginaw, Michigan [J]. Environmental Research, 2011, 111(1): 81-86.

[95] Cao J J, Xu X J, Hylkema M N, et al. Early-life exposure to widespread environmental toxicants and health risk: A focus on the immune and respiratory systems [J]. Annals of Global Health, 2016, 82(1): 119-131.

[96] Daston G P. Toxicity of minimal amounts of cadmium to the developing rat lung and pulmonary surfactant [J]. Toxicology letters, 1981, 9(2): 125-130.

[97] Daston G P, Grabowski C T. Toxic effects of cadmium on the developing rat lung. I. Altered pulmonary surfactant and the induction of respiratory distress syndrome [J]. Journal of toxicology and environmental health, 1979, 5(6): 973-983.

[98] Ramsey K A B A, McKenna K L. In Utero exposure to Arsenic alters lung development and genes related to immune and mucociliary function in mice [J]. 2013, 121(2): 244-250.

[99] Orisakwe O E, Frazzoli C, Ilo C E, et al. Public health burden of e-waste in Africa [J]. Journal of Health and Pollution, 2019, 9(22): 190610.

[100] Zoeller R T, Brown T R, Doan L L, et al. Endocrine-disrupting chemicals and public health protection: A statement of principles from the endocrine society [J]. Endocrinology, 2012, 153(9): 4097-4110.

[101] Okeme J O, Arrandale V H. Electronic waste recycling: Occupational exposures and work-related health effects [J]. Current Environmental Health Reports, 2019, 6(4): 256-268.

[102] Boas M, Feldt-Rasmussen U, Main K M. Thyroid effects of endocrine disrupting chemicals [J]. Molecular and Cellular Endocrinology, 2012, 355(2): 240-248.

[103] 居颖, 陈兰, 苏萍, 等. 电子垃圾拆解区居民血清中甲状腺激素和性激素水平研究 [J]. 环境与健康杂志, 2008, (6): 499-503.

[104] Tang J, Ma S T, Hu X, et al. Handwipes as indicators to assess organophosphate flame retardants exposure and thyroid hormone effects in e-waste dismantlers [J]. Journal of Hazardous Materials, 2023, 443: 130248.

[105] Carvalho D P, Dupuy C. Thyroid hormone biosynthesis and release [J]. Molecular and Cellular Endocrinology, 2017, 458(C): 6-15.

[106] Mughal B B, Fini J B, Demeneix B A. Thyroid-disrupting chemicals and brain development: An update [J]. Endocrine Connections, 2018, 7(4): R160-R186.

[107] Katarzynska D, Hrabia A, Kowalik K, et al. Comparison of the in vitro effects of TCDD, PCB 126 and PCB 153 on thyroid-restricted gene expression and thyroid hormone secretion by the chicken thyroid gland [J]. Environmental Toxicology and Pharmacology, 2015, 39(2): 496-503.

[108] McKinney J D, Waller C L. Polychlorinated biphenyls as hormonally active structural analogues [J]. Environmental health perspectives, 1994, 102(3): 290-297.

[109] Collins W T, Jr. , Capen C C. Fine structural lesions and hormonal alterations in thyroid glands of perinatal

[109] rats exposed in utero and by the milk to polychlorinated biphenyls [J]. The American journal of pathology, 1980, 99 (1): 125-142.

[110] Palkovicova L, Patayova H, Rausova K, et al. Prenatal PCB exposure and thyroid function at birth [J]. Epidemiology, 2011, 22 (1): S170-S170.

[111] Park C B, Kim G E, Kim Y J, et al. Reproductive dysfunction linked to alteration of endocrine activities in zebrafish exposed to mono-(2-ethylhexyl) phthalate (MEHP) [J]. Environmental Pollution, 2020, 265: 114362.

[112] You H H, Song G. Review of endocrine disruptors on male and female reproductive systems [J]. Comparative Biochemistry and Physiology C-Toxicology & Pharmacology, 2021, 244: 109002.

[113] Adegoke E O, Rahman M S, Park Y J, et al. Endocrine-disrupting chemicals and infectious diseases: From endocrine disruption to immunosuppression [J]. International Journal of Molecular Sciences, 2021, 22 (8): 3939.

[114] Pinson A, Bourguignon J P, Parent A S. Exposure to endocrine disrupting chemicals and neurodevelopmental alterations [J]. Andrology, 2016, 4 (4): 706-722.

[115] Seegal R F, Brosch K O, Okoniewski R J. Coplanar PCB congeners increase uterine weight and frontal cortical dopamine in the developing rat: Implications for developmental neurotoxicity [J]. Toxicological Sciences, 2005, 86 (1): 125-131.

[116] Pessah I N, Cherednichenko G, Lein P J. Minding the calcium store: Ryanodine receptor activation as a convergent mechanism of PCB toxicity [J]. Pharmacology & Therapeutics, 2010, 125 (2): 260-285.

[117] Kuo C H, Yang S N, Kuo P L, et al. Immunomodulatory effects of environmental endocrine disrupting chemicals [J]. Kaohsiung Journal of Medical Sciences, 2012, 28 (7): S37-S42.

[118] Palacios-Arreola M I, Morales-Montor J, Cazares-Martinez C J, et al. Environmental pollutants: An immunoendocrine perspective on phthalates [J]. Frontiers in Bioscience-Landmark, 2021, 26 (3): 401-430.

[119] Clayton E M R, Todd M, Dowd J B, et al. The impact of Bisphenol A and Triclosan on immune parameters in the U.S. population, NHANES 2003-2006 [J]. Environmental Health Perspectives, 2011, 119 (3): 390-396.

[120] Harris S, Shubin S P, Wegner S, et al. The presence of macrophages and inflammatory responses in an in vitro testicular co-culture model of male reproductive development enhance relevance to in vivo conditions [J]. Toxicology in Vitro, 2016, 36: 210-215.

[121] Horvathova M, Jahnova E, Palkovicova L, et al. Dynamics of lymphocyte subsets in children living in an area polluted by polychlorinated biphenyls [J]. Journal of Immunotoxicology, 2011, 8 (4): 333-345.

[122] McMurray R W. Estrogen, prolactin, and autoimmunity: Actions and interactions [J]. International Immunopharmacology, 2001, 1 (6): 995-1008.

[123] Ikhlas S, Usman A, Ahmad M. In vitro study to evaluate the cytotoxicity of BPA analogues based on their oxidative and genotoxic potential using human peripheral blood cells [J]. Toxicology in Vitro, 2019, 60: 229-236.

[124] Tan S, Chen Z, Wang R, et al. Emission characteristics of polybrominated diphenyl ethers from the thermal disassembly of waste printed circuit boards [J]. Atmospheric Environment, 2020, 226: 117402.

[125] Giammona C J, Sawhney P, Chandrasekaran Y, et al. Death receptor response in rodent testis after mono-(2-ethylhexyl) phthalate exposure [J]. Toxicology and Applied Pharmacology, 2002, 185 (2): 119-127.

[126] Wang W, Hafner K S, Flaws J A. In utero bisphenol a exposure disrupts germ cell nest breakdown and reduces fertility with age in the mouse [J]. Toxicology and Applied Pharmacology, 2014, 276 (2): 157-164.

[127] Kahn L G, Philippat C, Nakayama S F, et al. Endocrine-disrupting chemicals: Implications for human health [J]. The Lancet Diabetes & Endocrinology, 2020, 8 (8): 703-718.

[128] Vandenberg L N, Maffini M V, Schaeberle C M, et al. Perinatal exposure to the xenoestrogen bisphenol-A induces mammary intraductal hyperplasias in adult CD-1 mice [J]. Reproductive Toxicology, 2008, 26 (3-4): 210-219.

[129] De Coster S, van Larebeke N. Endocrine-disrupting chemicals: associated disorders and mechanisms of action [J]. Journal of environmental and public health, 2012, 2012: 713696.

[130] Ogunseitan O A. The Basel Convention and e-waste: translation of scientific uncertainty to protective policy [J]. Lancet Global Health, 2013, 1(6): E313-E314.

[131] Song Q B, Li J H. A systematic review of the human body burden of e-waste exposure in China [J]. Environment International, 2014, 68: 82-93.

[132] Huo X, Peng L, Xu X J, et al. Elevated blood lead levels of children in Guiyu, an electronic waste recycling town in China [J]. Environmental Health Perspectives, 2007, 115(7): 1113-1117.

[133] Xu X J, Chen X J, Zhang J, et al. Decreased blood hepatitis B surface antibody levels linked to e-waste lead exposure in preschool children [J]. Journal of Hazardous Materials, 2015, 298: 122-128.

[134] Lu X L, Xu X J, Zhang Y, et al. Elevated inflammatory Lp-PLA2 and IL-6 link e-waste Pb toxicity to cardiovascular risk factors in preschool children [J]. Environmental Pollution, 2018, 234: 601-609.

[135] Hou R K, Huo X, Zhang S C, et al. Elevated levels of lead exposure and impact on the anti-inflammatory ability of oral sialic acids among preschool children in e-waste areas [J]. Science of the Total Environment, 2020, 699: 134380.

[136] Zhang Y F, Xu J W, Yang Y, et al. The association between body lead levels and childhood rickets: A meta-analysis based on Chinese cohort [J]. Medicine (Baltimore), 2019, 98(8): e14680.

[137] Xue K B, Qian Y, Wang Z Y, et al. Cobalt exposure increases the risk of fibrosis of people living near e-waste recycling area [J]. Ecotoxicology and Environmental Safety, 2021, 215: 6.

[138] Decharat S. Urinary mercury levels among workers in e-waste shops in nakhon Si thammarat province, Thailand [J]. Journal of preventive medicine and public health = Yebang Uihakhoe chi, 2018, 51(4): 196-204.

[139] Liu W, Huo X, Liu D, et al. S100beta in heavy metal-related child attention-deficit hyperactivity disorder in an informal e-waste recycling area [J]. Neurotoxicology, 2014, 45: 185-191.

[140] Zhang R B, Huo X, Ho G Y, et al. Attention-deficit/hyperactivity symptoms in preschool children from an e-waste recycling town: Assessment by the parent report derived from DSM-IV [J]. Bmc Pediatrics, 2015, 15: s12887.

[141] Xu L, Huo X, Liu Y, et al. Hearing loss risk and DNA methylation signatures in preschool children following lead and cadmium exposure from an electronic waste recycling area [J]. Chemosphere, 2020, 246: 125829.

[142] Zeng X, Xu X J, Boezen H M, et al. Decreased lung function with mediation of blood parameters linked to e-waste lead and cadmium exposure in preschool children [J]. Environmental Pollution, 2017, 230: 838-848.

[143] Liu Y, Huo X, Xu L, et al. Hearing loss in children with e-waste lead and cadmium exposure [J]. Science of the Total Environment, 2018, 624: 621-627.

[144] Zheng X B, Xu X J, Lu F F, et al. High serum IgG subclass concentrations in children with e-waste Pb and Cd exposure [J]. Science of the Total Environment, 2021, 764: 142806.

[145] Wang X F, Miller G, Ding G Q, et al. Health risk assessment of lead for children in tinfoil manufacturing and e-waste recycling areas of Zhejiang province, China [J]. Science of the Total Environment, 2012, 426: 106-112.

[146] Wang Y, Sun X H, Fang L Y, et al. Genomic instability in adult men involved in processing electronic waste in northern China [J]. Environment International, 2018, 117: 69-81.

[147] Zhang Q, Lu M Y, Wang C, et al. Characterization of estrogen receptor alpha activities in polychlorinated biphenyls by in vitro dual-luciferase reporter gene assay [J]. Environmental Pollution, 2014, 189: 169-175.

[148] Lauby-Secretan B, Loomis D, Grosse Y, et al. Carcinogenicity of polychlorinated biphenyls and polybrominated biphenyls [J]. Lancet Oncology, 2013, 14(4): 287-288.

[149] Florencia P. Polychlorinated Biphenyls as a Cardiovascular Health Risk: A New Threat from an Old Enemy? [J]. Environmental Health Perspectives, 2020, 128(11): 114003.

[150] Wu K S, Xu X J, Peng L, et al. Association between maternal exposure to perfluorooctanoic acid (PFOA) from electronic waste recycling and neonatal health outcomes [J]. Environment International, 2012, 48: 1-8.

[151] Chan J K Y, Xing G H, Xu Y, et al. Body loadings and health risk assessment of polychlorinated dibenzo-*p*-dioxins and dibenzofurans at an intensive electronic waste recycling site in China [J]. Environmental Science & Technology, 2007, 41(22): 7668-7674.

[152] Leung A O W, Chan J K Y, Xing G H, et al. Body burdens of polybrominated diphenyl ethers in childbearing-aged women at an intensive electronic-waste recycling site in China [J]. Environmental Science and Pollution Research, 2010, 17(7): 1300-1313.

[153] Zhang Q, Hu M J, Wu H B, et al. Plasma polybrominated diphenyl ethers, urinary heavy metals and the risk of thyroid cancer: A case-control study in China [J]. Environmental Pollution, 2021, 269: 116162.

[154] Zhang S C, Huo X, Zhang Y, et al. Ambient fine particulate matter inhibits innate airway antimicrobial activity in preschool children in e-waste areas [J]. Environment International, 2019, 123: 535-542.

[155] Wang Q H, Xu X J, Zeng Z J, et al. PAH exposure is associated with enhanced risk for pediatric dyslipidemia through serum SOD reduction [J]. Environment International, 2020, 145: 106132.

[156] Kuang H X, Li Y H, Li L Z, et al. Four-year population exposure study: Implications for the effectiveness of e-waste control and biomarkers of e-waste pollution [J]. Science of the Total Environment, 2022, 842: 156595.

第 7 章 电子垃圾拆解排放水体和沉积物中有机污染物的转化与风险消减

在电子垃圾拆解过程中，一些阻燃剂、塑料其他助剂、多环芳烃(PAHs)及其他持久性有机污染物(POPs)会被释放迁移到环境中，并进入水体和沉积物中，然后通过食物链进入生物体内，从而对环境和生物产生健康风险。本章首先对电子垃圾拆解处理过程排放的水体及沉积物等环境介质中阻燃剂和塑料其他助剂等典型有机物的赋存特征和释放迁移进行了概述，然后探讨了水体和沉积物中典型有机物的生物转化、水解及光化学转化机制，最后以电子垃圾拆解排放水体和沉积物中的溴系阻燃剂(BFRs)为例，评估分析了其健康风险，并提出相应的风险消减方法。

7.1 污染物在水体和沉积物中分布特征

电子垃圾拆解区产生的有机污染物，经过直接排放、地表径流、大气输送及干湿沉降等过程进入水体，由于其在自然条件下难以降解，容易在水体(河流、海洋、湖泊、水库等)中蓄积，且大部分电子垃圾拆解区附近水体有机物污染随距离变远污染程度减小。一般来说，持久性有机污染物在水中溶解度较低，水体中有机污染物会经过吸附等物理化学过程迁移至沉积物中，并且沉积物中胶体等吸附作用使有机污染物在沉积物中大量富集。当水中的有机污染物浓度下降时，沉积物中的有机污染物也会通过悬浮等过程再次释放进入水体，在水体中形成二次污染。这些有机污染物在水体和沉积物两种介质中多次循环，最终大部分有机污染物在沉积物中蓄积。因此，沉积物在水体中有机污染物的迁移转化过程中发挥着重要作用，且沉积物中的有机污染物浓度更能反映该区域中有机污染物的污染状况。

7.1.1 阻燃剂

添加型阻燃剂主要通过物理填充方式加入到电子电器产品中，该过程没有稳定化学键的生成，因此在电子垃圾拆解的处理处置过程中容易通过多种方式释放到水体和沉积物中造成污染。

1. 多溴联苯醚

由于多溴联苯醚(PBDEs)在具有较高的辛醇-水分配系数(K_{ow}),在水中溶解度较低,且 PBDEs 具有亲脂性,更倾向于吸附在沉积物上,因此环境水体中 PBDEs 的浓度相对较低,沉积物中 PBDEs 的浓度相对较高。PBDEs 同系物的 K_{ow} 大都随着溴原子取代数的增加而增大,其在水中的溶解度随着溴原子取代数的增加而减小,一般来说水体中 PBDEs 含量较低并且以低溴代 PBDEs 为主,而沉积物中主要以高溴代 PBDEs 为主。目前,国内外对于电子垃圾拆解区周边水体和沉积物中的 PBDEs 已经开展大量相关研究。

(1)水体:水体中的 PBDEs 主要存在于水相和颗粒相(悬浮颗粒物)中,两者中 PBDEs 的分配系数受季节、盐度和悬浮颗粒物等因素的影响。一般来说国内外所报道的水体中 PBDEs 的浓度比沉积物中 PBDEs 的浓度小。对我国 BFRs 主要生产区域莱州湾东部海域的 PBDEs 研究表明,总 PBDEs 在水中溶解相(0.29~0.76 ng/L)和颗粒相中(1.79~3.60 ng/L)的含量远小于在表层沉积物(31.37~44.39 ng/g)含量[1]。对长江流域重庆段水体上、中、下游水体及沉积物中 PBDEs 的分布特征研究表明,在水体中并未检到 PBDEs,沉积物中总 PBDEs 的含量为 1.22~15.66 ng/g[2]。Wang 等[3]报道了常年从事电子垃圾拆解活动的浙江峰江水样中 PBDEs 总浓度范围为 8.8~36.53 ng/L。目前,我国已经实现了对河流等水体中常规参数指标的长期监测,但是对水体中,尤其是电子垃圾拆解区附近水体中 PBDEs 的监测研究还有待进一步完善。

(2)沉积物:沉积物由于具有复杂的孔隙结构,其中 PBDEs 含量相对较高,成为 PBDEs 在水-沉积物两相中迁移与转化的重要载体和储库。不仅在各种电子垃圾拆解区附近沉积物中检出了 PBDEs 的存在,而且在各种河流和河口以及海洋沉积物中也普遍检出了 PBDEs 的存在,甚至在北极地区的沉积物中也有少量 PBDEs 的存在[4]。就过去几十年的报道来看,欧洲地区沉积物中的 PBDEs 含量相对较高。例如,2014 年西班牙河流沉积物中 PBDEs 检测的含量范围为 1.5~44.3 ng/g[5],与世界其他地区相比,该地区的 PBDEs 污染较严重,高于美国密歇根湖(1.7~4 ng/g dw)[6]、我国珠江(0.3~21.8 ng/g dw)[7]、长江(1.98~3.23 ng/g dw)[8]和莱州湾白浪河(1.3~24.9 ng/g dw)[9]等水域沉积物的浓度。我国对淮河、黄河和巢湖三个沉积柱芯中 PBDEs 的研究发现,巢湖含量最高,其次是淮河、黄河,并且 BDE-209 的检出率最高[10]。

电子垃圾拆解处理带来的 PBDEs 污染尤为普遍,电子电器产品中的 PBDEs 在后期的储存、使用、维护和回收等活动过程中容易被释放到环境,从而对人体健康和生态环境造成危害。研究表明,在我国主要电子垃圾集中拆解场地附近土壤和水体中均存在 PBDEs 的污染问题。电子垃圾拆解场地附近水体中的 PBDEs 浓度较高,比非电子垃圾拆解区域高出一到两个数量级。例如,2005 年报道我国广东某典型电子垃圾拆解地检测到 PBDEs 在土壤和沉淀样品浓度为 0.26~824 ng/g dw[11]。2007 年报道检测该地具有单溴至七溴和十溴取代的联苯醚同系物[12]。2019 年报道了该电子垃圾回收区、河流的源头、河口以及相关支流中有机阻燃剂的含量及分布特征,其中 BDE-209 在上游占主导地位,但四溴联苯醚 BDE-49 在该地和许多下游站点占主导地位[13]。2014 年从清远电子垃圾拆解场地附近大雁河的 11 个地点收集了表层沉积物样本,测得总 PBDEs 含量为 0.052~126.640 ng/g,其中 BDE-47 和 BDE-99 是主要同系物[14]。2016 年台州电子垃圾拆解区沉积物中测得总 PBDEs 含量为 3.70~28 755.00 ng/g(平均值为 2779 ng/g),BDE-209 均为主要成分。与国内外其他地区相比,我国台州某电子垃圾拆解工业园的 PBDEs 污染较为严重,且电子垃圾拆解活动是 PBDEs 污染的主要来源。以该园区为圆心的 16 km 范围内,沉积物中 PBDEs 含量和距电子垃圾拆解工业园区中心的距离呈极显著负相关($P<0.01$)[15]。

BDE-209 是国内外电子垃圾拆解区附近沉积物中 PBDEs 的主要同系物。在 2012~2014 年对越南某一电子垃圾处理区域附近沉积物的研究表明,其 PBDEs 的浓度范围为 1.6~62 ng/g dw[16]。电子垃圾拆解处理车间附近河流沉积物中 PBDEs 的浓度中值和浓度范围分别为 660 ng/g 和 100~3800 ng/g,上游和下游场地河流沉积物中 PBDEs 的浓度中值(浓度范围)分别为 36 ng/g(2.8~38 ng/g)和 44 ng/g(<0.1~220 ng/g)。BDE-209 是最主要的 PBDEs 同系物,占电子垃圾处理车间地表土壤和河流沉积物中 PBDEs 同系物总量的 51%~95%。BDE-209 的高浓度与其广泛应用相关,根据先前对废弃电子电器产品中混合塑料中 PBDEs 浓度的研究结果,BDE-209 主要存在于电子垃圾的塑料成分中,如阴极射线管显示器、电视机和制热电子设备等的塑料外壳[17]。就目前国内外电子垃圾拆解地 PBDEs 污染的报道而言,其污染水平超过非电子垃圾拆解地,且 PBDEs 比其他 BFRs 的污染更为普遍。

2. 多氯联苯

多氯联苯(PCBs)是早期电子垃圾拆解区的典型持久性有机污染物之一。PCBs 污染会造成严重危害,例如日本发生的 PCBs 污染米糠油中毒事件。虽然国内外已经禁止 PCBs 的生产及使用,但其具有难降解和蓄积性等特点,PCBs 在水体、大气、土壤和沉积物等环境介质中仍广泛检出。

(1)水体:PCBs 在水体中的半衰期大约 2~6 年。对英国环境中海水、沉积物及土壤中的 PCBs 分布规律等研究发现,土壤中 PCBs 约占比 93.1%,约有 3.5% 进入海水中,底泥中约有 4.1%[18]。我国对水体环境中 PCBs 的监测研究主要集中在东部的一些河流、河口以及沿海近岸地带,如长江、珠江和海河等流域。总体上看，除海河附近有较为严重的 PCBs 污染外，其他的河流大多呈现轻污染状态，这种结果与 PCBs 的疏水性密切相关[19]。目前，国内外研究已经报道，很多电子垃圾拆解区的水体都存在不同程度的 PCBs 污染。1995 年台州电子垃圾拆解区附近水样的 PCBs 总含量为 2.97 ng/L，水体中二氯、三氯代 PCBs 含量相对较高，其原因在于低氯代 PCBs 具有相对较高的水溶性[20]。2014 年报道我国台州某电子垃圾拆解区焚烧场和手工拆解场不同环境介质(饮用水、大气、土壤、食物等)中 15 种 PCBs 的污染状况，与国际已有报道 PCBs 浓度相比，台州区域各介质中 PCBs 的污染程度相对较高，其中 PCB-47 和 PCB-171 浓度最高，其次含量较高的为 PCB-47、PCB-1 和 PCB-123。在饮用水中检测到的 PCB-171 含量最高，在手工拆解场和焚烧场检出分别为 13.15 mg/L、31.87 mg/L，其次为 PCB-47 含量分别为 12.81 mg/L、13.35 mg/L[21]。

(2)沉积物：由于 PCBs 具有低水溶性，且黏度较大，有亲脂性，容易被沉积物中有机质的吸附作用而进入沉积物，因此沉积物中 PCBs 含量一般是水中含量的数倍。关于沉积物中 PCBs 的污染，大规模的研究始于 20 世纪 90 年代，目前已基本完成我国各主要港口、海岸、江河沉积物中的 PCBs 污染状况调查。早在 20 世纪 80 年代末期，我国东南沿海某地部分居民从事随意拆解含 PCBs 的电力设备，导致发生了 PCBs 污染事件，对河流、土壤等造成污染。1995 年对该地区土壤及河流沉积物的 PCBs 污染调查发现，高氯代 PCBs 所占比例较高，且得出 PCBs 在沉积物中的迁移作用比较弱[22]。2012 年对台州某电子垃圾拆解区沉积物中 144 种 PCBs 的污染水平、分布特征及可能来源的研究表明，除 PCB-1、PCB-2、PCB-3 其余 141 种 PCBs 均有检出，总 PCBs 浓度为 1.66~5930 ng/g，对比国内外研究该地区污染处于中高水平。其中，tri-CBs、tetra-CBs 和 penta-CBs 检出浓度较大，占总量的 80.7%。通过聚类分析，37 个样品中与 8 种 PCBs 同族体有关系，其中 22 个样品中 PCBs 污染与工业品 Arl248 相关[23]。2015 年对广东清远龙塘底泥研究表明，PCBs 的含量随着深度增加而减小，三氯~五氯 PCBs 所占比例较高，占 PCBs 总量的 70%左右[24]。各 PCBs 同族物的水溶性、在颗粒相中的吸附能力差异较大，所以不同 PCBs 在环境介质中具有不同的浓度组成和迁移转化特征。

3. 四溴双酚 A

随着 TBBPA 在电子电器产品中的大量使用，TBBPA 可能在电子电器产品废弃后的处置过程中通过各种等途径进入到水体和沉积物中，造成电子垃圾拆解区中水体、沉积物等环境介质中 TBBPA 的检出率较高。

(1)水体：TBBPA 具有较强的疏水性，在水中溶解度为 4.16 mg/L，因此一般在水体中检出浓度较低，但是其在水体中溶解度会随着温度、pH 的等条件变化，例如随 pH 增大而显著增大。干旱地区的土壤一般是强碱性，因此 TBBPA 会从土壤中迁移转化对地下水造成水体污染。从国内外各地区各种水体中检测到 TBBPA 的浓度范围来看，我国水体中 TBBPA 的污染较为严重，特别是巢湖流域检测到的 TBBPA 浓度最高可达 4870 ng/L[25]以及环渤海区域检测到的 TBBPA 浓度为 38.97~672.64 ng/L[26]，远高于世界上其他国家的含量水平。污水处理厂并不能对 TBBPA 进行完全降解，且污水处理厂出水中也检测到 TBBPA，从而造成 TBBPA 随着污水处理厂出水进入环境水体中。

(2)沉积物：对比国内外检测到的沉积物中 TBBPA 的浓度范围，电子垃圾拆解区沉积物中 TBBPA 的含量要高于其他地方，这是由于对含 TBBPA 废弃物处置不当造成的。在典型污染地区河流沉积物中 TBBPA 浓度较高且相差较大，英国 BFRs 生产工厂所在流域的沉积物中达到 9750 ng/g dw[27]。在 2012~2014 年对越南某电子垃圾拆解处理车间附近的河流沉积物中 TBBPA 的浓度范围为 6.0~2400 ng/g(中值为 380 ng/g)[16]。2013 年在我国广东某电子垃圾拆解厂附近流域甚至高达 4.12×10^4 ng/g dw[28]，2019 年检测该拆解区附近沉积物样品 TBBPA 的浓度范围为 19.8~1.52×10^4 ng/g dw[29]。但是，近几年来该区域一直在开展相关的环境整治，减少了污染物向附近河流的排放。Wang 等[30]报道了我国清远电子垃圾拆解地区流域沉积物中 TBBPA 浓度为 24.7~914 ng/g dw，流经我国东部主要 BFRs 生产厂区的小清河沉积物浓度(2.08~1.35×10^3 ng/g dw)也要低于电子垃圾拆解区附近沉积物 TBBPA 含量[31]。虽然近年来研究发现 TBBPA 在沉积物中浓度呈现下降趋势[32]，但由于沉积物中本身浓度较高，所以仍需对 TBBPA 在沉积物中的分布及浓度水平进行监测。

4. 得克隆

得克隆(dechlorane plus，DP)已经使用了几十年，对各种环境介质中 DP 的分布和含量的研究表明，DP 已经在电子垃圾拆解场地周边的水、空气、土壤等环境介质被频繁检出，说明电子垃圾拆解行为可能造成了环境中的 DP 污染[33,34]。Wu 等[35]对我国广东清远电子垃圾拆解处理企业周边水库的 6 个地表水和 6 个沉积物进行分析检测得出，DP 在溶解性水、悬浮颗粒物和表层沉积物中的平均浓度分别为 0.80 ng/L、3930 ng/g dw 和 7590 ng/g dw，且发现早期的电子垃圾拆解处理给当地水生生物造成了 DP 污染。在对越南电子垃圾拆解区土壤及沉积物中 PBDEs、TBBPA、有机磷酸酯阻燃剂(OPFRs)、DP 等的研究分析发现，DP 在该电子垃圾处理车间地表土壤和河流沉积物中浓度相对较低[16]。

5. 有机磷阻燃剂

(1)水体：与传统阻燃剂 PBDEs 相比，含有磷酸酯结构的 OPFRs 具有强的水溶性,更容易进入水生系统中。OPFRs 在地下水和地表水(饮用水、河流、湖泊、地表径流、城镇污水等)中均有检出。目前普遍认为污水是地表水环境中 OPFRs 的主要来源,进入水生系统的 OPFRs 主要分配在水相或者沉积物中[36],其中磷酸三(丁氧基乙基)酯(TBEP)、磷酸三(2-氯乙基)酯(TCEP)、磷酸三(2-氯异丙基)酯(TCPP)和 TnBP 是检出率较高的化合物。由于 TCEP 和 TCPP 具有较低的 log K_{ow},即水溶性大和较低的蒸汽压即难挥发的特性,因而常在水体检测 OPFRs 中占主要成分。对我国(珠江三角洲、南海、黄河口)和日本(东京湾)水生环境中 OPFRs 的研究发现,TPPO 是黄海河口最主要的同系物,TEP 和 TCEP 在南海水中占主导地位,TEP 和 TPPO 是珠江三角洲水中主要的同系物,而 TCPP 和 TCEP 对东京湾水体贡献最大[37]。我国水体中 OPFRs 的浓度(85.1~325 ng/L,均值:165 ng/L)[38,39]略高于韩国(74.0~342 ng/L,中值:151 ng/L)[40,41]和美国(3.02~366 ng/L,中值:41.6 ng/L)[42]水体中浓度。

一般地,河流、湖泊等水体中 OPFRs 平均浓度略高于海水,例如纽约州湖泊(160 ng/L)[42]、德国易北河(260.3 ng/L)[43]、注入渤海河流(300 ng/L)[44]、意大利 Volcanic 湖(61.5 ng/L)[45]略高于黄海渤海海域(23.7 ng/L)[46]和纽约州附近海域(53.6 ng/L)[42],造成该种现象的主要原因是海水的稀释效应。另外,我国渤海黄海流域海水中的 OPFRs 由近岸水域至近海水域逐渐降低[46]。

由于受到工业区和人为活动的影响,城镇、工厂聚集区水体中 OPFRs 的浓度常常较高。由于未处理废水的排入,英国亚耳河中 OPFRs 的浓度高达 113~26300 ng/L(均值:6350 ng/L)[47]。受到港口汽车、轮船的运输压力,我国连云港附近的海域水体中 OPFRs 的平均浓度为 738.4 ng/L,高于厦门(227.7 ng/L)和青岛(129.5 ng/L)[48]。另外,水体中 OPFRs 还受到季节性的变化影响,例如珠海三角洲海域水体中枯水季(OPFRs:2040~3120 ng/L,均值:2620 ng/L)的浓度大于雨季(OPFRs:1080~2500 ng/L,均值:1520 ng/L)[49]。纽约哈德逊河水体中 OPFRs 的变化趋势为高温季节大于低温季节[42];北京城镇地表水中 OPFRs 分别在夏季和冬季达到最高浓度和最低浓度[50]。

(2)沉积物:在沉积物中 OPFRs 主要的同系物为 TPhP、TCPP、TEHP、TCEP 和 TBEP。Tan 等[51]通过调查珠三角地区沉积物中 OPFRs 的空间分布,发现城市地区和电子垃圾回收地区的 OPFRs 含量都很高,其中北江站点 OPFRs 含量最高(470 ng/g dw),因为该站点流经珠三角最大的电子垃圾拆解区。尽管 OPFRs 比 BFRs 的水溶性要大得多,但 Li 等[13]研究发现受到电子垃圾拆解活动污染区域大多数沉积物中的 OPFRs 浓度都超过了 BFRs。练江上游河流沉积物中 TCPP 是主要的同系物(TCPP 占比 15.3%~79.7%),而存在严重污染的该地段及其下游则以 TPhP 为主(67.3%~90.3%),TPhP 浓度为 4260~1 710 000 ng/g,OPFRs 浓度为 6010~2 120 000 ng/g,该地沉积物中的 OPFRs 浓度是迄今为止世界上报告的最高浓度之一。对越南北部某电子垃圾拆解区调研发现，该地的河流沉积物中 OPFRs 浓度为 7.3~38 ng/g dw，其中 TPhP 含量最高。我国南方某电子垃圾拆解区附近河流沉积物中 OPFRs 浓度为 11.75 ng/g dw，其中 TCP 含量较高。

7.1.2 塑料其他助剂

1. 邻苯二甲酸酯

邻苯二甲酸酯(PAEs)作为电子产品常用的一种增塑剂，随着电子垃圾处理过程逐渐释放入环境中，广泛存在于水体和沉积物中。PAEs 在水体中有少量以溶解状态存在，大部分被吸附在水中的悬浮颗粒物上。

(1)水体：地表水及地下水中的 PAEs 主要源于固体废弃物的堆放、工业废水的排放、PVC 塑料缓慢释放、雨水淋洗及土壤浸润等。此外，大气中的 PAEs 也可通过干、湿沉降转入到水环境中。工业废水是水体中 PAEs 的主要来源，很多水体中 PAEs 直接来自电子垃圾拆解区排放的废水，并且垃圾及渗滤液也是地表水体中 PAEs 输入的一个重要来源。水环境中已检测出多种类的 PAEs，并且含量较高，其中以 DnBP 和 DEHP 最为典型。生产 PAEs 的废水中 PAEs 含量最高可达 μg/L 级别，而地表水中 PAEs 的含量一般为 ng/L 级别[52-54]。

(2)沉积物：印度的一项研究采集了电子垃圾拆解区、工业区和住宅区附近河流的沉积物，研究其中 PAEs 的分布特征。结果表明，电子垃圾拆解区河流沉积物中 PAEs 的总浓度(407 ng/g)是工业区(150 ng/g)和住宅区(125 ng/g)河流沉积物的 2~3 倍，并且有显著的差异。在所有沉积物样品中，DEHP 和 DnBP 贡献了总浓度的 80%以上，并且近一半的 DEHP 来自电子垃圾拆解[55]。Chakraborty 等[56]在印度的四个大城市的电子垃圾拆解车间和附近露天垃圾场的表层土壤中检测到 6 种目标 PAEs 及其替代物，其中电子垃圾拆解车间附近土壤中 DEHP 占主导地位。

2. 双酚类化合物

对中国、日本、韩国及印度等国家地表水中双酚类化合物浓度研究表明，双酚 A(BPA)的浓度大致在几十到几百 ng/L，并且双酚 F(BPF)的浓度比 BPA 高 1~2 个数量级，是地表水中主要的双酚类化合物；其中双酚 S(BPS)也检测到具有较高的浓度，部分区域 BPS 浓度达到 μg/L 级别[57]。此外还有对美国、日本、韩国等国家部分工业区中底泥中的双酚类物质浓度进行了调查研究，所有样品中总双酚类物质的浓度平均水平为 201 ng/g dw，其中一个样本中双酚类物质总浓度高达到 25 300 ng/g dw[58]。

3. 全氟化合物

不同于上述的疏水性典型污染物，大部分的全氟化合物(PFCs)既疏水又疏油。在生产和使用的过程中，大部分 PFCs 会进入地表水中，沉积物和深海则是 PFCs 在环境中最重要的汇，PFCs 会在环境中停留较长的时间。PFCs 的水溶性较高，导致其在水体中的含量较高，而存在于沉积物含量相对较低。但在受到电子垃圾拆解处理影响的沉积物中依然含高浓度的 PFCs。

(1) 水体：水体是 PFCs 在环境中存在的一个重要介质，有研究显示，在电子垃圾拆解区水体中检测出了大量 PFCs，若处理不当，这些 PFCs 可能通过废水排放等进入到环境中。除此之外，电子垃圾拆解处理处置过程中的直接污染也是水体等环境中 PFCs 的重要来源。近年来，一些地区也发生了由于 PFCs 而导致的污染事件，环境水体中的 PFCs 得到了人们广泛的关注。PFCs 水溶性相对较高，故水体一直是 PFCs 的重要传播介质之一，其中全氟辛酸(PFOA)和全氟辛烷磺酸(PFOS)是两种最常检出的 PFCs。研究表明，PFOS 和 PFOA 广泛存在于不同国家和地区的海水、河水、湖水、饮用水、地表水以及地下水中，但是不同国家和地区水体中 PFOS 和 PFOA 的浓度水平差异较大，其浓度基本上都在 ng/L 级别。

(2) 沉积物：位于偏远地区的沉积物中也存在 PFCs 污染，其 PFCs 以 PFOS 和 PFOA 为主，长链的 PFCs 更倾向于累积在沉积物中。Stock 等[59]对位于加拿大北极圈努勒维特康沃利斯岛的三个湖泊研究表明，PFCs 浓度分别为 5 ng/g dw、7 ng/g dw 和 100 ng/g dw，PFOA 最高浓度为 1.7 ng/g dw。全氟庚酸(perfluoroheptanoic acid，PFHpA)，最高浓度为 3.9 ng/g dw，PFOS 的浓度最高，浓度范围为 24~85 ng/g dw，高浓度水平的 PFOS 可能来自湖畔军用机场的废水排放或在流域内使用水性灭火泡沫过程中 PFCs 的直接或间接排放。我国沉积物中 PFCs 的研究主要集中于受到污染较重的河流，研究发现黄浦江沉积物中 PFOA 和 PFOS 的浓度最高，分别为 203 ng/g dw 和 8.78 ng/g dw。在 Bao 等[60,61]的研究中，黄浦江沉积物中 PFOA 和 PFOS 的浓度则分别介于 0.2~0.64 ng/g dw 和 0~0.46 ng/g dw 之间。来自同一区域沉积物样品中 PFCs 浓度之间的显著差异可能是样品个体间的差异造成的[62,63]，其中 PFOA 和 PFOS 是沉积物中主要的 PFCs。

7.1.3 多环芳烃

目前已知的 PAHs 大约有 200 多种，其已经广泛存在于人类生活的自然环境，如大气、土壤、水体和沉积物中。在电子垃圾拆解区所有的污染物中，PAHs 因其在环境广泛分布、对人类健康具有重大威胁而成为国内外研究的重点。PAHs 溶解度较低、辛醇-水分配系数较高及蒸汽压小，因此相对于水体更易被沉积物吸附。PAHs 最初是通过大气或水体进入环境当中，无论是何种介质、无论在地球的哪一个角落都可以检测到 PAHs，它们遍布在各种环境介质中。PAHs 和 PAHs 衍生物在环境中虽然含量较低，但在环境中分布广泛，人们会通过多种途径摄入，对人体健康产生长期潜在危害，水体和沉积物中 PAHs 的污染备受关注。

1. 多环芳烃

(1) 水体：在过去的二十年中，针对来自电子垃圾拆解场地 PAHs 的环境发生和归宿、转化以及潜在的不良健康后果进行了大量研究。各个国家的研究起步不同，主要是由于经济发展速度不同而导致 PAHs 污染特征的研究程度差异。欧洲和北美研究起步较早，南美和非洲研究相对较晚。随着发达国家向发展中国家转移电子垃圾，例如菲律宾、印度、越南和加纳等出现了大量电子垃圾拆解点。在我国广东和浙江等典型的电子垃圾回收区域，经常对空气颗粒物样品中的 PAHs 进行调查。总体而言，与十年前的水平相比，我国广东等典型电子垃圾拆解场地大气中 PAHs 含量已大幅下降超过一个数量级。我国电子垃圾拆解场地土壤中的 PAHs 浓度也普遍低于印度、巴基斯坦和加纳等国的电子垃圾拆解场地的浓度值。目前，关于电子垃圾拆解区水体与沉积物中 PAHs 的研究程度远不及大气及土壤中 PAHs。水体中 PAHs 的研究主要集中在湖泊和近海，我国对东部沿海地区水体中 PAHs 的研究较多。就三大水系而言，对长江和珠江的研究多于黄河。三大水系主要河流中的 PAHs 污染对水生生物的风险较低，风险排序从大到小依次为：蒽＞芘＞苯并[a]蒽＞荧蒽＞苯并[a]芘＞菲＞䓛[64,65]。对于饮用水中 PAHs 也开展了研究，李桂英等[66]调查了珠江三角洲 9 个地级市 16 个饮用水源地的 PAHs 污染程度，结果表明 16 种美国优控的多环芳烃(萘除外)的总量在丰水期水相中为 32.03~754.76 ng/L；颗粒相中为 13.35~3 017.82 ng/L；枯水期水相中为 48.09~113.63 ng/L；颗粒相中为 8.56~69.60 ng/L。总之，我国各水体存在着不同程度的 PAHs 污染，其中电子垃圾拆解区、河口以及港口水体中浓度较高，而河流及湖泊水体中浓度较低。由于 PAHs 在水中溶解度及挥发性不同，水体中 PAHs 呈现溶解于水、乳化或吸附在悬浮颗粒不同状态，其中水体中萘的检出率较高，然后是菲、芘等组分。

(2) 沉积物：由于 PAHs 在水中溶解度较低，进入水体后大部分 PAHs 最终被沉积物所吸附。和水体一样，沉积物中 PAHs 的含量高低与人类活动的强弱密切相关。一般说来，不同类型的水体中，以海湾、港口及河口地区的沉积物中 PAHs 含量较高，而河流、湖泊及海洋沉积物中相对较低，影响沉积物中 PAHs 的迁移转化主要是沉积物的有机碳含量和粒径分布这两个因素。我国对电子垃圾拆解区沉积物中 PAHs 的研究较早，例如，有学者在 2004~2005 年间从广东某电子垃圾拆解区几种不同的土地用途采集 21 个沉积物样本,特别是废印刷电路板场地、鸭塘和稻田的酸浸,以及河流支流和水沟,研究表明该地沉积物中 PAHs 浓度由高到低的总体趋势为:酸淋＞水沟≈鸭塘＞径流沟＞河流支流＞水库,沉积物中 PAHs 的浓度范围为 143~534 ng/g,高于河流[67]。不同电子垃圾场的 PAHs 的组成特征差异很大。大部分研究发现,沉积物中 PAHs 主要以四环为主,二至三环的含量相对较少。电子垃圾拆解区沉积物中 PAHs 为点源污染,主要来源可能是燃烧排放。16 种优控 PAHs 的浓度介于 59.3~3180 µg/kg 之间,平均值为 722 µg/kg(标准偏差为 817),表明采样区 PAHs 污染离散程度高,各个采样点 PAHs 污染差异较大[68]。另外,台州沉积物中 PAHs 组分含量由高到低依次为:菲＞蒽＞芘＞苯并[b]荧蒽＞荧蒽＞苯并[k]荧蒽[69-71]。

2. 多环芳烃衍生物

电子垃圾拆解区和沉积物样本中已经发现 PAHs 衍生物,例如氯化/溴化多环芳烃(X-PAH)和甲基多环芳烃(Me-PAH),浓度低于 PAHs,但比不受电子垃圾拆解活动影响区域的样品中的检出率要高。但与母体 PAHs 相比,关于 PAHs 衍生物的产生和环境归宿的报道很少。实地观察和模拟研究报告了几种与电子垃圾拆解活动相关的 PAHs 衍生物,包括 Br-PAHs、Cl-PAHs、NPAHs、OPAHs 和烷基化 PAHs 等。这些衍生物可以从初级来源排放,如塑料聚合物的不完全燃烧,或者可能源自大气二次形成。此外,还在越南北部电子垃圾场的土壤和沉积物样本中检测到了 MeBaA 和 MePh,这些 PAHs 衍生物的浓度比其母体 PAHs 低两个或三个数量级。电子垃圾回收区的河流沉积物和表层土壤中的主要化合物是 1-Cl-Pyr 和 9-氯菲(9-Cl-Phe),研究已证实,电子垃圾的热解和燃烧处理是各种 PAHs 衍生物排放的主要原因之一[72]。

7.1.4 其他持久性有机污染

二噁英类化合物具有较高的辛醇-水分配系数,几乎不溶于水,因此湖泊、海洋、河水中的二噁英大多吸附在沉积物中,因此对于沉积物为研究对象的二噁英类报道较为多见。

1. 多氯二苯并二噁英及呋喃

目前关于电子垃圾拆解区附近环境中多氯二苯并二噁英及呋喃(PCDD/Fs)的研究集中于大气和土壤两种介质,对我国浙江和广东等电子垃圾拆解区附近的大气和土壤研究表明该地区已受到 PCDD/Fs 的污染。但由于二噁英类化合物溶解度低,水体中二噁英类含量较低研究相对较少。一般来说未受污染的水体中二噁

英类含量低于 μg/L 级。例如，瑞典 Eman 河中二噁英类含量为 0.022 pg/L[73]。但是污水中 PCDD/Fs 的含量要高，某造纸厂废水中二噁英的含量是 239 pg/L[74]。在休斯敦船道工业区和城区附近的水体检测到 PCDD/Fs 的含量 26.73~461.87 pg/L[75]，PCDD/Fs 的总含量明显高于其他离工业区较远或者没有被污染的水体，说明工业污染是 PCDD/Fs 的一个重要来源。美国 Hudson 和 Rantin Bay 同样在工业区附近，得到的 PCDD/Fs 的总含量分别是 36.2 pg/L 和 14.85 pg/L，这同样印证了 PCDD/Fs 的污染来源[76]。

二噁英不易溶于水，因此环境中的二噁英多会吸附在区域沉积物中，但是随着远距离传输，沉积物中的二噁英呈现广泛分布。早期对欧洲二噁英的调查结果表明，几乎所有欧洲国家的沉积物中都有二噁英的存在[77]，其中 OCDD 在所有检出的同类物中占比最高。这与我国部分地区沉积物中二噁英的同类物检测结果相一致。广东某电子垃圾拆解区早期落后的回收方式造成了严重的环境污染，在露天焚烧区域检测出 PCDD/Fs 在内的多种 POPs，其中酸浸场地和燃烧残渣混合的土壤样品的总 PCDD/Fs 浓度最高，范围分别为 12 500~89 800 pg/g dw 和 799 020~967 500 pg/g dw[67,78]。2022 年的文献报道了广东某电子垃圾回收区附近的练江 27 个表层沉积物中的 PCDD/Fs 浓度水平，其浓度由高到低依次为：该地＞该地下游＞该地上游的空间趋势，说明电子垃圾回收活动是该流域最主要的二噁英贡献源。来自该地不同村庄的沉积物在 PCDD/Fs 浓度和同系物组成方面存在很大差异，其原因归因于这些村庄开展的电子垃圾回收活动多种多样。电子垃圾露天焚烧区附近的沉积物中 PCDD/F 浓度极高且具有独特的 PCDD/Fs 分布，以低氯 PCDFs（四到六）为特征。练江沉积物中 PCDD/F 浓度的地理分布和剖面表明，该地沉积物中大量 PCDD/Fs 主要保留在当地和附近的水体中[79]。

2. 多溴二苯并二噁英及呋喃

根据最近几十年的调查研究发现，多溴二苯并二噁英及呋喃（PBDD/Fs）广泛存在于河流湖泊和沉积物[80,81]、土壤、生物有机体，甚至人体内，大气环境中也检测到 PBDD/Fs 的存在，它们可以通过大气沉降再次进入水体和沉积物中。有关研究对城市废水处理厂污泥样品中的 PBDD/Fs 进行分析，PBDFs 在所有污泥样品中均被检出，结果表明相关 PBDFs 的质量浓度甚至最高达到了 3 μg/kg[82]。

含 BFRs 的家用电器在使用过程中能够释放出 PBDD/Fs[83]，例如在电子垃圾中塑料聚合体碎屑的热处理过程中，研究人员检测到沉积物中 PBDD/Fs 的浓度达到了 1.2~13.9 μg/L[81]。Wang 等[84]对从电子垃圾回收站附近收集的河流沉积物样品进行分析，检测到 0.025~0.92 ng/g dw 的 PBDFs，表明不受控的电子垃圾回收会导致 PBDFs 的环境污染。Liu 等[79]对广东某电子垃圾回收区沉积物中 PBDD/Fs 的研究表明，电子垃圾拆解回收活动是二噁英的最大贡献者之一。

7.2 污染物在水体和沉积物中的转化机制

电子垃圾拆解过程中排放的阻燃剂、塑料添加剂、PAHs 及其他 POPs 等污染物大多具有大辛醇-水分配系数、低水溶性、亲脂性和热稳定性等特点，易进入水环境并于沉积物中富集。在自然条件下，这些污染物会经历各种环境过程，从而引起自身的迁移转化。为了评价在水体/沉积物中的残留污染水平，推断其在各水体和沉积物中的转化过程可能产生的潜在产物，需要对各污染物的转化过程、转化产物及其动力学行为进行研究。总体上来说，这些污染物在水体和沉积物中的转化主要有生物降解、光解、水解等过程。

7.2.1 生物转化机制

污染物进入水体和沉积物后容易被植物、鱼体吸收并进入食物链转移，水体和沉积物中存在许多微生物，这些微生物在污染物的降解中发挥着重要作用。吸附在微生物细胞膜表面的污染物进入细胞内后与相关特异性降解酶结合发生酶促反应从而被去除。目前，已有研究发现了许多具有相应降解能力的微生物，并且对于污染物在水体和沉积物生物降解过程中酶催化步骤、污染物转化机理有了深入了解。大多污染物的微生物转化又分为好氧氧化过程和厌氧还原过程，不同物质的降解机制也不尽相同。

1. 阻燃剂

1）多溴联苯醚

鱼体中多溴联苯醚的代谢转化机制：吴江平等[85]对电子垃圾拆解地及对照区野生鲮鱼和鲫鱼样品中的ΣPBDEs、ΣPBBs、ΣDPs 和ΣABFRs（非 PBDEs 类，与 PBDEs 相似的替代型溴系阻燃剂）进行了定量分析（表 7-1），发现电子垃圾拆解区的污染物含量均明显高于对照区，表明电子垃圾拆解活动使得水中鱼类受到卤代阻燃剂的污染。在该电子垃圾拆解地内，鲮鱼和鲫鱼体内的ΣPBDEs 在 2010~2016 年间无明显变化[71,86,87]。对于不同鱼类样品中，除 DP 外，鲮鱼体内卤代阻燃剂的含量显著高于鲫鱼。鲮鱼是一种底层鱼类，主要以藻类、有机碎屑为食，可能导致了其体内卤代阻燃剂的高蓄积。此外，由于卤代阻燃剂具有亲脂性，鱼体内的脂肪含量也可能影响其生物蓄积，相对于鲫鱼，鲮鱼体内脂肪含量显然要更高。另外，鲫鱼的食量及消化能力较弱，这也可能导致其体内卤素阻燃剂浓度较低。研究表明，鲮鱼和鲫鱼体内的 PBDEs 以 BDE-28、BDE-47、BDE-100、BDE-153、BDE-154 和 BDE-209 为主，约占总ΣPBDEs 含量的 97%[85]，

PBDEs 组成模式与其他相关研究相符[71,86,87]。这种组成模式与电子电器工业生产中使用的 PBDEs 阻燃剂种类、环境和生物体内降解及富集能力有关[71,86-88]。与鲮鱼相比，鲫鱼体内 BDE-100 和 BDE-154 的相对含量显著升高，而 BDE-28 的相对含量显著降低[85]，可能与 BDE-100 和 BDE-154 水生生物网内的生物放大能力较高而 BDE-28 较低有关[71,85-87]。

表 7-1 电子垃圾拆解地和对照区野生鲮鱼和鲫鱼脂肪(%)及卤代阻燃剂含量(ng/g dw)[85]

项目	鲮鱼		鲫鱼	
	电子垃圾区	对照区	电子垃圾区	对照区
脂肪含量	5.80 (5.61~5.99)	4.53 (2.87~5.51)	1.53 (1.09~1.77)	2.24 (1.46~5.51)
ΣPBDEs	1440 (1990~1610)	8.76 (4.49~14.0)	348 (110~587)	29.9 (9.31~79.3)
ΣPBBs	29.5 (26.8~34.5)	0.27 (0.17~0.45)	5.38 (2.47~9.45)	0.63 (0.18~1.31)
ΣDPs	15.1 (8.47~21.3)	0.31 (0.08~0.50)	24.9 (11.1~49.0)	0.45 (0.22~0.83)
ΣABFRs	8.93 (4.88~13.0)	2.84 (0.18~5.50)	6.20 (3.17~11.7)	3.90 (2.21~8.38)

从之前的体内/体外研究表明，PBDEs 在某些鱼类(如虹鳟鱼和鲤鱼)可能生物转化形成 OH-BDEs 和 MeO-BDEs[89-91]。Haglund 等[92]通过肠道菌群和沉积物中的微生物将 PBDEs 甲基化为 MeO-BDEs 的可能性。Kim 等[93]在一条河流的七种代表性鱼类中测定了 PBDEs 及其羟基化和甲氧基化衍生物(OH-BDEs 和 MeO-BDEs)的浓度和分布，发现 PBDEs 及其衍生物在鱼的内脏中不同程度地积累，其中 PBDEs 结构类似物的浓度比 PBDEs 的浓度高几倍。结构类似物主要为 MeO-BDEs，占总浓度的 60%~75%，而不是 OH-BDEs，五溴化 MeO-BDEs 是主要的同系物组，其次是四溴化 MeO-BDEs≈四溴化 OH-BDEs＞三溴化 OH-BDEs。在大多数鱼类中检测到四溴二苯醚至七溴二苯醚，其生物利用度高于溴化程度更高的八溴二苯醚至十溴二苯醚。而在大鳍刺鳚鲌和鲶鱼中发现了较多的高溴二苯醚，特别是十溴二苯醚，因为它们生活在河流下游的底栖栖息地存在沉积物的重新悬浮，造成它们比其他鱼类暴露于更高浓度的 PBDEs[94]。MeO-BDEs 浓度根据相对营养水平而增加，表明与淡水食物网中的母体 PBDEs 相比，PBDEs 衍生物的生物放大程度更高。与使用半透膜装置测量的水中溶解的分析物浓度相比，鱼类对非邻位取代的 MeO-BDEs 的吸收量更大。

PBDEs 的厌氧微生物降解：从电子垃圾回收点[95]、厌氧湿地沉积物[96,97]、河床底泥[98,99]等研究中得到了多种 PBDEs 厌氧降解菌，涉及脱卤球菌属（Dehalococcoides），脱卤单胞菌属（Dehalogenimonas）、脱卤杆菌属（Dehalobacter）、不动杆菌属（Acinetobacter）、硫磺单胞菌属（Sulfurospirillum）、脱卤脱亚硫酸菌属（Desulfitobacterium）等微生物种群。厌氧微生物降解转化 PBDEs 的主要途径是还原脱溴加氢过程。由于此过程是酶促作用下的电子传递引起的，苯环上的电子云密度和氧化还原电位较低，易发生亲核取代反应生成低溴化产物[100]。不同取代位点的反应难易程度不同，低电位的位置如间位溴原子更容易受到攻击，其次为对位，最后为邻位[101]。以 BDE-209 为例，大多数厌氧微生物首先对 BDE-209 进行一次脱溴，生成 BDE-206、BDE-207 等物质[102]，其中根据溴的取代位置依次进行，间溴和对溴通常首先被去除，然后是邻位[103]，如图 7-1 所示；也有一些研究发现，在海洋沉积物中 BDE-209 的厌氧降解过程中，溴的消除没有偏好[104]。因此，BDE-209 厌氧还原脱溴能力的差异可能与不同沉积物环境中微生物群落组成不同有关[104-106]。研究表明，高溴化 PBDEs 的脱溴速度比低溴化 PBDEs 慢[99, 107]，这可能由于辛醇-水分配系数高，高溴化 PBDEs 难溶于水[108]，从而限制了它们的生物利用度。然而，在某些情况下，高溴化 PBDEs 的化学性质可能不稳定，导致厌氧微生物脱溴率增加，例如 BDE-209 中含有更多的低电位溴原子，更易被还原。在河流沉积物中添加维生素 B_{12} 或零价铁可使 BDE-209 的脱溴率分别提高 36.4%和 86.4%[109]。高溴化 PBDEs 的厌氧降解率高于低溴化 PBDEs[105,106]。因此，水体和沉积物中成分的差异可能会对不同溴化程度的 PBDEs 降解产生很大影响。此外，厌氧微生物以 88%~100%的水平将 BDE-47 脱溴降解[110,111]。

PBDEs 的好氧微生物降解：与 PBDEs 的厌氧降解相比，好氧降解更彻底，过程更短[112,113]。但是好氧微生物难以直接对高溴代联苯醚进行开环降解，PBDEs 增加的尺寸和疏水性降低了它们的生物利用度和碳对羟基化的敏感性；因此，酶的攻击受到空间位阻，导致需氧降解与溴化程度成反比[112]，在水体和沉积物中，高溴代 PBDEs 一般先经过厌氧微生物作用或光解等途径脱溴为低溴代化合物后再进行被好氧降解为小分子物质[101]。Rhodococcus jostii RHA1、Burkholderia xenovorans LB400、Rhodococcus sp. RR1、乙醚降解 Pseudonocardia dioxanivorans CB1190 和其他各种菌株可以对 PBDEs 进行有氧降解[114]。苯环的羟基化和单氧合或双氧合开环导致在三羧酸循环中使用降解产物作为底物，不会产生高毒性中间体，例如 BDE-209 的部分降解途径（图 7-2）[115-118]。这种氧化产生的产物包括 OH-PBDEs、溴酚和溴邻苯二酚。由联苯双加氧酶基因编码的联苯双加氧酶是催化 PBDEs 有氧降解的关键酶[119]。其中，2,3-二羟基联苯双加氧酶攻击 2,3-碳键生成 2,3-二羟基联苯醚，然后在邻位或间位裂解开环，它在低溴化 PBDEs 的开环中起重要作用[120-122]。黄婷等[123]和 Cao 等[124]对 BDE-47 好氧降解产物进行了测定，生成 OH-PBDEs 及 2,4-

DBP，并推测可进一步分解为 CO_2、H_2O 及 Br。研究发现，对于 PBDEs 完全脱溴的产物二苯醚(DE)，可被好氧降解成苯酚及 2-吡喃酮-6-羧酸酯[120]，或者进一步降解为顺式黏康酸后进入三羧酸循环[115]，最终生成 CO_2 和 H_2O。

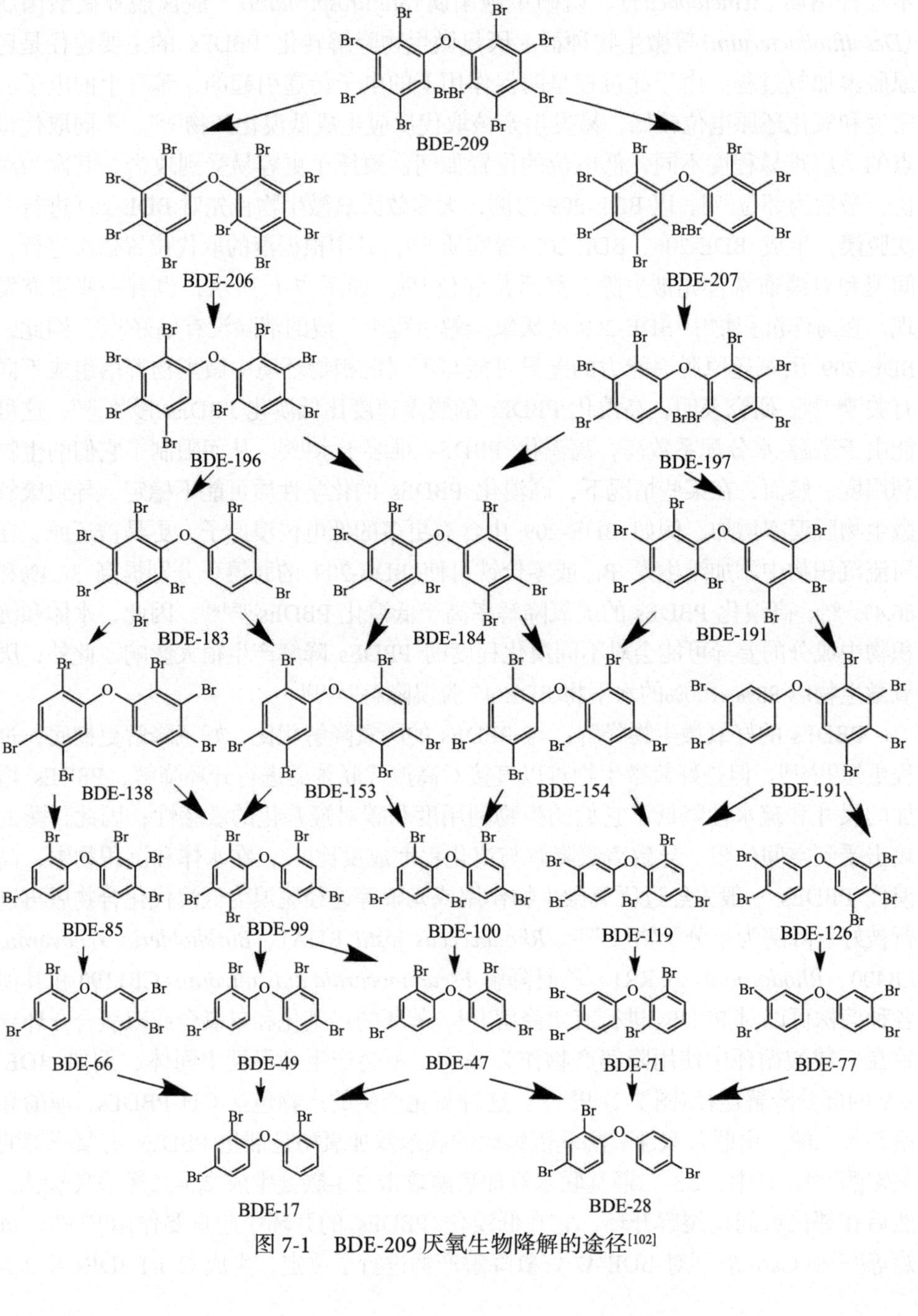

图 7-1　BDE-209 厌氧生物降解的途径[102]

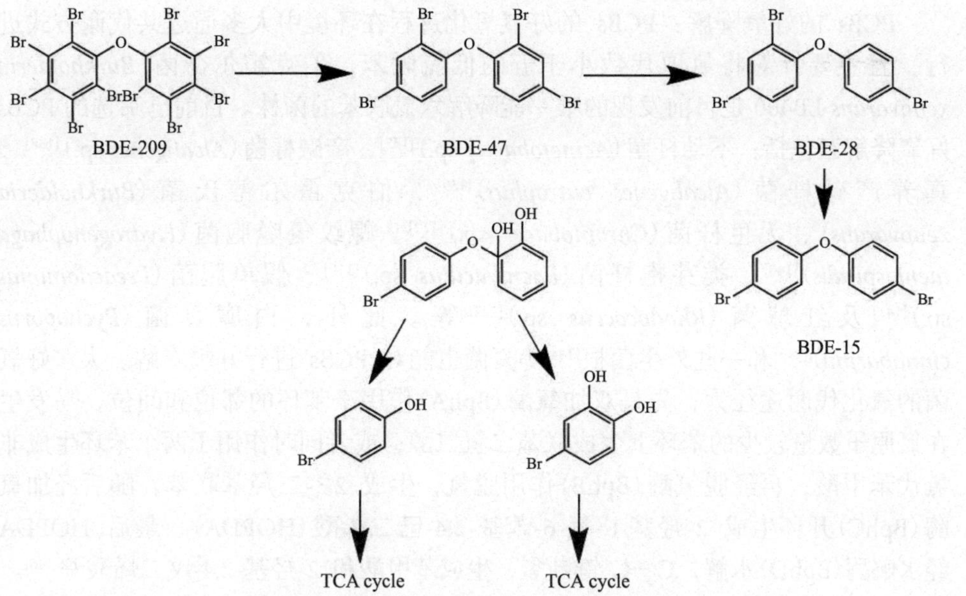

图 7-2　BDE-209 的部分好氧降解途径[125]

2) 多氯联苯

鱼体中 PCBs 的代谢转化机制：已有不少研究证明在鱼体内存在 PCBs 的生物转化过程[126,127]。北美五大湖区的不同鱼类以及从电子垃圾回收地区采集的乌鳢和鲮鱼体内均检测到了 OH-PCBs[128,129]以及 MeSO$_2$-PCBs[130,131]。Buckman 等[132]通过虹鳟鱼暴露实验证实，OH-PCBs 来自 PCBs 的生物转化，且依赖氯取代模式，在间/对位取代且具有邻近氢原子 PCBs 更易发生生物转化。尽管 OH-PCBs 添加了羟基基团增加了水溶性，但仍具有较强疏水性易在水生生物中积累[133]。

PCBs 的厌氧微生物降解：目前已有大量研究发现了水体/沉积物中 PCBs 的厌氧微生物还原过程，是环境中 PCBs 消减的关键环节。Brown 等[134]最早提出了 PCBs 的厌氧还原脱氯，并发现氯化程度较高的同系物更易被还原，而脱氯加氢生成的低氯代同系物则被好氧菌生物降解。Quensen 等[135]后来通过对美国哈德逊河沉积物的微生物在厌氧条件下能对大多数 PCBs 进行还原脱氯，脱氯主要发生在间位、对位或仅有邻位取代的同系物中。目前已分离得到多种厌氧脱氯菌，包括 *Dehalospirillum multivoram*、*Dehalococcoides etrtenogenes*、*Desulforomonas chloroethenica*、*Dehalobacter restrictus*、*Desulf tobacterium*、*DesulfonTOnile tiedjei*、*Enterobacter MS-1* 和 *Enterobacter agglomerans* 等[136]。具有不同脱卤酶的细菌具有不同的脱氯特性和脱氯途径，因此环境中 PCBs 的脱氯速度、程度、途径受到微生物群落组成的影响。

PCBs 的好氧降解：PCBs 的好氧氧化过程在环境中大多通过共代谢方式进行，且主要针对将氯取代数小于五的低氯联苯，伯克霍尔德菌 *Burkholderia xenovorans* LB400 是目前发现的唯一能降解六氯联苯的菌株。目前已筛选的 PCBs 好氧降解菌包括：不动杆菌(*Acinetobacter* sp.)[137]、产碱杆菌(*Alcaligenes* sp.)[138]、真养产碱杆菌(*Alcaligenes eutrophus*)[139]、伯克霍尔德氏菌(*Burkholderia xenovorans*)、无色杆菌(*Chromobacer* sp.)[112]、螺纹氢噬胞菌(*Hydrogenophaga taeniospiralls*)[140]、类芽孢杆菌(*Paenibacillus* sp.)[119]、假单胞菌(*Pseacdomonas* sp.)[141]及红球菌(*Rhodococcus* sp.)[142]等。此外，白腐真菌(*Pycnoporus cinnabarinu*)[143]和一些外生菌根[144]等真菌也能对 PCBs 进行好氧降解。大多好氧菌的氧化代谢途径为：先是双加氧酶(BphA)作用于苯环的邻位和间位，易发生在氯原子数量较少的苯环上生成联苯二氢二醇，或者同时作用于两个苯环生成非氯代苯甲酸；再经脱氢酶(BphB)作用脱氢，生成 2,3-二羟基联苯，随后经加氧酶(BphC)开环生成 2-羟基-6-氧-6-苯基-2,4-己二烯酸(HOPDA)，最后 HOPDA 经水解酶(BphD)水解，C—C 键断裂，生成苯甲酸和 2-羟基-2,4-戊二烯酸[145,146]。前期研究发现，PCBs 的好氧微生物降解还具有以下规律[147]：①低氯代 PCBs 更易被降解，降解速率高于高氯代同系物；②邻位和间位取代的 PCBs 易发生降解；③具有相邻的氯取代位点的 PCBs 难易发生降解；④由于邻位效应的影响，如果其中一个苯环上有两个邻位都被氯原子取代，或者两个苯环上各有一个邻位被取代，则更难被降解；⑤PCBs 的开环反应更容易发生在没有氯取代或氯取代数量少的苯环上[136,148,149]。

3) 四溴双酚 A

鱼体中 TBBPA 的转化降解：张梦迪等[29]于 2019 年对采集自广东练江的某电子垃圾拆解区附近河流的沉积物、罗非鱼肉、螺中的 TBBPA 及其脱溴产物、甲基化产物进行了定性定量分析(采样点分布见图 7-3)，浓度见表 7-2 至表 7-4，沉积物中其浓度分别为 $19.8\sim1.52\times10^4$ ng/g、$8.05\sim1.84\times10^3$ ng/g 和 $0.08\sim11.9$ ng/g，在生物样品中为 $6.96\sim1.97\times10^5$ ng/g、$3.84\sim7.07\times10^3$ ng/g、$3.42\sim472$ ng/g，且靠近拆解地区的浓度显著高于远离地区。沉积物中普遍检测出所有的脱溴产物，但是甲基化产物检出率较低，表明在沉积物中 TBBPA 更易发生脱溴降解，可能与沉积物中的厌氧微生物降解有关。据 Sun 等[150]报道，TBBPA 在厌氧环境下更易发生脱溴反应，其甲基化则主要发生在有氧环境中。研究中还发现 TBBPA 及其转化产物能在生物体内发生富集，且甲基化产物的生物富集性更高。TBBPA 与脱溴产物浓度具有相关性而与甲基化产物浓度不相关，每一种脱溴产物与对应的甲基化产物相关，说明鱼、螺等水生生物体内能对脱溴产物发生甲基化转化，甲基化是生物体内代谢的主要途径，而脱溴产物由于环境中也大量存在，无法证明其是否能在生物体内转化。对 TBBPA 及其产物的在沉积物、螺肉和鱼体组织中的组成特征研究发现(图 7-4)，TBBPA 所占的比例最高，而 tirBBPA 被发现是最主要的脱溴产物，diMeO-TBBPA 是最主要的甲基化产物，其次是 diMeO-triBBPA。总的来说，目前有关鱼体内的转化机制的研究较少，其转化机理还需要进一步研究。

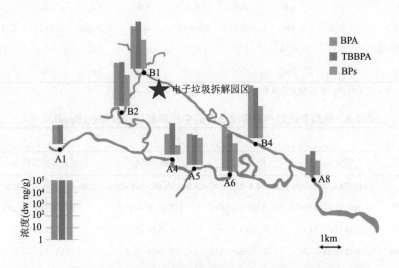

图 7-3 2019 年各采样点 TBBPA、BPA、BPs 浓度对比图
A 点为干流采样点，B 点为支流采样点

表 7-2 典型电子垃圾拆解地区沉积物中 TBBPA 及其转化产物特征

化合物	A1	A4	A5	A6	A8	B1	B2	B4
BPA	35.9	50.1	$3.65×10^3$	$4.68×10^3$	52.0	$3.46×10^3$	$4.06×10^3$	$2.57×10^4$
monoBBPA	3.33	1.18	4.30	20.2	1.47	103	74.9	401
2,2'-diBBPA	7.53	0.98	3.26	16.0	0.89	68.1	25.5	182
2,6-diBBPA	2.04	1.27	3.52	20.5	1.26	82.4	32.1	230
triBBPA	4.72	17.3	33.5	118	12.0	596	218	$1.03×10^3$
TBBPA	19.8	413	253	$1.27×10^3$	89.1	$7.72×10^3$	$4.16×10^3$	$1.52×10^4$
diMeO-monoBBPA	0.12	0.20	0.18	n.d.	n.d.	0.65	1.17	1.48
diMeO-2,2'-diBBPA	n.d.	n.d.	0.17	1.92	n.d.	0.16	0.83	1.17
diMeO-2,6-diBBPA	n.d.	0.12	0.31	n.d.	0.26	0.32	0.52	n.d.
diMeO-triBBPA	n.d.	n.d.	0.31	n.d.	n.d.	0.89	1.70	2.80
diMeO-TBBPA	n.d.	0.40	0.48	0.82	0.25	5.92	3.95	3.11

续表

化合物	A1	A4	A5	A6	A8	B1	B2	B4
2-monoBP	0.10	0.40	5.22	3.32	1.70	36.6	12.0	41.0
2,4-diBP	0.35	1.02	5.64	8.84	2.50	47.3	16.5	135
2,4,6-triBP	0.56	4.19	167	171	12.2	162	357	348
2,3,4,6-TBP	n.d.	n.d.	n.d.	n.d.	n.d.	n.d.	n.d.	n.d.
PBP	n.d.	n.d.	n.d.	n.d.	n.d.	n.d.	n.d.	n.d.
ΣBBPA	37.4	434	297	1.44×10^3	105	8.57×10^3	4.51×10^3	1.71×10^4
diMeO-ΣBBPA	0.30	0.76	1.77	2.75	0.54	8.82	9.87	11.4
ΣBPs	1.01	5.62	178	183	16.6	246	386	524

注：A点为干流采样点，B点为支流采样点，具体采样点布置见图7-3

表 7-3　鱼类各组织中目标化合物的中位浓度(范围)(ng/g lipid)

化合物	鱼			
	肌肉($n=13$)	鳃($n=13$)	内脏($n=10$)	肠道及其内容物($n=12$)
BPA	482 (74.4~1.26×10^3)	328 (68.8~2.31×10^3)	3.26×10^4 (481~1.07×10^5)	2.96×10^4 (1.69×10^3~7.70×10^4)
monoBBPA	8.61 (1.17~32.8)	3.03 (0.74~11.0)	173 (2.10~811)	209 (10.8~1.45×10^3)
2,6-diBBPA	4.38 (1.77~18.1)	1.37 (0.43~5.33)	134 (5.89~297)	127 (11.3~891)
2,2'-diBBPA	5.42 (0.98~15.2)	1.29 (0.56~6.01)	177 (3.03~362)	115 (19.0~702)
triBBPA	23.8 (7.45~134)	6.77 (1.94~27.3)	748 (17.1~2.74×10^3)	614 (140~4.85×10^3)
TBBPA	253 (26.3~749)	57.0 (6.96~126)	8.34×10^3 (127~1.47×10^5)	1.78×10^4 (1.50×10^3~1.97×10^5)
diMeO-monoBBPA	7.46 (0.55~16.4)	2.15 (0.60~5.71)	16.9 (0.79~38.0)	6.94 (1.82~18.7)
diMeO-2,2'-diBBPA	3.42 (1.22~9.16)	1.20 (0.21~7.37)	8.50 (0.74~13.0)	8.15 (1.04~22.6)
diMeO-2,6-diBBPA	9.14 (1.89~13.5)	2.37 (0.40~16.1)	21.3 (1.81~31.0)	33.7 (1.70~45.2)
diMeO-triBBPA	11.3 (2.37~24.4)	2.42 (0.58~5.19)	14.6 (1.73~90.6)	8.63 (3.51~60.2)
diMeO-TBBPA	20.5 (8.41~49.2)	15.9 (1.63~114)	43.3 (1.55~83.0)	32.7 (3.78~73.8)
2-monoBP	28.9 (2.96~133)	14.7 (1.06~34.6)	66.3 (6.41~109)	32.0 (7.20~65.9)
2,4-diBP	59.3 (7.53~140)	48.7 (5.57~74.7)	351 (9.07~935)	93.0 (51.6~787)
2,4,6-triBP	108 (23.6~514)	41.0 (23.1~126)	2.08×10^3 (95.4~3.85×10^3)	2.19×10^3 (412~3.86×10^3)
2,3,4,6-TBP	n.d.	n.d.	n.d. (n.d.~16.1)	3.45 (n.d.~9.21)
PBP	n.d.	n.d.	n.d.	n.d.
ΣBBPA	396 (38.5~930)	69.6 (10.8~177)	9.62×10^3 (152~151×10^5)	2.83×10^4 (1.69×10^3~205×10^5)
ΣdiMeO-ΣBBPA	48.5 (14.4~89.3)	24.0 (3.42~148)	104 (6.62~256)	90.2 (70.7~221)
ΣBPs	188 (50.6~787)	104 (29.7~235)	2.51×10^3 (111~4.91×10^3)	2.31×10^3 (471~4.72×10^3)

表 7-4 螺中目标化合物的中位浓度(ng/g lipid)

化合物	AW-1	AW-2	AW-3	AW-4	BW-3	BW-4	BW-7
BPA	259	229	222	98.4	445	694	650
monoBBPA	5.28	4.21	5.52	3.06	7.31	9.70	11.5
diBBPA	5.44	4.30	10.8	3.25	12.0	11.0	11.3
triBBPA	10.5	10.2	22.6	7.06	18.5	23.9	27.5
TBBPA	49.2	40.4	82.6	39.4	137	148	212
diMeO-monoBBPA	1.04	1.75	7.95	5.66	2.62	9.86	14.6
diMeO-2,2'-diBBPA	5.85	1.28	17.2	3.56	6.20	7.29	16.8
diMeO-2,6-diBBPA	4.54	2.51	7.47	6.70	5.73	6.61	6.36
diMeO-triBBPA	11.4	12.2	26.3	21.9	20.1	21.6	146
diMeO-TBBPA	18.7	9.50	31.0	10.6	28.2	29.6	289
2-monoBP	29.5	21.7	49.5	28.6	23.9	71.2	70.1
2,4-diBP	48.0	53.8	13.1	9.26	34.1	48.1	31.3
2,4,6-triBP	228	295	52.0	37.2	117	169	88.1
TBP	n.d.	n.d.	n.d.	n.d.	n.d.	n.d.	n.d.
PBP	n.d.	n.d.	n.d.	n.d.	n.d.	n.d.	n.d.
ΣBBPA	70.4	59.1	122	52.8	175	193	472
diMeO-ΣBBPA	41.5	21.8	90.0	48.0	62.8	75.0	72.2
ΣBPs	306	371	115	75.5	175	289	189

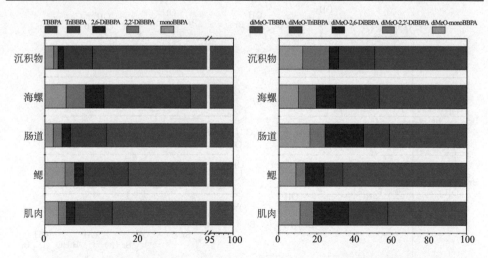

图 7-4 沉积物、海螺、鱼类各组织样品中 TBBPA 及其代谢物的组成特征

TBBPA 的厌氧微生物转化机制：厌氧还原脱溴是 TBBPA 在河床和底泥中的重要降解途径。在厌氧条件下，由于 TBBPA 苯环中的溴取代基电负性较强，易

发生还原脱溴，TBBPA 逐步脱溴生成 BPA，但 BPA 很难被进一步降解，易发生积累。文竣平等[151]筛选出肠杆菌 Enterobacter sp. NJUST20 能够脱溴还原降解 TBBPA，6 天内对 10 mg/L 的 TBBPA 的降解率为 81.9%。肠杆菌 NJUST20 对 BPA 也具有降解效果，推测其降解途径为：四溴双酚 A 首先脱掉 4 个溴原子，变成 BPA；之后 BPA 两个苯环之间断裂变成 2,4-二羟基苯乙酮和 2,5-二羟基苯甲酸。2,4-二羟基苯乙酮可能脱掉一个碳原子变成 2,4-二羟基苯甲酸；然后 2,5-二羟基苯甲酸和 2,4-二羟基苯甲酸发生开环反应变成 4-羟基-5-甲基-2-己酮；4-羟基-5-甲基-2-己酮最终被降解为无毒且可以被微生物利用的草酸和异丁酸。除此之外，Peng 等[152]从厌氧污泥体系中经过筛选分离出一株丛毛单胞菌 Comamonas sp. JXS-2-02，能将 TBBPA 还原为 BPA，并进一步开环降解为苯乙酮和草酰乙酸，从而完成产物的无毒化。

TBBPA 的好氧微生物转化机制：TBBPA 的好氧降解主要依赖于菌株的好氧脱溴能力。An 等[153]从电子垃圾拆解场地污泥中筛选出一株苍白杆菌属的菌株 (Ochrobactrum sp.)，其能使 TBBPA 完全脱溴，鉴定中间代谢产物并得出两条代谢途径：一条是 TBBPA 分子在异丙基处断裂，然后脱水氧化，进一步脱溴直至矿化；另一条是 TBBPA 逐步脱溴生成 BPA，然后 C—C 键断裂生成对羟基苯酚和开环产物，进一步氧化直至生成 CO_2 和 H_2O。范真真[154]则筛选出的假单胞菌株 Pseudomonas sp. Fz 能通过共代谢降解 TBBPA，通过降解产物推测出了 TBBPA 的降解途径(图 7-5)。

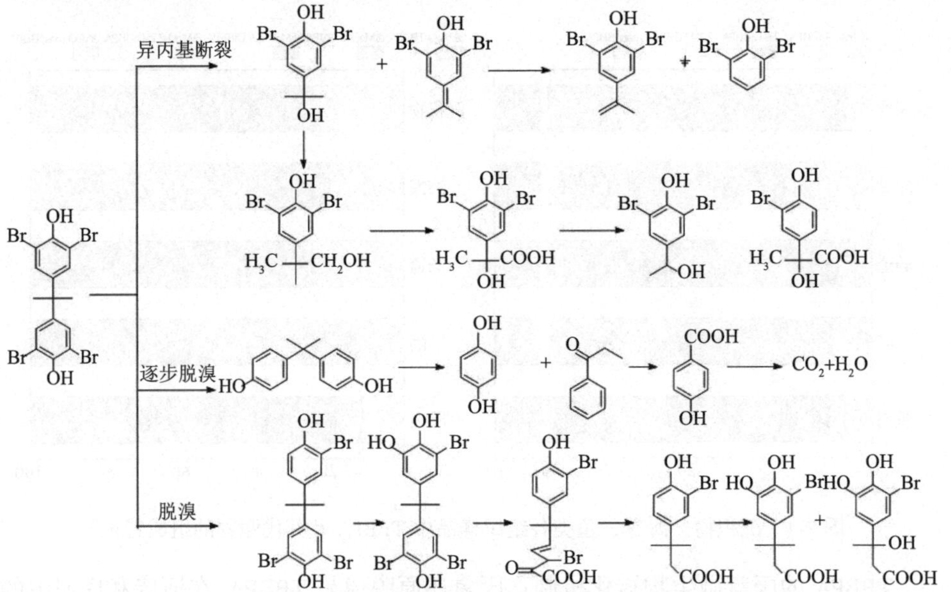

图 7-5 四溴双酚 A 的好氧降解途径[154,155]

综上所述，各种微生物对水体和沉积物中 TBBPA 的降解机制是：TBBPA 主要在无氧条件下脱溴降解为 BPA，在好氧条件下则经过甲基化、本位取代，以及烷基链断裂生成多种代谢产物。

4) 有机磷阻燃剂

OPFRs 的生物降解机制研究较少，目前发现的均为好氧降解菌，且多为鞘氨醇菌属。好氧降解机制主要靠磷酸水解酶能够依次水解脱去支链最终生成无机磷(Pi)，如图 7-6 所示。Takahashi 等[156]于 2008 年首次报道了能降解磷酸三(2-氯乙基)酯(TCEP)和磷酸三(1,3-二氯丙基)酯(TDCPP)的菌群，并筛选出了两个 TCEP 的降解菌 Sphingobium sp.TCM 1 和 TDK1，确定了两种能参与 TCEP 降解的磷酸单酯酶基因[157]。Rangu 等[158]于 2014 年发现鞘氨醇单胞菌(Sphingobium sp. strain RSMS)能将磷酸三丁酯(TBP)可作为其唯一的碳源和磷源，并且具有高效矿化高浓度 TBP 的能力，还确定了主要降解酶磷酸三酯酶(PTE)、磷酸二酯酶(PDE)和磷酸单酯酶(PME)。除好氧细菌外，研究发现白腐真菌中的血红密孔菌(Pycnoporus sanguineus)对磷酸三苯酯(TPHP)也具有降解能力，生成氧化裂解、甲基化和羟基化产物[159]。

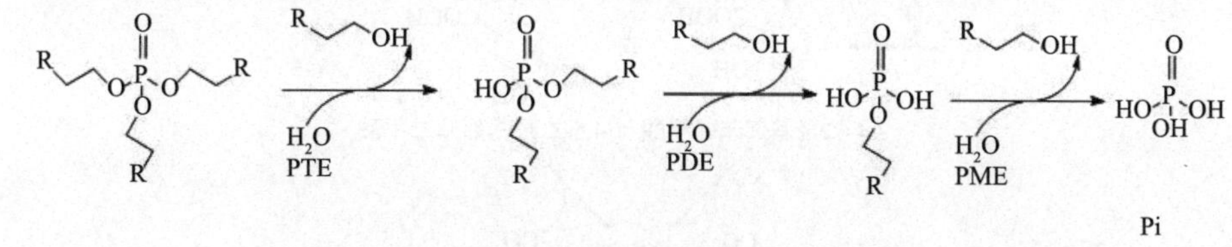

图 7-6 有机磷酸酯的生物降解示意图[157]

2. 塑料其他助剂

1) 邻苯二甲酸酯

厌氧微生物降解：关于 PAEs 化合物的厌氧生物降解途径主要包括两个步骤，即发生脱羧反应生成邻苯二甲酸(PA)和相应的醇，以及苯甲酸进一步开环降解成乙酸和甲烷[160]。目前筛选得到了多种优势降解菌，包括严格厌氧和兼性厌氧菌，如假单胞菌属(Pseudomonas)、芽孢杆菌属(Bacillus)、发酵菌如嗜热互营杆菌属(Syntrophorhabdus)、硝化菌如固氮弧菌属(Azoarcus)等[161-163]。

好氧微生物降解：在环境中分出的典型好氧 PAEs 降解菌从形态上讲大部分都是杆状，主要集中鞘氨醇单胞菌属(Sphingomonas)[164]、伯克霍尔德氏菌属(Burkhoideria)[165]、假单胞菌属(Pseudomonas[166])、节杆菌属(Arthobacter[167,168])和不动菌属(Acinetobacter[167])等。PAEs 的好氧降解的首要步骤与厌氧类似均为侧链的断裂，生成 PA，而后 PA 继续被氧化、开环，直至被矿化(图 7-7)[169]。侧链的水解过程中有三种途径，包括脱脂作用、β 氧化和转酯化作用。脱脂作用即在酯酶作用下，侧链的逐步水解，生成 PA 的过程，也是主要作用途径；当 PAEs 侧链酯基碳原子数较高时，β 氧化过程能够同时脱去双链烷基；转酯化作用则会脱去单侧链烷基。Tang 等[170]对 *Rhizobium sp.* LMB-1 降解邻苯二甲酸二丁酯(DBP)的产物进行了测定，厘清了 DBP 的代谢途径，邻苯二甲酸二丁酯(DBP)经 β 氧化逐步生成邻苯二甲酸二乙酯(DEP)，DEP 继续 β 氧化会生成邻苯二甲酸二甲酯(DMP)，后经转酯化作用生成 3-甲氧基丙酸甲酯(MMP)，最终生成 PA；DEP 也能发生侧链酯基水解生成邻苯二甲酸单乙酯(MEP)，MEP 经脱脂化作用最终生成 PA。PA 的降解主要为双加氧酶作用下氧化而后脱氢生成二羟基邻苯二甲酸，再脱羧生成原儿茶酸[171,172]。原儿茶酸经对位或间位开环形成有机酸，进一步参与进而转化成丙酮酸、琥珀酸、草酰乙酸等进入三羧酸循环，直至被完全降解[173,174]。

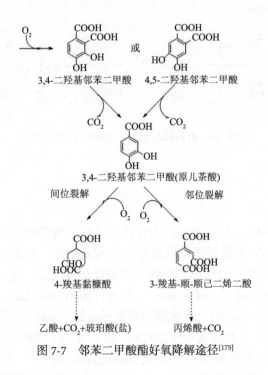

图 7-7 邻苯二甲酸酯好氧降解途径[175]

2) 双酚类化合物

厌氧生物降解：Sarmah 等[176]在河水-沉积物模型中发现了 BPF、DHBP、

BPA、TDP、BPS、BPE 和 BPB 能够得到不同程度的去除，然而他们发现这种降解受到硫酸盐、硝酸盐、铁还原等非生物因素的影响。Li 等[177]分离出一株兼性厌氧菌芽孢杆菌 GZB，可以利用 Fe^{3+} 为电子受体使 BPA 厌氧降解。但以上研究均未检测出相应的降解产物。虽然菌株 *Comamonas* sp. strain JXS-2-02[152]和 *Enterobacter* sp.NJUST20[151]可以在厌氧条件下将 TBBPA 降解为 BPA，并进一步降解转化。但由于没有进行以 BPA 单一为底物的实验，其转化机制仍不明确，关于 BPs 的厌氧生物降解途径亟待研究。

好氧生物降解：目前已分离出许多 BPA 好氧降解菌，大多属于 Sphingomonadaceae 科，该科可分为 *Sphangomonas*[178]、*Sphangobaum*[179]、*Sphangopyxas*[180]和 *No-vosphangobaum*[180] 4 个菌属，*Bacallus*[181]、*Pseudomonas*、*Streptomyces*[181]、*Enterobacter*[181]等菌属均有报道含有 BPA 降解菌。不同的菌株的酶组成不同，具有不同的代谢途径。Sasaki 等[182]发现菌株 *Sphingomona* sp AO1 中由细胞色素 P450 和铁氧化还原蛋白组成的细胞色素 P450 单加氧酶系统参与了 BPA 的降解。*Sphangobaum* sp.strain BiD32 则是通过羟基苯甲酸羟化酶来降解 BPA[183]。此外，Hirano 等[184]研究证实氨单加氧酶也与 BPA 的降解过程有关。

细菌对 BPA 的降解途径主要分为有 3 种（图 7-8），分别为氧化骨架重排、本位取代及酚环羟基化-间位裂解。氧化骨架重排是连接两个苯环的脂肪链发生氧

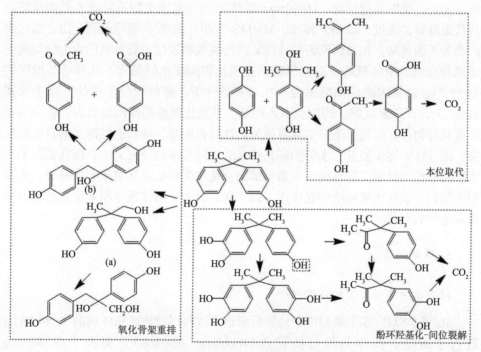

图 7-8 BPA 的好氧降解途径[191]

化重排，从而进一步脱水断键生成对羟基苯乙酮与对羟基苯甲酸(主要方式)或者被氧化羟基化。菌株 MV 1[178]、*Pseudomonas paucamobilis* strain FJ-4、AO1[182]和 *Sphingomonas* sp. strain BP-7[185]等通过此途径降解 BPA。本位取代途径是先将 C4 位置的 C 原子羟基化，然后连接酚环和异丙基的 C—C 键氧化断裂，生成对苯二酚和 4-异丙醇苯酚，从而被进一步分解代谢。菌株 *Sphangomonas* sp.strain TTNP3[186]、*Sphangobaum xenophagum* Bayram[187]和 *Cupraavadus basalensa*s JF1[188]通过本位取代降解 BPA。酚环羟基化—间位裂解途径的菌株则包括有 *Sphangobaum fulaganas* OM[189]和 *Sphingobium fulaginas* TIK1[190]。

3) 全氟化合物

由于 C—F 共价键键强度高和碳中心保护作用，PFCs 在自然界中很难降解(即 PFCs 中 C—F 键很难断裂)，一般会随着食物链富集。目前对全氟烷基化合物(PFCA)的微生物降解机制研究较少，主要集中在好氧降解。相关研究表明全氟辛酸(PFOA)和全氟辛烷磺酸(PFOS)对土壤、沉积物、水体、污泥等环境中微生物的群落结构产生影响[181,192,193]。Kwon 等[194]筛选出能够好氧降解全氟辛烷磺酸 PFOS 的铜绿假单胞菌 HJ4，检测出了全氟丁烷磺酸(PFBS)和全氟已烷磺酸(PFHxS)等一系列代谢产物，还根据该实验检测到的一些未知产物提出了降解过程中 C—C 键断裂的可能。Dinglasan 等[195]利用好氧微生物菌群对 4-氟肉桂酸进行代谢降解，通过 GC-MS 和 LC-MS-MS 气相色谱-质谱联用仪测定出主要代谢产物为不饱和酸。相关研究表明 PFCs 进行脱氟和氧化主要是利用微生物细胞中还原脱卤酶、单加氧酶、双加氧酶、过氧化物酶或水解酶等，从而将其彻底矿化[196,197]。加氧酶的作用机制为：在好氧条件下，将氧气分子中的氧原子加到 PFCs 上，在原来氟原子的位点引入羟基。过氧化物酶的作用机制为：在一些介质或共底物存在的情况下产生具有强氧化性的自由基，从而引发链式的自由基反应，使 PFCs 逐步脱氟。水解酶则是将水分子羟基加到 PFCs 上，取代氟原子。水解酶的作用机制实际是酶作为亲核试剂直接攻击碳氟键导致了它的断裂。还原脱卤酶则一般在厌氧环境中起作用，以 PFCs 为电子受体发生脱氟加氢过程，很难发生彻底矿化。

3. 多环芳烃

1) 鱼体中多环芳烃的代谢转化

Shi 等[198]对广东某典型电子垃圾拆解地下游海门湾地区鱼体内的 16 种 PAHs 进行了测定，海洋鱼类包括带鱼(*Trichiurus haumela*)、褐蓝子鱼(*Siganus fuscescens*)、马面鲀(*Thamnaconus septentrionalis*)、沙丁鱼(*Sardina pilchardus*)、

青砧鱼(*Pneumatophorus japonicas*)、二长棘鲷(*Parargyrops edita*)和蓝圆鲹(*Decapterus maruadsi*)。所有海洋鱼类肌肉中均检测到PAHs，范围为388.5~5640 ng/g(平均值：1768 ng/g，中位数：1478 ng/g)，最低浓度在马面鲀中测得，而在沙丁鱼中的浓度最高。通过皮肤和鳃从水中生物富集以及摄入食物和/或沉积物是PAHs在鱼类体内积累的潜在途径，生物积累的比例主要取决于它们的一般行为、饮食习惯、营养水平、脂质含量和代谢能力[199-201]。沙丁鱼的PAHs含量最高，在所有鱼类中，沙丁鱼的脂质含量最高。二长棘鲷是一种近海底层鱼类，其积累量明显更高，浓度高于任何其他物种(沙丁鱼除外)，这表明它们有更多机会接触和吸收沉积物中的PAHs(沉积物是PAHs的重要汇)。具有最高营养水平(4.5)的带鱼是一种底栖鱼类，其PAHs的积累水平高于营养水平最低(2.0)的褐蓝子鱼。马面鲀虽然与青砧鱼、蓝圆鲹、沙丁鱼等具有相似的营养水平(3.4)，但其脂质含量(范围：0.59%~1.69%，平均：0.98%)是所有鱼类中最低的，这部分解释了在马面鲀中检测到的最低水平的PAHs。海洋鱼类中的PAHs，三环化合物在鱼肌肉中的分布占主导地位，其次是四环化合物。它们占PAHs总量的55.2%~65.9%。特别是低分子化合物，包括二环和三环PAHs，占PAHs总量的52.1%(范围：40.1%~63.6%)。菲是最普遍的单个化合物，占PAHs总量的12.6%(范围：8.7%~21.2%)。最主要的化合物(包括菲、蒽和萘)也占PAHs总量的35.4%(范围：31.7%~53.1%)。在不同的鱼种中，虽然五环化合物在二长棘鲷(25.5%)和蓝圆鲹(24.2%)中的总PAHs的比例高于其他物种，但PAHs同系物分布相似。

研究表明，PAHs生物转化始于细胞色素P450系统的激活[202,203]，导致氧化化合物的主要CYP1A酶上调。然后，氧化的PAHs受到与CYP450系统相关的酶的活性影响，例如环氧化物水解酶，这导致PAHs代谢物的形成(图7-9)，其中一些代谢物的毒性更大。这些生化过程是PAHs在无脊椎动物[204]、鱼类[205, 207]和哺乳动物[208,209]中生物转化的普遍途径。CYP介导的环氧化通常是PAHs生物转化的第一步，然后进一步羟基化产生活性代谢物(即PAH-二醇)，通过亲核攻击形成DNA加合物[210-212]。过氧化物酶和一些CYP450酶也可能催化PAH的单电子氧化，产生有毒自由基，可进一步氧化为醌自由基。

2)多环芳烃的微生物降解机制

各种微生物对不同PAHs的降解机理有所不同，萘、菲等低分子量的PAHs容易作为碳源被微生物利用，高分子量的PAHs由于其结构复杂、难溶于水的特性，大多数是以共代谢的方式进行降解。

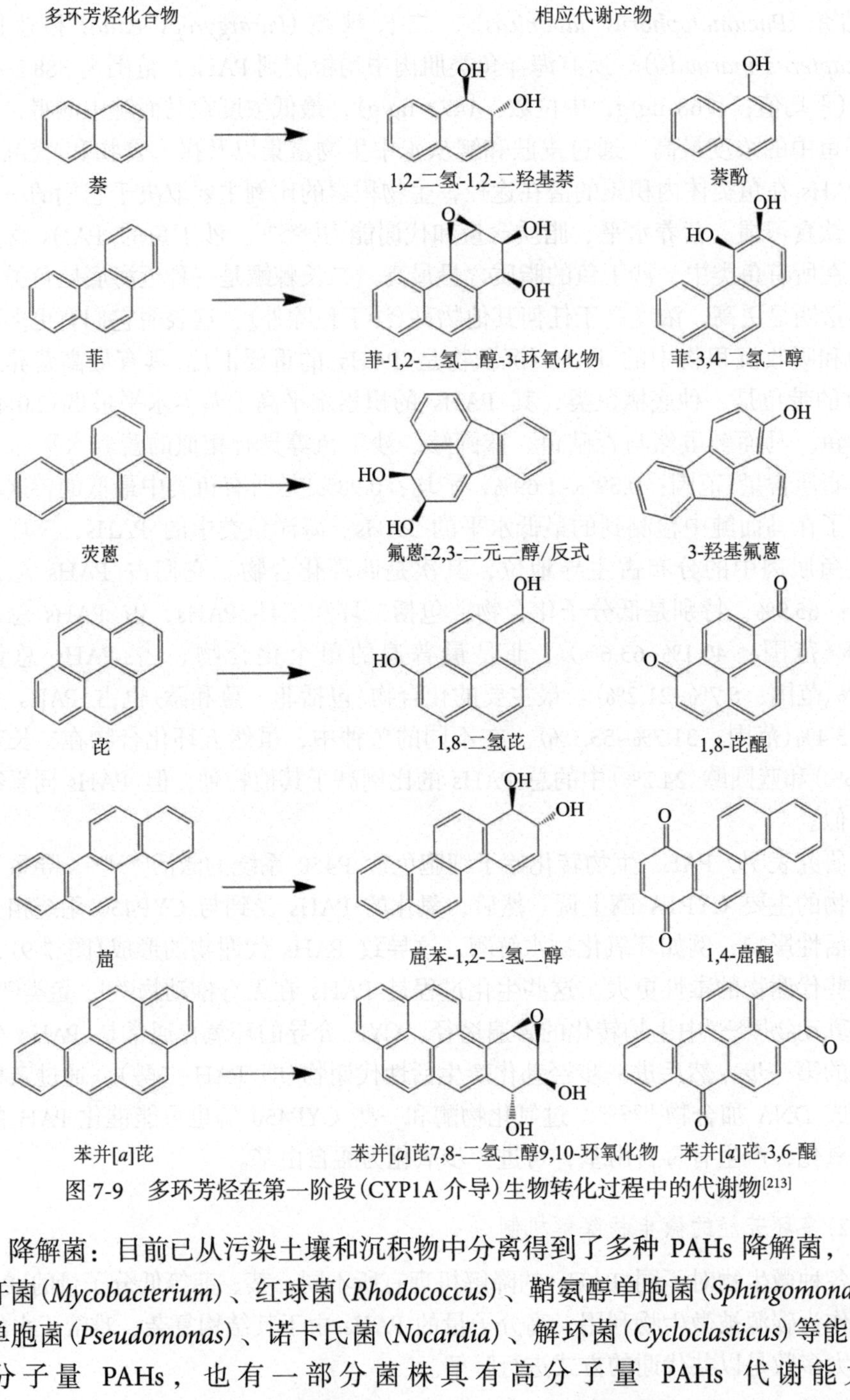

| 多环芳烃化合物 | 相应代谢产物 |

[图示:
- 萘 → 1,2-二氢-1,2-二羟基萘；萘酚
- 菲 → 菲-1,2-二氢二醇-3-环氧化物；菲-3,4-二氢二醇
- 荧蒽 → 荧蒽-2,3-二元二醇/反式；3-羟基荧蒽
- 芘 → 1,8-二氢芘；1,8-芘醌
- 䓛 → 䓛苯-1,2-二氢二醇；1,4-䓛醌
- 苯并[a]芘 → 苯并[a]芘7,8-二氢二醇9,10-环氧化物；苯并[a]芘-3,6-醌
]

图 7-9　多环芳烃在第一阶段（CYP1A 介导）生物转化过程中的代谢物[213]

降解菌：目前已从污染土壤和沉积物中分离得到了多种 PAHs 降解菌，如分枝杆菌（*Mycobacterium*）、红球菌（*Rhodococcus*）、鞘氨醇单胞菌（*Sphingomonas*）、假单胞菌（*Pseudomonas*）、诺卡氏菌（*Nocardia*）、解环菌（*Cycloclasticus*）等能降解低分子量 PAHs，也有一部分菌株具有高分子量 PAHs 代谢能力，*Mycobacterium*[214]属最多；芽孢杆菌（*Bacillus*）、气单胞菌（*Aeromonas*）、伯克氏菌

(*Burkholderia*)等能代谢低分子量 PAHs[215]。此外，土壤真菌、腐生真菌和木腐真菌等真菌[216]和藻类如羊角月牙藻(*Selenastrum capricornutum*)[217,218]也具有降解 PAHs 的能力。

降解途径：细菌和真菌对 PAHs 的好氧降解主要包括三种不同的降解机制(图 7-10)：一是芳香环被细胞色素 P450 单加氧酶氧化形成环氧化物，随后重新排列苯酚类物质或加氢形成反式二氢二醇物，该降解途径在原核生物和真核中均能发生，通常与微生物的解毒机制相关，但是往往并不能使 PAHs 完全矿化；二是芳香环被白腐真菌等木质素降解真菌分泌的木质素/锰过氧化酶和漆酶氧化形成醌类物质，并进一步环裂解直至完全矿化，该降解途径通常需要共代谢方式实现；三是芳香环被环羟化双加氧酶氧化形成二氢二醇类物质，随后脱氢、环裂解直至完全矿化，在降解过程中，微生物不仅能够以 PAHs(通常是低分子量 PAHs)为唯一碳源生长，还可以以共代谢的方式对高分子量 PAHs 进行降解[218,219]。

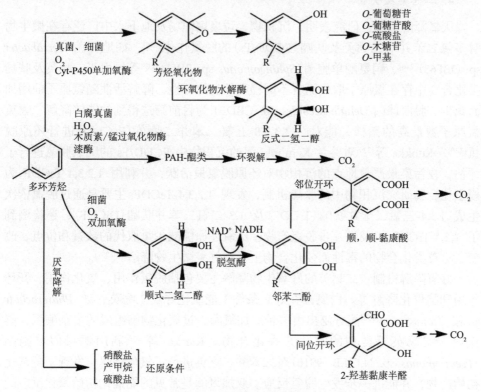

图 7-10 多环芳烃的降解机制[215]

Mycobacterium vanbaalenii PYR-1 降解菲时的 3 种途径，一是 3,4 位 C 被羟基化后通过邻苯二甲酸途径降解，二是 9,10 位 C 被双加氧酶攻击形成顺式-9,10-二

氢二醇菲，三是受单加氧酶的攻击形成 9,10-环氧化菲[220]。*Bacillus*、*Pseudomonas*、*Alcaligenes* 及 *Micrococcus* 属降解菌降解菲多采用该途径。*Mycobacterium* 属的高分子量 PAEs 降解菌(如 PYR-1、6PYR、PO1、PO2、czh-10)具有类似的转化途径，即环羟化双加氧酶作用于 1,2 位点生成 1,2-二氢二醇芘，并经过一系列酶簇反应形成 1,2-二甲氧基芘；或者作用与 4,5 位点生成 4,5-二氢二醇芘，经过脱氢酶、环裂解酶和脱羧酶的作用下形成 3,4-二羟基菲，通过菲的邻苯二甲酸降解途径进一步降解，直至被矿化[171, 220-222]。迄今为止，对于 5 环及 5 环以上 PAHs 降解途径的研究非常有限。

另外，研究发现 PAHs 能够在硫酸还原条件、硝酸还原条件和产甲烷还原条件下发生降解[219,223,224]，但 PAHs 的厌氧降解是一个非常缓慢的过程，其生化机制也尚未阐明[225]。

4. 二噁英及呋喃

厌氧降解机制：研究表明，沉积物、垃圾填埋场及地下水中广泛存在微生物对多氯二苯并对二噁英及呋喃(PCDD/Fs)的脱氯作用[226]，如地杆菌(*Terrabacter* sp. DBF63)[226]、鞘氨醇单胞菌(*Sphingomonas* sp. RW1)[227]等。由于二噁英及呋喃类化合物中存在氯原子取代而有不同程度的生物毒性，降解活性随氯原子的增加而减少。脱卤球菌 *Dehalococcoides* sp. CBDB1 为目前筛选得到的能够降解二噁英氯原子数最高的菌株，能对 1,2,3,7,8-五氯二苯并二噁英(PeCDD)进行还原脱氯[228]。Kuokka 等[229]对芬兰 Kymijoki 河的沉积物中 PCDD/Fs 的厌氧脱氯进行了评估，在污染最严重地点的沉积物对还原脱氯最活跃，并利用 1,2,3,4-TeCDF 为模型化合物研究沉积物中的脱氯机制，发现 1,2,3,4-TeCDF 主要是通过脱氯依次生成 1,3,4-三氯二苯并呋喃(TrCDF)及 1,3-二氯二苯并呋喃(DiCDF)，还检测到了 1,2,4-TrCDF 以及进一步脱氯产物。厌氧还原能减少氯取代的数量和位点、改变二噁英类化合物的毒性而使化合物更易被好氧微生物降解[230,231]。

好氧降解机制：二噁英的好氧生物降解主要包括脱氯作用、氧化作用、开环作用和酶催化降解等。白腐菌在好氧条件下能氧化降解二噁英，如 *Pharerochate* sp.和 *Trametes* sp[232]能分泌出由多种双加氧酶、过氧化物酶组成的复杂酶系，能作用于二噁英等类型的复杂芳香化合物。Klecka 等[233]利用铜绿假单胞菌(*Pseudomonas* sp. N.C.I.B. 9816)对二苯并二噁英进行了氧化降解，发现了羟基化初始产物，并能进一步发生脱氢反应。双加氧酶是常见的能催化初始氧化反应的酶。双加氧酶能攻击二噁英苯环上 1,2-(主要)和 2,3-取代位点，生成双羟基化合物，再脱氢成为烯醇，然后进一步开环氧化，最后通过三梭基循环代谢成 CO_2 和 H_2O[234]，也能作用在没有取代基的碳位上，生成儿茶酚和氯代水杨酸。二噁英氯代的位置和数量与进一步的降解将有密切的联系，例如在 1-CDD 和 4-CDF

的氧化过程中由于中间产物 3-氯儿茶酚的积聚会阻止醚键的进一步断裂；由 2,3-diCDF 和 2,3-diCDD 生成的 4,5-二氯儿茶酚则会抑制儿茶酚-1,2-双加氧酶的作用。

7.2.2 水解机制

大多电子垃圾拆解排放水体的污染物为化学性质稳定的 POPs，且在水相中的溶解度较低，不易水解，也不易与强酸强碱等发生反应，所以水解机制仅针对 OPFRs 和 PAHs 进行阐述。

1. 有机磷阻燃剂

鉴于 OPFRs 的基本分子结构(即酯键)，其非生物水解与它们在水环境中的稳定性密切相关。已有文献表明，OPFRs 水解过程中酯键的断裂主要与侧链分子结构组成有关，芳基类 OPFRs 最易发生水解反应，其次是氯代烷基 OPFRs，而烷基 OPFRs 稳定性最高，且烷基中碳原子数量越多稳定性越高[235]。OPFRs 的水解机制主要包括两种途径如图 7-11 所示：(a) H_2O 攻击碳原子；(b) OH^- 或 H_2O 进攻磷原子，此反应较易进行[236]。水解过程还受 pH 的影响，三甲基磷酸(TMP)能水解生成正磷酸盐和甲醇，但在酸性或中性条件下易发生 C—O 键的断裂，而在碱性条件下易发生 S_N2 亲核取代反应造成 P—O 键的断裂[237,238]。

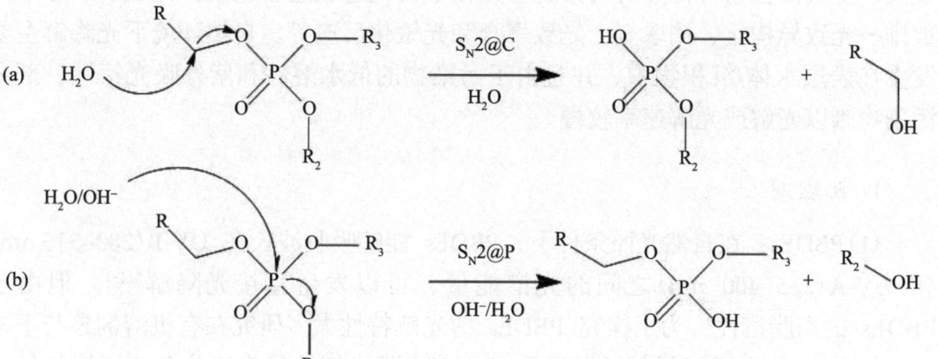

图 7-11 有机磷物质水解示意图[236]

近年来的研究发现矿物催化水解是实际土壤或水生环境中 Cl-OPFRs 降解的重要途径之一。以针铁矿为例，矿物表面与 OPFRs 的相互作用是促进水解反应的关键，主要包括：①表面金属原子与磷中心直接配位使 $S_N2@P$ 反应更易进行；②表面金属原子与酯基直接配位会削弱其与磷中心的键合从而促进裂解；③形成的表面 Fe(Ⅲ)羟基物种能提高矿物-水界面上的亲核剂浓度从而促进水解反应[239]。

2. 邻苯二甲酸酯

由于 PAEs 类物质的分子中含酯基及长短不同的脂肪链,所以其水解速率受酸、碱及碳链长度的影响。在天然水体中 PAEs 的水解速率很慢,半衰期从几年到几千年不等[240]。PAEs 的水解过程主要是 H 逐步置换侧链生成邻苯二甲酸,侧链则生成相应的醇,无法完全矿化 PAEs。PAEs 的水解过程受 pH 和烷基侧链碳原子数的应用,酸性和碱性环境下水解速率均要高于中性环境中[241],随着碳原子数量越多,半衰期越长[242],水解反应式如图 7-12 所示。

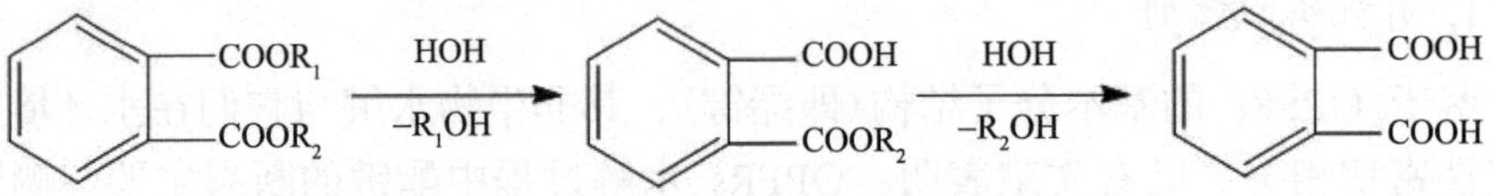

图 7-12 PAEs 的水解反应式[241]

7.2.3 光化学转化机制

自然界中的光解过程包括直接光解和间接光解两种。直接光解过程中污染物能吸收光子,导致其化学键断裂而降解;而间接光解过程中光子则被光敏化物质吸收,生成活性物种再降解污染物。光化学转化过程主要包括:光裂解、分子内重排、光致异构化、摘取氢、光致聚合和光敏化反应等。自然环境下光降解主要发生在表层水体/沉积物中,并且由于污染物的低水溶性和紫外吸光特性,部分污染物难以光解或光解速率较慢。

1. 阻燃剂

(1)PBDEs:在自然光照条件下,PBDEs 能够吸收波长在 UV-B(280~315 nm)和 UV-A(315~400 nm)之间的光谱能量,可以发生直接光降解[243]。但由于 PBDEs 的高脂溶性,为了探究 PBDEs 的光解特性大多研究在有机溶剂参与下进行。PBDEs 的光解特征包括以下几点:①主要光解途径为逐步还原脱溴加氢,同时伴随着分子内脱去溴化氢生成多溴二苯并呋喃和醚键断裂生成溴酚等途径;②PBDEs 的光解难易程度与溴取代基数量、取代位点有关。溴取代基越多的同系物越易受紫外照射发生光解;大部分 PBDEs 的不同取代位置对其自然光解难易程度的影响作用顺序为:对位<间位<邻位;苯环中溴取代基分布越集中的同系物越有利降解。Qu 等[244]利用氙灯降解水溶液中固体基质表面的 BDE-209 模拟其在天然水体中的光解,发现除了直接光解外,BDE-209 能被硅胶基质产生的羟基自由基氧化,并提出了连续脱溴、分子内消除 HBr、羟基加成和醚键裂解作

为降解途径。Roszko 等[245]研究水介质中 PBDEs 在紫外条件下的光解，发现由于 PBDEs 在水中的溶解度极低，脱溴反应只在固体颗粒表面进行，脱溴产物随后进一步降解为简单的化合物，如二苯醚或溴酚，但降解率很低甚至低于在无水相固体颗粒中的降解。在自然介质中，PBDEs 在实验条件下与自然条件下的降解产物的差别并不大，但在自然介质中的半衰期会更久。

(2) PCBs：在自然条件下，PCBs 能进行直接光解[246]，有机质、无机离子等在光照条件下生成自由基也可能使 PCBs 发生光敏化反应。PCBs 的光解过程包括脱氯、羟基化和异构化反应。为了提高多氯联苯在水中的溶解性，更加清晰地观察降解过程，在水中添加少量有机溶剂作为促溶剂和氢供体，可以提高光解速率[247]。在实验条件下的研究发现，PCBs 的光降解主要是逐步脱氯的过程，且遵循准一级反应动力学，氯原子数量越多光解速率越快，且脱氯优先发生在氯原子较多的苯环上，脱氯顺序呈现邻位＞对位＞间位的特点，除了直接脱氯途径，在不同条件下也有少量羟基化、乙基化、甲氧基化等反应发生[248,249]。Chen 等[250]在水中加入少量甲醇为促溶剂对 PCBs 进行光解观察到了脱氯和羟基化产物，并发现有机质腐殖酸能对光解起到促进作用，鉴定出起作用的主要活性氧物种为·OH 和单线态氧。

(3) TBBPA：对于 TBBPA 的光化学转化已经开展了一些研究。Wang 等[251]发现 TBBPA 在除去溶解氧的溶液中主要光转化途径以还原脱溴为主，在含有溶解氧的溶液中 O_2 可以引发 TBBPA 的氧化降解生成溴酚等单苯环产物，而在两种条件下均检出了羟基取代的产物。Bao 和 Niu[252]发现 TBBPA 可以敏化水中的溶解氧生成 ROS、·OH、$^1O_2$ 和 O_2^- 等，生成的 ROS（主要是 O_2^-）可以引发 TBBPA 的自敏化光解，主要途径为脱溴和 C—C 键断裂，而腐殖酸会与 TBBPA 竞争光子而抑制 TBBPA 的降解。

(4) OPFRs：由于部分 OPFRs 侧链烷基缺乏发色团，如毒性最大的氯代烷基磷酸酯，没有吸收波峰，不能直接光解。但是研究发现 H_2O_2、过硫酸盐等氧化剂在光照条件下能对 OPFRs 进行降解，在环境水体和沉积物中也有·OH 和过硫酸盐等氧化自由基存在，因此 OPFRs 在环境中也会发生光化学转化。OPFRs 的支链结构主要包括氯代烷烃、烷烃和芳香烃三种，支链结构会使降解速率呈现差异。相比于烷烃，氯代烷烃更难降解，且氯和氧原子数量越高降解速率越慢，如对于降解速率，TBEP＞磷酸三丁酯(TBP)＞磷酸三(2-氯乙基)酯(TCEP)＞磷酸三(2-氯丙基)酯(TCPP)[253]。以 TCEP 的光降解为例，在·OH 和过硫酸根离子存在条件下可能会发生三种转化途径，如图 7-13 所示。有过硫酸根离子参加反应时，分为两种途径[254]：(a) $SO_4^-·$ 进攻 P 原子，发生·OH/$SO_4^-·$ 加成反应，而后羟基取代支链生成产物 1；(b) $SO_4^-·$ 进攻支链末端的 C 原子，发生取代反应

$SO_4^-·$取代 Cl 原子，再经水解生成产物 4。·OH 也能直接作用于 TCEP，使其进行光解，同样也分为两种途径[237]：(a)·OH 加成至 P 原子，P—O 键断裂，生成产物 1；(b)·OH 进攻支链末端 C 原子，生成碳中心自由基，随后伴随着氧化反应、Russel 反应、水解反应，最终生成产物 2 和产物 3。此外，Moussa 等[255]还通过 TBP 的光降解实验发现，·OH 也可以通过进攻侧链丁基上 α 点位的 C 原子生成碳中心自由基致使 C—H 键断裂，而后被氧化生成烷氧基，最终产生二丁基磷酸(HDBP)。对于含苯环的有机磷[256]，其光解则要包括光致苯环结构的 C—H 键均裂以及光致水解反应。

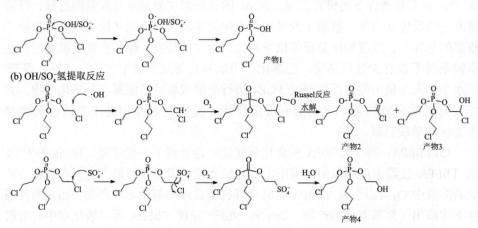

图 7-13　自由基氧化 TCEP 的降解途径[254]

2. 塑料其他助剂

（1）PAEs：PAEs 能吸收 290~400 nm 的紫外光，并发生光解。此外，天然水体和沉积物中存在的各种光敏物质产生的单线态氧($^1O_2$)和羟基自由基(·OH)，也能促进 PAEs 的降解。Meylan 等[257]提出了 PAEs 的光解机制，即两个酯基优先断裂，形成邻苯二甲酸，随后苯环上连接羧酸的 C—C 键断裂，最后再进一步降解形成 CO_2 和 H_2O。

（2）全氟化合物(PFCs)：在自然光条件下，全氟辛酸(PFOA)和全氟辛烷磺酸(PFOS)都不能吸收可见光，仅在 290~300 nm 内有较弱吸收[258]。因此，在自然水体和沉积物中 PFCs 难以发生直接光解，大多研究采用 UV 光源或添加催化剂、氧化剂或光敏剂来提高光降解效率，且仅发生断链生成短链的全氟同系物或发生脱氟作用，难以进一步降解[259-261]。

3. 多环芳烃

在大多数的天然水体表面中，波长大于 290 nm 的自然光能导致 PAHs 的直接光解。研究认为，水中的 PAHs 能被光诱导产生的 $^1O_2$、O_3 或 ·OH 氧化，从而被降解。其降解过程主要由光致电离，电子转移到分子氧(O_2)，活泼三重态与 O_2 反应三个主要反应引起，如图 7-14 所示。PAHs 吸收光子同时释放一个电子生成 P^+，电子与 O_2 反应生成超氧化物(O_2^-)，P^+ 与 H_2O 或者 OH^- 反应生成二级中间体，并进一步形成稳定产物(步骤 1~4)[262]。当 PAHs 失去电子后也能与 O_2^- 生成复合物[P^+-O_2^-]，而后分离或回归基态，或者与碰撞化合物反应生成复合物 [P-$^1O_2$]，或者激发生成单线态($^1P^*$)、三线态($^3P^*$)(步骤 5~14)。PAH 也能直接吸收光子形成激发单线态($^1P^*$)和激发三线态($^3P^*$)，$^3P^*$ 能与 O_2 反应产生复合物 [$^3P^*$-$^3O_2$] 或 [P^+-O_2^-]，随后进行降解[262]。

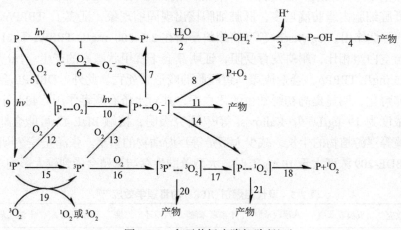

图 7-14　多环芳烃光降解路径[262]

4. 二噁英及呋喃

多氯二苯并二噁英(PCDDs)及呋喃(PCDFs)能吸收紫外光发生还原脱氯，生成低氯代产物[263]。研究发现，氯原子数目对 PCDD/Fs 的光解速率有影响，低氯取代的 PCDD/Fs 的降解速率要高于高氯原子的 PCDD/Fs，这与 PCBs 或 PBDEs 的光化学行为有所不同。PCDDs 由于其结构和溶解度的影响，在水中的光解速率较低；而 PCDFs 由于含有一个杂环 O 氧原子，在水中活性更强，且在水中的降解速率高于有机溶剂中[264]。然而，PCDD/Fs 在自然水体中的光解途径主要是间接光解或者光敏化反应。Friesen 等[265,266]对 PCDD/Fs 在自然水体中直接太阳光光解的产物分析结果表明，其光解途径包括还原脱氯、C—O 键断裂及羟基化作用，并且脱氯不是 PCDF 的主要光解途径。在水体中，PCDD/F 分子主要与·OH 反应导致 C—O 键的断裂；PCDDs 还能经光化学反应转化生成氯代二羟基联苯或 PCBs 等代谢产物，从而进行光解[266]。

7.3 污染物在水体和沉积物中健康风险消减

由于 BFRs 具有环境持久性、长距离迁移性及生物富集性，众多研究表明其对水生生物具有雌激素效应、神经毒性、甲状腺毒性、胚胎毒性、致畸、致突变和致癌性（表 7-5）等效应。例如，彭浩等[267]研究了 BFRs 对部分水生生物的半致死浓度，这些水生生物及其半致死浓度分别是：水蚤 0.96 mg/L、海藻 0.09 mg/L、蓝鳃太阳鱼 0.1 mg/L、虹鳟鱼 0.18 mg/L 以及胖头鲤鱼 0.26 mg/L。刘红玲等[268]通过鱼体测试发现，TBBPA 会造成实验鱼烦躁、抽搐、呼吸困难、变色、产卵量下降等现象。陈玛丽等[269]将红鲫暴露在 0.25 mg/L TBBPA 中，12 周后发现肝脏细胞内脂肪滴增多，肝脏细胞核出现固缩现象，证实了 TBBPA 对红鲫肝脏的毒性作用。Jagnytsch 等[270]将蝌蚪暴露在 2.5~500 mg/L TBBPA 下 21 天后发现，与空白组相比，蝌蚪发育受阻，证明其会干扰甲状腺素的产生。Hu 等[271]发现 0.75 mg/L TBBPA，会造成斑马鱼胚胎的畸形、死亡。此外，DBDPE 会抑制斑马鱼卵孵化，明显提高初孵幼虫的死亡率，并对水蚤有剧毒作用，48 h 半数有效致死浓度为 19 μg/L[272]。Kallqvist 等[273]研究表明，接触 BDE-47 可能会抑制中肋骨条藻等浮游植物的生长，减少大型蚤等浮游动物的繁殖。张泽光等[274]研究表明，当 BDE-209 浓度大于 10 mg/L 时，大型蚤的生存与繁殖会受到较大影响。

表 7-5 溴代阻燃剂（BFRs）的毒理学效应[275]

毒理学效应	五溴联苯醚	八溴联苯醚	十溴联苯醚	六溴环十二烷	四溴双酚 A	十溴二苯乙烷
浮游动物毒性	+		+	+		+
浮游植物毒性	+			+		
鱼类毒性	+		+	+	+	+
生长发育毒性	+					
内分泌毒性	+			+	+	+
急性毒性				+	+	
生殖毒性	+	+				
遗传毒性				+		
神经毒性		+		+	+	
肝脏毒性	+				+	

除了 BFRs，目前在各类水环境中的鱼类体内也已经检测到了 PAHs 的存在。蔡立哲等[276]发现九龙江口红树林区弹涂鱼和孔虾虎鱼肌肉 PAHs 的总含量分别为 40.65 ng/g 和 5.46 ng/g。我国北部白洋淀的四种鱼类(鲫、乌鳢、草鱼和鲢)的不同组织(肌肉、脑、鳃、肝)中 PAHs 的含量为 4.76~144.25 ng/g[277]。国外学者的研究发现，意大利亚得里亚大西洋的红鲻鱼、鲭鱼、蓝鳕和无须鳕体内 PAHs 总含量分别为 16.52 ng/g、63.33 ng/g、55.53 ng/g 和 44.14 ng/g[278]。但是，阿拉伯海峡不同种类鲨鱼中 PAHs 的湿重可高达 130~730 ng/g[279]。因此，不断有研究报道 PAHs 对浮游动植物和鱼类毒性的报道。吴玲玲等[280]将斑马鱼暴露于 100 μg/L 的菲溶液中 360 天后，发现其肝细胞形状变得不规则，细胞核固缩，胞质中空泡化程度加重，细胞核溶解，肝组织发生局部坏死。苗晶晶[281]用 10 μg/L 苯并[a]芘处理栉孔扇贝 10 天，发现雌性生殖腺中靠近被膜的卵母细胞细胞质疏松，核膜模糊，可见苯并[a]芘具有干扰内分泌的功能。上述研究再次证明，PAHs 作为一类具有"三致"作用的有机污染物，将会严重危害水生生物的健康。

此外，OPFRs 可被水体生物通过饮食及接触等方式摄入体内。据报道，在我国珠江三角洲地区，鲶鱼和草鱼的鱼肉中 TnBP、TCEP、TCPP 和 TBEP 浓度分别为 43.9~2946 ng/g lipid、82.7~4692 ng/g lipid、62.7~883 ng/g lipid 和 164~8842 ng/g lipid；清道夫、罗非鱼、鲮鱼和鲶鱼中 TiPP、TnPP、TnBP 和 TEHP 含量分别为 3.82~18.8 ng/g lipid、15.1~255 ng/g lipid、11.7~94.6 ng/g lipid 和 12.7~96.1 ng/g lipid[282,283]。在北美五大湖区的红点鲑和碧古鱼中，Σ_6OPFRs 平均含量为 11.2(密歇根湖)~97.4(伊利湖)ng/g lipid，其中 TPHP 为主要污染物(3.93~50.7 ng/g)，然后依次为 TCEP(6.55~35.6 ng/g)、TCPP(2.91~26.2 ng/g) 和 TNBP(0.18~5.04 ng/g)[284]。在菲律宾马尼拉湾四线鸡鱼和黄纹绯鮨鲤中 OPFRs 类物质的总浓度为 110~1900 ng/g lipid，并且底栖鱼类比浮游鱼类更易富集 OPFRs 类物质[285]。根据已有的毒理学研究发现，部分 OPFRs 添加剂具有明显的免疫毒性、内分泌干扰性、诱变性、致癌性、神经毒性等[286]。Wei 等[287]也发现暴露于 OPFRs，如 TPhP、TDCPP、TCEP 和 TBEP 等可能对鱼类等造成不利的生殖效应、内分泌干扰效应和神经毒性。相关研究也指出 TCP 对水生生物具有急性毒性、生殖毒性，且可对中枢神经系统产生毒害[288]。

综上所述，亟须对水体和沉积物中有机污染物的生态毒性和内分泌干扰活性进行评价研究，并消减其健康风险。一方面，由于 BFRs 的阻燃效率高、适用面广、耐热性好、水解稳定性优异等特点，BFRs 是电子电器产品中一类用量巨大的化合物。另一方面，BFRs 具有持久性、亲脂性和生物富集性，在水体和沉积物被广泛检测出，尤其是在水生生物中。Johnson-Restrepo 等[289]在美国佛罗里达州沿海海域的三种海洋生物中检测到了不同含量的 TBBPA(湿重)，其中牛鲨和剑吻鲨肌肉中分别含有(9.5±12.0)ng/g 和(0.87±0.5)ng/g，宽吻海豚脂肪中含有

(1.2±3.0)ng/g。此外，有学者在荷兰的鱼中检测出 0.06~3.4 ng/g 的 TBBPA，在贻贝样品中检测到 BDE-47(0.7~17 ng/g dw)、BDE-99(0.7~17 ng/g dw)和 BDE-153（＜0.1~1.5 ng/g dw）[290,291]；Whitfield 等[292,293]在澳大利亚的棕藻(14~38 ng/g)、红藻(4.5~68 ng/g)和苔藓虫门(24~27 ng/g)中检测出 TBP，虾中检测到 TBP 平均含量为 7.8~97 ng/g，鱼中 TBP 最高含量为 3.4 ng/g。Oliveira 等[294]在黄笛鲷和敏尾笛鲷的胃组织中分别检测到 15~171 ng/g 和 6~119 ng/g 的 TBP。因此，本章节以 BFRs 为例，阐明其在水体和沉积物中的毒性评价方法和健康风险消减。

7.3.1 阻燃剂的健康风险

1. 溴代阻燃剂的内分泌干扰活性评价

鉴于 TBBPA 及其代谢中间产物 BPA 具有内分泌干扰活性[295]，可以通过内分泌毒性来评估水体和沉积物的毒性，具体的评估方法如下：

(1) 对水体进行评估时，TBBPA 和 BPA 内分泌干扰活性 17β-雌二醇当量(E_2EQ)的计算公式为[296]

$$E_2EQ = E_2EF \times MEC \tag{7.1}$$

式中，E_2EQ 为 17β-雌二醇当量；E_2EF 为雌二醇当量因子，对 TBBPA 和 BPA，其分别为 0.45×10^{-6} 和 13.70×10^{-6}[295]；MEC 为有机物的环境浓度(ng/L)。

(2) 对沉积物进行评估时，主要基于孔隙水浓度 C_{pw}，其计算公式为[297]

$$C_{pw} = C_s / (f_{oc} K_{oc}) \tag{7.2}$$

式中，C_{pw} 和 C_s 分别为有机物在孔隙水(ng/L)和沉积物(ng/g)中的浓度；f_{oc} 和 K_{oc} 分别为沉积物中有机碳的馏分和ΣE分配系数[298]，其中 K_{oc} 通过 Advanced Chemistry Development(ACD/Labs)计算得到。

(3) TBBPA 和 BPA 总的 E_2EQ 值的计算公式为[296]

$$E_2EQ_{total} = \sum E_2EQ_i \tag{7.3}$$

依据欧盟定义，当环境中内分泌干扰化学物质高于阈值时，即 17β-雌二醇当量值>1 ng/E_2L，会对水生生物内分泌系统造成影响[299]。Xiong 等[300]对北江水体和沉积物的 TBBPA 和 BPA 污染进行了探究，发现水体中 17β-雌二醇当量值大于沉积物但小于 1 ng/E_2L，说明其对水生生物尚不存在危害。但是值得注意的是 BPA 的 17β-雌二醇当量数比 TBBPA 大 2~4 倍，说明 BPA 是雌激素活性的主要贡献者(图 7-15)。

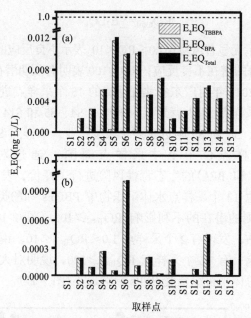

图 7-15 估算水体(a)和沉积物(b)中 TBBPA 和 BPA 的 17β-雌二醇当量数之和

2. 阻燃剂对水生生物的生态毒性评价

对阻燃剂水生生物的毒性评价主要通过生态毒性来评估，使用水生生物的风险商(RQ)来进行评估，其计算公式为[47]

$$RQ=MEC/PNEC=MEC/(EC_{50} 或 LC_{50}/f) \tag{7.4}$$

式中，MEC 为环境浓度(mg/L)；PNEC 代表预测无影响时的浓度(mg/L)。PNEC 可以通过毒理学相关浓度 EC_{50} 或者 LC_{50} 与 f 的商计算得到，其中 f 为安全系数，数值上等于 1000；EC_{50} 为生物半最大效应浓度(mg/L)，LC_{50} 为半数致死浓度(mg/L)，通过 ECOSAR 软件计算得到[298]。通过计算 TBBPA、八种 PBDEs(BDE-28、BDE-47、BDE-99、BDE-100、BDE-153、BDE-154、BDE-183 和 BDE-209)、TBP、PeBP 和 BPA 对三种不同营养级的水生生物(鱼、大型蚤和绿藻)的 EC_{50}，发现 BDE-28、BDE-47、BDE-99、BDE-100 对大型蚤的 EC_{50} 分别是 0.11071 mg/L、0.00789 mg/L、0.00261 mg/L 和 0.01112 mg/L[301]，而 ECOSAR 软件估算 BDE-153、BDE-154 和 BDE-183 对大型蚤的 EC_{50} 分别是 0.002 mg/L、0.004 mg/L 和 0.0002 mg/L。BDE-209 对绿藻(*Pseudokirchneriella subcapitata*)和鱼(*Oryzias latipes*)的 EC_{50} 为 0.00520 mg/L 和 0.00455 mg/L。

总的风险商的计算公式为[47]

$$RQ_{total} = \sum E_2 EQ_i \tag{7.5}$$

若 RQ<1.0 表示无显著风险；1.0≤RQ<10 表示不良反应的可能性很小；10≤RQ<100 表明有显著的潜在不良反应；RQ≥100 表明有预期潜在的不良反应。

Xiong 等[300]于 2015 年对广东省北江河流的 15 个采样点进行了沉积物采样分析，其中上游水库采样点 S1 为背景点，采样点 S4、S5 和 S14 附近有电子废物回收车间，采样点 S7、S8 和 S9 附近有多家服装行业，采样点 S2、S3 和 S13 有塑料工业，其他采样点位于上述采样点周围。通过对 PBDEs 和酚类 BFRs(TBBPA、TBP 和 BPA)的生态毒性风险进行了评价，发现北江水体除了 2 个采样点之外，其他 13 个采样点水体沉积物中 PBDEs 和酚类 BFRs 对鱼、大型蚤和绿藻存在可预测的潜在的不利影响(RQ_{Total}≥100)(图 7-16)。同时对沉积物的生态毒性进行评估，发现有 2 个采样点 1.0≤RQ_{Total}<10，说明对大型蚤和鱼存在小的潜在不利影响；有 12 个采样点 RQ_{Total}≥100，说明对大型蚤和鱼存在潜在的不良反应，而且总体来讲绿藻受 BFRs 影响最小(图 7-17)。

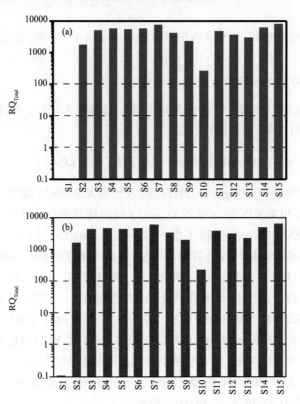

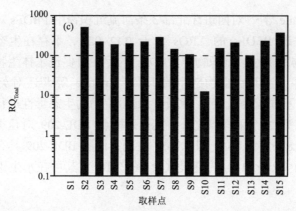

图 7-16 水体中 PBDEs、TBP、PeBP、TBBPA 和 BPA 对(a)鱼、(b)大型蚤和(c)绿藻的总风险商评价

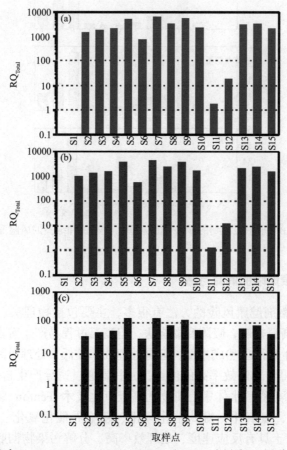

图 7-17 沉积物中 PBDEs、TBP、PeBP、TBBPA 和 BPA 对(a)鱼、(b)大型蚤和(c)绿藻的总风险商评价

此外，Cristale 等[298]对西班牙北部 3 条河流沉积物中 PBDEs 对大型蚤的毒性进行了评估，发现 PBDEs 的 ΣRQs 介于 0.12~0.71，不存在生态风险，但由于 PBDEs 具有生物蓄积性，沉积物中 PBDEs 的存在可促进水体生物对这些污染物的持续摄入，并促进其在食物链中的积累。Cristale 等[47]还对英国某一河流中 BDE-209 对三个不同营养级(鱼、水蚤、绿藻)生物的生态毒性进行了评估，发现除了接近河流源头 2 个采样点之外，其他采样点 BDE-209 对鱼类、蚤和藻类的 RQs 分别介于 3.8~65、3.6~62 和 3.3~57 之间，说明 BDE-209 具有较低(1.0≤RQ<10)和显著(10≤RQ<100)的健康风险(图 7-18)，其可在水生生物进行生物积累和生物放大。

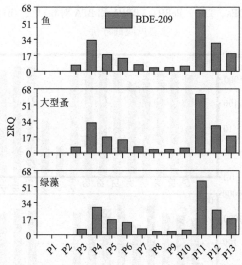

图 7-18　水中检测到的 BDE-209 对鱼、大型蚤和绿藻的风险商值

7.3.2　阻燃剂健康风险的消减方法

当前消减阻燃剂健康风险的方法有很多，主要包括物理法、化学法和生物法。物理法包括高温热解，吸附和燃烧法，但高温过程易产生有毒的中间产物，并且成本较高；生物法虽然成本低，但是处理效率较低、处理时间较长；化学法同时具备原料易得、降解效率高的优点，尤其是通过诱导产生自由基(羟基自由基和超氧自由基等)降解水体中有机物的高级氧化技术(Fenton 氧化、光催化氧化、电化学氧化、氯氧化、湿式催化氧化、超临界催化氧化、臭氧催化氧化等)。这些技术由于具有反应速度、处理效率高、分解污染物彻底等优点已经成为了目前难降解有机污染物领域的研究热点[302]。针对水体或沉积物中 BFRs 污染，以下将具体介绍化学和生物技术对 BFRs 健康风险的消减。

1. 化学法

近年来，许多科研学者利用高级氧化技术来去除水中 BFRs 的健康风险。如 Zhong 等[303]用钛磁铁催化氧化 20 mg/L TBBPA，发现在 pH=6.5 时，紫外线照射 240 min 后，TBBPA 的降解率可以达到 100%。Horikoshi 等[304]通过 UV/TiO_2 体系降解 TBBPA，发现在碱性(pH=12)条件下反应 300 min 后，其降解率可以达到 45%~60%。Zhang 等[305]利用 52.3 m/h 的 O_3 在 pH=9 条件下降解 50 mg/L 的 TBBPA 25 min，去除率可达到 99.3%。Lin 等[306]利用 MnO_2 对 TBBPA 进行氧化，发现在 21 ℃、pH=4.5，635 μmol/L MnO_2 体系中，TBBPA（3.5 μmol/L）在 5 min 内的降解率为 50%，并随着时间的延长有更高的降解效率。Zhang 等[307]发现 MnO_2/MWCNT 复合材料，可以去除 98.3%的 TBBPA。何勇[308]利用电化学氧化法降解 5 mg/L TBBPA，发现在 10 mA/cm^2 电流密度、5 mmol/L 电解质浓度、pH=7.62 条件下，其去除率可达 97.72%。可见高级氧化技术可大大地消减水体中 TBBPA 的健康风险。

此外，还有学者研究了水体和沉积物中 PBDEs、HBCD、TBP 的健康风险消减。董洪梅[309]采用 UV/O_3 对废水中的 BDE-47 进行处理，发现 60 min 后其降解率达到 83.8%。高亚杰等[310]发现在 UV 照射下，0.10 g/L TiO_2 在 120 min 内可以降解 70%的 HBCD。An 等[311]通过制备介孔 TiO_2 对 TBP 进行光催化降解研究，发现 TBP 在 1 h 内被完全降解，其还制备 TiO_2(Degussa P25)-蒙脱土复合材料 NLCT14，结合 UV 光谱阐明复合材料吸附降解水体 PBDEs 性能，发现复合材料吸附效率可以达到 80%~90%，光催化降解效率(97%)要高于直接光解（图 7-19）[312]。Peng 等[313]在 pH=3，0.2 mol/L 过硫酸钠下处理水中 BDE-209，发现 6 h 后其去除率为 53%。Pan 等[314]发现纳米零价铁能提高沉积物中 BDE-209 的去除率。Li 等[315]利用生物炭材料负载纳米零价铁后用以催化降解 BDE-209，发现当 pH=3 时，BDE-209 在 240 min 内的去除率为 82.06%。这说明水体和沉积中的其他 BFRs 的健康风险也可以通过高级氧化技术得到消减。

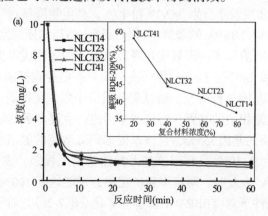

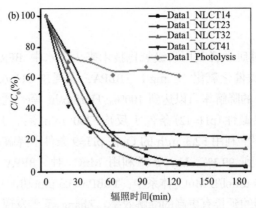

图 7-19　四种光催化剂吸附(a)和降解(b)水相 BDE-209 性能研究

2. 生物法

与化学和物理修复相比，生物修复技术具有成本低、可原位修复、不造成二次污染以及可在大范围的污染场地使用等优点，其在污染物修复方面有着广阔的应用前景。用于消减水体和沉积物中阻燃剂健康风险的生物方法/技术主要包括生物刺激、生物强化、根际过滤、植物萃取、植物固定化和植物降解技术等。但是目前大部分生物修复都集中在实验室研究阶段，主要利用实验室分离筛选、驯化得到的单一优势菌种、并利用基因工程和酶工程等手段获得基因工程菌和高效降解酶消减 BFRs 的健康风险。如 Peng 等[152]发现丛毛单胞菌属(Comamonas sp. strain)可在 10 天内降解 86% 0.5 mg/L TBBPA。钱艳园等[316]筛选并分离了四株 TBBPA 的好氧降解菌，发现其在腐殖酸和 K^+ 存在下，可以降解 81.6%的 TBBPA。

An 等[153]从电子垃圾拆解工业园污泥中获得一株能高效降解 TBBPA 的苍白芽孢杆菌(Ochrobactrum sp. T)，发现该菌株降解 TBBPA 的最佳反应条件为：接种量 25 mL/100 mL，温度 35 ℃，转速 200 r/min，pH 7.0，反应 72 h 后对 3 mg/L 的 TBBPA 降解率和脱溴率分别为 92%和 87%，其生物降解过程符合一级动力学反应模型。该菌株对 TBBPA 的降解机理包括氧化路径和还原路径，TBBPA 先被好氧氧化降解，同时消耗了一定氧气并释放出电子为还原脱溴提供条件，降解产物为 CO_2、H_2O 和 Br^- 等。Liang 等[317]进一步对苍白芽孢杆菌(Ochrobactrum sp. T)的 TBBPA 降解基因进行研究，通过全基因组序列，获得了 24 个可能与卤代化合物降解或者代谢相关的基因。对这些基因进行克隆和表达筛选到了一个含有编码 TBBPA 降解酶的基因 tbbpa，且该基因编码降解酶的氨基酸序列与卤代酸脱卤酶的序列相似性为 100%。利用该基因构建基因工程菌，发现该菌在 30℃和 pH 7.0 降解 TBBPA 96 h 时，降解率和矿化率分别能达到 100%和 37.8%，均要高于野生菌在同样条件下对 TBBPA 的降解和矿化(图 7-20)。对于不同添加剂对基因工程菌降解 TBBPA 影响研究，发现加入乙醇、NH_4NO_3 和葡萄糖的反应体系能够促进 TBBPA 的降解，但醋酸钠和柠檬酸钠的加入会使得 TBBPA 降解率从 91.1%下降到 67%和 62.8%，抑制 TBBPA 的降解，这可能是因为这两个添加剂不能作为重组菌的电子供体或者碳源(图 7-21)。

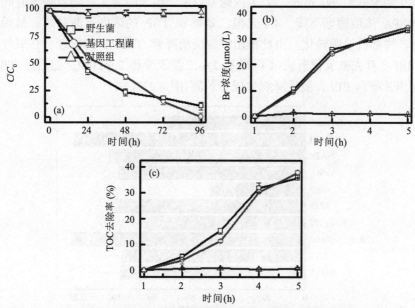

图 7-20 基因工程菌和野生菌对 TBBPA 降解特性的比较

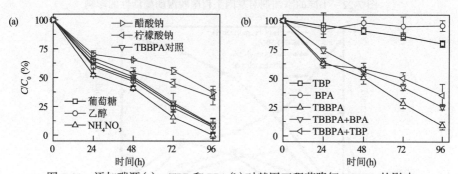

图 7-21 添加碳源(a)，TBP 和 BPA(b)对基因工程菌降解 TBBPA 的影响

加入含溴共代谢底物 TBP 可以促进重组菌降解 TBBPA，但内分泌干扰素 BPA 的加入对重组菌降解 TBBPA 具有抑制的作用，这可能是因为 BPA 对基因工程菌产生了雌激素毒性[317]。由于苍白芽孢杆菌 *Ochrobactrum* sp. T 对 TBBPA 的降解主要在脱卤酶/降解酶催化下进行，Liang 等[318]通过对基因工程菌所表达的脱卤酶进行镍离子亲和层析，获得了一个溴酚脱卤酶，分子量为 117 kDa，其在 30℃和 pH 6.5 降解 TBBPA 时，米氏常数(K_m)和最大反应速度(V_{max})可以达到

26.6 μM(μmol/L)和 0.133 μM/(min·mg)。还原型辅酶Ⅱ(nicotinamide adenine dinucleotide phosphate, NADPH)和甲基紫精的加入可以促进该脱卤酶降解 TBBPA，但加入 Cu^{2+}、细胞色素 C 和抗坏血酸抑制脱卤酶的降解作用(图 7-22)。该脱卤酶具有很宽的底物谱，能转化 2,6-DBP 和 TBP 等一系列的溴代底物。其在降解的前 60 min，以 4-丙烯基-2,6-二溴苯酚为主，随着降解时间的延长，TBBPA 浓度逐渐下降，4,4-二溴二苯醚和 TBP 的浓度逐渐升高，最后所有的中间产物都被该酶转化。由此推出该脱卤酶降解 TBBPA 的机理：在氧气存在下，TBBPA 首先被氧化形成 4-丙烯基-2,6-二溴苯酚和 2,6-DBP，随后经 TBP 转化成 4-溴苯酚(4-BP)，最后降解生成二苯醚(图 7-23)。

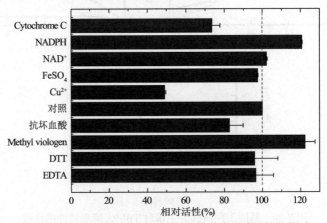

图 7-22 不同的添加剂对基因工程菌脱溴酶酶活性的影响

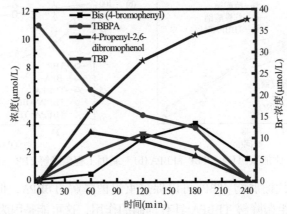

图 7-23 基因工程菌脱溴酶降解 TBBPA 的中间产物浓度变化曲线

TBP 作为 TBBPA 等 BFRs 的一个中间产物，已经在多种环境介质中被检测出，且不断有研究报道其在环境中的消减过程。范真真[154]发现假单胞菌属 *Pseudomonas* sp.可以以葡萄糖作为共代谢碳源，在好氧条件下降解 95.6%10 mg/L

的 TBP。此外，Zu 等[319,320]发现芽孢杆菌 Bacillus sp. GZT 可用于 TBP 的有氧降解，其在接种量 20 mL/100 mL，37℃，200 r/min，pH 7.0 时，降解 3 mg/L TBP 120 h，降解率和脱溴率可达 93%和 89%。Liang 等[321]进一步通过阴离子交换层析、硫酸铵沉淀和凝胶过滤等步骤从芽孢杆菌 Bacillus sp. GZT 中分离获得了一个 TBP 的脱卤酶，分子量约为 63.4 kDa，该脱卤酶与寡肽 ABC 转运寡肽结合蛋白和多肽 ABC 转运底物结合蛋白具有很高的同源性。在 pH 6.5，35℃下降解 TBP 120 min 后，降解率可达到 80%，并对 2-BP，2,4-DBP、2,6-DBP，3-BP 和 4-BP 有降解活性。其降解 TBP 的机理为：在 TBP 的邻位和对位上分别脱去一个溴原子生成 2,4-DBP 和 2,6-DBP，紧接着 2,4-DBP 和 2,6-DBP 通过生成 2-BP 而被转化生成苯甲酸和苯乙酸(图 7-24)。此外，通过对芽孢杆菌 Bacillus sp. GZT 全基因组进行测序，发现 21 个与酚类或卤代化合物的代谢或者降解相关的基因。将其进行克隆和表达，成功鉴定了 TBP、2,4-DBP、2-BP、苯酚和 2,6-二溴-4-甲基酚的降解基因 *tbpA*、*tbpB*、*tbpC*、*tbpD* 和 *tbpE*。具有 TBP 降解功能的基因工程菌可以将 TBP 转化为 2,4-DBP 和 2,6-二溴-4-甲基酚，对 TBP 的降解率为 88%；而具有 2,4-DBP 降解功能的基因工程菌可以将其快速转化为 2-BP，并在具有 2,6-二溴-4-甲基酚和 2-BP 降解功能的基因工程菌的降解产物中检测到了苯酚(图 7-25)。利用五种基因工程菌(1∶1∶1∶1∶1)的混合菌液对 TBP 进行降解，经过 40 h 后，TBP 降解率可以达到 90%，与野生菌的降解效果相当(图 7-26)。这些基因工程菌降解 TBP 的机理为：首先通过脱溴或者甲基化生成 2,4-DBP 或者 2,6-二溴-4-甲基酚，2,4-DBP 被进一步转化为 2-BP，而 2,6-二溴-4-甲基酚和 2-BP 脱溴生成苯酚[322]。

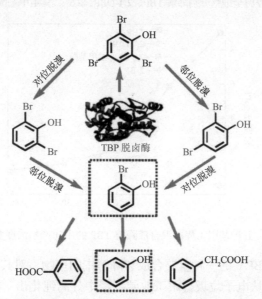

图 7-24 芽孢杆菌(*Bacillus* sp. GZT)溴酚脱卤酶降解 TBP 的路径

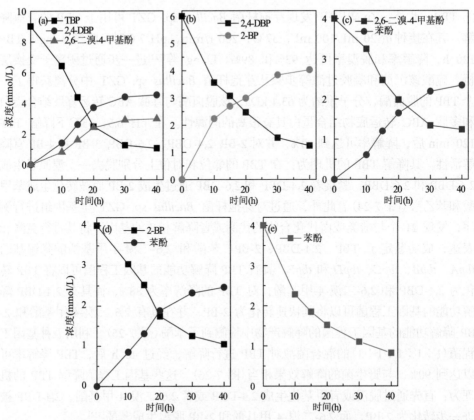

图 7-25 五种基因工程菌全细胞分别降解 TBP、2,4-DBP、2,6-二溴-4-甲基酚、2-BP 和苯酚的情况

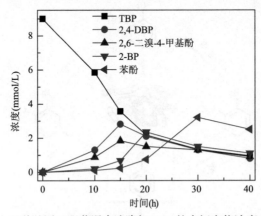

图 7-26 五种基因工程菌混合液降解 TBP 的中间产物浓度变化曲线

BPA 作为 TBBPA 的一种重要合成和降解中间产物，对其生物降解的研究报道较多。Li 等[323]从电子垃圾拆解工业园污泥中分离纯化出一株兼性厌氧芽孢杆菌 Bacillus sp. GZB，并发现该菌在接种量为 20 mL/100 mL，37℃，150 r/min， pH=7.0~8.5 时，对 5 mg/L BPA 的好氧降解率和矿化率为 100%和 51%(96 h)，且该生物降解过程符合一级动力学降解模型。在无氧条件下，Fe^{3+} 可以替代 O_2 作为电子受体进行 BPA 的氧化降解(图 7-27)。进一步测定了 5 mg/L、10 mg/L、15 mg/L、20 mg/L、30 mg/L BPA 溶液暴露下的酵母菌 β-半乳糖苷酶活性，并拟合 BPA 的雌激素剂量-效应曲线(图 7-28)，发现 β-半乳糖苷酶活性随着 BPA 的浓度增加而增大，BPA 的 EC_{50} 值为 31.73 mg/L(139 μmol/L)，其雌激素效应比 17β-雌二醇低 5 个数量级左右。如图 7-29 所示，不同初始浓度的 BPA 生物降解过程中雌激素效应的变化趋势基本一致，且 BPA 浓度越大，β-半乳糖苷酶酶活越高。在反应的 12 h 之内，30 mg/L BPA β-半乳糖苷酶酶活呈明显的下降趋势，随着反应时间延长至 24 h 和 60 h，β-半乳糖苷酶酶活先快速增加到 1.25 U，然后基本保持不变，60 h 之后开始迅速下降。而 10~20 mg/L BPA 在降解前 12 h，β-半乳糖苷酶酶活基本保持不变。在 12~24 h 之间，β-半乳糖苷酶酶活均呈现明显升高的趋势，并且较高的 β-半乳糖苷酶酶活一直持续至 60 h。96 h 后 β-半乳糖苷酶酶活比初始降解前低。这说明在 BPA 降解过程中产生了具有雌激素效应的中间产物，导致了 β-半乳糖苷酶酶活的升高，但最终可被该高效降解菌种降解，充分证明芽孢杆菌 *Bacillus* sp. GZB 是对 BPA 具有较强的降解脱毒能力。Das 等[324,325]进一步从芽孢杆菌 *Bacillus* sp. GZB 中纯化了一种编码 513 个氨基酸的孢子-漆酶，并鉴定了漆酶的编码基因 *cotA*，构建了基因工程菌，发现从野生菌分离的孢子-漆酶可以在 30 h 内完全去除 10 mg/L 的 BPA，此时酶活性变化率为 61.67%±1.44%，而从基因工程菌分离的漆酶在 14 h 内就可以达到 100%的去除率，此时酶活性变化率为 54.98%±4.98%(图 7-30)。进一步用发光菌评估漆酶降解 BPA 过程中的急性毒性，发现 10 mg/L BPA 会抑制 66%发光菌发光，而这一毒性会随着孢子-漆酶和漆酶降解 BPA 而减少，对比孢子-漆酶和漆酶两种条件毒性变化发现利用漆酶降解 BPA 可以有效降低 BPA 的毒性作用(图 7-31)。

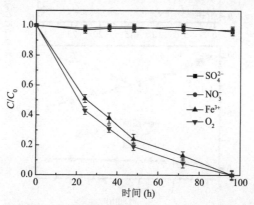

图 7-27　不同电子受体对芽孢杆菌 *Bacillus* sp. GZB 降解 BPA 的影响

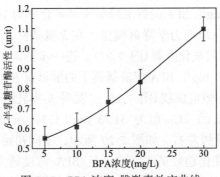

图 7-28　BPA 浓度-雌激素效应曲线

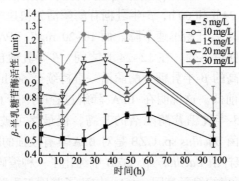

图 7-29　芽孢杆菌 *Bacillus* sp. GZB 降解 BPA 过程中雌激素效应变化

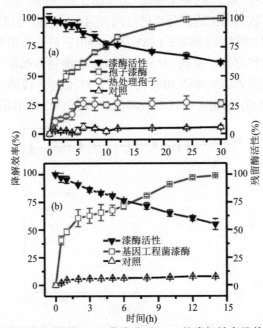

图 7-30　野生菌孢子-漆酶和基因工程菌漆酶对 BPA 的降解效率及其残留酶活性的变化

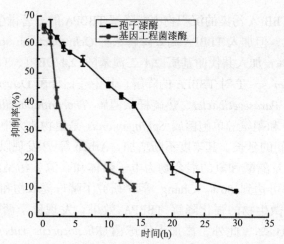

图 7-31　野生菌孢子-漆酶和基因工程菌漆酶降解 BPA 过程中毒性的变化

除了 TBBPA 及其中间产物的生物降解，对于 PBDEs 的生物降解的研究也很多。Lu 等[326]发现具有金属抗性的 *Bacillus cereus* JP12 在接种量 $OD_{600}=0.6$，pH=6.0，30℃，150 r/min 时降解 1 mg/L BDE-209 12 天，降解率可达到 88.7%。当 BDE-209 的浓度增加到 20 mg/L 时，降解率下降到 44.6%，但当 Cu^{2+} 的浓度≤8 mg/L 时，降解率可达到 51.8%，说明 Cu^{2+} 在一定程度上可以促进 BDE-209 的降解。Shi 等[113]发现绿脓杆菌 *Pseudomonas aeruginosa* 在葡萄糖浓度为 5 mg/L 时，可以在 7 天内降解 56.0% 1 mg/L BDE-209。Yu 等[327]发现以生丝微菌属 *Hyphomicrobium*，*Pseudomonas*，氨基杆菌属 *Aminobacter*，金黄杆菌属 *Chryseobacterium*，寡养单胞菌 *Stenotrophomonas*，鞘氨醇杆菌属 *Sphingobacterium* 和微杆菌属 *Microbacterium* 为主的微生物菌群 GY1 可以去除 57.2% 0.5 mg/L 的 BDE-209，但对于 10 mg/L 的 BDE-209 去除率只有 22.3%。Zhang 等[328]发现斯氏假单胞菌 *Pseudomonas stutzeri* 可以在 14 天内降解 97.94%的 BDE-47。Huang 等[329]发现河流沉积物中分离的假单胞菌 *Pseudomonas* sp.和芽孢杆菌 *Bacillus* sp.可以在 14 d 降解 90% BDE-15。

对于水体和沉积物 BFRs 的污染，主要是通过生物刺激或生物放大的方法来进行修复。生物刺激是指在不添加外来菌种的条件下，通过向水体或沉积物中投加特定的物质如电子供体或者营养物质从而强化原有菌群对污染物的降解能力，达到加速污染物去除效果。生物放大则指的是通过向水体或沉积物中投加高效降解菌种或者基因工程菌等，从而达到去除目标污染物的目的。虽然实验室分离的高效降解菌在实验室中能够高效、彻底地降解目标污染物，但当将其投加到自然环境后，可能由于外部条件的不可控性，难以达到期望的处理效果，且引进专门的微生物菌剂也可能由于不适应外场复杂的环境，导致存活率、生物降解活性低下。目前，生物修复法在 BFRs 污染与控制领域已经得到了一定的研究与应用。

如 Li 等[330]发现 TBBPA 污染的沉积物中能降解 TBBPA 的土著微生物在 10 周仅降解 3.4%的 TBBPA，但加入 TBBPA 高效降解菌种 Ochrobactrum sp. T 后 TBBPA 降解率提高到 52.1%，加入共代谢基质[2,4-二溴苯酚(2,4-DBP)、TBP 和 BPA]则会抑制 Ochrobactrum sp. T 对 TBBPA 的降解。其中苍白杆菌 Ochrobactrum sp.、珞珈山副土壤杆菌 Parasegetibacter、热硫杆状菌属 Thermithiobacillus、苯基杆菌属 Phenylobacterium 和鞘氨醇单胞菌属 Sphingomonas 是生物放大中的主要微生物菌属，随着实验时间的延长，其丰度不断增加。Arbeli 等[331]发现以 2%~3% NaCl、丙酮酸、乙醇以及醋酸和氢的混合物为电子供体和碳源，污染河床的沉积物对 TBBPA 的降解率可达到 86%。Chang 等[155]研究了酵母提取物和鼠李糖脂等对水体沉积物中土著微生物的强化降解 TBBPA 效果，发现鼠李糖脂的强化效果最佳，可以达到 94.8%。此外，添加赤红球菌(Rhodococcus ruber)、泡状枝动菌(Mycoplana bullata)、(Bacillus pumilus)、坚强芽孢杆菌(Bacillus firmus)可以促进 TBBPA 的降解，其中短小芽孢杆菌的生物放大效果最明显，在去除沉积物 TBBPA 过程中，优势菌群为红球菌和短小芽孢杆菌。Li 等[332]发现添加 BPA 优势降解菌鞘氨醇单胞菌 TTNP3(Sphingomonas sp. strain TTNP3)可以提高土著微生物对 TBBPA 的降解效率，并发现鞘氨醇单胞菌 TTNP3 对 TBBPA 的降解起代谢补偿作用。Wang 等[333]发现在共代谢底物鼠李糖存在的情况下，strain NLPSJ-22 对 BDE-99 的去除率可以达到 96%，此外，strain NLPSJ-22 可以强化 BDE-99 的降解，在生物膜反应器上降解率高达 80%，优势降解菌群为红球菌属 Rhodococcus，芽孢杆菌属 Bacillus，假单胞菌属 Pseudomonas，伯克氏菌属 Burkholderia 和鞘脂菌属 Sphingobium。在研究白腐真菌降解 BDE-209 时，发现其在 5d 内可以将 1 mg/L BDE-209 降解 63%，添加低浓度 Cu(≤1 mg/L)、Cd(≤0.5 mg/L)可促进白腐菌对 BDE-209 的降解；但重金属添加浓度＞1 mg/L 时，则会抑制 BDE-209 的降解。胞外酶对 BDE-209 去除率为 54.14%，菌种对 BDE-209 去除率为 60.17%。与纯水体系相比，由于土著微生物的作用，沉积物中白腐菌对 BDE-209 的降解能力高，不添加重金属时降解率为 71.1%。重金属对降解的影响作用较小，且不同重金属影响不同，在添加＞2 mg/L Cd、Pb 具有一定的抑制作用，且具有浓度依赖性，而添加 Cu 降解表现出促进的作用[334]。对于 BFRs 的代谢产物，Xiong 等[335]研究发现 TBP 降解菌 Bacillus sp. GZT 的加入水体和沉积物介质中可以强化降解 TBP，降解率增加到 40.7%。尽管 2-溴苯酚、2,6-二溴苯酚和 2,4-二溴苯酚等共代谢基质的加入对 TBP 降解没有促进作用，但是添加物酵母膏提取物、氯化钠、腐殖酸、乳酸钠、葡萄糖和丙酸能够明显促进 TBP 的生物降解速率，7 周内 TBP 的降解率分别可达 54.0%、46.6%、46.0%、44.3%、59.0%和 47.0%，可见酵母膏提取物和葡萄糖对 TBP 的生物刺激作用最明显。对河流沉积物中能够生物刺激土著微生物的化学物质进行研究，发现粗酶可以将

BPA 的降解率提高到 100%，主要降解微生物为假单胞菌 Pseudomonas sp.[336]。Yang 等[337]发现河流沉积物好氧降解 BPA 4 天后，BPA 降解率高达 96%，优势菌群为变形菌门(Proteobacteria)。此外，加入 BPA 优势降解菌 Bacillus sp. GZB 后，表层沉积物中 BPA 的降解明显增加，7 周内 BPA 降解率达到 35.9%；加入酵母膏提取物和葡萄糖对生物强化降解 BPA 效果最明显，7 周内 BPA 的降解率可分别达到 56.3%和 56.6%，其中变形菌门 Proteobacteria 的丰度为 37.54%~59.05%、厚壁菌门 Firmicutes 的丰度为 11.59%~33.97%[338]。可见，生物刺激和生物强化可以很好地消减水体和沉积物中 BFRs 的健康风险。

除了生物刺激和生物强化，根际过滤、植物萃取、植物固定化和植物降解技术等植物修复也可以大大消减 BFRs 的健康风险。植物修复主要利用植物吸收作用，植物分泌物和酶的刺激作用以及生物转化作用等去除和降解沉积物中的 BFRs。杨雷峰等[339]对河涌底泥中 PBDEs 进行原位生物修复研究，发现与空白组相比，再力花处理组可以有效提高底泥中 BDE-209 的去除率，经过 390 天，1.33 mg/kg BDE-209 去除率为 27%。Farzana 等[340]发现红树林在一年内可以去除沉积物中 60%的 BDE-209。Jiang 等[341]也发现红树植物可以提高沉积物中 TBBPA 的去除效果。此外，也有学者采用植物和微生物联合的方法，杨雷峰等[342]利用河边常见植物再力花(Thalia dealbata)作为修复植物，同时向体系中投加铅黄肠球菌(Enterococcus casseliflavus)和短芽孢杆菌(Brevibacillus brevis)，发现投加菌种后 BDE-209 的去除率高于相应的对照组，其中加入 B. brevis 根际处理组 BDE-209 的去除率(66%)大于非根际处理组(37.93%)和无植物处理的对照组(39.27%)，说明外源微生物可以强化植物根系修复 BDE-209 污染的底泥。也有研究指出丛植菌根真菌能够强化黑麦草对土壤中 BDE-209 的去除[333]。Pan 等[314]利用纳米零价铁和微生物联合处理被 BDE-209 污染的红树林沉积物，发现与单一 10%纳米零价铁(降解率为 41.99%)和微生物处理(降解率为 15.80%)相比，联合处理 12 月 BDE-209 的降解率可以更高(66.9%)。可见，通过植物和微生物联合修复的方法可以有效地解决单种修复方法过程中所出现的问题，并且可以提高降解效果，从而达到较为快速高效去除污染物的目的。

总之，可以采用物理法、化学法及生物降解法等对水体和沉积物中有机污染物进行风险消减，各种方法各有利弊和适用范围，其作用机制也有所不同。从防止二次污染的角度出发，相比物理法和化学法，生物修复法具有一定的优势，且生物法还具有成本低、操作简便等特点，所以生物修复法的应用场景正在不断拓宽。但考虑到现实环境污染的复杂性，在选择有机污染物消减技术方法过程中，应根据环境污染具体情况，并综合考虑各种技术的优缺点，选择最适宜的技术或技术组合来消减健康风险。

(熊举坤 梁志梳 李桂英 安太成)

参 考 文 献

[1] 牟亚南, 王金叶, 张艳, 等. 莱州湾东部海域多溴联苯醚的污染特征及生态风险评价[J]. 环境化学, 2019, 38(1): 131-141.

[2] 赵敬敬, 向新志, 王正虹, 等. 长江流域重庆段饮用水水源地多溴联苯醚的分布[J]. 环境与健康杂志, 2018, 35(10): 893-895.

[3] Wang J, Lin Z, Lin K, et al. Polybrominated diphenyl ethers in water, sediment, soil, and biological samples from different industrial areas in Zhejiang, China [J]. Journal of Hazardous Materials, 2011, 197: 211-219.

[4] Herrero H, Rodado G M, Mucientes A E. Computational techniques applied to the study of the oxidation kinetics of iron and molybdenum cyanocomplexes by peroxynitrous acid [J]. Journal of Chemometrics, 2008, 22(9-10): 556-562.

[5] Baron E, Santin G, Eljarrat E, et al. Occurrence of classic and emerging halogenated flame retardants in sediment and sludge from Ebro and Llobregat river basins (Spain) [J]. Journal of Hazardous Materials, 2014, 265: 288-295.

[6] Song W L, Li A, Ford J C, et al. Polybrominated diphenyl ethers in the sediments of the great lakes. 2. Lakes Michigan and Huron [J]. Environmental Science & Technology, 2005, 39(10): 3474-3479.

[7] Mai B X, Chen S J, Luo X J, et al. Distribution of polybrominated diphenyl ethers in sediments of the Pearl River Delta and adjacent South China Sea [J]. Environmental Science & Technology, 2005, 39(10): 3521-3527.

[8] Shen M, Yu Y, Zheng G J, et al. Polychlorinated biphenyls and polybrominated diphenyl ethers in surface sediments from the Yangtze River Delta [J]. Marine Pollution Bulletin, 2006, 52(10): 1299-1304.

[9] Jin J, Liu W, Wang Y, et al. Levels and distribution of polybrominated diphenyl ethers in plant, shellfish and sediment samples from Laizhou Bay in China [J]. Chemosphere, 2008, 71(6): 1043-1050.

[10] 都烨. 多溴联苯醚在不同区域沉积物中的分布特征[J]. 蚌埠学院学报, 2019, 8(5): 26-32.

[11] Wang D L, Cai Z W, Jiang G B, et al. Determination of polybrominated diphenyl ethers in soil and sediment from an electronic waste recycling facility [J]. Chemosphere, 2005, 60(6): 810-816.

[12] Luo Q, Cai Z W, Wong M H. Polybrominated diphenyl ethers in fish and sediment from river polluted by electronic waste [J]. Science of the Total Environment, 2007, 383(1-3): 115-127.

[13] Li H R, La Guardia M J, Liu H H, et al. Brominated and organophosphate flame retardants along a sediment transect encompassing the Guiyu, China e-waste recycling zone [J]. Science of the Total Environment, 2019, 646: 58-67.

[14] Hai Z, Guocheng H, Zhencheng X, et al. Characterization and distribution of heavy metals, polybrominated diphenyl ethers and perfluoroalkyl substances in surface sediment from the Dayan River, south China [J]. Bulletin of Environmental Contamination and Toxicology, 2015, 94(4): 503-510.

[15] 陈香平, 彭宝琦, 吕素平, 等. 台州电子垃圾拆解区水和沉积物中多溴联苯醚污染特征与生态风险[J]. 环境科学, 2016, 37(5): 1771-1778.

[16] Matsukami H, Suzuki G, Someya M, et al. Concentrations of polybrominated diphenyl ethers and alternative flame retardants in surface soils and river sediments from an electronic waste-processing area in northern Vietnam, 2012-2014 [J]. Chemosphere, 2017, 167: 291-299.

[17] Wager P A, Schluep M, Mueller E, et al. RoHS regulated substances in mixed Plastics from Waste Electrical and Electronic Equipment [J]. Environmental Science & Technology, 2012, 46(2): 628-635.

[18] Harrad S J, Sewart A P, Alcock R, et al. Polychlorinated biphenyls (PCBs) in the British environment: Sinks, sources and temporal trends [J]. Environmental pollution (Barking, Essex : 1987), 1994, 85(2): 131-146.

[19] 王泰, 张祖麟, 黄俊, 等. 海河与渤海湾水体中溶解态多氯联苯和有机氯农药污染状况调查[J]. 环境科学, 2007, (4): 4730-4735.

[20] 储少岗, 徐晓白, 童逸平. 多氯联苯在典型污染地区环境中的分布及其环境行为[J]. 环境科学学报, 1995, (4): 423-432.

[21] 杨彦, 王宗庆, 王琼, 等. 电子垃圾拆解场多环境介质多氯联苯(PCBs)污染特征及风险评估[J]. 生态毒理学报, 2014, 9(1): 133-144.

[22] 储少岗, 杨春, 徐晓白, 等. 典型污染地区底泥和土壤中残留多氯联苯(PCBs)的情况调查[J]. 中国环境科学, 1995, (3): 199-203.

[23] 王学彤, 李元成, 缪绎, 等. 电子废物拆解区河流沉积物中多氯联苯的污染水平、分布及来源[J]. 环境科学, 2012, 33(7): 2347-2351.

[24] 林娜娜, 单振华, 朱崇岭, 等. 清远某电子垃圾拆解区河流底泥中重金属和多氯联苯的复合污染[J]. 环境化学, 2015, 34(9): 1685-1693.

[25] 曹洋, 李莉, 杨苏文, 等. 高效液相色谱法分析湖泊环境介质中的四溴双酚 A [J]. 吉林广播电视大学学报, 2010, (4): 9-10+55.

[26] 张琳, 云霞, 那广水, 等. 环境水体中四溴双酚 A 的 HPLC-MS/MS 分析方法的建立与应用[J]. 环境工程学报, 2011, 5(5): 1077-1080.

[27] Morris S, Allchin C R, Zegers B N, et al. Distribution and fate of HBCD and TBBPA brominated flame retardants in north Sea estuaries and aquatic food webs [J]. Environmental Science & Technology, 2004, 38(21): 5497-5504.

[28] Li H, Guardia M L, Liu H, et al. Brominated and organophosphate flame retardants along a sediment transect encompassing the Guiyu, China e-waste recycling zone [J]. Science of the Total Environment, 2019, 646(PT. 1-1660): 58-67.

[29] 张梦迪. 四溴双酚 A(TBBPA)及其转化产物在沉积物-螺-鱼体中富集、分布及转化规律研究[D]. 广州: 广东工业大学, 2022.

[30] Wang J X, Liu L L, Wang J F, et al. Distribution of metals and brominated flame retardants (BFRs) in sediments, soils and plants from an informal e-waste dismantling site, South China [J]. Environmental Science and Pollution Research, 2015, 22(2): 1020-1033.

[31] Zhu A, Liu P, Gong Y, et al. Residual levels and risk assessment of tetrabromobisphenol A in Baiyang Lake and Fuhe river, China [J]. Ecotoxicology & Environmental Safety, 2020, 200: 110770.

[32] Kotthoff M, Rüdel H, Jürling H. Detection of tetrabromobisphenol A and its mono- and dimethyl derivatives in fish, sediment and suspended particulate matter from European freshwaters and estuaries [J]. Analytical & Bioanalytical Chemistry, 2017, 409(3): 3685-3694.

[33] Sun H, Li Y, Wang P, et al. Atmospheric levels and distribution of dechlorane Plus in an e-waste dismantling region of east China [J]. Science China Chemistry, 2016, 60(2): 305-310.

[34] 任玥, 许鹏军, 齐丽, 等. 典型电子废物处置场地表层土壤中的德克隆阻燃剂研究[J]. 中国环境科学, 2013, 33(8): 1420-1425.

[35] Wu J P, Zhang Y, Luo X J, et al. Isomer-specific bioaccumulation and trophic transfer of dechlorane plus in the freshwater food web from a highly contaminated site, South China [J]. Environmental Science & Technology, 2010, 44(2): 606-611.

[36] Martinez-Carballo E G-B C, Sitka A, et al. Determination of selected organophosphate esters in the aquatic environment of Austria [J]. Science of the Total Environment, 2007, 388(1-3): 290-299.

[37] Lai N L S, Kwok K Y, Wang X H, et al. Assessment of organophosphorus flame retardants and plasticizers in aquatic environments of China (Pearl River Delta, south China Sea, Yellow River Estuary) and Japan (Tokyo Bay) [J]. Journal of Hazardous Materials, 2019, 371: 288-294.

[38] Li J, Yu N, Zhang B, et al. Occurrence of organophosphate flame retardants in drinking water from China [J]. Water Research, 2014, 54: 53-61.

[39] Ding J, Shen X, Liu W, et al. Occurrence and risk assessment of organophosphate esters in drinking water from eastern China [J]. Science of the Total Environment, 2015, 538: 959-965.

[40] Park H, Choo G, Kim H, et al. Evaluation of the current contamination status of PFASs and OPFRs in south Korean tap water associated with its origin [J]. Science of the Total Environment, 2018, 634: 1505-1512.

[41] Lee S, Jeong W, Kannan K, et al. Occurrence and exposure assessment of organophosphate flame retardants (OPFRs) through the consumption of drinking water in Korea [J]. Water Research, 2016, 103: 182-188.

[42] Un-Jung K, Kurunthachalam K. Occurrence and distribution of organophosphate flame retardants/plasticizers in surface waters, tap water, and rainwater: Implications for human exposure [J]. Environmental Science & Technology, 2018, 52(10): 5625-5633.

[43] Bollmann U E, Moeler A, Xie Z, et al. Occurrence and fate of organophosphorus flame retardants and plasticizers in coastal and marine surface waters [J]. Water Research, 2012, 46(2): 531-538.

[44] Wang R, Tang J, Xie Z, et al. Occurrence and spatial distribution of organophosphate ester flame retardants and plasticizers in 40 rivers draining into the Bohai Sea, north China [J]. Environmental Pollution, 2015, 198: 172-178.

[45] Bacaloni A, Cucci F, Guarino C, et al. Occurrence of organophosphorus flame retardant and plasticizers in three volcanic lakes of Central Italy [J]. Environmental Science & Technology, 2008, 42(6): 1898-1903.

[46] Zhong M, Tang J, Wu H. Occurrence and spatial distribution of organophosphorus flame retardants and plasticizers in the Chinese Bohai and Yellow Seas [J]. Abstracts of Papers of the American Chemical Society, 2018, 256.

[47] Cristale J, Katsoyiannis A, Sweetman A J, et al. Occurrence and risk assessment of organophosphorus and brominated flame retardants in the River Aire (UK) [J]. Environmental Pollution, 2013, 179: 194-200.

[48] Hu M, Li J, Zhang B, et al. Regional distribution of halogenated organophosphate flame retardants in seawater samples from three coastal cities in China [J]. Marine Pollution Bulletin, 2014, 86(1-2): 569-574.

[49] Wang X, He Y, Lin L, et al. Application of fully automatic hollow fiber liquid phase microextraction to assess the distribution of organophosphate esters in the Pearl River Estuaries [J]. Science of the Total Environment, 2014, 470: 263-269.

[50] Shi Y, Gao L, Li W, et al. Occurrence, distribution and seasonal variation of organophosphate flame retardants and plasticizers in urban surface water in Beijing, China [J]. Environmental Pollution, 2016, 209: 1-10.

[51] Tan X X, Luo X J, Zheng X B, et al. Distribution of organophosphorus flame retardants in sediments from the Pearl River Delta in south China [J]. Science of the Total Environment, 2016, 544: 77-84.

[52] Zhang Z-M, Yang G-P, Zhang H-H, et al. Phthalic acid esters in the sea-surface microlayer, seawater and sediments of the east China Sea: Spatiotemporal variation and ecological risk assessment [J]. Environmental Pollution, 2020, 259: 113802.

[53] 王俊安, 李冬, 郑晓英, 等. 城市污水中邻苯二甲酸酯的研究与防治对策 [J]. 给水排水, 2007, (S1): 170-173.

[54] Sha Y, Xia X, Yang Z, et al. Distribution of PAEs in the middle and lower reaches of the Yellow River, China [J]. Environmental Monitoring and Assessment, 2007, 124(1-3): 277-287.

[55] Mukhopadhyay M, Sampath S, Munoz-Arnanz J, et al. Plasticizers and bisphenol A in Adyar and Cooum riverine sediments, India: Occurrences, sources and risk assessment [J]. Environmental Geochemistry and Health, 2020, 42(9): 2789-2802.

[56] Chakraborty P, Sampath S, Mukhopadhyay M, et al. Baseline investigation on plasticizers, bisphenol A, polycyclic aromatic hydrocarbons and heavy metals in the surface soil of the informal electronic waste recycling workshops and nearby open dumpsites in Indian metropolitan cities [J]. Environmental Pollution, 2019, 248: 1036-1045.

[57] Yamazaki E, Yamashita N, Taniyasu S, et al. Bisphenol A and other bisphenol analogues including BPS and BPF in surface water samples from Japan, China, Korea and India [J]. Ecotoxicology and Environmental Safety, 2015, 122: 565-572.

[58] Liao C Y, Liu F, Moon H B, et al. Bisphenol analogues in sediments from industrialized areas in the United States, Japan, and Korea: Spatial and temporal distributions [J]. Environmental Science & Technology, 2012, 46(21): 11558-11565.

[59] Stock N L, Furdui V I, Muir D C G, et al. Perfluoroalkyl contaminants in the Canadian arctic: Evidence of atmospheric transport and local contamination [J]. Environmental Science & Technology, 2007, 41(10): 3529-3536.

[60] Bao J, Liu W, Liu L, et al. Perfluorinated compounds in urban river sediments from Guangzhou and Shanghai of China [J]. Chemosphere, 2010, 80(2): 123-130.

[61] Bao J, Liu W, Liu L, et al. Perfluorinated compounds in the environment and the blood of residents living near fluorochemical plants in fuxin, China [J]. Environmental Science & Technology, 2011, 45(19): 8075-8080.

[62] Bao J, Liu L, Wang X, et al. Human exposure to perfluoroalkyl substances near a fluorochemical industrial park in China [J]. Environmental Science and Pollution Research, 2017, 24(10): 9194-9201.

[63] Yang L P, Zhu L Y, Liu Z T. Occurrence and partition of perfluorinated compounds in water and sediment from Liao River and Taihu Lake, China [J]. Chemosphere, 2011, 83(6): 806-814.

[64] 冯承莲, 夏星辉, 周追, 等. 长江武汉段水体中多环芳烃的分布及来源分析 [J]. 环境科学学报, 2007, (11): 1900-1908.

[65] 冯承莲, 雷炳莉, 王子健. 中国主要河流中多环芳烃生态风险的初步评价 [J]. 中国环境科学, 2009, 29(6): 583-588.

[66] 李桂英, 乔梦, 孙红卫, 等. 珠江三角洲地区饮用水源水中多环芳烃污染现状及人体健康危害的评价研究 [Z]. 第四届广东省分析化学研讨会. 中国广东恩平. 2010: 16-18+24.

[67] Wong M H, Wu S C, Deng W J, et al. Export of toxic chemicals - A review of the case of uncontrolled electronic-waste recycling [J]. Environmental Pollution, 2007, 149(2): 131-140.

[68] 王学彤, 贾英, 孙阳昭, 等. 典型污染区河流表层沉积物中 PAHs 的分布、来源及生态风险 [J]. 环境科学, 2010, 31(1): 153-158.

[69] Chen L, Yu C N, Shen C F, et al. Study on adverse impact of e-waste disassembly on surface sediment in East China by chemical analysis and bioassays [J]. Journal of Soils and Sediments, 2010, 10(3): 359-367.

[70] Hong S, Kim Y, Lee Y, et al. Distributions and potential sources of traditional and emerging polycyclic aromatic hydrocarbons in sediments from the lower reach of the Yangtze River, China [J]. Science of the Total Environment, 2022, 815: 152831.

[71] Jiang J H. Concentrations and characteristics of polycyclic aromatic hydrocarbons in seawater and sediment from mixed-aquiculture ponds in Taizhou, China [J]. Fresenius Environmental Bulletin, 2009, 18(5A): 749-756.

[72] Hoa N T Q, Anh H Q, Tue N M, et al. Soil and sediment contamination by unsubstituted and methylated polycyclic aromatic hydrocarbons in an informal e-waste recycling area, northern Vietnam: Occurrence, source apportionment, and risk assessment [J]. Science of the Total Environment, 2020, 709: 135852.

[73] Rivera J, Eljarrat E, Espadaler I, et al. Determination of PCDF/PCDD in sludges from a drinking water treatment plant influence of chlorination treatment [J]. Chemosphere, 1997, 34(5-7): 989-997.

[74] 青宪, 黄锦琼, 余小巍, 等. 某造纸厂废水中二噁英含量及其电子束辐照降解 [J]. 环境科学, 2014, 35(7): 2645-2649.

[75] Suarez M P, Rifai H S, Palachek R, et al. Distribution of polychlorinated dibenzo-p-dioxins and dibenzofurans in suspended sediments, dissolved phase and bottom sediment in the Houston Ship Channel [J]. Chemosphere, 2006, 62(3): 417-429.

[76] Lohmann R, Nelson E, Eisenreich S J, et al. Evidence for dynamic air-water exchange of PCDD/Fs: A study in the Raritan Bay/Hudson River Estuary [J]. Environmental Science & Technology, 2000, 34(15): 3086-3093.

[77] Meijer S N, Steinnes E, Ockenden W A, et al. Influence of environmental variables on the spatial distribution of PCBs in Norwegian and UK soils: Implications for global cycling [J]. Environmental Science & Technology, 2002, 36(10): 2146-2153.

[78] Leung A O W, Luksemburg W J, Wong A S, et al. Spatial distribution of polybrominated diphenyl ethers and polychlorinated dibenzo-*p*-dioxins and dibenzofurans in soil and combusted residue at Guiyu, an electronic waste recycling site in southeast China [J]. Environmental Science & Technology, 2007, 41(8): 2730-2737.

[79] Liu M Y, Li H R, Chen P, et al. PCDD/Fs and PBDD/Fs in sediments from the river encompassing Guiyu, a typical e-waste recycling zone of China [J]. Ecotoxicology and Environmental Safety, 2022, 241: 113730.

[80] Ueno D, Darling C, Alaee M, et al. Hydroxylated Polybrominated diphenyl ethers (OH-PBDEs) in the abiotic environment: Surface water and precipitation from Ontario, Canada [J]. Environmental Science & Technology, 2008, 42(5): 1657-1664.

[81] Ren M, Peng P A, Chen D Y, et al. PBDD/Fs in surface sediments from the East River, China [J]. Bulletin of Environmental Contamination and Toxicology, 2009, 83(3): 440-443.

[82] Unger M, Asplund L, Haglund P, et al. Polybrominated and mixed brominated/chlorinated dibenzo-*p*-dioxins in Sponge (Ephydatia fluviatilis) from the Baltic Sea [J]. Environmental Science & Technology, 2009, 43(21): 8245-8250.

[83] Birnbaum L S, Staskal D F, Diliberto J J. Health effects of polybrominated dibenzo-*p*-dioxins (PBDDs) and dibenzofurans (PBDFs) [J]. Environment International, 2003, 29(6): 855-860.

[84] Wang D, Jiang G, Cai Z. Method development for the analysis of polybrominated dibenzo-*p*-dioxins, dibenzofurans and diphenyl ethers in sediment samples [J]. Talanta, 2007, 72(2): 668-674.

[85] 吴江平, 冯文露, 吴思康, 等. 废弃电子垃圾拆解地野生鱼类卤系阻燃剂残留 [J]. 中国环境科学, 2021, 41(4): 1886-1892.

[86] Sun R, Luo X, Tang B, et al. Persistent halogenated compounds in fish from rivers in the Pearl River Delta, south China: Geographical pattern and implications for anthropogenic effects on the environment [J]. Environmental Research, 2016, 146: 371-378.

[87] Tao L, Zhang Y, Wu J-P, et al. Biomagnification of PBDEs and alternative brominated flame retardants in a predatory fish: Using fatty acid signature as a primer [J]. Environment International, 2019, 127: 226-232.

[88] Wu J-P, Luo X-J, Zhang Y, et al. Biomagnification of polybrominated diphenyl ethers (PBDEs) and polychlorinated biphenyls in a highly contaminated freshwater food web from south China [J]. Environmental Pollution, 2008, 157(3): 904-909.

[89] Liu F Y, Wiseman S, Wan Y, et al. Multi-species comparison of the mechanism of biotransformation of MeO-BDEs to OH-BDEs in fish [J]. Aquatic Toxicology, 2012, 114: 182-188.

[90] Shen M N, Cheng J, Wu R H, et al. Metabolism of polybrominated diphenyl ethers and tetrabromobisphenol A by fish liver subcellular fractions in vitro [J]. Aquatic Toxicology, 2012, 114: 73-79.

[91] Zeng Y H, Luo X J, Chen H S, et al. Gastrointestinal absorption, metabolic debromination, and hydroxylation of three commercial polybrominated diphenyl ether mixtures by common carp [J]. Environmental Toxicology and Chemistry, 2012, 31(4): 731-738.

[92] Haglund P S, Zook D R, Buser H R, et al. Identification and quantification of polybrominated diphenyl ethers and methoxy-polybrominated diphenyl ethers in baltic biota [J]. Environmental Science and Technology, 1998, 31(11): 3281-3287.

[93] Kim U J, Jo H, Lee I S, et al. Investigation of bioaccumulation and biotransformation of polybrominated diphenyl ethers, hydroxylated and methoxylated derivatives in varying trophic level freshwater fishes [J]. Chemosphere, 2015, 137: 108-114.

[94] Covaci A, Losada S, Roosens L, et al. Anthropogenic and naturally occurring organobrominated compounds in two deep-sea fish species from the Mediterranean Sea [J]. Environ Sci Technol, 2008, 42(23): 8654-8660.

[95] Zhao S Y, Ding C, Xu G F, et al. Diversity of organohalide respiring bacteria and reductive dehalogenases

that detoxify polybrominated diphenyl ethers in e-waste recycling sites [J]. Isme Journal, 2022, 16(9): 2123-2131.

[96] Wang G G, Jiang N, Liu Y, et al. Competitive microbial degradation among PBDE congeners in anaerobic wetland sediments: Implication by multiple-line evidences including compound-specific stable isotope analysis [J]. Journal of Hazardous Materials, 2021, 412: 125233.

[97] 黄晨晨. 基于单体多维稳定同位素技术的沉积物中持久性有机污染物微生物厌氧降解研究[D]. 广州: 中国科学院大学(中国科学院广州地球化学研究所), 2021.

[98] Rodenburg L A, Meng Q Y, Yee D, et al. Evidence for photochemical and microbial debromination of polybrominated diphenyl ether flame retardants in San Francisco Bay sediment [J]. Chemosphere, 2014, 106: 36-43.

[99] Robrock K R, Korytar P, Alvarez-Cohen L. Pathways for the anaerobic microbial debromination of polybrominated diphenyl ethers [J]. Environmental Science & Technology, 2008, 42(8): 2845-2852.

[100] Waaijers S L, Parsons J R. Biodegradation of brominated and organophosphorus flame retardants [J]. Curr Opin Biotechnol, 2016, 38: 14-23.

[101] 张馨予, 王继华. 多溴联苯醚的降解途径研究进展 [J]. 环境科学与技术, 2019, 42(10): 186-196.

[102] Chang Y T, Lo T, Chou H L, et al. Anaerobic biodegradation of decabromodiphenyl ether (BDE-209)-contaminated sediment by organic compost [J]. International Biodeterioration and Biodegradation, 2016, 113: 228-237.

[103] Gerecke A C, Hartmann P C, Heeb N V, et al. Anaerobic degradation of decabromodiphenyl ether [J]. Environmental Science & Technology, 2005, 39(4): 1078-1083.

[104] Zhu X F, Zhong Y, Wang H L, et al. New insights into the anaerobic microbial degradation of decabrominated diphenyl ether (BDE-209) in coastal marine sediments [J]. Environmental Pollution, 2019, 255: 113151.

[105] Yang C W, Huang H W, Chang B V. Microbial communities associated with anaerobic degradation of polybrominated diphenyl ethers in river sediment [J]. Journal of Microbiology Immunology and Infection, 2017, 50(1): 32-39.

[106] Yang C W, Lee C C, Ku H, et al. Bacterial communities associated with anaerobic debromination of decabromodiphenyl ether from mangrove sediment [J]. Environmental Science and Pollution Research, 2017, 24(6): 5391-5403.

[107] Shih Y H, Chou H L, Peng Y H. Microbial degradation of 4-monobrominated diphenyl ether with anaerobic sludge [J]. Journal of Hazardous Materials, 2012, 213: 341-346.

[108] Cruz R, Cunha S C, Marques A, et al. Polybrominated diphenyl ethers and metabolites—An analytical review on seafood occurrence [J]. Trac-Trends in Analytical Chemistry, 2017, 87: 129-144.

[109] Wei X P, Yin H, Peng H, et al. Reductive debromination of decabromodiphenyl ether by iron sulfide-coated nanoscale zerovalent iron: Mechanistic insights from Fe(II) dissolution and solvent kinetic isotope effects [J]. Environmental Pollution, 2019, 253: 161-170.

[110] Lee L K, Ding C, Yang K L, et al. Complete debromination of tetra- and penta-brominated diphenyl ethers by a coculture consisting of dehalococcoides and desulfovibrio species [J]. Environmental Science & Technology, 2011, 45(19): 8475-8482.

[111] Ding C, Chow W L, He J Z. Isolation of acetobacterium sp strain AG, which reductively debrominates Octa- and pentabrominated diphenyl ether technical mixtures [J]. Applied and Environmental Microbiology, 2013, 79(4): 1110-1117.

[112] Robrock K R, Coelhan M, Sedlak D L, et al. Aerobic biotransformation of polybrominated diphenyl ethers (PBDEs) by bacterial isolates [J]. Environmental Science & Technology, 2009, 43(15): 5705-5711.

[113] Shi G Y, Yin H, Ye J S, et al. Aerobic biotransformation of decabromodiphenyl ether (PBDE-209) by Pseudomonas aeruginosa [J]. Chemosphere, 2013, 93(8): 1487-1493.

[114] Chou H L, Chang Y T, Liao Y F, et al. Biodegradation of decabromodiphenyl ether (BDE-209) by bacterial mixed cultures in a soil/water system [J]. International Biodeterioration & Biodegradation, 2013, 85: 671-682.

[115] Kim Y M, Nam I H, Murugesan K, et al. Biodegradation of diphenyl ether and transformation of selected brominated congeners by Sphingomonas sp PH-07 [J]. Applied Microbiology and Biotechnology, 2007, 77(1): 187-194.

[116] Tokarz J A, Ahn M Y, Leng J, et al. Reductive debromination of polybrominated diphenyl ethers in anaerobic sediment and a biomimetic system [J]. Environmental Science & Technology, 2008, 42(4): 1157-1164.

[117] Lv Y C, Zhang Z, Chen Y C, et al. A novel three-stage hybrid nano bimetallic reduction/oxidation/biodegradation treatment for remediation of 2, 2′4, 4′-tetrabromodiphenyl ether [J]. Chemical Engineering Journal, 2016, 289: 382-390.

[118] Lv Y C, Li L H, Chen Y C, et al. Effects of glucose and biphenyl on aerobic cometabolism of polybrominated diphenyl ethers by Pseudomonas putida: Kinetics and degradation mechanism [J]. International Biodeterioration & Biodegradation, 2016, 108: 76-84.

[119] Sakai M, Ezaki S, Suzuki N, et al. Isolation and characterization of a novel polychlorinated biphenyl-degrading bacterium, Paenibacillus sp KBC101 [J]. Applied Microbiology and Biotechnology, 2005, 68(1): 111-116.

[120] Pfeifer F, Truper H G, Klein J, et al. Degradation of diphenylether by Pseudomonas cepacia Et4: Enzymatic release of phenol from 2, 3-dihydroxydiphenylether [J]. Archives of microbiology, 1993, 159(4): 323-329.

[121] Jiang L F, Luo C L, Zhang D Y, et al. Biphenyl-metabolizing microbial community and a functional operon revealed in e-waste-contaminated soil [J]. Environmental Science & Technology, 2018, 52(15): 8558-8567.

[122] Pan Y, Chen J, Zhou H C, et al. Degradation of BDE-47 in mangrove sediments under alternating anaerobic-aerobic conditions [J]. Journal of Hazardous Materials, 2019, 378: 120709.

[123] 黄婷, 段星春, 陶雪琴, 等. 2, 2′, 4, 4′-四溴联苯醚高效好氧降解菌的鉴定及其降解路径 [J]. 环境科学学报, 2017, 37(12): 4705-4714.

[124] Cao Y J, Yin H, Peng H, et al. Biodegradation of 2, 2′, 4, 4′-tetrabromodiphenyl ether (BDE-47) by Phanerochaete chrysosporium in the presence of Cd^{2+} [J]. Environmental Science and Pollution Research, 2017, 24(12): 11415-11424.

[125] Zhang Y F, Xi B D, Tan W B. Release, transformation, and risk factors of polybrominated diphenyl ethers from landfills to the surrounding environments: A review [J]. Environment International, 2021, 157: 106780.

[126] Wong C S, Lau F, Clark M, et al. Rainbow trout (Oncorhynchus mykiss) can eliminate chiral organochlorine compounds enantioselectively [J]. Environmental Science & Technology, 2002, 36(6): 1257-1262.

[127] Wong C S, Mabury S A, Whittle D M, et al. Organochlorine compounds in Lake Superior: Chiral polychlorinated biphenyls and biotransformation in the aquatic food web [J]. Environmental Science & Technology, 2004, 38(1): 84-92.

[128] Campbell L M, Muir D C G, Whittle D M, et al. Hydroxylated PCBs and other chlorinated phenolic compounds in lake trout (Salvelinus namaycush) blood plasma from the Great Lakes Region [J]. Environmental Science & Technology, 2003, 37(9): 1720-1725.

[129] Zeng Y H, Luo X J, Zheng X B, et al. Species-specific bioaccumulation of halogenated organic pollutants and their metabolites in fish Serum from an e-waste site, south China [J]. Archives of Environmental Contamination and Toxicology, 2014, 67(3): 348-357.

[130] Stapleton H M, Letcher R J, Baker J E. Metabolism of PCBs by the deepwater sculpin (Myoxocephalus thompsoni) [J]. Environmental Science & Technology, 2001, 35(24): 4747-4752.

[131] Zhang Y, Wu J P, Luo X J, et al. Methylsulfonyl polychlorinated biphenyls in fish from an electronic waste-

recycling site in south China: Levels, congener profiles, and chiral signatures [J]. Environmental Toxicology and Chemistry, 2012, 31(11): 2507-2512.

[132] Buckman A H, Wong C S, Chow E A, et al. Biotransformation of polychlorinated biphenyls (PCBs) and bioformation of hydroxylated PCBs in fish [J]. Aquatic Toxicology, 2006, 78(2): 176-185.

[133] McKim J, Schmieder P, Veith G. Absorption dynamics of organic chemical transport across trout gills as related to octanol-water partition coefficient [J]. Toxicology and applied pharmacology, 1985, 77(1): 1-10.

[134] Brown J F, Jr. , Bedard D L, Brennan M J, et al. Polychlorinated biphenyl dechlorination in aquatic sediments [J]. Science, 1987, 236(4802): 709-712.

[135] Quensen J F, 3rd, Tiedje J M, Boyd S A. Reductive dechlorination of polychlorinated biphenyls by anaerobic microorganisms from sediments [J]. Science, 1988, 242(4879): 752-754.

[136] Borja J, Taleon D M, Auresenia J, et al. Polychlorinated biphenyls and their biodegradation [J]. Process Biochemistry, 2005, 40(6): 1999-2013.

[137] Kohler H P, Kohler-Staub D, Focht D D. Cometabolism of polychlorinated biphenyls: Enhanced transformation of Aroclor 1254 by growing bacterial cells [J]. Applied and Environmental Microbiology, 1988, 54(8): 1940-1945.

[138] Commandeur L C, May R J, Mokross H, et al. Aerobic degradation of polychlorinated biphenyls by Alcaligenes sp. JB1: Metabolites and enzymes [J]. Biodegradation, 1996, 7(6): 435-443.

[139] Bedard D L, Wagner R E, Brennan M J, et al. Extensive degradation of Aroclors and environmentally transformed polychlorinated biphenyls by Alcaligenes eutrophus H850 [J]. Applied and Environmental Microbiology, 1987, 53(5): 1094-1102.

[140] Lambo A J, Patel T R. Isolation and characterization of a biphenyl-utilizing psychrotrophic bacterium, Hydrogenophaga taeniospiralis IA3-A, that cometabolize dichlorobiphenyls and polychlorinated biphenyl congeners in Aroclor 1221 [J]. Journal of Basic Microbiology, 2006, 46(2): 94-107.

[141] Master E R, Mohn W W. Psychrotolerant bacteria isolated from Arctic soil that degrade polychlorinated biphenyls at low temperatures [J]. Applied and Environmental Microbiology, 1998, 64(12): 4823-4829.

[142] Bruhlmann F, Chen W. Transformation of polychlorinated biphenyls by a novel BphA variant through the meta-cleavage pathway [J]. FEMS Microbiology Letters, 1999, 179(2): 203-208.

[143] Kubatova A, Erbanova P, Eichlerova I, et al. PCB congener selective biodegradation by the white rot fungus Pleurotus ostreatus in contaminated soil [J]. Chemosphere, 2001, 43(2): 207-215.

[144] 陈瑞蕊, 林先贵, 尹睿, 等. 有机污染土壤中菌根的作用 [J]. 生态学杂志, 2005, (2): 176-180.

[145] Seah S Y K, Labbe G, Nerdinger S, et al. Identification of a serine hydrolase as a key determinant in the microbial degradation of polychlorinated biphenyls [J]. Journal of Biological Chemistry, 2000, 275(21): 15701-15708.

[146] 蔡慧. PCBs 降解菌的筛选鉴定及降解特性的研究 [D]. 苏州: 苏州科技学院, 2015.

[147] 贾凌云. 多氯联苯降解菌的筛选及其降解性能研究 [D]. 大连: 大连理工大学, 2008.

[148] Bedard D L, Haberl M L. Influence of chroline substitution pattern on the degradation of polychlorinated biphenyls by eight bacterial strains [J]. Microbial Ecology, 1990, 20(1): 87-102.

[149] Field J A, Sierra-Alvarez R. Microbial transformation and degradation of polychlorinated biphenyls [J]. Environmental Pollution, 2008, 155(1): 1-12.

[150] Sun F F, Kolvenbach B A, Nastold P, et al. Degradation and metabolism of tetrabromobisphenol A (TBBPA) in submerged soil and soil-plant systems [J]. Environmental Science & Technology, 2014, 48(24): 14291-14299.

[151] 文竣平. 四溴双酚 A 厌氧降解菌的筛选及其降解特性研究 [D]. 南京: 南京理工大学, 2015.

[152] Peng X X, Zhang Z L, Luo W S, et al. Biodegradation of tetrabromobisphenol A by a novel Comamonas sp strain, JXS-2-02, isolated from anaerobic sludge [J]. Bioresource Technology, 2013, 128: 173-179.

[153] An T C, Zu L, Li G Y, et al. One-step process for debromination and aerobic mineralization of

tetrabromobisphenol-A by a novel Ochrobactrum sp T isolated from an e-waste recycling site [J]. Bioresource Technology, 2011, 102(19): 9148-9154.

[154] 范真真. 四溴双酚 A 好氧共代谢降解特性及其机理研究[D]. 大连: 大连理工大学, 2013.

[155] Chang B V, Yuan S Y, Ren Y L. Aerobic degradation of tetrabromobisphenol-A by microbes in river sediment [J]. Chemosphere, 2012, 87(5): 535-541.

[156] Takahashi S, Kawashima K, Kawasaki M, et al. Enrichment and characterization of chlorinated organophosphate ester-degrading mixed bacterial cultures [J]. Journal of Bioscience and Bioengineering, 2008, 106(1): 27-32.

[157] Takahashi S, Katanuma H, Abe K, et al. Identification of alkaline phosphatase genes for utilizing a flame retardant, tris(2-chloroethyl) phosphate, in Sphingobium sp strain TCM1 [J]. Applied Microbiology and Biotechnology, 2017, 101(5): 2153-2162.

[158] Rangu S S, Muralidharan B, Tripathi S C, et al. Tributyl phosphate biodegradation to butanol and phosphate and utilization by a novel bacterial isolate, Sphingobium sp strain RSMS [J]. Applied Microbiology and Biotechnology, 2014, 98(5): 2289-2296.

[159] 周家华. 血红密孔菌对磷酸三苯酯的降解机制及其降解产物毒性研究 [D]. 桂林: 桂林理工大学, 2021.

[160] O'Connor O A, Rivera M D, Young L Y. Toxicity and biodegradation of phthalic acid esters under methanogenic conditions [J]. Environmental Toxicology and Chemistry, 1989, 8(7): 569-576.

[161] Shariati S, Pourbabaee A A, Alikhani H A, et al. Anaerobic biodegradation of phthalic acid by an indigenous Ralstonia pickettii strain SHAn2 isolated from Anzali international wetland [J]. International Journal of Environmental Science and Technology, 2022, 19(6): 4827-4838.

[162] 田田. 白洋淀岸边带中降解 PAEs 的兼性厌氧菌的分离筛选与活性研究 [D]. 保定: 河北大学, 2020.

[163] Junghare M, Spiteller D, Schink B. Anaerobic degradation of xenobiotic isophthalate by the fermenting bacterium Syntrophorhabdus aromaticivorans [J]. Isme Journal, 2019, 13(5): 1252-1268.

[164] Gu J G, Han B P, Duan S S, et al. Degradation of the endocrine-disrupting dimethyl phthalate carboxylic ester by Sphingomonas yanoikuyae DOS01 isolated from the south China sea and the biochemical pathway [J]. International Biodeterioration & Biodegradation, 2009, 63(4): 450-455.

[165] Wang Y, Yin B, Hong Y G, et al. Degradation of dimethyl carboxylic phthalate ester by Burkholderia cepacia DA2 isolated from marine sediment of south China sea [J]. Ecotoxicology, 2008, 17(8): 845-852.

[166] Liao C S, Chen L C, Chen B S, et al. Bioremediation of endocrine disruptor di-n-butyl phthalate ester by Deinococcus radiodurans and Pseudomonas stutzeri [J]. Chemosphere, 2010, 78(3): 342-346.

[167] Yang X, Zhang C, He Z, et al. Isolation and characterization of two n-butyl benzyl phthalate degrading bacteria [J]. International Biodeterioration & Biodegradation, 2013, 76: 8-11.

[168] Wen Z D, Gao D W, Wu W M. Biodegradation and kinetic analysis of phthalates by an Arthrobacter strain isolated from constructed wetland soil [J]. Applied Microbiology and Biotechnology, 2014, 98(10): 4683-4690.

[169] 韩永和, 何睿文, 李超, 等. 邻苯二甲酸酯降解细菌的多样性、降解机理及环境应用 [J]. 生态毒理学报, 2016, 11(2): 37-49.

[170] Tang W J, Zhang L S, Fang Y, et al. Biodegradation of phthalate esters by newly isolated Rhizobium sp LMB-1 and its biochemical pathway of di-n-butyl phthalate [J]. Journal of Applied Microbiology, 2016, 121(1): 177-186.

[171] Kim S J, Kweon O, Jones R C, et al. Complete and integrated pyrene degradation pathway in mycobacterium vanbaalenii PYR-1 based on systems biology [J]. Journal of bacteriology, 2007, 189(2): 464-472.

[172] Habe H, Miyakoshi M, Chung J, et al. Phthalate catabolic gene cluster is linked to the angular dioxygenase gene in Terrabacter sp strain DBF63 [J]. Applied Microbiology and Biotechnology, 2003, 61(1): 44-54.

[173] 沈萍萍, 王莹莹, 顾继东. 活性污泥中细菌对邻苯二甲酸酯的降解及其途径 [J]. 应用与环境生物学报, 2004, (5): 643-646.

[174] Kamimura N, Takamura K, Hara H, et al. Regulatory system of the protocatechuate 4, 5-cleavage pathway genes essential for lignin downstream catabolism [J]. Journal of Bacteriology, 2010, 192(13): 3394-3405.

[175] Ejlertsson J, Meyerson U, Svensson B H. Anaerobic degradation of phthalic acid esters during digestion of municipal solid waste under landfilling conditions [J]. Biodegradation, 1996, 7(4): 345-352.

[176] Sarmah A K, Northcott G L. Laboratory degradation studies of four endocrine disruptors in two environmental media [J]. Environmental Toxicology and Chemistry, 2008, 27(4): 819-827.

[177] Li G Y, Zu L, Wong P K, et al. Biodegradation and detoxification of bisphenol A with one newly-isolated strain Bacillus sp GZB: Kinetics, mechanism and estrogenic transition [J]. Bioresource Technology, 2012, 114: 224-230.

[178] Lobos J H, Leib T K, Su T M. Biodegradation of bisphenol A and other bisphenols by a gram-negative aerobic bacterium [J]. Applied and Environmental Microbiology, 1992, 58(6): 1823-1831.

[179] Zhou N A, Lutovsky A C, Andaker G L, et al. Cultivation and characterization of bacterial isolates capable of degrading pharmaceutical and personal care products for improved removal in activated sludge wastewater treatment [J]. Biodegradation, 2013, 24(6): 813-827.

[180] Toyama T, Sato Y, Inoue D, et al. Biodegradation of bisphenol A and bisphenol F in the rhizosphere sediment of Phragmites australis [J]. Journal of Bioscience and Bioengineering, 2009, 108(2): 147-150.

[181] Matsumura Y, Hosokawa C, Sasaki-Mori M, et al. Isolation and characterization of novel bisphenol - A-degrading bacteria from soils [J]. Biocontrol Science, 2009, 14(4): 161-169.

[182] Sasaki M, Maki J, Oshiman K, et al. Biodegradation of bisphenol A by cells and cell lysate from Sphingomonas sp strain AO1 [J]. Biodegradation, 2005, 16(5): 449-459.

[183] Roh H, Subramanya N, Zhao F M, et al. Biodegradation potential of wastewater micropollutants by ammonia-oxidizing bacteria [J]. Chemosphere, 2009, 77(8): 1084-1089.

[184] Hirano T, Honda Y, Watanabe T, et al. Degradation of bisphenol a by the lignin-degrading enzyme, manganese peroxidase, produced by the white-rot basidiomycete, Pleurotus ostreatus [J]. Bioscience Biotechnology and Biochemistry, 2000, 64(9): 1958-1962.

[185] Sakai K, Yamanaka H, Moriyoshi K, et al. Biodegradation of bisphenol A and related compounds by Sphingomonas sp strain BP-7 isolated from seawater [J]. Bioscience Biotechnology and Biochemistry, 2007, 71(1): 51-57.

[186] Tanghe T, Dhooge V, Verstraete W. Isolation of a bacterial strain able to degrade branched nonylphenol [J]. Applied and Environmental Microbiology, 1999, 65(2): 746-751.

[187] Gabriel F L P, Cyris M, Giger W, et al. Ipso-substitution: A general biochemical and biodegradation mechanism to cleave alpha-quaternary alkylphenols and bisphenol A [J]. Chemistry & Biodiversity, 2007, 4(9): 2123-2137.

[188] Fischer J, Kappelmeyer U, Kastner M, et al. The degradation of bisphenol A by the newly isolated bacterium Cupriavidus basilensis JF1 can be enhanced by biostimulation with phenol [J]. International Biodeterioration & Biodegradation, 2010, 64(4): 324-330.

[189] Yuka O, Shohei G, Tadashi T, et al. The 4-tert-butylphenol-utilizing bacterium Sphingobium fuliginis OMI can degrade bisphenols via phenolic ring hydroxylation and meta-cleavage pathway [J]. Environmental Science & Technology, 2013, 47(2): 1017-1023.

[190] Toyama T, Ojima T, Tanaka Y, et al. Sustainable biodegradation of phenolic endocrine-disrupting chemicals by Phragmites australis-rhizosphere bacteria association [J]. Water Science and Technology, 2013, 68(3): 522-529.

[191] 马力超, 吕红, 魏浩, 等. 双酚类化合物的生物降解研究进展 [J]. 工业水处理, 2017, 37(12): 11-16.

[192] Xu R, Tao W, Lin H Z, et al. Effects of perfluorooctanoic acid (PFOA) and perfluorooctane sulfonic acid

(PFOS) on soil microbial community [J]. Microbial Ecology, 2022, 83(4): 929-941.
[193] 王滢. 典型全氟烷基化合物在人工湿地的生态效应研究 [D]. 南京: 东南大学, 2021.
[194] Kwon B G, Lim H-J, Na S-H, et al. Biodegradation of perfluorooctanesulfonate (PFOS) as an emerging contaminant [J]. Chemosphere, 2014, 109: 221-225.
[195] Dinglasan M J A, Ye Y, Edwards E A, et al. Fluorotelomer alcohol biodegradation yields poly- and perfluorinated acids [J]. Environmental Science & Technology, 2004, 38(10): 2857-2864.
[196] Hasan S A, Ferreira M I M, Koetsier M J, et al. Complete biodegradation of 4-fluorocinnamic acid by a consortium comprising arthrobacter sp. strain G1 and ralstonia sp. strain H1 [J]. Applied and Environmental Microbiology, 2011, 77(2): 572-579.
[197] Chan P W Y, Yakunin A F, Edwards E A, et al. Mapping the reaction coordinates of enzymatic defluorination [J]. Journal of the American Chemical Society, 2011, 133(19): 7461-7468.
[198] Shi J C, Zheng G J S, Wong M H, et al. Health risks of polycyclic aromatic hydrocarbons via fish consumption in Haimen bay (China), downstream of an e-waste recycling site (Guiyu) [J]. Environmental Research, 2016, 147: 233-240.
[199] Cheung K C, Leung H M, Kong K Y, et al. Residual levels of DDTs and PAHs in freshwater and marine fish from Hong Kong markets and their health risk assessment [J]. Chemosphere, 2007, 66(3): 460-468.
[200] van der Oost R, Beyer J, Vermeulen N P E. Fish bioaccumulation and biomarkers in environmental risk assessment: A review [J]. Environmental Toxicology and Pharmacology, 2003, 13(2): 57-149.
[201] Meador J P, Stein J E, Reichert W L, et al. Bioaccumulation of polycyclic aromatic hydrocarbons by marine organisms [J]. Reviews of Environmental Contamination and Toxicology, 1995, 143: 79-165.
[202] Frasco M F, Guilhermino L. Effects of dimethoate and beta-naphthoflavone on selected biomarkers of Poecilia reticulata [J]. Fish Physiology and Biochemistry, 2002, 26(2): 149-156.
[203] Santana M S, Sandrini-Neto L, Neto F F, et al. Biomarker responses in fish exposed to polycyclic aromatic hydrocarbons (PAHs): Systematic review and meta-analysis [J]. Environmental Pollution, 2018, 242: 449-461.
[204] Stiborova M, Moserova M, Cerna V, et al. Cytochrome b5 and epoxide hydrolase contribute to benzo[a]pyrene-DNA adduct formation catalyzed by cytochrome P450 1A1 under low NADPH: P450 oxidoreductase conditions [J]. Toxicology, 2014, 318: 1-12.
[205] Pangrekar J, Kole P L, Honey S A, et al. Metabolism of chrysene by brown bullhead liver microsomes [J]. Toxicological Sciences, 2003, 71(1): 67-73.
[206] Zhou H L, Wu H F, Liao C Y, et al. Toxicology mechanism of the persistent organic pollutants (POPs) in fish through AhR pathway [J]. Toxicology Mechanisms and Methods, 2010, 20(6): 279-286.
[207] Pangrekar J, Kole P L, Honey S A, et al. Metabolism of phenanthrene by brown bullhead liver microsomes [J]. Aquatic toxicology, 2003, 64(4): 407-418.
[208] Ramesh A, Hood D B, Inyang F, et al. Comparative metabolism, bioavailability, and toxicokinetics of benzo[a]pyrene in rats after acute oral, inhalation, and intravenous administration [J]. Polycyclic Aromatic Compounds, 2002, 22(3-4): 969-980.
[209] Ramesh A, Walker S A, Hood D B, et al. Bioavailability and risk assessment of orally ingested polycyclic aromatic hydrocarbons [J]. International Journal of Toxicology, 2004, 23(5): 301-333.
[210] Pampanin D M, Le Goff J, Skogland K, et al. Biological effects of polycyclic aromatic hydrocarbons (PAH) and their first metabolic products in in vivo exposed Atlantic cod (Gadus morhua) [J]. Journal of Toxicology and Environmental Health-Part a-Current Issues, 2016, 79(13-15): 633-646.
[211] Pampanin D M, Brooks S J, Grosvik B E, et al. DNA adducts in marine fish as biological marker of genotoxicity in environmental monitoring: The way forward [J]. Marine Environmental Research, 2017, 125: 49-62.
[212] Pirsaheb M, Irandost M, Asadi F, et al. Evaluation of polycyclic aromatic hydrocarbons (PAHs) in fish: A review and meta-analysis [J]. Toxin Reviews, 2020, 39(3): 205-213.

[213] Franco M E, Lavado R. Applicability of in vitro methods in evaluating the biotransformation of polycyclic aromatic hydrocarbons (PAHs) in fish: Advances and challenges [J]. Science of the Total Environment, 2019, 671: 685-695.

[214] Hennessee C T, Li Q X. Effects of polycyclic aromatic hydrocarbon mixtures on degradation, gene expression, and metabolite production in four mycobacterium species [J]. Applied and Environmental Microbiology, 2016, 82(11): 3357-3369.

[215] 孙姗姗. Mycobacterium sp. WY10 和 Rhodococcus sp. WB9 的 PAHs 降解机制及其对污染土壤的修复作用研究 [D]. 杭州: 浙江大学, 2021.

[216] Silva I S, Grossman M, Durranta L R. Degradation of polycyclic aromatic hydrocarbons (2-7 rings) under microaerobic and very-low-oxygen conditions by soil fungi [J]. International Biodeterioration & Biodegradation, 2009, 63(2): 224-229.

[217] Luo S S, Chen B W, Lin L, et al. Pyrene degradation accelerated by constructed consortium of bacterium and microalga: Effects of degradation products on the microalgal growth [J]. Environmental Science & Technology, 2014, 48(23): 13917-13924.

[218] Cerniglia C E. Recent advances in the biodegradation of polycyclic aromatic hydrocarbons by Mycobacterium species; proceedings of the NATO Advanced Research Workshop on the Utilization of Bioremediation to Reduce Soil Contamination, Prague, Czech Republic, F Jun 14-19, 2000 [C]. Springer: DORDRECHT, 2003.

[219] Haritash A K, Kaushik C P. Biodegradation aspects of polycyclic aromatic hydrocarbons (PAHs): A review [J]. Journal of Hazardous Materials, 2009, 169(1-3): 1-15.

[220] Wagner-Dobler I, Bennasar A, Vancanneyt M, et al. Microcosm enrichment of biphenyl-degrading microbial communities from soils and sediments [J]. Applied and Environmental Microbiology, 1998, 64(8): 3014-3022.

[221] Moody J D, Freeman J P, Doerge D R, et al. Degradation of phenanthrene and anthracene by cell suspensions of mycobacterium sp strain PYR-1 [J]. Applied and Environmental Microbiology, 2001, 67(4): 1476-1483.

[222] Kim Y H, Freeman J P, Moody J D, et al. Effects of pH on the degradation of phenanthrene and pyrene by mycobacterium vanbaalenii PYR-1 [J]. Applied Microbiology and Biotechnology, 2005, 67(2): 275-285.

[223] Rockne K J, Chee-Sanford J C, Sanford R A, et al. Anaerobic naphthalene degradation by microbial pure cultures under nitrate-reducing conditions [J]. Applied and Environmental Microbiology, 2000, 66(4): 1595-1601.

[224] Tsai J C, Kumar M, Lin J G. Anaerobic biotransformation of fluorene and phenanthrene by sulfate-reducing bacteria and identification of biotransformation pathway [J]. Journal of Hazardous Materials, 2009, 164(2-3): 847-855.

[225] Coates J D, Anderson R T, Lovley D R. Oxidation of polycyclic aromatic hydrocarbons under sulfate-reducing conditions [J]. Applied and Environmental Microbiology, 1996, 62(3): 1099-1101.

[226] Habe H, Ide K, Yotsumoto M, et al. Degradation characteristics of a dibenzofuran-degrader Terrabacter sp. strain DBF63 toward chlorinated dioxins in soil [J]. Chemosphere, 2002, 48(2): 201-207.

[227] Wilkes H, Wittich R, Timmis K N, et al. Degradation of chlorinated dibenzofurans and dibenzo-p-dioxins by Sphingomonas sp. strain RW1 [J]. Applied and Environmental Microbiology, 1996, 62(2): 367-371.

[228] Bunge M, Adrian L, Kraus A, et al. Reductive dehalogenation of chlorinated dioxins by an anaerobic bacterium [J]. Nature, 2003, 421(6921): 357-360.

[229] Kuokka S, Rantalainen A L, Haggblom M M. Anaerobic reductive dechlorination of 1,2,3,4-tetrachlorodibenzofuran in polychlorinated dibenzo-p-dioxin- and dibenzofuran-contaminated sediments of the Kymijoki River, Finland [J]. Chemosphere, 2014, 98: 58-65.

[230] Adriaens P, Fu Q, Grbic-Galic D. Bioavailability and transformation of highly chlorinated dibenzo-p-

dioxins and dibenzofurans in Anaerobic soils and sediments [J]. Environmental Science & Technology, 1995, 29(9): 2252-2260.

[231] Barkovskii A L, Adriaens P. Microbial dechlorination of historically present and freshly spiked chlorinated dioxins and diversity of dioxin-dechlorinating populations [J]. Applied and Environmental Microbiology, 1996, 62(12): 4556-4562.

[232] Schmid A, Rothe B, Altenbuchner J, et al. Characterization of three distinct extradiol dioxygenases involved in mineralization of dibenzofuran by *Terrabacter* sp. strain DPO360 [J]. Journal of Bacteriology, 1997, 179(1): 53-62.

[233] Klecka G M, Gibson D T. Metabolism of dibenzo [1, 4]dioxan by a Pseudomonas species. [J]. The Biochemical Journal, 1979, 180(3): 639-645.

[234] Wittich R M. Degradation of dioxin-like compounds by microorganisms [J]. Applied Microbiology and Biotechnology, 1998, 49(5): 489-499.

[235] Su G Y, Letcher R J, Yu H X. Organophosphate flame retardants and plasticizers in aqueous solution: pH-dependent hydrolysis, kinetics, and pathways [J]. Environmental Science & Technology, 2016, 50(15): 8103-8111.

[236] Wu T, Gan Q, Jans U. Nucleophilic substitution of phosphorothionate ester pesticides with bisulfide (HS-) and polysulfides (S-n(2-)) [J]. Environmental Science & Technology, 2006, 40(17): 5428-5434.

[237] Wu L P, Chladkova B, Lechtenfeld O J, et al. Characterizing chemical transformation of organophosphorus compounds by C-13 and H-2 stable isotope analysis [J]. Science of the Total Environment, 2018, 615: 20-28.

[238] Wanamaker E C, Chingas G C, McDougal O M. Parathion hydrolysis revisited: *In situ* aqueous kinetics by H-1 NMR [J]. Environmental Science & Technology, 2013, 47(16): 9267-9273.

[239] Fang Y D, Kim E, Strathmann T J. Mineral- and base-catalyzed hydrolysis of organophosphate flame retardants: Potential major fate-controlling sink in soil and Aquatic environments [J]. Environmental Science & Technology, 2018, 52(4): 1997-2006.

[240] 潘水红. 水环境中邻苯二甲酸酯光降解机理的研究 [D]. 温州: 温州大学, 2018.

[241] 纪秀. 邻苯二甲酸二丁酯降解菌株的筛选及相关降解特性的研究 [D]. 镇江: 江苏大学, 2017.

[242] 李海涛, 黄岁樑. 水环境中邻苯二甲酸酯的迁移转化研究 [J]. 环境污染与防治, 2006, (11): 853-858.

[243] Bezares-Cruz J, Jafvert C T, Hua I. Solar photodecomposition of decabromodiphenyl ether: Products and quantum yield [J]. Environmental Science & Technology, 2004, 38(15): 4149-4156.

[244] Qu R J, Li C G, Pan X X, et al. Solid surface-mediated photochemical transformation of decabromodiphenyl ether (BDE-209) in aqueous solution [J]. Water Research, 2017, 125: 114-122.

[245] Roszko M, Szymczyk K, Jedrzejczak R. Photochemistry of tetra- through hexa-brominated dioxins/furans, hydroxylated and native BDEs in different media [J]. Environmental Science and Pollution Research, 2015, 22(23): 18381-18393.

[246] 陈蕾. 天然有机质介导的多氯联苯环境转化与降解机制 [D]. 杭州: 浙江大学, 2012.

[247] Crosby D G, Moilanen K W. Photodecomposition of chlorinated biphenyls and dibenzofurans [J]. Bulletin of Environmental Contamination and Toxicology, 1973, 10(6): 372-377.

[248] Miao X S C S G, Xu X B. Degradation pathways of PCBs upon UV irradiation in hexane [J]. Chemosphere, 1999, 39(10): 1639-1650.

[249] Ruzo L O, Zabik M J, Schuetz R D. ChemInform abstract: Photochemistry of bioactive compounds, photochemical processes of polychlorinated biphenyls [J]. Chemischer Informationsdienst, 1974, 5(33): 33211.

[250] Chen L, Shen C F, Zhou M M, et al. Accelerated photo-transformation of 2,2',4,4',5,5'-hexachlorobiphenyl (PCB 153) in water by dissolved organic matter [J]. Environmental Science and Pollution Research, 2013, 20(3): 1842-1848.

[251] Wang X W, Hu X F, Zhang H, et al. Photolysis kinetics, mechanisms, and pathways of tetrabromobisphenol

A in water under simulated solar light irradiation [J]. Environmental Science & Technology, 2015, 49(11): 6683-6690.

[252] Bao Y, Niu J. Photochemical transformation of tetrabromobisphenol A under simulated sunlight irradiation: Kinetics, mechanism and influencing factors [J]. Chemosphere, 2015, 134: 550-556.

[253] Watts M J, Linden K G. Advanced oxidation kinetics of Aqueous trialkyl phosphate flame retardants and plasticizers [J]. Environmental Science & Technology, 2009, 43(8): 2937-2942.

[254] Ou H S, Liu J, Ye J S, et al. Degradation of tris(2-chloroethyl) phosphate by ultraviolet-persulfate: Kinetics, pathway and intermediate impact on proteome of Escherichia coli [J]. Chemical Engineering Journal, 2017, 308: 386-395.

[255] Moussa D, Brisset J L. Disposal of spent tributylphosphate by gliding arc plasma [J]. Journal of Hazardous Materials, 2003, 102(2-3): 189-200.

[256] 孙世宾. 水环境中磷酸三甲苯酯的光解研究 [D]. 大连: 大连理工大学, 2019.

[257] Meylan W M, Howard P H. Computer estimation of the Atmospheric gas-phase reaction rate of organic compounds with hydroxyl radicals and ozone [J]. Chemosphere, 1993, 26: 2293-2299.

[258] Cheng J-h, Liang X-y, Yang S-w, et al. Photochemical defluorination of aqueous perfluorooctanoic acid (PFOA) by VUV/Fe^{3+} system [J]. Chemical Engineering Journal, 2014, 239: 242-249.

[259] Hori H, Hayakawa E, Einaga H, et al. Decomposition of environmentally persistent perfluorooctanoic acid in water by photochemical approaches [J]. Environmental Science & Technology, 2004, 38(22): 6118-6124.

[260] Chen J, Zhang P Y, Liu J. Photodegradation of perfluorooctanoic acid by 185 nm vacuum ultraviolet light [J]. Journal of Environmental Sciences, 2007, 19(4): 387-390.

[261] Panchangam S C, Lin A Y C, Tsai J H, et al. Sonication-assisted photocatalytic decomposition of perfluorooctanoic acid [J]. Chemosphere, 2009, 75(5): 654-660.

[262] 徐香. 海洋环境中有机污染物降解机理及构效关系的理论研究 [D]. 青岛: 中国海洋大学, 2012.

[263] 任曼. 环境与生物样品中 PCDD/Fs 和 DL-PCBs 的分析方法与环境行为初步研究 [D]. 广州: 中国科学院研究生院(广州地球化学研究所), 2006.

[264] Kim M K, O'Keefe P W. Photodegradation of polychlorinated dibenzo-p-dioxins and dibenzofurans in aqueous solutions and in organic solvents [J]. Chemosphere, 2000, 41(6): 793-800.

[265] Friesen K J, Muir D C G, Webster G R B. Evidence of sensitized photolysis of polychlorinated dibenzo-p-dioxins in natural waters under sunlight conditions [J]. Environmental Science & Technology, 1990, 24: 1739-1744.

[266] Friesen K J, Foga, M. M, et al. Aquatic photodegradation of polychlorinated dibenzofurans: Rates and photoproduct analysis [J]. Environmental Science & Technology, 1996, 30: 2504-2510.

[267] 彭浩, 金军, 王英, 等. 四溴双酚-A 及其环境问题 [J]. 环境与健康杂志, 2006, (6): 571-573.

[268] 刘红玲, 刘晓华, 王晓祎, 等. 双酚 A 和四溴双酚 A 对大型溞和斑马鱼的毒性 [J]. 环境科学, 2007, (8): 1784-1787.

[269] 陈玛丽, 瞿璟琰, 刘青坡, 等. 四溴双酚-A 和五溴酚对红鲫肝脏组织和超微结构的影响 [J]. 安全与环境学报, 2008, 8(4): 8-12.

[270] Jagnytsch O, Opitz R, Lutz I, et al. Effects of tetrabromobisphenol A on larval development and thyroid hormone-regulated biomarkers of the amphibian Xenopus laevis [J]. Environmental Research, 2006, 101(3): 340-348.

[271] Hu J, Liang Y, Chen M J, et al. Assessing the toxicity of TBBPA and HBCD by zebrafish embryo toxicity assay and biomarker analysis [J]. Environmental Toxicology, 2009, 24(4): 334-342.

[272] Nakari T, Huhtala S. *In vivo* and *In vitro* toxicity of decabromodiphenyl ethane, a flame retardant [J]. Environmental Toxicology, 2010, 25(4): 333-338.

[273] Kallqvist T, Grung M, Tollefsen K-E. Chronic toxicity of 2, 4, 2′, 4′-tetrabromodiphenyl ether on the marine alga Skeletonema costatum and the crustacean Daphnia magna [J]. Environmental Toxicology and

Chemistry, 2006, 25(6): 1657-1662.
[274] 张泽光, 黄满红, 陈东辉, 等. 十溴联苯醚对大型蚤和发光菌的毒性研究 [J]. 环境工程, 2013, 1: 299-302.
[275] 黄铸颖, 李海燕, 吴启航, 等. 溴代阻燃剂环境污染及毒性研究进展 [J]. 环境与健康杂志, 2014, 31(11): 1026-1032.
[276] 蔡立哲, 马丽, 袁东星. 九龙江红树林区底栖动物体内的 PAHs [J]. 海洋学报, 2005, 27(5): 112-118.
[277] Xu F L, Wu W J, Wang J J, et al. Residual levels and health risk of polycyclic aromatic hydrocarbons in freshwater fishes from Lake Small Bai-Yang-Dian, northern China [J]. Ecological Modelling, 2011, 222(2): 275-286.
[278] Perugini M, Visciano P, Giammarino A, et al. Polycyclic aromatic hydrocarbons in marine organisms from the Adriatic Sea, Italy [J]. Chemosphere, 2007, 66(10): 1904-1910.
[279] Al-Hassan J M, Afzal M, Rao C V N, et al. Petroleum hydrocarbon pollution in sharks in the Arabian Gulf [J]. Bulletin of Environmental Contamination and Toxicology, 2000, 65(3): 391-398.
[280] 吴玲玲, 陈玲, 张亚雷. 菲对斑马鱼鳃和肝组织结构的影响 [J]. 生态学杂志, 2007, 26(5): 688-692.
[281] 苗晶晶. 多环芳烃化合物对栉孔扇贝致毒机理的研究 [D]. 青岛: 中国海洋大学, 2006.
[282] Ma Y Q, Cui K Y, Zeng F, et al. Microwave-assisted extraction combined with gel permeation chromatography and silica gel cleanup followed by gas chromatography-mass spectrometry for the determination of organophosphorus flame retardants and plasticizers in biological samples [J]. Analytica Chimica Acta, 2013, 786: 47-53.
[283] Liu Y E, Huang L Q, Luo X J, et al. Determination of organophosphorus flame retardants in fish by freezing-lipid precipitation, solid-phase extraction and gas chromatography-mass spectrometry [J]. Journal of Chromatography A, 2018, 1532: 68-73.
[284] Guo J H, Venier M, Salamova A, et al. Bioaccumulation of dechloranes, organophosphate esters, and other flame retardants in Great Lakes fish [J]. Science of the Total Environment, 2017, 583: 1-9.
[285] Kim J W, Isobe T, Chang K H, et al. Levels and distribution of organophosphorus flame retardants and plasticizers in fishes from Manila Bay, the Philippines [J]. Environmental Pollution, 2011, 159(12): 3653-3659.
[286] van der Veen I, de Boer J. Phosphorus flame retardants: Properties, production, environmental occurrence, toxicity and analysis [J]. Chemosphere, 2012, 88(10): 1119-1153.
[287] Wei G-L, Li D-Q, Zhuo M-N, et al. Organophosphorus flame retardants and plasticizers: Sources, occurrence, toxicity and human exposure [J]. Environmental Pollution, 2015, 196: 29-46.
[288] Bolgar M, Hubball J A, Groeger J H, et al. Handbook for the Chemical Analysis of Plastic and Polymer Additives, Second Edition [M]. CRC Press, 2015.
[289] Johnson-Restrepo B, Adams D H, Kannan K. Tetrabromobisphenol A (TBBPA) and hexabromocyclododecanes (HBCDs) in tissues of humans, dolphins, and sharks from the United States [J]. Chemosphere, 2008, 70(11): 1935-1944.
[290] de W-S R M B, van D G, et al. Dietary intake of brominated flame retardants by the Dutch population [Z]. Inname van gebromeerde brandvertragers door deNederlandse bevolking via de voeding Rijksinstituut voor Volksgezondheid en Milieu RIVM. 2003.
[291] de Boer J, Wester P G, van der Horst A, et al. Polybrominated diphenyl ethers in influents, suspended particulate matter, sediments, sewage treatment plant and effluents and biota from the Netherlands [J]. Environmental Pollution, 2003, 122(1): 63-74.
[292] Whitfield F B, Shaw K J, Walker D I. The source of 2, 6-dibromophenol: Cause of an iodoform taint in Australian prawns [J]. Water Science and Technology, 1992, 25(2): 131-138.
[293] Whitfield F B, Helidoniotis F, Svoronos D, et al. The source of bromophenols in some species of Australian ocean fish [J]. Water Science and Technology, 1995, 31(11): 113-120.
[294] Oliveira A S, Silva V M, Veloso M C C, et al. Bromophenol concentrations in fish from Salvador, BA, Brazil [J].

Anais Da Academia Brasileira De Ciencias, 2009, 81(2): 165-172.

[295] Kitamura S, Suzuki T, Sanoh S, et al. Comparative study of the endocrine-disrupting activity of bisphenol A and 19 related compounds [J]. Toxicological Sciences, 2005, 84(2): 249-259.

[296] Sun Y, Huang H, Sun Y, et al. Ecological risk of estrogenic endocrine disrupting chemicals in sewage plant effluent and reclaimed water [J]. Environmental Pollution, 2013, 180: 339-344.

[297] Toro D M D Z C S, Hansen D J. Technical basis for establishing sediment quality criteria for nonionic organic chemicals using equilibrium partitioning: Annual review [J]. Journal Environmental Toxicology and Chemistry, 1991, 10(12): 1541-1583.

[298] Cristale J, Vazquez A G, Barata C, et al. Priority and emerging flame retardants in rivers: Occurrence in water and sediment, Daphnia magna toxicity and risk assessment [J]. Environment International, 2013, 59: 232-243.

[299] van Leeuwen K. Technical guidance document in support of commission directive 93/67/EEC on risk assession regulation (EC) no. 1488/94 on risk assessment for existing substances [Z]. Office for Official Publications of the European Community. 1996.

[300] Xiong J K, An T C, Zhang C S, et al. Pollution profiles and risk assessment of PBDEs and phenolic brominated flame retardants in water environments within a typical electronic waste dismantling region [J]. Environmental Geochemistry and Health, 2015, 37(3): 457-473.

[301] Davies R, Zou E. Polybrominated diphenyl ethers disrupt molting in neonatal Daphnia magna [J]. Ecotoxicology, 2012, 21(5): 1371-1380.

[302] 孙家宁, 孙韶华, 宋娜, et al. 高级氧化技术去除水中溴系阻燃剂的研究进展 [J]. 工业水处理, 2022, 42(2): 19-26.

[303] Zhong Y H, Liang X L, Zhong Y, et al. Heterogeneous UV/Fenton degradation of TBBPA catalyzed by titanomagnetite: Catalyst characterization, performance and degradation products [J]. Water Research, 2012, 46(15): 4633-4644.

[304] Horikoshi S, Miura T, Kajitani M, et al. Photodegradation of tetrahalobisphenol-A (X = Cl, Br) flame retardants and delineation of factors affecting the process [J]. Applied Catalysis B-Environmental, 2008, 84(3-4): 797-802.

[305] Zhang J, He S L, Ren H X, et al. Removal of tetrabromobisphenol-A from wastewater by ozonation; proceedings of the International Conference on Mining Science and Technology, Xuzhou, China, F, 2009 [C]. Elsevier Science Bv.

[306] Lin K D, Liu W P, Gan J. Reaction of Tetrabromobisphenol A (TBBPA) with Manganese Dioxide: Kinetics, Products, and Pathways [J]. Environmental Science & Technology, 2009, 43(12): 4480-4486.

[307] Zhang Y M, Chen Z, Zhou L C, et al. Efficient electrochemical degradation of tetrabromobisphenol A using MnO_2/MWCNT composites modified Ni foam as cathode: Kinetic analysis, mechanism and degradation pathway [J]. Journal of Hazardous Materials, 2019, 369: 770-779.

[308] 何勇. MIEX 树脂吸附—电解两步法去除水中溴代阻燃剂研究 [D]. 赣州: 江西理工大学, 2017.

[309] 董洪梅. O_3/UV 工艺降解 2, 2′, 4, 4′-四溴联苯醚废水的实验研究 [D]. 长沙: 湖南师范大学, 2013.

[310] 高亚杰, 张娴, 颜昌宙. 水环境中六溴环十二烷的光降解研究 [J]. 环境化学, 2011, 30(3): 598-603.

[311] An T C, Liu J K, Li G Y, et al. Structural and photocatalytic degradation characteristics of hydrothermally treated mesoporous TiO_2 [J]. Applied Catalysis A: General, 2008, 350(2): 237-243.

[312] An T, Chen J, Li G, et al. Characterization and the photocatalytic activity of TiO_2 immobilized hydrophobic montmorillonite photocatalysts: Degradation of decabromodiphenyl ether (BDE 209) [J]. Catalysis Today, 2008, 139(1): 69-76.

[313] Peng H J, Zhang W, Liu L, et al. Degradation performance and mechanism of decabromodiphenyl ether (BDE209) by ferrous-activated persulfate in spiked soil [J]. Chemical Engineering Journal, 2017, 307: 750-755.

[314] Pan Y, Leung P Y, Li Y Y, et al. Enhancement effect of nanoscale zero-valent iron addition on microbial

degradation of BDE-209 in contaminated mangrove sediment [J]. Science of the Total Environment, 2021, 781: 146702.

[315] Li H H, Zhu F, He S Y. The degradation of decabromodiphenyl ether in the e-waste site by biochar supported nanoscale zero-valent iron/persulfate [J]. Ecotoxicology and Environmental Safety, 2019, 183: 109540.

[316] 钱艳园, 刘莉莉, 于晓娟. 四溴双酚 A 好氧降解菌的筛选及其降解特性研究 [J]. 环境科学, 2012, 33(11): 3962-3966.

[317] Liang Z S, Li G Y, Mai B X, et al. Application of a novel gene encoding bromophenol dehalogenase from Ochrobactrum sp. T in TBBPA degradation [J]. Chemosphere, 2019, 217: 507-515.

[318] Liang Z S, Li G Y, Xiong J K, et al. Purification, molecular characterization and metabolic mechanism of an aerobic tetrabromobisphenol A dehalogenase, a key enzyme of halorespiration in Ochrobactrum sp. T [J]. Chemosphere, 2019, 237: 124461.

[319] Zu L, Li G Y, An T C, et al. Biodegradation kinetics and mechanism of 2, 4, 6-tribromophenol by Bacillus sp GZT: A phenomenon of xenobiotic methylation during debromination [J]. Bioresource Technology, 2012, 110: 153-159.

[320] Zu L, Li G Y, An J B, et al. Kinetic optimization of biodegradation and debromination of 2, 4, 6-tribromophenol using response surface methodology [J]. International Biodeterioration & Biodegradation, 2013, 76: 18-23.

[321] Liang Z S, Li G Y, An T C. Purifying, cloning and characterizing a novel dehalogenase from Bacillus sp GZT to enhance the biodegradation of 2, 4, 6-tribromophenol in water [J]. Environmental Pollution, 2017, 225: 104-111.

[322] Liang Z S, Li G Y, Mai B X, et al. Biodegradation of typical BFRs 2, 4, 6-tribromophenol by an indigenous strain Bacillus sp. GZT isolated from e-waste dismantling area through functional heterologous expression [J]. Science of the Total Environment, 2019, 697: 134159.

[323] Li G, Zu L, Wong P-K, et al. Biodegradation and detoxification of bisphenol A with one newly-isolated strain Bacillus sp. GZB: Kinetics, mechanism and estrogenic transition [J]. Bioresource Technology, 2012, 114: 224-230.

[324] Das R, Li G Y, Mai B X, et al. Spore cells from BPA degrading bacteria Bacillus sp GZB displaying high laccase activity and stability for BPA degradation [J]. Science of the Total Environment, 2018, 640: 798-806.

[325] Das R, Liang Z S, Li G Y, et al. Genome sequence of a spore-laccase forming, BPA-degrading Bacillus sp. GZB isolated from an electronic-waste recycling site reveals insights into BPA degradation pathways [J]. Archives of Microbiology, 2019, 201(5): 623-638.

[326] Lu M, Zhang Z-Z, Wu X-J, et al. Biodegradation of decabromodiphenyl ether (BDE-209) by a metal resistant strain, Bacillus cereus JP12 [J]. Bioresource Technology, 2013, 149: 8-15.

[327] Yu Y Y, Yin H, Peng H, et al. Biodegradation of decabromodiphenyl ether (BDE-209) using a novel microbial consortium GY1: Cells viability, pathway, toxicity assessment, and microbial function prediction [J]. Science of the Total Environment, 2019, 668: 958-965.

[328] Zhang S, Xia X, Xia N, et al. Identification and biodegradation efficiency of a newly isolated 2, 2′, 4, 4′-tetrabromodiphenyl ether (BDE-47) aerobic degrading bacterial strain [J]. International Biodeterioration & Biodegradation, 2013, 76: 24-31.

[329] Huang H W, Chang B V, Cheng C H. Biodegradation of dibromodiphenyl ether in river sediment [J]. International Biodeterioration & Biodegradation, 2012, 68: 1-6.

[330] Li G, Xiong J, Wong P K, et al. Enhancing tetrabromobisphenol A biodegradation in river sediment microcosms and understanding the corresponding microbial community [J]. Environmental Pollution, 2016, 208: 796-802.

[331] Arbeli Z, Ronen Z, Díaz-Báez M C. Reductive dehalogenation of tetrabromobisphenol-A by sediment from

a contaminated ephemeral streambed and an enrichment culture [J]. Chemosphere, 2006, 64(9): 1472-1478.

[332] Li F, Wang J, Nastold P, et al. Fate and metabolism of tetrabromobisphenol A in soil slurries without and with the amendment with the alkylphenol degrading bacterium Sphingomonas sp. strain TTNP3 [J]. Environmental Pollution, 2014, 193: 181-188.

[333] Wang S, Zhang S, Huang H, et al. Behavior of decabromodiphenyl ether (BDE-209) in soil: Effects of rhizosphere and mycorrhizal colonization of ryegrass roots [J]. Environmental Pollution, 2011, 159(3): 749-753.

[334] 熊士昌. 重金属对白腐菌降解水/沉积物体系十溴联苯醚的影响 [D]. 广州: 暨南大学, 2012.

[335] Xiong J K, Li G Y, An T C. The microbial degradation of 2, 4, 6-tribromophenol (TBP) in water/sediments interface: Investigating bioaugmentation using Bacillus sp GZT [J]. Science of the Total Environment, 2017, 575: 573-580.

[336] Chang B V, Liu J H, Liao C S. Aerobic degradation of bisphenol-A and its derivatives in river sediment [J]. Environmental Technology, 2014, 35(4): 416-424.

[337] Yang Y, Wang Z, Xie S. Aerobic biodegradation of bisphenol A in river sediment and associated bacterial community change [J]. Science of the Total Environment, 2014, 470-471: 1184-1188.

[338] Xiong J, An T, Li G, et al. Accelerated biodegradation of BPA in water-sediment microcosms with Bacillus sp. GZB and the associated bacterial community structure [J]. Chemosphere, 2017, 184: 120-126.

[339] 杨雷峰, 尹华, 叶锦韶, 等. 再力花对河涌底泥中多溴联苯醚的去除 [J]. 2015, 34(1): 130-136.

[340] Farzana S, Zhou H, Cheung S G, et al. Could mangrove plants tolerate and remove BDE-209 in contaminated sediments upon long-term exposure? [J]. Journal of Hazardous Materials, 2019, 378: 120731.

[341] Jiang Y C, Lu H L, Xia K, et al. Effect of mangrove species on removal of tetrabromobisphenol A from contaminated sediments [J]. Chemosphere, 2020, 244: 125385.

[342] 杨雷峰, 尹华, 彭辉, 等. 外源微生物对植物根系修复十溴联苯醚污染底泥的强化作用[J]. 环境科学, 2017, 38(2): 721-727.

第 8 章 电子垃圾拆解排放大气污染物的控制与健康风险消减

电子垃圾拆解处理所排放的气体污染物包括颗粒物(PM)、挥发性有机物(VOCs)、半挥发性有机物(SVOCs)和不完全燃烧产物多环芳烃(PAHs)等组分。早期分散的家庭作坊式电子垃圾拆解处理基本没有大气治理措施,气体污染物直接排放,对大气环境造成严重污染,对相关暴露人群造成健康风险。随着电子垃圾流入有处理资质企业或工业园区的集中拆解回收,企业或园区可以通过安装配套的大气污染控制装置,如负压集气罩收集烟气、气体净化装置降解气体污染物等,从而减少气体污染物对环境的排放和降低对暴露人群的健康风险。本章基于电子垃圾拆解处理排放大气污染物的污染控制实际应用案例,开展了相关大气污染物的控制与健康风险消减研究,为电子垃圾拆解处理过程的污染控制和绿色回收提供技术支撑。

8.1 颗粒物的控制与健康风险消减

8.1.1 颗粒物的控制与减排

电子垃圾拆解处理过程不可避免会释放出大量的颗粒物,而这些颗粒物会附带着大量的持久性有机污染物,因此暴露于电子垃圾拆解废气时会对人体造成严重的伤害。颗粒物在大气环境中呈现不同的粒径分布,有机污染物在不同粒径的颗粒物上分布也不尽相同。研究表明,部分有机物主要分布在粗颗粒物上[1,2],有些有机物在细颗粒物上所占的比重较大,而有些有机物则平均分布在不同粒径的颗粒物上。世界范围内已有大量关于不同粒径颗粒物上有机污染物分布特征的研究[3-7],但是缺乏电子垃圾拆解现场的数据。在实际的气体污染物控制净化过程中,颗粒物由于粒径大小不同,导致其在净化技术的各个单元停留时间不同和处理机制不同,最终去除效率也会相应地发生变化。因此,研究电子垃圾拆解处理排放的废气中不同粒径颗粒物以及颗粒相污染物的控制与减排具有重要的理论和实际意义,为高效安全地去除有机废气提供指导。

颗粒物污染的控制过程包括收集去除和尾气治理净化两个方面:①负压操作台可以将大部分的烟气颗粒物收集去除,减轻拆解处理过程中排放颗粒物对操作工人和车间环境的影响；②对电子垃圾拆解处理过程所产生的烟气，可以进行收集尾气的净化控制，从而减少对外部环境的排放。

1. 负压集气罩对颗粒物的收集

废电路板(WPCB)是废旧家电等拆解处理得到的主要拆解产物之一，由电子元器件和基板组成，且电子元器件通过焊料的黏附而附着在基板上。因此，在回收过程中的拆卸步骤中需要热处理来熔化焊料以去除电子部件。WPCB 脱焊产生的烟气，包括颗粒物和持久性毒害有机物，对环境质量和人类健康有许多负面影响。我国政府已经禁止在无污染控制的 WPCB 加热脱焊工艺，但按照《废弃电器电子产品规范拆解处理作业及生产管理指南(2015 年版)》规定，允许以加热等方式拆解电路板上元器件、零部件、汞开关等，但必须使用负压工作台，设置能够有效收集铅烟尘、有害气体的废气收集处理系统。在上海某电子垃圾回收企业的 WPCB 加热-人工拆解车间，利用荷电低压撞击器(ELPI)对不同位置颗粒物进行现场实时监测（图 8-1）[8]。WPCB 主要来自废阴极射线管电视机的电路板，其加热人工脱焊步骤如下：①锡锭在 300 ℃的加热炉中熔化成液态；②将具有焊锡面的 WPCB 样品浸入液态锡中，20~30 s 后，WPCB 上的焊锡熔化；③工人用钳子、刮刀等工具将电子元件（散热片、变压器、电池、电容器和电阻等）从基板上拔除。该车间加热脱焊操作台配有负压集气罩、管道收集及尾气处置装置。ELPI 可以将 0.03~10 μm 粒径范围的颗粒物分为 12 个通道，根据空气动力学直径的几何平均值(D_i)将不同粒径颗粒物分为 3 类：粗颗粒(PM$_{2.5~10}$，2.5<D_i<10 μm，通道 10~12)、细颗粒(PM$_{2.5}$，D_i<2.5 μm，通道 1~9)和超细颗粒(PM$_{0.1}$，D_i<0.1 μm，通道 1~2)。

图 8-1　废电路板电烤锡炉加热-人工拆解元器件过程

如前所述，ELPI 的气体入口分别在 3 个位置采样（图 8-2）：①采样点

1(0~100 s)：采取了车间的气体背景样本，并且电流值相对稳定，没有出现明显的峰。这意味着车间的背景 PM 值是稳定的。②采样点 2(100~370 s)：对位于负压集气罩外部(靠近工人鼻子)的气体样本进行采样，并将这些数据用于工人的健康风险评估。在此时间段内，可以看到四个电流值峰值(A~D)，代表四块 WPCB 的人工脱焊过程。ELPI 的电流值与所收集颗粒的数量和质量浓度成正比，电流峰值表示颗粒数量的实时增加。WPCB 的加热人工脱焊过程主要包括两个步骤：第一步是熔化焊料的加热过程，第二步是通过拆卸工人用钳子敲打和拉动来手动移除电子组件。在加热过程中，WPCB 基板中包含的树脂基体和有机添加剂也被液态焊锡加热，从而产生烟雾。在加热 20~30 s 后，WPCB 上的焊料熔化，工人开始用钳子、刮刀等工具去除电路板上电子元器件。同时，从基板释放的烟雾被负压集气罩收集或泄漏到罩外。每块电路板电子元器件的拔除和切割过程持续约 20 s，这与每个峰值的持续时间一致。拆卸工人用钳子夹住电路板，敲击操作台面使焊锡脱落，这个操作过程搅动了负压集气罩内的空气流动，导致烟雾泄漏到集气罩外部，即出现了如图 8-2 所示的峰 A~D。③采样点 3(370~520 s)：将 ELPI 气体入口移至负压集气罩内(加热炉顶部)，可以看出，采样点位置 3 的电流峰值明显高于采样点位置 1 和位置 2 的峰值。

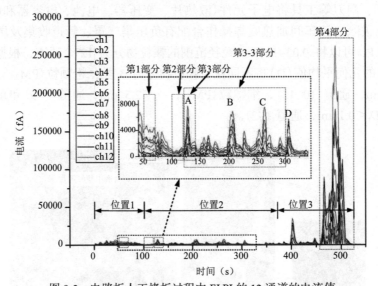

图 8-2　电路板人工烤板过程中 ELPI 的 12 通道的电流值

ELPI 的 3 个采样进口：位置 1–车间大气，可作为车间背景颗粒物浓度；位置 2–集气罩外部，靠近操作工人口鼻处；位置 3–集气罩内部，熔锡炉顶部

为了比较颗粒物数量浓度和质量浓度分布，选择了四个时间段，如第 1 部分(Part 1，50~70 s)，第 2 部分(Part 2，100~120 s)，第 3 部分(Part 3，121~141 s)

和第 4 部分(Part 4,460~520 s)。第 1 部分代表车间中 PM 的背景值,第 2 部分和第 3 部分代表负压罩外 WPCB 脱焊过程之前和之后的 PM 水平,第 4 部分代表负压罩内部的 PM 排放。图 8-3 显示了第 1 部分,第 2 部分,第 3 部分和第 4 部分中 PM 的数量和质量浓度分布。

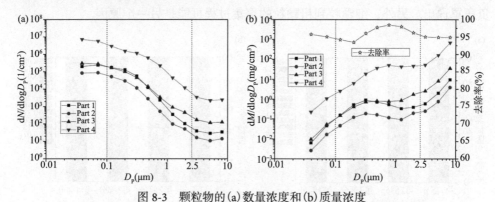

图 8-3　颗粒物的(a)数量浓度和(b)质量浓度

Part 1 表示车间颗粒物的背景值;Part 2 和 Part 3 表示人工烤板前后集气罩外部的颗粒物浓度;
Part 4 表示集气罩内部颗粒物浓度

通常,PM 的数量浓度随粒径增加而降低,而质量浓度随粒径增加而增加。但是,质量浓度分布显示出单峰分布,第 1 部分和第 2 部分的峰位于 0.31 μm。第 2 部分中的 PM 受罩内负压的影响,因此在 WPCB 脱除之前,罩外 PM 浓度不高,并且低于车间内浓度(第 1 部分)。第 3 部分和第 4 部分中的质量浓度分布($dM/d\log D_p \sim D_p$)没有明显的峰值[图 8-3(b)],并且形状相似。$PM_{0.1}$ 的数量和质量浓度从 8.67×10^4 count/cm^3 至 7.09×10^6 count/cm^3、2.66×10^{-3} mg/m^3 至 1.07 mg/m^3 不等。$PM_{2.5 \sim 10}$ 的数量和质量浓度从 11 count/cm^3 到 3445 count/cm^3,从 0.25 mg/m^3 到 679 mg/m^3 不等。第 4 部分中的 PM 数量和质量浓度值比第 1 部分,第 2 部分和第 3 部分高约 2~3 个数量级。

负压集气罩用于收集 WPCB 加热脱焊过程中产生的烟雾,第 3 部分中的 PM 是 PM 从负压罩中泄漏出来。因此,可以通过比较第 3 部分和第 4 部分来计算负压集气罩对 PM 的去除率。与第 4 部分中的 PM 相比,第 3 部分中的总 PM(从通道 1 到 12)质量浓度从 203.6 mg/m^3 降低至 9.4 mg/m^3,相当于总去除率为 95.4%。图 8-3(b)中显示了负压集气罩对不同粒径颗粒物的去除率,其去除率可达到 93.4%~98.5%,并且负压集气罩对 D_i 为 0.76 μm 和 0.20 μm 的颗粒物具有最高和最低的去除率。当工人敲击电路板而造成气流紊动时,颗粒物的流动状态受到许多因素的影响,例如惯性碰撞,拦截,扩散碰撞和重力等。

在第 1 部分,第 2 部分,第 3 部分和第 4 部分中不同粒径颗粒物的数量和质量占比如图 8-4 所示,4 个部分具有类似的分布模式,其中第 1、2、3 和 4 部分

[图 8-4(a)]中超细颗粒($PM_{0.1}$)的数量占比分别为 59.0%、68.1%、68.3% 和 71.4%。第 1、2、3 和第 4 部分[图 8-4(b)]中的粗颗粒($PM_{2.5\sim10}/PM_{10}$)的质量占比分别为 75.7%、82.9%、88.6% 和 80.4%，即负压集气罩外的第 3 部分中粗颗粒的累积占比最高，可能由于粗颗粒容易受到湍流空气的影响，导致更多的粗颗粒从负压罩逸出。另外，细颗粒到粗颗粒的聚集过程可能是另一个原因。

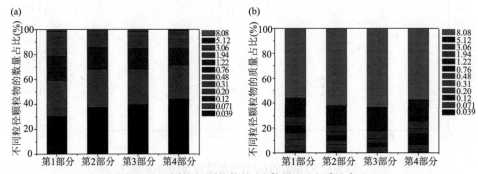

图 8-4　不同粒径颗粒物的(a)数量和(b)质量占比

Part 1 表示车间颗粒物的背景值；Part 2 和 Part 3 表示人工烤板前后集气罩外部的颗粒物浓度；
Part 4 表示集气罩内部颗粒物浓度

对 ELPI 采集的不同粒径颗粒物进行了扫描电镜图与能谱分析等表征。图 8-5(a)显示了 1.0~1.6 μm 颗粒物的形态分布。左上和右上图分别是图 8-5(a)的能谱分析和 Sn 的面扫描分布。用红色虚线圆圈圈出的一些细小颗粒的粒径约为 1 μm。同时，图 8-5(a)发现一些不规则形状颗粒物的松散聚集，其粒径较大的颗粒主要是由电路板上焊锡熔化过程中被捕获在采样滤膜上。在焊料熔化和手动去除元器件过程中，ELPI 泵抽取了细小的金属液滴颗粒。从图 8-5 的左上图的能谱分析得出，Al 元素的占比最大，达到了 88.04%，这主要由于采样滤膜为铝基滤膜。另外，扫描结果中还包括其他几种金属元素，如 Mn(2.07%)，Fe(0.59%)，Zn(0.36%)，Sn(0.81%) 和 Au(2.18%) 等。根据图 8-5 右上图的 Sn 空间分布图，Sn 分布与扫描电镜图 8-5(a)一致，并且 0.81% 的 Sn 元素分散在整个区域，特别是粗颗粒表面。在 WPCB 脱焊过程中，炉子的固定温度为 300℃，而 Sn 元素的熔点为 232 ℃。因此，WPCB 上的锡锭和焊料会熔化成液态，并产生 Sn 小液滴或蒸气。在集气罩处于负压的情况下，Sn 小液滴或蒸气会漂浮到烟雾中。当烟雾被抽入 ELPI 采样器时，Sn 小液滴或蒸气冷却并吸附在颗粒物表面。图 8-5(b)表明，粒径为 2.5 μm 的细颗粒主要以单个球形颗粒的形式存在。在图 8-5(c)中，可以看到在 6.8~10 μm 范围内具有不规则形状的粗颗粒，WPCB 基板受热释放的烟气颗粒物是其主要来源。

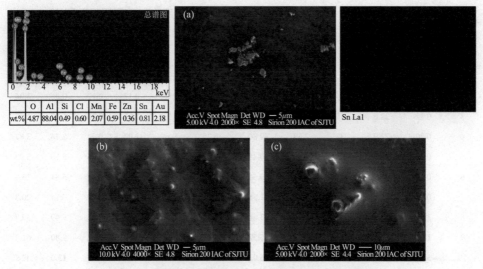

图 8-5　不同粒径颗粒物的扫描电镜及能谱分析图
(a) $1.0<D_i<1.6$ μm；(b) $2.5<D_i<4.4$ μm；(c) $6.8<D_i<9.97$ μm
左上图和右上图分别为图 8-5(a)的能谱图和 Sn 元素的面扫描图

为了评估在 WPCB 脱焊过程中通过吸入引起的颗粒物暴露，选择了图 8-2 所示的第 3-3 部分(从 120 s 到 300 s)作为实际暴露场景，将 ELPI 进样口位置放置在负压集气罩外部靠近拆解工人的鼻子处，其第 3-3 部分代表处理 3 块 WPCB 的所用时间。图 8-2 中的峰(A、B、C)代表在 WPCB 人工去除元器件过程中从负压集气罩逸出到车间的烟雾，而一些较低的峰则代表在 WPCB 加热过程中从负压集气罩泄漏的烟雾。因此，将第 3-3 部分中颗粒物质量浓度的平均值作为操作工人吸入颗粒物的质量浓度。表 8-1 显示了人体呼吸系统头部(HA)、气管和支气管(TB)和肺泡(AR)中颗粒物的沉积分数和通量的数据。根据国际防辐射委员会的 ICRP 模型和 ELPI 实时检测颗粒物的质量浓度，可以算出人体呼吸系统 HA、TB 和 AR 中颗粒物的总沉积通量分别为 1930 μg/h、74.0 μg/h 和 123 μg/h。HA 中 2.5~10 μm 的粗颗粒沉积通量最大(1830 μg/h)，占呼吸系统中颗粒物沉积总量的 86.1%。相反，HA、TB 和 AR 中超细颗粒($D_i<0.1$ μm，通道 1 和 2)的沉积通量分别为 0.16 μg/h、0.27 μg/h 和 1.30 μg/h，这表明超细颗粒物主要沉积在 AR 区域，占呼吸系统中超细颗粒物的 75.1%。颗粒物可通过撞击、沉降或扩散而沉积在呼吸系统的不同区域中。人体吸入空气气流后，由于颗粒物的惯性撞击和重力沉降，大粒径颗粒物($D_i>5$ μm)容易在 HA 区域被去除。对于 TB 区域，对于粒径大于 2.5 μm 的颗粒物，撞击仍然是重要的沉积机理。另外，在吸入的气流进入支气管后，气道容积增加而速率降低，从而提供了更多的时间来更有效的沉积颗粒物。但是粒径较小的颗粒物($D_i<0.5$ μm)的主要沉积机理是扩散，因为重力可以忽略不计。

表 8-1　颗粒物质量平均浓度、吸入因子(IF)、沉积部分(DF)和沉积通量(F)

通道	D_i μm	均值 μg/m³	IF	DF-HA	DF-TB	DF-AR	DF	F-HA	F-TB	F-AR	F
						%				μg/h	
1	0.039	1.70	1	4.79	8.91	37.0	50.7	0.04	0.08	0.34	0.47
2	0.071	8.23	1	2.64	4.30	21.5	28.4	0.12	0.19	0.96	1.26
3	0.12	22.6	1	2.02	2.02	11.2	15.2	0.25	0.25	1.37	1.87
4	0.20	72.9	1	2.58	0.86	6.25	9.69	1.02	0.34	2.46	3.81
5	0.31	98.8	1	4.68	0.48	5.93	11.0	2.50	0.26	3.16	5.92
6	0.48	136	1	9.31	0.66	7.94	17.9	6.85	0.49	5.84	13.1
7	0.76	122	1	19.1	1.65	10.8	31.5	12.6	1.09	7.13	20.8
8	1.22	138	1	36.8	3.64	12.7	53.1	27.5	2.72	9.49	39.7
9	1.94	150	0.998	58.0	5.57	12.1	75.6	46.9	4.51	9.80	61.2
10	3.06	256	0.991	76.0	6.03	9.40	91.4	105	8.35	13.0	126
11	5.12	829	0.966	86.6	4.39	5.58	96.5	387	19.6	24.9	432
12	8.08	2900	0.896	85.4	2.31	2.84	90.5	1340	36.1	44.4	1420
合计	—	4730	—	—	—	—	—	1930	74.0	123	2120

自从 2016 年初，我国电子垃圾的拆解处理模式从分散的家庭作坊式转变为工业园区集中拆解处理，但拆解的技术并没有发生改变，还是采用加热-人工脱焊的方法进行 WPCB 上元器件的拆解去除。负压集气罩可以一定程度将颗粒物等气体污染物收集起来，减轻对操作工人的暴露剂量。但是，还需采用大气污染控制及净化技术，将电子垃圾处理过程产生烟气进行了尾气治理，从而大幅度减少电子垃圾回收过程大气污染物的排放。以下将对排放气体污染物的控制净化技术及实际案例进行介绍。

2. 颗粒物的控制净化

颗粒物常见的控制净化技术包括：①喷淋塔(spray tower，ST)。ST 是一种广泛使用、易于建造和操作的除尘装置，主要用于去除工业废气中不同尺寸的颗粒物[9,10]。这项技术还可以去除一些水溶性大气污染物。经过 ST 预处理后，后续的有机物净化系统可以实现高效去除气态污染物[11]。②静电除尘(electrostatic precipitator，EP)。EP 能有效去除工业废气中的超细颗粒(包括微米和纳米尺寸)[12,13]。通过 EP 有效地截留悬浮颗粒物，从而提高后续处理系统去除废气有机成分的能力。然而，除超细颗粒外，尺寸较大的颗粒(如电子垃圾表面的聚集灰尘)必须在到达 EP 装置之前清除。这是因为较大的颗粒会损坏 EP 装置的放电电极，降低 EP 装置去除超细颗粒的效率。

以下将介绍电子垃圾拆解过程排放颗粒物控制方面的两个应用实例：①采用生物滴滤塔(BTF)和"静电除尘预处理(EP)-光催化反应器(PC)-臭氧深度氧化"等单元串联组成的一体化(EPO)净化装置两部分组成，开展了 BTF-EPO 组合工艺净化广东省某电子垃圾线路板拆解车间内排放总悬浮颗粒物(TSP)的性能评价；②采用 ST-EP-PC 联用反应器处理广东省某废旧电视机拆解车间内排放的 TSP、PM_{10} 和 $PM_{2.5}$。

1) 案例应用一

颗粒物净化装置：废气流量为 1000 m^3/h，收集的废气依次进入 BTF 和 EPO，通过 BTF-EPO 组合工艺对实际电子垃圾拆解过程中排放的废气进行污染控制[14]。图 8-6 给出了设备示意图。

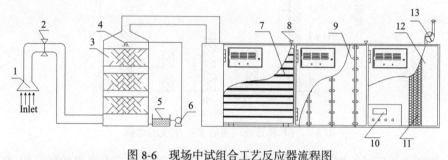

图 8-6　现场中试组合工艺反应器流程图

1-集气罩；2-阀门；3-填料；4-喷淋头；5-循环水槽；6-水泵；7-EP；8-取样口；9-PC；10-臭氧发生器；
11-蜂窝状活性炭；12-深度氧化室；13-风机

BTF 为玻璃钢材质，内径 1.2 m，高 2.2 m，有效填料体积约 1.4 m^3。BTF 中装填商业化陶粒(粒径为(20.00±2.00) mm；孔隙率为 (41.25±2.01)%；堆积密度为(210.36±14.62) kg/m^3；持水能力为 (12.53±0.12) g/g)。BTF 中使用的微生物菌群由降解二甲苯、甲苯和苯乙烯的菌群组成[15]。三个菌群培养物先在实验室培养至生长对数期，然后接入到 BTF 中，同时把营养液在 BTF 内进行循环流动，每天维持 8 小时。1 天后通入废气进行驯化。在驯化前 15 天，培养基更换频率为：三天换一次，而驯化 15 天以后，培养基更换频率为：一天换一次。无机盐培养基主要成分为(g/L)：2.0 KNO_3，0.6 $Na_2HPO_4·12H_2O$，0.25 $MgSO_4·7H_2O$，0.02 $CaCl_2$，0.005 $FeSO_4·7H_2O$，0.005 NaH_2PO_4。

TSP 采样方法：在第 1 天、第 15 天、第 30 天、第 45 天和第 60 天于 EP 装置的进出口连续采集 5 个时间点的样本，以准确评估综合废气去除技术的污染物水平和稳定性；在中试开始的第 1 天、第 7 天和第 15 天于 BTF-EPO 联用装置的进口处和出口处采集 TSP 样品。利用 TSP 采样器采集联用设备的进口和出口处废气。采样器中放置石英微孔纤维滤膜(20.3 cm × 25.4 cm)。采样周期为 8 小时，采样总体积约 140 m³。将覆盖有 TSP 样品的滤膜用预先焙烧过的铝箔纸包好保存在-20℃冷冻条件。将样品在 25℃、50%的相对湿度下平衡 24 小时以上，然后通过在采样前后称量平衡过滤器，确定 TSP 质量。

颗粒物净化效率：如图 8-7 所示，入口 TSP 浓度非常高（范围为 $3.8×10^3$ ~ $1.3×10^4$ μg/m³），远高于我国现行环境空气质量标准二级值（300 μg/m³，GB 3095—2012）。在 EP 处理后 TSP 浓度明显降低，在 60 天期间，平均去除效率达到 47.2%。同时吸附在 TSP 上的重金属[16]和 SVOCs[17]等污染物也可以有效地同时去除。

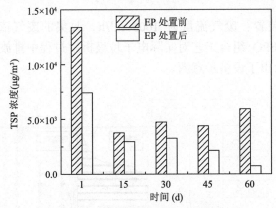

图 8-7 EP 装置处理前后 TSP 浓度的比较

如图 8-8 所示，联用装置进口处第一次、第二次和第三次采集的样品中 TSP 的质量浓度分别是 6783.0 μg/m³，3792.5 μg/m³ 和 7387.9 μg/m³。而经过装置处理后，出口处第一次、第二次和第三次采集样品中 TSP 质量浓度分别降至 1437.5 μg/m³、747.4 μg/m³ 和 1750.9 μg/m³。通过计算得到相应的去除率为 78.8%，80.2%和 76.3%，说明联用装置可以较好地去除电子垃圾拆解过程中产生的高浓度 TSP。

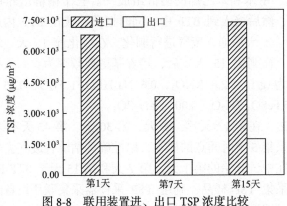

图 8-8 联用装置进、出口 TSP 浓度比较

虽然有近 80%的 TSP 已经被去除，但是出口 TSP 浓度仍然较高，可能造成这一现象的原因是：①由于中试现场主要是采用焚烧的方法来进行对电子垃圾拆解的处理，所以拆解过程产生的烟尘有不同大小粒径的颗粒物，这些颗粒物均被收集后进入到组合联用设备中，而在随后除尘过程中，因为大颗粒物(如动力学当量直径大于 100 μm 的颗粒物)较易被去除，会被联用设备优先去除，因此对小的颗粒物(比如直径小于 100 μm 的颗粒物)的去除效果造成了一定影响，导致出口 TSP 浓度仍然较高；②因为联用设备放置的位置比较靠近废气源，并且产生的 TSP 浓度过高，使得产生的高浓度 TSP 在联用设备中停留的时间过短，进而导致了 TSP 来不及被完全去除就已经排出。

2) 应用案例二

颗粒物净化装置：现场中试设备为 ST-EP-PC 联用反应器，如图 8-9 所示。电子垃圾拆解过程产生的废气首先进入 ST 单元(4500 mm×3000 mm，高×直径)底部，与塔内由上而下的喷淋水充分接触，而后进入 EP-PC 单元(长×宽×高为 2260 mm×2500 mm×3590 mm)。ST 和 PC 单元后分别装有离心机以确保达到稳定的流量 10000 m³/h。气体在反应器内停留时间为 9.3 s。

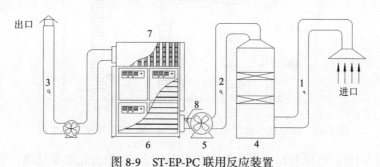

图 8-9 ST-EP-PC 联用反应装置

1~3：采样点；4-ST；5-离心泵；6-EP；7-PC；8-聚丙烯棉

ST 的材质为聚丙烯，内部循环喷淋水的流速为 0.4 m³/h。塔内被平均分隔成相邻贯通的 3 层，每一层内部装有商业拉西环(12 mm × 12 mm× 3 mm，堆积密度：0.69~0.73 g/cm³，SBET：0.42~0.48 m²/g，孔隙率≥70%)。EP-PC 单元下部为 EP 单元，内部对称分布有 16 个静电除尘器。上部的 PC 单元中有 20 块泡沫镍(1000 mm×500 mm，长×宽)，每块泡沫镍喷涂 TiO_2 催化剂(商业 P25，颗粒直径：300 nm，比表面积：50 m²/g)。泡沫镍两侧等距平行固定 4 支 30 W 的真空紫外灯(最大波长 254 nm，最小波长 185 nm)。

TSP、PM_{10} 和 $PM_{2.5}$ 采用大流量采样器(300 L/min)中的玻璃纤维过滤器(GFFs)(10.2 cm×12.7 cm)收集。在 0.3 m³/min 流量下抽取体积为 94.5~146.1 m³

的空气样品。取样前，GFFs 在 450 ℃下烘烤 6 h，然后在干燥器中平衡 24 h，温度为 25 ℃，湿度为 50%。取样后，用 450 ℃预焙铝箔纸包裹 GFFs，用双层聚乙烯袋密封，然后运输至实验室，在−20℃下储存。

颗粒物净化效率：图 8-10 给出了联用技术处理前后 TSP、PM_{10} 和 $PM_{2.5}$ 的浓度变化。可以看出，电烤锡炉拆解电视机线路板过程释放出大量的颗粒物，其 TSP、PM_{10} 和 $PM_{2.5}$ 的浓度分别是 $3.3×10^3$ μg/m³、$2.8×10^3$ μg/m³ 和 $2.8×10^3$ μg/m³。经过 ST 处理后，三种类型的颗粒物浓度没有降低反而增加，分别达到了 $7.1×10^3$ μg/m³、$9.0×10^3$ μg/m³ 和 $1.6×10^3$ μg/m³，表明 ST 技术对小粒径颗粒物没有去除效果，这可能是因为与进口废气大流量相比(10000 m³/h)，ST 内喷淋水循环速度太低(0.4 m³/h)。有研究表明在最优的操作条件下，ST 才能近乎完全地去除小粒径颗粒物[18]。此外，废气中携带的部分松脂油也会不断地积聚沉淀在喷淋塔内，导致塔内填料发生堵塞，影响去除效率。因此，在大流量有机废气的净化中如何有效地提高 ST 对小粒径颗粒物的去除效率还需要进一步研究和优化。

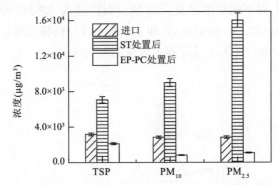

图 8-10　联用技术处理前后 TSP、PM_{10} 和 $PM_{2.5}$ 浓度变化情况

从图 8-10 还可以看出，颗粒物的粒径越大，ST 的去除效率越高。而相较于 ST，EP-PC 对颗粒物表现出高效稳定的去除能力，其中对 TSP、PM_{10} 和 $PM_{2.5}$ 的去除率分别达到 70.1%、90.9%和 93.3%，这主要归功于 EP 技术。前期研究表明，EP 可有效去除电子垃圾拆解过程中产生的 TSP，其最高去除效率可达 87.7%[19]。这是由于 ST 对大颗粒进行了有效的预处理，极大地保护了 EP 放电电极免受大颗粒的干扰损坏，达到 EP 单元高效去除小颗粒的目的。总的来说，在经过 ST-EP-PC 处理后，TSP、PM_{10} 和 $PM_{2.5}$ 的平均去除率可分别达到了 33.3%、79.9%和 61.8%。然而，仍有 6.7%~29.9%的颗粒物进入大气环境中，对工人的健康造成潜在威胁。因此，在后续的工业应用中，需要进一步改进颗粒物去除技术。

除了对颗粒物本身净化之外，也进一步考察了颗粒物上附着有机物的去除效率。从表 8-2 可以看出，联用技术对 TSP、PM_{10} 和 $PM_{2.5}$ 中 PAHs 的平均去除率分别为 19.2%、31.0%和 13.6%，相应的 PBDEs 的平均去除率分别为 72.8%、77.3%和 63.0%。相比较来说，联用技术对 PBDEs 的去除效率要高于 PAHs，这种差异主要归因于 PAHs 和 PBDEs 的气-固相分配不同。绝大多数的 PAHs 分布在气相中，而 PBDEs 主要富集在颗粒相中；颗粒相中的污染物主要依靠 EP 对颗粒物的去除，气相中的污染物主要依靠 PC 处理。此外，该联用技术对颗粒相 PAHs 和 PBDEs 的去除率显著高于气相中的 PAHs 和 PBDEs。第一次采样期间，联用技术的去除效率最高，TSP、PM_{10} 和 $PM_{2.5}$ 上 PAHs 的去除率分别达到了 57.4%、63.8%和 72.4%；而对 PBDEs 呈现出更高的去除效果，去除率分别为 99.0%、98.4%和 99.5%。随着中试时间的进行，颗粒物在联用技术的不同去除单元不断累积，在降低各单元性能的同时发生颗粒物的二次释放，导致去除效率明显降低。

表 8-2　不同采样期间内联用技术对 PAHs 和 PBDEs 的去除率(%)

		1st			2nd			3rd		
		TSP	PM_{10}	$PM_{2.5}$	TSP	PM_{10}	$PM_{2.5}$	TSP	PM_{10}	$PM_{2.5}$
PAHs	G	51.8	85.6	90.1	−114.3	−13.2	15.0	−5.6	11.5	−228.1
	P	59.8	57.3	65.9	13.2	18.5	19.1	19.3	33.4	3.5
	G+P	57.4	63.8	72.4	14.3	12.1	18.4	14.6	29.3	−13.0
PBDEs	G	50.0.0	73.1	78.1	−4055.3	−2044.6	−1579.6	−662.6	−533.9	15.9
	P	99.5	99.9	99.9	99.6	99.9	99.4	93.0	99.1	48.2
	G+P	99.0	98.4	99.5	41.1	55.9	60.0	78.3	76.5	29.5

注：G 代表气相，P 代表颗粒相，G+P 代表气相和颗粒相之和

8.1.2　颗粒物健康风险消减

工业生产过程排放的颗粒物会通过不同途径进入人体，尤其是在一些特殊的职业环境(如电子垃圾拆解)中，工作者会暴露在高浓度的颗粒物环境中，造成气体污染物的暴露情景和潜在健康风险。因此以前面所提到的利用 ST-EP-PC 联用工艺净化电子垃圾拆解排放废气前后颗粒物的研究为例，进一步探究了 ST-EP-PC 联用工艺净化电子垃圾拆解排放废气前后颗粒物上 PAHs 和 PBDEs 的健康风险变化情况。

1. 风险指数计算公式

PAHs 的暴露致癌风险采用毒性当量因子法进行计算，即 PAHs 中毒性最强的苯并[a]芘为 1，其他 PAHs 的毒性因子见表 8-3 所示。根据不同的暴露途径，

通过式(8.1)至式(8.4)估算呼吸($CDI_{inhalation}$)、皮肤吸收(CDI_{dermal})的长期暴露量以及相关的暴露风险。对于 PBDEs 的暴露非致癌风险，风险指数(HI)定义为长期的摄入量(CDI)与相应化合物参考剂量的比值(RfD)(表 8-4)。对于致癌性物质，一般不存在剂量阈值，微量的物质即会对人体造成危害，其致癌风险值(LCR)通过人体长期暴露浓度与相应致癌因子(CSF)及暴露时间获得，其风险指数用式(8.5)和式(8.6)计算。

$$CDI_{inhalation} = \frac{C_i \times IR_i \times T_i \times EF \times ED}{AT \times BW} \quad (8.1)$$

$$CDI_{dermal} = \frac{C_i \times k_{p,d} \times SA \times f_{sa} \times T_i \times EF \times ED}{AT \times BW} \quad (8.2)$$

$$CDI_{inhalation} = \frac{C_i \times TEF_i \times IR_i \times T_i \times EF \times ED}{AT \times BW} \quad (8.3)$$

$$CDI_{dermal} = \frac{C_i \times TEF_i \times k_{p,d} \times SA \times f_{sa} \times T_i \times EF \times ED}{AT \times BW} \quad (8.4)$$

$$HI_i = \frac{CDI_i}{RfD_i} \quad (8.5)$$

$$LCR_i = CDI_i \times CSF_i \quad (8.6)$$

式中，C_i 为化合物组分 i 的暴露浓度($\mu g/m^3$)；IR_i 为成年人吸入空气的速率(m^3/h)；T_i 为车间内拆解工人每日持续暴露的时间(h/d)；ED 为拆解工人的平均工作时间(a)；EF 为暴露频率(d/a)；$k_{p,d}$ 为颗粒态化合物的皮肤渗透系数(m/h)；SA 为皮肤表面积(m^2)；f_{sa} 为皮肤暴露的百分比；AT 为非致癌和致癌时间(d)；BW 为拆解工人的平均体重(kg)；CSF_i 为暴露组分 i 的致癌斜率因子(mg/kg·d))。

一般认为，成年人的吸入空气速率为 1.5 m^3/h。由于电子垃圾拆解过程包括常规的烧板过程和比较剧烈的切除电子元器件过程，因此拆解车间内工人的呼吸速率按照 2 h 为 2.05 m^3/h 和 8 h 为 1.39 m^3/h 计算[20]；按照实际的拆解工作时间，T_i 取 10 h；假设拆解工作时间为 24 年，每年工作 360 天；污染物导致非致癌风险和致癌风险的时间分别为 8650 天和 25500 天；正常人的体重为 60 kg；不同年龄段和不同性别人群皮肤表面积按照平均值 0.33 m^2 进行计算[21]；正常情况下，拆解工人仅胳膊、头部及颈部暴露在外面，其暴露百分比参照文献选取 25%[22]；所有 $k_{p,d}$ 均采用 PAHs 的 0.007 m/h[22]。BDE-209 的 CSF_i 为 7×10^{-4} kg/(d·mg)[23,24]。PAHs 的吸入和皮肤接触 CSF_i 分别为 3.14 kg/(d·mg) 和 30.5 kg/(d·mg)[25]。

表 8-3　PAHs 毒性当量因子

化学物	缩写	TEF	化学物	缩写	TEF
Naphthalene	NAP	0.001	Benz[a]anthracene	BaA	0.100
Acenaphthylene	AC	0.001	Chrysene	CHR	0.010
Acenaphthene	ACE	0.001	Benzo[b]fluoranthene	BbF	0.100
Fluorene	FL	0.001	Benzo[k]fluoranthene	BkF	0.100
Phenanthrene	PHE	0.001	Benzo[a]pyrene	BaP	1.000
Anthracene	ANT	0.010	Indeno[1,2,3-c,d]pyrene	Ind	0.100
Fluoranthene	FLU	0.001	Dibenzo[a,h]anthracene	DiB	1.000
Pyrene	PYR	0.001	Benzo[g,h,i]perylene	BghiP	0.010

表 8-4　PBDEs 毒性参考剂量

化学物	RfD	化学物	RfD	化学物	RfD	化学物	RfD
BDE-17	1×10^2	BDE-100	1×10^2	BDE-138	2×10^2	BDE-196	3×10^3
BDE-28	1×10^2	BDE-99	1×10^2	BDE-183	2×10^2	BDE-208	3×10^3
BDE-71	1×10^2	BDE-85	1×10^2	BDE-190	2×10^2	BDE-207	3×10^3
BDE-47	1×10^2	BDE-154	2×10^2	BDE-197	3×10^3	BDE-206	3×10^3
BDE-66	1×10^2	BDE-153	2×10^2	BDE-203	3×10^3	BDE-209	7×10^3

2. 不确定性分析

健康风险评价所采用的模型假设各种污染物之间不存在协同或拮抗作用，并且所有的污染物对人体产生的健康影响一致，造成评估在一定程度上存在不确定性。此外，由于有机物存在于不同粒径的颗粒物中，因此吸入暴露评估应充分考虑颗粒物的粒径分布[26]。而污染物的浓度采用 TSP 的浓度，一定程度上高估了污染物的摄入量，进而高估了其相应的风险值。

对于颗粒相中相关污染物的健康风险评估，皮肤的渗透系数和暴露面积的选择可直接影响健康风险评估的准确性，导致吸入和皮肤暴露风险值的差异。例如，皮肤暴露所摄入的污染物的量等于甚至大于通过呼吸途径的摄入量[27]。此外，一般皮肤暴露系数选择 25%，即假设拆解工人所穿的衣服对污染物的传输起到一定程度的阻挡作用，而这可能低估污染物的皮肤暴露量。研究表明，与干净的衣物相比，吸附污染物的衣服可增强皮肤对污染物的吸收[28]。因此，关于皮肤暴露评估还需深入的研究，以期得到准确、全面的风险评估。

3. 健康风险消减评估

图 8-11(a)给出了联用技术处理前 PAHs 的吸入致癌风险。TSP、PM_{10} 和

PM$_{2.5}$ 中 PAHs 的吸入致癌风险平均值分别为 $1.7×10^{-6}$、$1.3×10^{-6}$ 和 $8.7×10^{-7}$，其中 TSP 和 PM$_{10}$ 上 PAHs 超出了可接受的致癌风险值 $1.0×10^{-6}$，对拆解工人造成一定程度的健康危害。与颗粒相相比，相应的气相 PAHs 所致的吸入致癌风险要低（TSP：$2.1×10^{-7}$，PM$_{10}$：$1.6×10^{-7}$，PM$_{2.5}$：$1.3×10^{-7}$），都处于安全范围内。由于高毒性的 PAHs 主要集中在颗粒相，从而导致了低含量的 PAHs 表现出较高的致癌风险。因此，对于 PAHs 而言，对颗粒相 PAHs 的去除至关重要。经过联用技术处理之后[图 8-11(b)]，TSP、PM$_{10}$ 和 PM$_{2.5}$ 上 PAHs 的吸入致癌风险平均值分别下降到 $1.6×10^{-6}$、$8.9×10^{-7}$ 和 $6.6×10^{-7}$。虽然风险值均有所降低，但是对于 TSP 所携带的 PAHs 而言，其仍会对拆解工人造成吸入致癌风险。

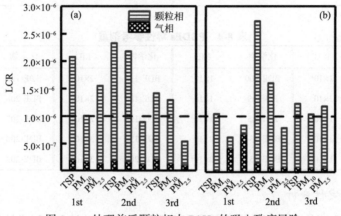

图 8-11　处理前后颗粒相中 PAHs 的吸入致癌风险
(a) 进口；(b) 出口

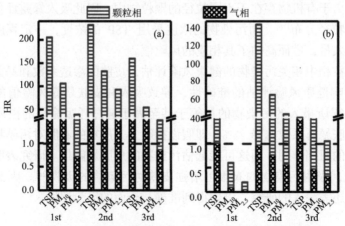

图 8-12　处理前后颗粒相中 PBDEs 的吸入非致癌风险
(a) 进口；(b) 出口

图 8-12 给出了联用技术处理前后 PBDEs 通过吸入导致的非致癌风险。处理前的有机废气中，其 TSP、PM_{10} 和 $PM_{2.5}$ 中 PBDEs 引起的吸入非致癌风险值分别高达 199.4、98.2 和 44.9。由此可见，电视机线路板拆解有机废气中的 PBDEs 可引起显著的吸入非致癌风险，必须采取相应的治理和减排措施，其中相应的颗粒相 PBDEs 导致的吸入非致癌风险值分别为 196.6、96.0 和 43.6，占总的吸入非致癌风险值的 98.6%、97.8%和 97.2%。对于电子垃圾拆解有机废气中主要的 PBDEs 成分 BDE-47 和 BDE-99，其处理前 TSP、PM_{10} 和 $PM_{2.5}$ 中 PBDEs 的吸入非致癌风险值分别为 78.8、37.8、17.9 和 51.6、24.4、12.2，均超过了确定非致癌风险值(1.0)。

而经过联用技术处理后，PBDEs 所导致的吸入非致癌风险明显降低，TSP、PM_{10} 和 $PM_{2.5}$ 中 PBDEs 的平均吸入非致癌风险值分别降至 57.9、24.1 和 12.8，降低了 71.0%、75.5%和 71.4%，可见经过处理后，颗粒物中的 PBDEs 引起的健康风险有所降低，说明该联用技术可有效地消减颗粒相中 PBDEs 的健康风险。

8.2 VOCs 的控制与健康风险消减

8.2.1 VOCs 的控制与减排

除了排放各种粒径的颗粒物之外，在电子垃圾拆解过程中会释放出大量的 VOCs(详细内容已在第 3 章阐述)，这些 VOCs 具有很强的毒性，它们可能会对车间内工人甚至拆解区附近的居民的生存环境以至人体健康产生一定的毒害作用。因此，了解电子垃圾拆解车间内不同拆解工艺排放 VOCs 的污染特征及其健康风险，对研发电子垃圾拆解车间内 VOCs 污染源排放的安全高效消减控制技术具有非常重要的现实意义。

1. 电子垃圾拆解过程排放 VOCs 控制技术

电子垃圾拆解过程排放 VOCs 的浓度、组成和流量随着拆解工艺、电子垃圾类型等不同呈现较大差异，导致治理及控制方面有很大的难度[19,29-32]。目前，电子垃圾拆解过程排放 VOCs 的控制技术分为前端控制技术和末端处理技术两类。其中前端控制技术主要是在生产各种电子元器件时尽量少地使用含有毒害性物质的材质，如有机溶剂等。然而由于这些产品良好的性能并且在短时间内很难找到合适的替代品，限制其使用还需要很长一段时间才能实现，因此关于电子垃圾拆解过程排放 VOCs 的治理一般仍采用末端处理技术，而常用的末端处理技术分为三大类：物理法、化学法和生物法[33]。如图 8-13 所示，物理法主要包括吸附法、吸收法、冷凝法等，其中以活性炭吸附法应用最为广泛。但是物理法并不能彻底去除 VOCs，只是将 VOCs 从气相转移到液相或固相。与物理法相对应的是能将 VOCs 彻底降解去除的化学法和生物法，其主要是利用光、热、催化剂或微生物等将 VOCs 有效转化成为二氧化碳、水和无机物等。

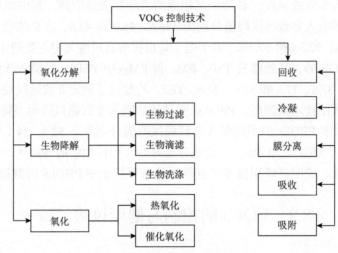

图 8-13　常用 VOCs 控制技术

常用 VOCs 控制工艺的基本情况如表 8-5 所述[34]。

表 8-5　常用 VOCs 控制工艺的基本情况

工艺	去除率(%)	产生二次污染	优点	缺点
热氧化	95~99	焚烧产物	可回收能量	卤代或氧化产物需后续深度处理
催化氧化	90~98	焚烧产物	可回收能量	催化剂易中毒，需后续深度处理
生物过滤	60~95	生物量	二次污染少，且无毒害	具有选择性，速率慢，填料回收难
冷凝	70~85	有机物	回收产品补偿操作成本	维护操作严格，不适宜高沸点物质
吸收	90~98	废水	回收产品补偿操作成本	维护操作严格，VOCs 需预处理
活性炭吸附	80~90	炭和有机物	回收产品补偿操作成本	受湿度影响大，孔易堵塞，脱附难
沸石吸附	90~96	沸石和有机物	回收产品补偿操作成本	沸石成本高，应用受限
膜分离	90~99	废膜	回收产品补偿操作成本	膜成本太高

下面对主要的 VOCs 处理技术进行简要的介绍。

1) 吸附法

含 VOCs 的气体在与多孔结构的吸附剂进行接触时，由于吸附剂的表面存在未平衡的分子间吸引力或化学键力作用，使得混合气体中的 VOCs 组分会被吸附在吸附剂的表面，从而能有效地将气体中的 VOCs 去除，以达到净化气体的效果。基于吸附剂与吸附质之间的不同相互作用可分为化学吸附与物理吸附。物理吸附是一个可逆过程，该吸附现象的作用力主要是由吸附质分子与吸附剂表面间的范德华力、静电作用力构成；而化学吸附则是指吸附质与吸附剂之间电子交换、转移或共有(形成共价键)的过程，通常为不可逆过程[35,36]。研究表明，在需要实现分离处理的情况下，物理吸附更加有效。吸附设备简单，操作灵活，是有效、经济的 VOCs 回收技术之一。目前，商业化的活性炭净化器对 VOCs 去除率可达 90%[36]，最高可以达到 95%~99%[37]。吸附法常被用来处理中低浓度的 VOCs 废气，其对 VOCs 去除率的高低取决于吸附剂的性质、VOCs 的组分和浓度、操作的条件(如吸附系统的操作温度、湿度、压力等)等因素。一般情况下，不饱和化合物的吸附会更完全，而环状化合物也比直链结构的物质更容易被吸附。然而，吸附剂的吸附能力会随着使用时间的延长而逐渐降低，吸附剂往往需要再生利用，但却存在吸附剂再生回收成本高的难题。

2) 吸收法

溶剂吸收法是采用低挥发性或不挥发性溶剂对 VOCs 进行吸收，再利用吸收剂物理性质和 VOCs 分子的差异进行分离。其对 VOCs 的吸收效果主要取决于吸收剂设备的结构特征和吸收性能。其工艺流程如下：含 VOCs 的废气由底部进入吸收塔，在上升的过程中与来自塔顶的吸收剂逆流并充分接触而被吸收，净化后的气体再由塔顶排出，在吸收了 VOCs 的吸收剂通过热交换器后，会进入汽提塔的顶部，在温度高于吸收温度或压力低于吸收压力的时候能得以解吸，吸收剂经冷凝器冷凝后再进入吸收塔循环使用。解吸出的 VOCs 气体经过冷凝器、气液分离器后会以纯 VOCs 气体的形式离开汽提塔，并可以进一步回收利用[33]。

3) 冷凝法

冷凝法是基于有机物在不同温度下的饱和蒸气压不同，通过提高系统压力、降低温度或者在降低温度同时提高系统压力的方法，使处于蒸气状态的 VOCs 冷凝并与气体进行分离，以实现废气净化的一种 VOCs 回收方法。常用的冷凝方法有表面冷凝和接触冷凝。表面冷凝中，含 VOCs 气体不与冷凝剂直接接触；而接触冷凝则是冷却介质与被冷凝气体在接触冷凝器中直接进行接触，使得气体中的 VOCs 组分进行降温冷凝，冷凝液与冷却介质会以废液形式排出冷凝器[38]。冷凝技术的设备和操作过程都比较简单，并且回收物质纯度较高。当 VOCs 含量较低时，冷凝法需要采取更进一步的冷冻措施，这会使得运行成本提高。因此，冷凝法不适宜用于处理低浓度的 VOCs 废气，而常作为使用其他方法时净化高浓度 VOCs 废气的前处理方法，用来降低有机负荷，并回收有机物。

4) 燃烧法

将有害气体、蒸气、液体或烟尘通过燃烧而转化为无害物质的过程称为燃烧净化法，亦称焚烧法。燃烧法净化废气的实质是废气中的有害成分在高温下燃烧氧化和热分解，以实现无害化的过程。因此，燃烧法只适用于那些含可燃的或在高温下可以分解的有害物质的废气净化。然而，在燃烧和焚化过程中，由于燃烧不完全，在燃烧过程中会产生许多有害的中间产物，如乙醛、二噁英、呋喃、多环芳烃等，造成二次污染。目前常用的燃烧法有直接燃烧法、催化燃烧法和热力燃烧法，其中尤以催化燃烧应用最为广泛。

5) 光催化氧化法

光催化氧化法是基于利用光催化剂在紫外或可见光照射下促进反应物发生反应的催化氧化还原反应。光催化氧化技术因具有原材料无毒、价廉易得、反应条件温和、降解产物为二氧化碳、水和无机物等，反应设备简单安全且容易和其他工艺联用及能适用于多种有机物分解等优点，近年来广泛应用于有机废气的降解[39]。光催化降解有机物的基本原理是：当光催化剂受到能量相当于或高于其禁带宽度的光照辐射时，价带上的电子会被激发，进而跃迁到高能的导带，并产生电子-空穴对，价带上具有强氧化性的空穴会将吸附在催化剂表面的水(或 OH^-)氧化成羟基自由基，吸附在光催化剂表面的氧气则会与导带上的电子结合，生成 O_2^-，再经过一系列反应生成反应活性很高的羟基自由基($OH·$)或 $HOO·$。这些自由基具有很强的氧化能力，能破坏有机物中的 C—C、C—H、C—O、C—N、N—H 等化学键，对光催化氧化过程起到决定作用。氧化作用可以通过表面羟基的间接氧化作用，也可以在粒子内部或颗粒表面经价带空穴进行直接氧化；或者这两种氧化机理同时起作用进行催化氧化[40]。光催化机理如图 8-14 所示[41]。光催化氧化还原反应多以半导体为催化剂，比较常用的光催化剂有 TiO_2、ZnO、CdS、Fe_2O_3、SnO_2、WO_3 等。由于 TiO_2 的化学性质和光化学性质十分稳定，且无毒、价廉易得，其应用最为广泛。

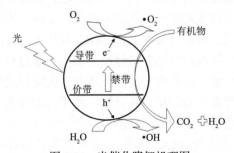

图 8-14 光催化降解机理图

目前，光催化技术在降解甲苯[42,43]、乙苯[44]、二甲苯[45]以及苯乙烯和苯酚[46]等均有研究，这些物质广泛存在于电子垃圾拆解过程释放的 VOCs 中[29, 47-49]。此外，光催化氧化可有效地降解 PAHs 和 PBDEs[50]，比如 BDE-209 和菲通过光催化过程发生开环，生成毒性较低的有机酸等中间产物[51,52]，而 Ag/TiO_2 催化剂可有效地降解 BDE-47[53]。

6) 生物法

生物法来净化 VOCs 的过程主要是通过附着在滤料介质上的微生物在合适的温度、pH 等环境条件下，吸收废气中的 VOCs 作为碳源和能源，用以维持其生命活动，并将 VOCs 进行分解为二氧化碳和水的过程。气相主体中的 VOCs 首先经历由气相到固相/液相的传质过程，然后在固/液相中被微生物降解。生物法处理 VOCs 废气的过程中，pH 值、水分含量、滤床温度、填料类型、营养物质以及微生物菌群等因素都会对净化效率产生影响。常用的生物法处理 VOCs 工艺有生物洗涤塔、生物滴滤塔和生物过滤塔。生物法相较常规处理方法具有设备简单、运行费用低、较少形成二次污染等优点，其广泛应用于有机废气的净化，尤其在处理低浓度、生物降解性好的气态污染物时效果更好。然而，生物法也存在设备占地面积较大、驯化周期长、抗冲击负荷差的弱点。

7) 等离子体法

等离子体是指电离度大于 0.1%，由离子、电子以及未电离的中性粒子混合组成，整体呈中性的电离气体。因为等离子体中含有大量的活性粒子，如高能电子、离子、自由基以及激发态分子等，而这些活性粒子的能量比气体分子的键能要高，可以破坏 VOCs 分子的化学键，实现 VOCs 的分解。等离子体技术对 VOCs 中的丁酮、苯及甲苯的去除率高于 90%[54]，且对控制各浓度范围的 VOCs 尾气都表现出了较好的处理效果。然而，等离子体技术仅仅只能将那部分靠近电极与高能等离子体接触的气体污染物分解，容易导致污染物分解不完全。由于设备扩大后产生等离子体平均能量相对较低(不到 20 eV)，限制了等离子设备规模的扩大。

8) 膜分离法

膜分离法是指在分离膜的供给侧与渗透侧压差作用下，利用有机蒸气能选择性透过无孔的气体分离膜以实现有机蒸气与气流分离的技术。通常利用真空泵在分离膜渗透侧抽气以维持该侧比供给侧更低的压力。这种膜对有机分子是全透性的，而对空气则是半透性的。当有机物通过分离膜后，有机蒸气被冷凝而实现回收，净化后的空气作为残留物而去除。通过膜分离技术可以回收的有机物包括乙醛、丙酮、苯、氯仿、甲苯、二甲苯、三氯乙烯、正己烷以及氯乙烯。该技术不会分解污染物而产生有害的副产物,并且还可以实现污染物的回收再利用。然而,膜分离技术需要真空泵和冷凝设备,这会增加 VOCs 回收过程中的基建和运行成本。另外,为了获得较高的 VOCs 透过浓度,往往还需要多级分离系统,这些都会增加运行成本。

9) 臭氧氧化法

由于臭氧具有强氧化性,且臭氧还原产物为无害的氧气而受到广泛的关注。臭氧氧化法现已广泛应用于污废水的深度净化处理过程。目前,国内外研究仍主要集中在利用臭氧氧化法对液相有机物的去除上,臭氧氧化分解空气中有机废气的研究,还仍然处于起始阶段[55]。

2. 电子垃圾拆解过程 VOCs 减排控制应用

1) 案例应用一

选择某旋转灰化炉烧板车间作为 VOCs 污染控制的中试现场,开展了 EPO 净化 VOCs 的中试研究,对该工艺的处理效果及设备的性能进行了评价。旋转灰化炉结构如图 8-15 所示,其工作原理如下:首先将已预先回收的电子元器件线路板投入旋转炉烧板机中进行深度拆解(以回收利用含有贵金属的集成芯片及焊接用的锡)并回收铜;其次,通过电镀来提炼集成芯片中的贵金属,与此同时,从裸板中提取铜;旋转炉在旋转的同时通过电力或者燃烧液化气来加热。最后,线路板被加热后,焊接锡被熔化,从而实现了锡、裸板和其他组件(如集成芯片等)的分离。

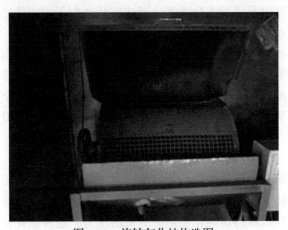

图 8-15 旋转灰化炉构造图

VOCs 净化装置:在一体化反应器(额定功率:1.5 kW;尺寸:3300 mm × 1200 mm × 1200 mm)中进行了中试试验,现场装置示意图如图 8-16 所示。中试现场废气以 720 m³/h 的速度直接进入 EPO 一体化反应器进行处理。装置被平均分隔成 3 间相邻贯通的小单元，依次为 EP、PC 及臭氧深度氧化室。EP 单元中并排放置 3 个静电除尘器(720 mm × 360 mm × 850 mm)，每个静电除尘器中含 4× 11 个静电场，每个静电场的放电电压为 6000~14000 kV，放电电流约为 0.8 A，每个静电场的直径为 50 mm，长度为 265 mm。PC 单元中放置 3 块涂有 TiO_2 的泡沫镍(1000 mm × 1000 mm)和八个 30W 低压真空紫外灯(ZY30S19W，可以产生臭氧)[56]，平行固定在泡沫镍的两侧。紫外灯管到泡沫镍的间距为 50 mm，此时泡沫镍处光强为 2.5 mW/cm²。臭氧深度氧化室内放置 1 台臭氧发生器(尺寸：350 mm×320 mm×600 mm，臭氧产生量：15 g/h)和置于其后端的蜂窝状活性炭改性臭氧分解催化剂(比表面积：800 m²/g；体积密度：0.40~0.48 g/mL；孔密度：16 孔/平方厘米)。

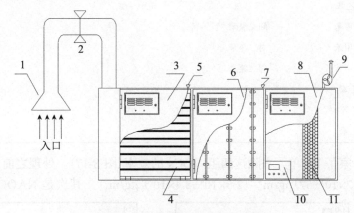

图 8-16　一体化净化设备示意图
1-集气罩；2-阀门；3-除尘室；4-静电除尘器；5-取样口；6-光催化反应室；7-取样口；8-深度氧化单元；
9-风机；10-臭氧发生器；11-蜂窝状活性炭

在泡沫镍上涂覆 TiO_2 的程序：通过喷涂方法制备负载型光催化剂，首先将 TiO_2 在剧烈搅拌下分散到蒸馏水中(1 g/100 mL)，持续 30 分钟，直到形成均匀的 TiO_2 悬浮液。三块(1000 mm×1000 mm)多孔泡沫镍(孔隙率≥95%；孔密度：80~100 ppi)用作基质，因为它们具有高度开放的多孔结构和优良的流体力学性能，可以在高流速下降低空气阻力和光催化剂的损失。用乙醇清洗去除油脂后，使用实验室喷枪喷涂制备的 TiO_2 悬浮液，对泡沫镍进行涂层。涂层在室温下自然干燥 24 小时，并在以下光催化反应器中获得用于光催化剂的重量为 5.19 g/m² 的 TiO_2 膜。

VOCs 净化效率：Chen 等[19]发现在使用旋转焚烧炉拆除线路板的过程中，共检测到 43 种 VOCs。按照它们的结构可以分为四类：芳香烃(AHs)、脂肪族烃(AlHs)、卤代烃(HHs)以及含氮和含氧化合物(NAOCC) (表 8-6)。

表 8-6　VOCs 分类

芳香烃(AHs)	卤代烃(HHs)	脂肪族烃(AlHs)	含氮和含氧化合物(NAOCCs)
苯	氯甲烷	3-甲基-1-丁烯	丙烯腈
甲苯	溴甲烷	α-蒎烯	甲基丙烯酸甲酯
乙苯	三氯甲烷	1-己烯	苯甲醛
间/对二甲苯	三氯乙烯	甲基氯乙烷	甲基异丁基甲酮
邻二甲苯	四氯乙烯	异戊烷	
苯乙烯	1,2-二氯乙烷	环戊烯	
异丙苯	1,2-二氯丙烷	正庚烷	
正丙苯	1,1,1-三氯乙烷	β-蒎烯	
3-甲基乙苯	氯苯	正癸烷	
2-甲基乙苯	苄基氯	正十一烷	
1,3,5-三甲苯	间二氯苯		
1,2,4-三甲苯	对二氯苯		
1,2,3-三甲苯	邻二氯苯		
间二乙苯			
邻二乙苯			
对二乙苯			

首先考察了 EP 单元去除这四组 VOCs 的效率(图 8-17)。处理之前 AHs 的浓度最高((1.7×10^3±57)μg/m^3~(2.5×10^4±8.4×10^2) μg/m^3)，其次是 NAOOCs((1.3±

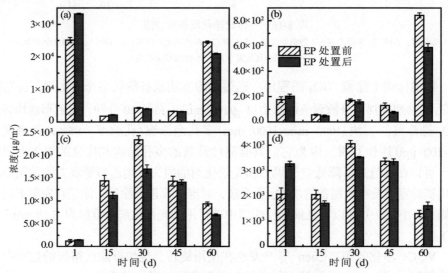

图 8-17　EP 处理前后 VOCs 的浓度变化
(a) AHs；(b) AlHs；(c) HHs；(d) NAOCCs

$0.13) \times 10^3$ μg/m³ ~ $(3.4 \pm 0.13) \times 10^3$ μg/m³)和 HHs（$(1.2 \pm 0.32) \times 10^2$ μg/m³ ~ $(2.3 \pm 0.86) \times 10^3$ μg/m³），而 AlHs 的浓度最低（$(51.2 \pm 0.32) \times 10^2$ μg/m³ ~ $(2.3 \pm 0.86) \times 10^3$ μg/m³）。

不同类型 VOCs 浓度的显著差异是电子垃圾的性质不同造成的。众所周知，线路板主要是由含有阻燃剂的环氧树脂或酚醛树脂作为基板树脂原料，如四溴双酚A[57]。在较高的燃烧温度下（旋转焚烧炉中温度>400℃），溴化环氧或酚醛树脂中的碳原子和苯环之间的单键在加热过程中可能被其他自由基破坏或取代，释放出大量 AHs[58]。尽管 AlHs 也会被释放[59]，但它们更容易被氧化并可能转化为 NAOCCs，导致 AHs 和 NAOCCs 的浓度远高于 AlHs。当这些 VOCs 经过 EP 单元处理后，其浓度和成分都没有明显变化，表明 EP 技术对 VOCs 的去除效率很低。但是 EP 可有效去除废气中的颗粒物[60]，保护后续 PC 单元中的光催化剂不受这些颗粒物的影响，从而实现高稳定的 VOCs 去除能力。由图 8-18 所示可知，在 PC 和臭氧氧化协

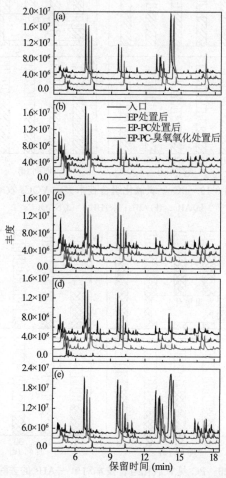

图 8-18　第 1 天(a)、15 天(b)、30 天(c)、45 天(d)和 60 天(e)检测的 VOCs 总离子色谱图

同处理后，总离子色谱图上所有 VOCs 的峰强迅速下降，表明使用高级氧化技术可以有效去除 VOCs。图 8-19 至图 8-21 分别给出了联用高级氧化技术对四类 VOCs 的去除效率。由图 8-19 可知，PC-臭氧氧化组合技术对 AHs、AlHs、HHs 和 NAOCCs(不含苯甲醛)的平均分辨率分别为 95.7%、95.4%、87.4%和 97.5%。进一步观察发现，PC 单元对组合工艺的总去除效率具有不同的贡献。PC 单元仅能够去除 33.0%的 HHs[图 8-19(c)]，而对于 AHs[图 8-19(a)]和 NAOCC[图 8-19(d)]，则去除效率分别增加到 45.2%和 57.6%。但是 PC 单元对 AlHs 显示出最高的去除效率，平均去除率达到了 74.0%[图 8-19(b)]，这可能是由于 AlHs 的浓度最低(图 8-17)。

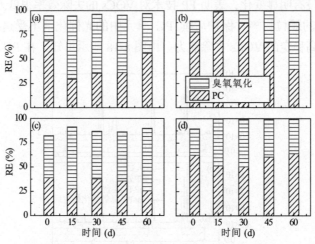

图 8-19 PC-臭氧氧化组合技术去除 VOCs 效率
(a) AHs；(b) AlHs；(c) HHs；(d) NAOCCs

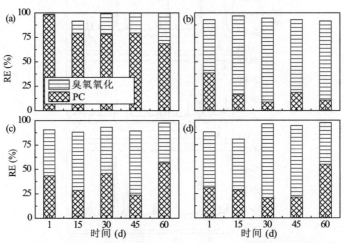

图 8-20 PC-臭氧氧化组合技术对单一 AHs 的去除效率
(a)苯乙烯；(b)苯；(c)二甲苯；(d)甲苯

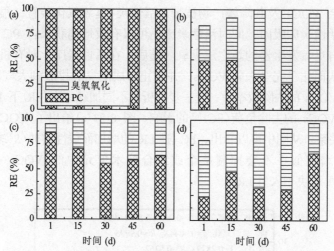

图 8-21　PC-臭氧氧化组合技术对其他单一 VCOs 的去除效率
(a) α-蒎烯；(b) 氯苯；(c) 甲基丙烯酸甲酯；(d) 丙烯腈

PC 单元对苯的平均去除效率只有 18.5%，这可能是因为它是其他大分子量 AHs 的降解产物[61]。然而，在经过后续臭氧氧化单元处理后，苯的平均去除效率达到 93.8%（图 8-20）。由于苯甲醛也是 AHs 的氧化产物[62-64]，在 PC 单元出口处也观察到苯甲醛浓度的增加（图 8-22）。同样地，经随后的臭氧氧化单元处理后苯甲醛浓度迅速下降，出口浓度远远低于进口浓度。这一结果再次说明了 PC-臭氧氧化组合技术对所有 VOCs 都具有较高的去除效率，主要净化机制如下：PC 会产生大量的活性氧物种，能够有效降解气态有机污染物[65]，同时由于使用的紫外灯可以额外产生另一种强氧化剂——臭氧，因此活性氧物种和臭氧协同作用可以更高效地去除 VOCs[62]。值得注意的是，由于使用真空紫外灯后具有优异的再生能力，因此可以保持光催化剂的高光活性并提高其使用寿命[66]。经过 PC

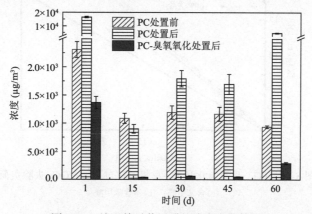

图 8-22　处理前后苯甲醛的浓度变化情况

处理后的残余 VOCs 及其降解产物被带入活性炭+臭氧氧化装置，很容易被该单元中布置的活性炭床吸附，同时活性炭床还可以有效地捕获来自 PC 单元剩余的臭氧以及来自臭氧发生器的新生成臭氧。随后，在活性炭床上对有机物进行臭氧催化降解，从而几乎完全去除 VOCs，并防止形成的臭氧排放到大气环境中。

在比较 VOCs 的去除效率后，进一步分析了不同加载速率(LR)下 PC-臭氧氧化组合技术对 VOCs 的去除负荷(EC)[67]的影响。图 8-23(a)给出了 VOCs 的 LRs 和 ECs 之间的关系。从图中可以看出，EC 随 LR 的增加而线性增加($R^2 \geqslant 0.9262$)，表明入口 LR 增加似乎不会抑制单个或组合技术单元中的 VOCs 去除；最大 EC(EC_{max})主要取决于入口 LR。

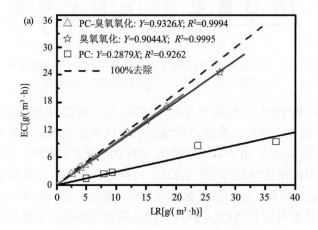

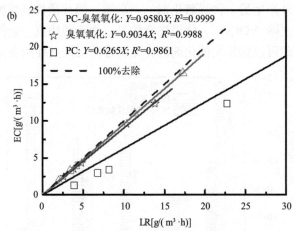

图 8-23　PC、臭氧氧化和 PC-臭氧氧化对不同 VOCs 去除负荷
(a)含苯甲醛；(b)不含苯甲醛

对于 PC 和臭氧单元，当最大 LR 为 36.8 g/(m³·h) 和 27.4 g/(m³·h) 时，EC_{max} 分别为 9.4 g/(m³·h) 和 24.5 g/(m³·h)。PC-臭氧氧化组合技术的 EC_{max} 不高于 16.9 g/(m³·h)，主要原因是反应器体积较大(是 PC 或臭氧装置的 2 倍)。而 EC_{max} 在单独臭氧处理时最大，其次是 PC-臭氧处理，然后是 PC 单独处理。虽然该结果不能定义为实际 EC_{max}，因为没有测试到较高的 LRs，但可以得出结论：在研究条件下，臭氧氧化在组合技术中的贡献要大于 PC。此外，拟合曲线斜率表明 PC 装置、臭氧装置和 PC-臭氧联合装置分别实现了 28.8%、90.4%和 93.3%的 VOCs 去除效率，其中 PC-臭氧联合装置超过了单独臭氧装置和单独 PC 装置，且使用 PC 装置的 VOCs 平均去除效率远小于使用臭氧装置，这可能是 PC 技术的 LRs 更高以及产生高浓度中间产物——苯甲醛所致。

因此进一步研究了 PC-臭氧氧化组合技术的 LRs 和 ECs 与除苯甲醛外的其他 VOCs 之间的关系[图 8-23(b)]。发现 PC 装置的最大 LR 略有降低(从 36.8 g/(m³·h)降到 34.5 g/(m³·h))，臭氧装置的最大 LR 大幅降低(从 27.4 g/(m³·h)降到 10.8 g/(m³·h))，导致 EC_{max} 相反(PC 装置从 9.4 g/(m³·h)增至 23.6 g/(m³·h)，臭氧装置从 24.5 g/(m³·h) 降到 9.3 g/(m³·h))，PC 装置大于 PC-臭氧氧化装置，后者大于单独的臭氧氧化装置。此外，PC 和 PC-臭氧氧化装置的 VOCs 平均去除效率分别增加至 62.7%和 95.8%，而臭氧氧化装置的 VOCs 的平均去除效率略有下降(90.3%)，但 PC-臭氧氧化装置仍然大于臭氧氧化装置，臭氧氧化装置大于单独使用 PC 装置。这些结果表明，在 PC-臭氧氧化系统中，PC 单元预处理具有高 LRs 和高 EC，而臭氧氧化可有效去除 VOCs，使该组合技术能够高效、稳定地去除电子垃圾拆解过程排放的 VOCs。

在 EPO 净化 VOCs 研究的基础上，进一步提出了将 EPO 工艺与 BTF 联用组合工艺，并开展了对该组合工艺净化电子垃圾拆解车间内 VOCs 的性能评价。表 8-7 列出了 BTF-EPO 联用组合工艺中试研究期间第 1 天，第 7 天和第 15 天电子垃圾拆解过程产生 VOCs 组分及浓度变化情况。由表中可知，三天采集的样品中 TVOC 的浓度分别为 (6604.3±448.0) μg/m³、(5499.1±854.7) μg/m³ 和 (26834.0±447.0) μg/m³，其浓度远远高于城市垃圾压缩过程产生的 TVOC[68]，但是低于油漆生产加工现场产生的 TVOC[15]。

表 8-7 VOCs 成分及其对应含量(μg/m³)

VOCs	第 1 天	第 7 天	第 15 天
芳香烃类	3119.60±216.98	2369.91±359.77	24419.58±229.50
苯	927.16 ± 42.75	1071.42 ± 167.01	1881.56 ± 95.88
甲苯	310.87 ± 12.81	317.10 ± 65.20	3561.79 ± 49.62
乙苯	83.15 ± 1.38	56.33 ± 5.97	2649.53 ± 27.23

续表

VOCs	第 1 天	第 7 天	第 15 天
对/间二甲苯	153.52±7.02	105.50±11.24	3541.93±126.51
邻二甲苯	73.58 ± 2.51	45.67 ± 6.63	514.55 ± 17.39
苯乙烯	1314.89 ± 140.27	660.27 ± 98.48	10803.23 ± 357.83
异丙苯	4.91 ± 0.28	5.87 ± 0.82	105.70 ± 1.58
3-甲基乙苯	20.17 ± 0.68	9.58 ± 0.27	133.07 ± 1.02
4-甲基乙苯			697.92 ± 5.77
2-甲基乙苯	21.48 ± 0.63	10.55 ± 0.11	
1,3,5-三甲苯	29.28 ± 2.86	16.23 ± 0.70	132.55 ± 2.77
1,2,4-三甲苯	135.61 ± 6.66	36.69 ± 1.95	397.75 ± 1.02
1,2,3-三甲苯	27.50 ± 3.46	10.35 ± 0.52	
1,4-二乙苯		24.36 ± 0.86	
1,3-二乙苯	17.45 ± 0.68		
脂肪烃类	44.58 ± 0.81	174.37 ± 0.53	102.00 ± 3.43
1-丁烯		146.74 ± 2.31	
环戊烯	2.86 ± 0.13		6.18 ± 0.12
1-己烯	29.80 ± 0.92	21.26 ± 1.19	79.53 ± 3.96
α-蒎烯	11.92 ± 0.25	7.16 ± 0.16	16.29 ± 0.41
卤代烃类	1295.90 ± 62.68	1936.59 ± 353.32	1170.64 ± 146.46
三氯甲烷			7.04 ± 0.21
1,1,1-三氯乙烷	120.82 ± 31.93	24.90 ± 4.72	
1,2-二氯乙烷		13.32 ± 3.08	56.77 ± 1.69
三氯乙烯	30.13 ± 1.13	9.04 ± 0.34	53.38 ± 3.47
1,2-二氯丙烷	14.45 ± 2.48		49.40 ± 0.02
一溴甲烷	497.34 ± 57.25	1827.69 ± 356.49	414.22±23.58
四氯乙烯	30.06 ± 2.71	17.60 ± 1.23	398.73 ± 13.69
氯苯	80.20 ± 0.60	23.80 ± 1.87	117.54 ± 5.90
对二氯苯		20.25 ± 1.88	
间二氯苯			19.11 ± 0.92
苄基氯	522.91 ± 37.06		153.58 ± 6.81
邻二氯苯			45.99 ± 3.01

续表

VOCs	第1天	第7天	第15天
含氮含氧VOCs	2144.19±167.50	1018.22±142.11	1141.79±74.51
苯甲醛	1493.75±92.86	577.99±65.28	212.77±4.09
2-丙烯腈	230.43±28.08		131.18±25.10
2-甲基丙烯酸甲酯	237.48±21.76	440.23±76.83	797.84±53.50
异己酮	182.54±24.80		
总挥发性有机物	6604.28±447.97	5499.10±854.66	26834.01±447.04

注：在第1天、第7天、第15天分别拆解电脑WPCB、硬盘光驱WPCB、电源电话等各种混杂的WPCB

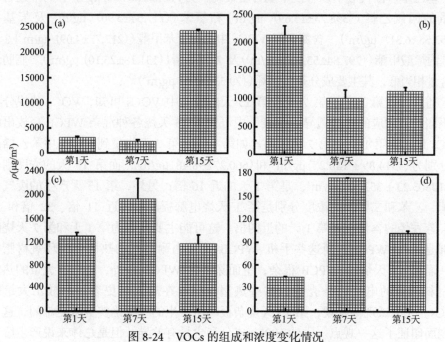

图 8-24 VOCs 的组成和浓度变化情况
(a)芳香烃类；(b)含氮含氧类；(c)卤代烃类；(d)脂肪烃类

中试期间内芳香烃类 VOCs、卤代烃类 VOCs、含氮含氧 VOCs 及脂肪烃类 VOCs 浓度如图 8-24 所示。由于在采样期内电子垃圾拆解现场产生的电子垃圾的种类和数量有所不同，这四类 VOCs 的浓度在三次采样时间有一定的变化[14]，但是总体来看，这四类 VOCs 的浓度大小呈现出如下的规律：其中芳香烃类 VOCs 的浓度最高，其第 1 天、第 7 天和第 15 天采集的样品对应的浓度分别为 $(3119.6±217.0)\mu g/m^3$、$(2369.9±359.8)\mu g/m^3$ 和 $(24419.6±229.5)\mu g/m^3$；接着是含氮含氧类 VOCs，其浓度分别为 $(2144.2±167.5)\mu g/m^3$、$(1018.2±142.1)\mu g/m^3$ 和

$(1141.8±74.5)\mu g/m^3$；而卤代烃类 VOCs 的浓度与含氮含氧类 VOCs 较为接近，其浓度分别是$(1295.9±62.7)\mu g/m^3$、$(1936.6±353.3)\mu g/m^3$ 和$(1170.6±146.5)\mu g/m^3$；而脂肪烃类的 VOCs 的浓度最低，分别是$(44.6±0.8)\mu g/m^3$、$(174.4±0.5)\mu g/m^3$ 和$(102.0±3.4)\mu g/m^3$。

从表 8-7 中可以看出，中试期间以第 15 天时 VOCs 成分最为复杂，含量为最高。因此选定第 15 天产生的废气中 VOCs 来进行详细分析。芳香烃类中主要成分为苯乙烯、二甲苯、甲苯、乙苯、苯为主要成分，其浓度分别为$(10803.23±357.83)\mu g/m^3$、$(4056.48±143.90)\mu g/m^3$、$(3561.79±49.62)\mu g/m^3$、$(2649.53±27.23)\mu g/m^3$、$(1881.56±95.88)\mu g/m^3$，另外还含有少量的三甲苯、二乙苯及甲基乙苯。卤代烃中以一溴甲烷含量最高，为$(414.22±23.58)\mu g/m^3$，另外还含有大量的四氯乙烯$((398.73±13.69)\mu g/m^3)$和氯苯$((117.54±5.90)\mu g/m^3)$及苄基氯$((153.58±6.81)\mu g/m^3)$。含氮含氧 VOCs 中主要有苯甲醛$((212.77±4.09)\mu g/m^3)$、2-甲基丙烯酸甲酯$((797.84±53.50)\mu g/m^3)$及 2-丙烯腈$((131.18±25.10)\mu g/m^3)$。脂肪烃类含量相当低，其主要成分是 1-己烯$((79.53±3.96)\mu g/m^3)$。

通过比较第 1 天、第 7 天及第 15 天的废气中 VOCs 可知，VOCs 的成分有所不同并且对应的含量迥异。从表 1 中可知第 15 天烧各种混杂 WPCB 释放出的 VOCs 量远比第 1 天和第 7 天的高。如第 1 天及第 7 天收集到的样品中苯乙烯浓度分别为$(1314.89±140.27)\mu g/m^3$ 和$(660.27±98.48)\mu g/m^3$，而第 15 天的浓度则高达$(10803.23±357.83)\mu g/m^3$，是第 1 天的近 16 倍；另外，第 15 天产生的废气中甲苯、乙苯和二甲苯的浓度分别是第 1 天烧电脑板产生的近 11 倍、32 倍和 18 倍，芳香烃的含量也是第 1 天的近四倍。这可能主要是因为第 1 天和第 7 天烧的全部是手机 WPCB，而这些手机 WPCB 均是由耐高温的热固性的环氧树脂构成，而第 15 天烧的 WPCB 很杂，里面除了含 WPCB 板外，还含有大量的热塑性材料构成的电线、外壳等，这些热塑性材料在高温时更容易释放出大量的VOCs。在中试现场发现，第 15 天烧杂板时所产生的气味确实更浓更难闻，这也从侧面印证了这一观点。虽然 VOCs 成分和含量有差异，但是总体来说产生废气的 VOCs 以芳香烃类化合物为主，其组分包括苯乙烯、苯、甲苯、乙苯、二甲苯、苯甲醛、一溴甲烷、2-甲基-丙烯酸甲酯等。

中试期间采用 BTF-EPO 工艺处理电子垃圾拆解过程产生的 VOCs，处理前后采集的 VOCs 的气样经预浓缩-气相色谱-质谱仪的 TIC 图如图 8-25 所示。在对该综合工艺设备对 VOCs 去除率效果考察时，首先考察了 BTF 或者 EPO 对 TVOC 的去除情况。从图 8-26 可以看出，虽然 TVOC 的进口浓度有明显较大幅度的波动，但 BTF 对 TVOC 的去除率在逐步地增加(由 60.4%增加到 79.1%)，这是因为 BTF 中的微生物生长逐渐稳定，使得其生物量也在逐渐增加，因此其去除率都有所增加。但 BTF 对 TVOC 的去除率却低于 EPO 组合工艺。从图中还可以看出,虽然 EPO 对 TVOC 的去除率随着中试的进行有所降低,但是去除率仍保持在 84.6%以上。然而,相较单独使用 BTF 或者 EPO,BTF-EPO 联用工艺对 TVOC 表现出更强和更稳定的去除能力,其对 TVOC 的去除率能稳定在 95.4%以上,这一结果表明 BTF-EPO 联用工艺在去除电子垃圾拆解现场排放的 VOCs 具有高效的处理能力。

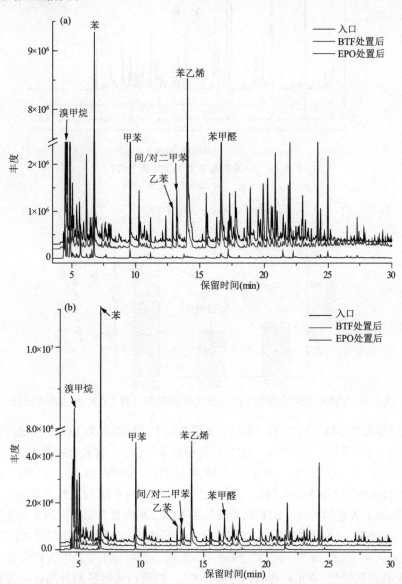

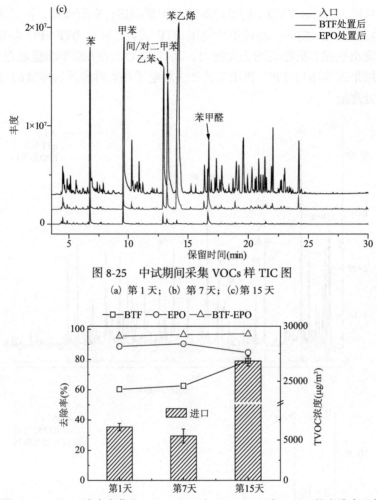

图 8-25 中试期间采集 VOCs 样 TIC 图
(a) 第1天；(b) 第7天；(c) 第15天

图 8-26 TVOC 浓度变化和 BTF、EPO 和 BTF-EPO 对 TVOC 的去除率比较

进一步考察 BTF、EPO 和 BTF-EPO 工艺对中试现场四类 VOCs 的处理效果，结果如图 8-27 所示。由图可知，BTF 对芳香烃类 VOCs、含氮含氧类 VOCs、卤代烃类 VOCs 和脂肪烃类 VOCs 的去除率分别在 62.9%~79.2%、54.1%~90.1%、29.5%~71.8%和 48.3%~86.2%之间。通过比较不同时间内 BTF 对这几类 VOCs 去除率发现，从第 1 天至第 7 天，BTF 对芳香烃类 VOCs 和含氮含氧类 VOCs 的去除率则分别从 62.9%和 54.1%增加到 65.2%和 90.1%，而对卤代烃类和脂肪烃类 VOCs 的去除率分别从 47.9%和 66.2%降低到 29.5%和 48.3%。这可能主要是因为相较第 1 天，第 7 天的脂肪烃类 VOCs 和卤代烃类 VOCs 的进口浓度急剧升高，而芳香烃类 VOCs 和含氮含氧类 VOCs 的进口浓度有所降低所致。与第 7 天相比，在第 15 天时，BTF 对芳香烃类 VOCs、卤代烃类 VOCs 和脂肪烃类 VOCs 的去除率明显升

高，然而，对含氮含氧类 VOCs 的去除率则有所降低。这是因为相比第 7 天，第 15 天时卤代烃类 VOCs 和脂肪烃类 VOCs 的进口浓度有所降低，而含氮含氧类 VOCs 的进口浓度则有所增加所导致。对比分析可知，BTF 对脂肪烃类 VOCs、含氮含氧类 VOCs 和卤代烃类 VOCs 去除率的变化与污染物的进口浓度成正相关关系。然而对芳香烃类 VOCs 而言，虽然其进口浓度显著升高（从（2369.9±359.8）μg/m³ 升至 （24419.6±229.5）μg/m³），但是 BTF 对其去除率并没有降低，反而还有所提高，这可能与其选用的优势菌种有关，因为 BTF 所使用的功能微生物菌群是由降解甲苯、二甲苯和苯乙烯的菌群组成，故对芳香烃类 VOCs 去除效果较为显著。

同时也考察了单独使用 EPO 工艺对这四类 VOCs 的去除能力。从图 8-27 中还可以看出，相比于 BTF，虽然 EPO 对这四类 VOCs 的去除率也会有所浮动，但是总体上均表现为 EPO 的去除率更高：对芳香烃类 VOCs、含氮含氧类 VOCs 和卤代烃类 VOCs 的去除率分别在 85.9%~94.0%、72.0%~91.9% 和 49.5%~76.1% 之间。这可能是因为经 BTF 中微生物降解后，进入到 EPO 的废气中 VOCs 浓度已大大下降，导致光催化氧化的有机负荷降低。其中特别需要指出的是经 EPO 处理后可以完全去除脂肪烃类 VOCs。由图 8-27 可以发现，相较单独使用 BTF 或者 EPO 工艺的处理结果，BTF-EPO 联用技术对 VOCs 的去除率明显提高：其中卤代烃类 VOCs、脂肪烃类 VOCs、芳香烃类 VOCs 和含氮含氧类 VOCs 的去除率分别大于或者等于 83.4%、100%、97.0% 和 92.4%。

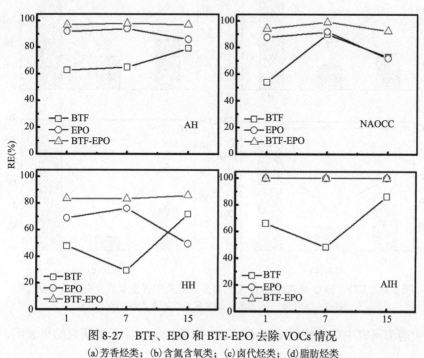

图 8-27　BTF、EPO 和 BTF-EPO 去除 VOCs 情况
(a)芳香烃类；(b)含氮含氧类；(c)卤代烃类；(d)脂肪烃类

BTF、EPO 及 BTF-EPO 联用工艺对芳香烃类中各主要成分的去除率如图 8-28 所示。从图中可以看出，尽管芳香烃类各主要组分如苯、甲苯、乙苯、二甲苯、苯乙烯及三甲苯的浓度在剧烈变化，尤其是在第 15 天时各组分浓度达到最高，但 BTF 工艺对它们的去除率总体呈现上升趋势。而 EPO 工艺对除了苯外的其他组分去除率则相对比较稳定，只是略微下降。尽管单独 BTF 工艺和 EPO 工艺对这些组分的去除率有起伏，但 BTF-EPO 联用工艺对这些组分的去除率基本上都达到 90%以上，特别是对苯乙烯，更是达到 100%的去除。

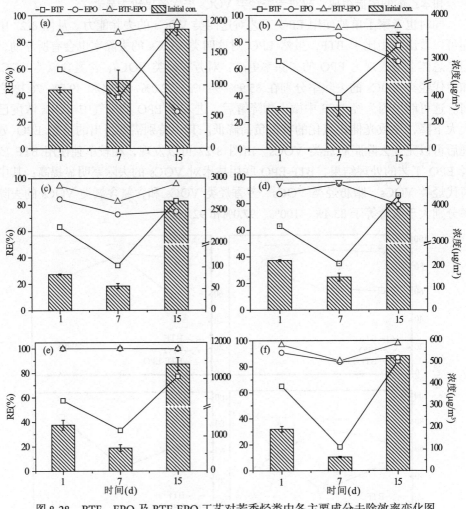

图 8-28　BTF、EPO 及 BTF-EPO 工艺对芳香烃类中各主要成分去除效率变化图
(a)苯；(b)甲苯；(c)乙苯；(d)二甲苯；(e)苯乙烯；(f)三甲苯

由前面研究中单独使用 BTF 和 EPO 技术处理 VOCs 的研究结果表明，BTF

和 EPO 技术均是去除电子垃圾拆解废气中 VOCs 的高效实用单元技术，但是二者在单独应用于处理实际工业废气时都还是存在一定的缺点：对 BTF 而言，由于微生物生长比较缓慢，挂膜需要一定的时间才能完成[69]；而 EPO 在使用时，经过长时间较高处理负荷后，催化剂表面会积聚细颗粒物，因而掩蔽了紫外光与光催化剂的接触，使得其效率也会逐渐降低[70]。相较单一 BTF 或 EPO 工艺，BTF 与 EPO 的顺序式串联工艺就具有明显的优势：BTF 在前面的处理中可以发挥处理高浓度 VOCs 的作用，而后续的 EPO 技术则可以进一步对于微量毒害性 VOCs 进行深度氧化达标排放。

BTF、EPO 和 BTF-EPO 工艺的去除负荷关系和有机负荷如图 8-29 所示。虚线和实线的斜率分别代表 TVOC 理论上完全降解和 TVOC 实际去除率。由图可知，各种工艺的去除负荷随着 TVOC 的有机负荷的增长而呈现出线性增长的关系（线性系数 $R^2 > 0.987$），表明在进口处有机负荷的增加对单一或联用工艺对 VOCs 的去除率并没有明显的抑制作用，其中最大去除负荷值主要是取决于进口的有机负荷值。而对 BTF 工艺来说，当有机负荷最大值为 15.8 g/(m³·h)时，其去除负荷值也达到最大值 12.5 g/(m³·h)，但 EPO 工艺的最大去除负荷仅为 4.4 g/(m³·h)左右，因为其有机负荷较小，最大仅为 5.2 g/(m³·h)。另外值得指出的是，BTF-EPO 联用工艺的最大去除负荷值也只有 9.4 g/(m³·h)，比单独的 BTF 工艺要低，这是因为结合 BTF 较高的去除负荷和 EPO 较高的去除率，使用 BTF-EPO 联用工艺可以高效去除 VOCs。

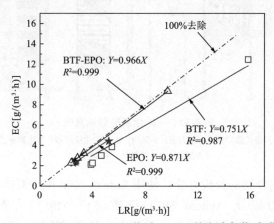

图 8-29　BTF、EPO 和 BTF-EPO 工艺对 TVOC 的去除负荷-负荷率对比图

2）案例应用二

Chen 等[50]采用袋式除尘器（BF）与光催化-生物滴滤器（PC-BTF）相结合的除尘净化一体化技术来净化电子垃圾拆解车间内的 VOCs，并对该设备对 VOCs 的

净化性能进行评估。

VOCs 净化装置：如图 8-30 所示，用于这些实验的集成反应器包括三个顺序连接的部分：BF、PC 和 BTF 反应器。不锈钢袋式除尘器高为 900 mm，直径为 600 mm，不锈钢光催化反应器为 450 mm × 450 mm × 1000 mm 的长方体，其中十个紫外灯(36 W，最大波长 254 nm)等距固定在五排，并配有两块泡沫镍，用 TiO_2 涂层平行固定在每个灯的顶部和底部，泡沫镍与紫外灯之间的距离约为 50 mm；生物滴滤系统主要包括一个由透明有机玻璃(高 2220 mm，内径 400 mm)制成的 BTF 和一个循环营养单元。循环营养单元由 96 升营养液罐、蠕动泵和流量计组成，为了加快 BTF 的启动时间，采用两步接种法将微生物接种到填料上。首先，在实验室进行了生物膜在填料上的定殖实验：即将包装材料浸入含有 B350 和自分离蜡样芽孢杆菌混合物的营养液中，每天从桶底曝气 20 小时，然后去除上清液，加入新鲜营养液和混合菌群，第一步接种持续 13 天，然后在生物滴滤器中进行生物膜的形成。电子垃圾拆解过程排放的废气首先通过集气罩收集到高炉中，使用安装在高炉和 PC 反应器之间的风机，速度为 200 m^3/h。然后将气体依次引入 PC 反应器(上流式)和 BTF(下流式)。实验持续了 30 天。

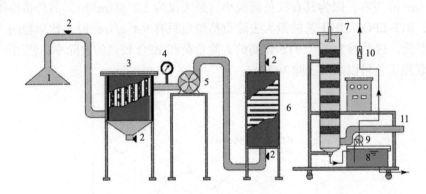

图 8-30 一体化反应器示意图

1-集气罩；2-阀门；3-袋式除尘器；4-压力表；5-风扇；6-光催化反应器；7-生物滴滤器；8-营养液罐；9-蠕动泵；10-流量计；11-通风口

挥发性有机化合物的分析方法：使用 Entech 7100 预浓缩器和 Agilent 6890 N 型气相色谱(GC)以及 Agilent 5973 型质谱仪(MS)结合美国环境保护署(USEPA)TO-14 方法对挥发性有机物进行定性和定量检测。使用 Entech 7100，从苏码罐中提取 250 毫升废气，并将其收集在 1/4 英寸液氮冷阱中。除去 H_2O 和 CO_2 后，样品扫入 GC 进行分离，然后进入 MS 进行分析。

VOCs 净化效率：表 8-8 列出了电子垃圾回收过程中产生的 VOCs 的浓度和成分。苯、甲苯、乙苯、间/对二甲苯和邻二甲苯是前五位的主要挥发性有机化合物。总 VOC(TVOC)浓度范围为 48~9.2×10² μg/m³，甲苯平均占 TVOC 的 57.9%（图 8-31）。

表 8-8 处理前主要 VOCs 的浓度和组成

采样周期	浓度（μg/m³）					
	苯	甲苯	乙苯	间/对二甲苯	邻二甲苯	TVOC
1st	1.0×10²	6.4×10²	68	79	34	9.2×10²
2nd	16	1.7×10²	77	1.1×10²	32	4.1×10²
3rd	3.6	27	7.6	8.7	5.7	52
4th	4.5	32	3.4	5.4	1.9	48

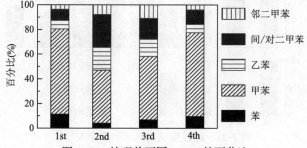

图 8-31 处理前不同 VOCs 的百分比

废电视机中 WPCB 主要由含有阻燃剂的环氧树脂或酚醛树脂制成，其中碳原子和苯环之间的单键很容易断裂，然后在加热时释放出芳香烃。应用综合技术可在 30 天的运行期内平均去除 83.7% 的 TVOC（图 8-32）。具体而言，PC 技术去除了约 65.3% 的 TVOC，其次是 BTF(14.6%) 和 BF(3.8%) 技术。这一结果表明 PC 因为其芳香烃降解能力对 VOC 去除的贡献最大，而除尘技术对 VOCs 的去除效果较差。

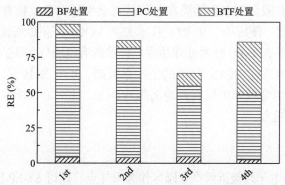

图 8-32 使用综合技术在四个采样样本中 TVOC 消除情况

对于单个挥发性有机化合物也获得了类似的结果。图 8-33 显示，采用综合技术，苯、甲苯、乙苯、间/对二甲苯和邻二甲苯的平均去除率分别为 75.2%、84.8%、82.2%、80.2%和 83.2%。其中苯是最难消除的，因为当其他大分子量芳烃（如甲苯）降解时，苯也是降解的中间产物。这五种挥发性有机物中约有 56.9%~67.4%通过 PC 技术去除，其次是 BTF 技术(14.8%)和 BF 技术(4.2%)，这与 TVOC 去除结果一致。

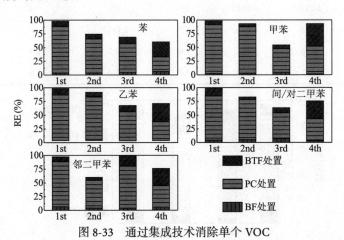

图 8-33 通过集成技术消除单个 VOC

以甲苯为例，随着处理时间的推移，上述处理技术去除甲苯的能力发生了变化。如 PC 对甲苯的去除率由 89.4%下降到 51.3%，BTF 的去除率由 6.4%上升到 40.5%。这可能是因为在处理初期，PC 反应器中的催化剂和紫外光灯处于峰值状态，对 VOCs 的去除能力很强。然而，在复杂条件下长时间处理后，细小颗粒和难降解的中间产物会污染催化剂和灯的表面，降低催化剂和灯之间的有效接触，进而降低去除效率。TVOC 的去除率从 87.4%下降到 46.3%，BTF 的去除率从 6.8%上升到 37.1%，再次支持上述假设。相反，在处理期开始时，BTF 处于启动模式，VOCs 去除能力有限。随着生物膜在包装材料上生长，微生物有助于减少挥发性有机化合物。因此，使用 PC 和 BTF 技术去除 VOCs 可获得最佳的总体效果。这一趋势与之前应用 BTF-PC 技术处理油漆厂排放的有机废气的研究一致[15]。

综上所述，PC 对 VOCs 气体的去除率较高；BTF 加快了 VOCs 被生物降解。BF-PC-BTF 组合技术可以有效地消除 VOCs，证明了该技术净化电子垃圾拆解车间内大气的能力。

3) 案例应用三

Liu 等[71]选择电子垃圾拆解产业园区作为采样点，探讨 ST-EP-PC 联用中试设备对电子垃圾拆解回收过程中释放的 VOCs 的消除能力，评估该联用设备的性能。

VOCs 净化装置：选取我国南方某个电子垃圾回收小镇作为研究地点。在电子垃圾集中拆解园区内某一建筑楼顶放置自发研究的 ST-EP-PC 联用设备。此建筑内共有 20 个电子垃圾回收车间，主要采用电烤锡炉拆解回收电视机线路板。现场中试设备如图 8-34 所示，喷淋塔后和出口处安装有风机，以确保稳定的有机废气流量 60000 m^3/h。喷淋塔(6000 mm × 3200 mm，高×直径)一共有四层，每一层装有 500 mm 高的商业拉西环(12 mm × 12 mm × 3 mm)填料。静电除尘-光催化联用反应器是密封的不锈钢长方体结构反应箱(13000 mm × 2500 mm × 2500 mm)，箱体分为静电除尘单元和光催化反应器。静电除尘单元内安装了 28 个静电除尘器，光催化单元内平行安装了 128 盏真空紫外灯(最大波长 254 nm，最小波长 185 nm)和 20 张负载 TiO_2 的泡沫镍网(1000 mm × 1250 mm)。两灯之间的垂直距离为 200 mm，真空紫外灯与泡沫镍板距离 10 cm。联用设备的详细参数如表 8-9 所示。有机废气在喷淋塔内塔的有效保留时间为 0.72 s，在静电除尘-光催化单元为 4.9 s，整个联用设备的保留时间为 5.62 s。中试时间为 2015 年 7 月至 2016 年 1 月。分别在联用设备的进口和出口使用 2.7 L 内表面经过硅烷化处理的苏码罐进行样品的采集，每隔约 90 天进行一次样品的采集，每个采样点采集三个样品作为废气污染浓度分析。

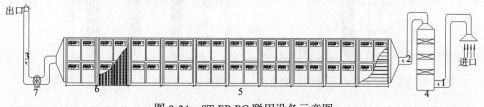

图 8-34　ST-EP-PC 联用设备示意图
1~3-采样点；4-ST；5-EP；6-PC；7-风机

表 8-9　联用设备参数

处理单元	参数	数值
喷淋塔	拉西环（$D × H × \delta$, mm）	12 × 12 × 3
	密度（g/cm^3）	0.69 ~ 0.73
	比表面积 S_{BET} (m^2/g)	0.42 ~ 0.48
	孔隙率	>70%
静电除尘	放电电压（V）	220
	放电功率（W）	9600
光催化反应	TiO_2 比表面积 S_{BET} (m^2/g)	50

VOCs 净化效率：共检测到 39 种 VOCs，包括 15 种 AHs，10 种 HHs，10 种 AlHs 和 4 种 OVOCs。图 8-35(a)给出了 180 天中试期内不同采样期间有机废气

中 VOCs 的浓度变化。从图中可以看出，电子垃圾拆解有机废气中的 VOCs 以 AHs 为主，浓度范围为 $1.3×10^2$~$6.4×10^2$ μg/m³（平均浓度 $3.0×10^2$ μg/m³），占 TVOCs 总含量的 55.9%，其次是 HHs（58~$3.8×10^2$ μg/m³，平均浓度 $1.8×10^2$ μg/m³），AlHs（12~72 μg/m³，平均浓度 35 μg/m³）和 OVOCs（6.3~55 μg/m³，平均浓度 24 μg/m³），分别占 TVOCs 总含量的 33.3%、6.51%和 4.23%[图 8-35(d)]。这与前期的研究结果一致[29, 72]，即采用电烤锡炉处理电视机线路板时，AHs 和 HHs 是拆解有机废气 VOCs 中含量最高的两类，主要由于酚醛树脂和氯代阻燃剂的加热降解行为造成的[29, 72]。

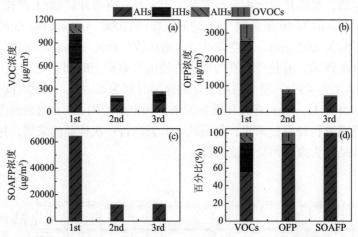

图 8-35　联用技术处理前 VOCs 的浓度(a)、OFP (b)、SOAFP (c) 和相应的比例(d)

电子垃圾拆解过程排放的 VOCs 进入大气环境后，在太阳光照射下与空气中的 NO_x 发生反应，导致臭氧浓度升高及光化学烟雾等光化学污染，对人类的眼睛和呼吸道等造成危害。因此可通过臭氧生成潜势（OFP）来评价电子垃圾工业源排放的 VOCs 对光化学污染形成的贡献值，如图 8-35(b)所示。从图可以看出，电子垃圾拆解行业存在严重的光化学污染的可能，其中 AHs 为臭氧生成的主要贡献者，贡献平均浓度为 $1.1×10^3$ μg/m³，占了总臭氧生成潜势的 86.5%，远远高于 OVOCs（$2.1×10^2$ μg/m³）、AlHs（39 μg/m³）和 HHs（28 μg/m³），分别占总的臭氧生成潜势的 9.1%、2.4%和 2.0%。虽然 OVOCs 的释放浓度比 HHs 和 AlHs 低 7.41 和 1.42 倍，但是其臭氧生成潜势分别是 HHs 和 AlHs 的 7.45 和 5.35 倍，说明 OVOCs 是大气中 O_3 的重要来源。OVOCs 的低释放和高臭氧生成潜势主要归因于 OH 自由基较高的损失率和化学反应活性[73,74]。与 OFP 相似，AHs 也是二次有机气溶胶生成潜势（SOAFP）的主要贡献者，贡献平均浓度为 11.8 μg/m³，占了总二次有机气溶胶生成潜势的 99.9%[图 8-35(c)]，表明电子垃圾拆解排放的 VOCs 中 AHs 对二次有机气溶胶的生成占绝对优势。通过图 8-35(b)和(c)可以看出，

AHs 比例越大，其 O_3 和 SOA 的生成潜势越大，这与 Ou 等的研究结果一致[75]。

图 8-36 为 VOCs 浓度、OFP 和 SOAFP 中排名前十的 VOCs 组分。从图中可以看出，这十种 VOCs 分别占 TVOCs、OFP 和 SOAFP 总量的 73.1%、84.9% 和

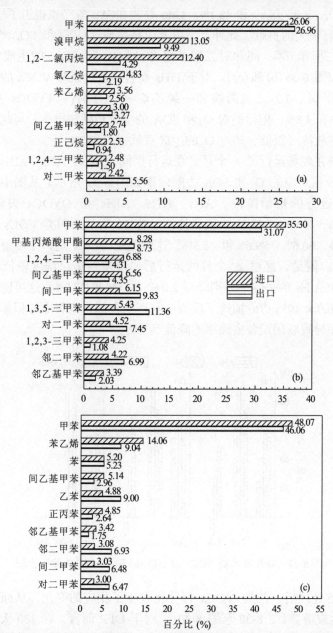

图 8-36　联用设备处理前 VOCs 的浓度(a)、臭氧生成潜势(b)和二次有机气溶胶生成潜势(c)排名前十的 VOCs 组分

94.7%，其中 AHs 类 VOCs 占据了一半以上，更加印证了 AHs 对 O_3 和 SOA 生成潜势的显著贡献。在电子垃圾拆解过程中，排放的 VOCs 中甲苯的含量最高，占 TVOCs 总含量的 26.1%，其对 O_3 和 SOA 的生成潜势的贡献也最大，分别贡献了总 OFP 和 SOAFP 的 35.3% 和 48.1%。Guo 等[76]和 Wu 等[77]也报道了甲苯对 O_3 和 SOA 生成占有很大的贡献。此外，二甲苯也是生成 O_3[78,79] 和 SOA[80,81]的重要贡献者。同时二甲苯（邻、间和对二甲苯之和）对 O_3 和 SOA 的生成分别贡献了 14.9% 和 9.1%[图 8-36(b) 和 (c)]。对于 HHs 而言，虽然某些 VOCs 的排放浓度很大，如一溴甲烷、1,2-二氯丙烷和一氯乙烷等，分别占 TVOCs 总排放量的 13.0%、12.4% 和 4.8%，但他们对 O_3 和 SOA 的生成贡献却很小。与此相反，甲基丙烯酸甲酯排放浓度很低，却对 O_3 的生成贡献很大(8.3%)。

联用设备在现场运行了 6 个月。在运行期间，AHs、HHs、AlHs 和 OVOCs 四类 VOCs 及相对应的 O_3 和 SOA 去除效率如图 8-37 所示。从图中可以看出，在联用设备运行的初始阶段，AHs、HHs、AlHs 和 OVOCs 去除率分别为 55.7%、70.1%、40.8% 和 25.0%；设备连续运行 90 天后，四类 VOCs 的去除率分别增至 71.9%、80.6%、89.6% 和 65.5%，这可能是 VOCs 进口浓度显著降低所致[图 8-37(a)]。但是，经过 6 个月的运行后，即使在进口浓度较低的情况下，AHs、HHs 和 AlHs 的去除率分别降至 2.0%、51.0% 和 27.0%。这可能是由于高速的进口风量(6.0×10^4 m^3/h)加速了反应中间产物在催化剂表面的积累，引起催化剂失活，进而导致联用设备去除率的降低[64, 82-84]。

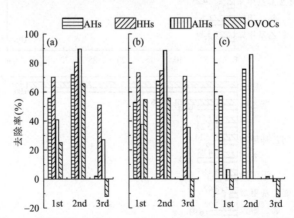

图 8-37 联用技术对 VOCs(a)、O_3(b) 和 SOA(c) 去除率比较

联用设备处理后，四类 VOCs 百分比发生了相应的变化，从而导致不同的 O_3 和 SOA 生成潜势(图 8-38 至图 8-40)。对于 OFP 而言，在 180 天的采样周期内，HHs 的去除对于预防 O_3 的形成效果最显著，可抑制 70.6%~74.5% 的臭氧生成。另外，AlHs 和 OVOCs 也分别可抑制 35.5%~88.6% 和 12.4%~55.6% 的 O_3 形

成。这与 VOCs 的去除率顺序高度一致，说明 VOCs 的消减可有效地预防 O_3 的形成。废气经过治理之后，AHs 和 OVOCs 仍是 O_3 生成贡献最大的两类 VOCs，分别贡献了 87.0%和 10.1%，比处理前分别增加 0.53%和 1.03%。而 AHs 的去除可抑制 44.7%的 SOA 生成，其次为 AlHs(30.0%)。然而，处理后的有机废气仍可造成严重的 SOA 污染(926.07 g/h)，影响区域气候变化。

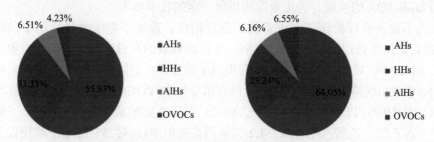

图 8-38　联用设备处理前后四类 VOCs 百分比

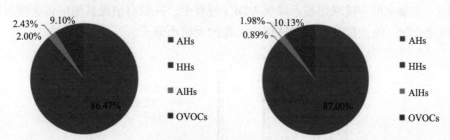

图 8-39　联用设备处理前后 OFP 中四类 VOCs 百分比

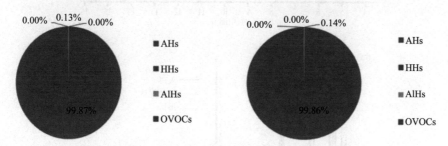

图 8-40　联用设备处理前后 SOAFP 中四类 VOCs 百分比

接着进一步研究了处理前后十种占比最高的 VOCs 组分的变化。从图 8-36 可以看出，甲苯仍然是主要的 VOCs 组分，占处理后 TVOCs 总含量的 26.96%，而且是最主要的 O_3 和 SOA 诱导物，分别占 OFP 和 SOAFP 的 31.1%和 46.1%。虽然甲苯在 TVOCs 中的占比增加 0.90%，但是其 OFP 和 SOAFP 分别降低 4.23%和 2.01%，说明其他 VOCs 组分的贡献有所增加，尤其是 AHs 类 VOCs。例如，对二甲苯的百分含量从 2.42%增加到 5.56%，比处理前增加 2.3 倍；对臭氧和二次有机气溶胶生成潜势的贡献分别增加 1.65 倍和 2.16 倍。此外邻二甲苯、间二甲苯和 1,3,5-三甲苯也发现了相同的现象。

综上所述，电子垃圾拆解工业释放的大量 VOCs 可诱导严重的 O_3 和 SOA 污染，尤其是 AHs。ST-EP-PC 联用技术可有效地去除废气中的 VOCs，从而有效抑制 O_3 和 SOA 的生成，为工业应用提供一定的技术参考。

为了进一步评价该工艺设备的工程应用性，选取了电子垃圾拆解中试现场（电烤锡炉加热处理回收电视机线路板）车间内释放的有机废气作为研究对象，开展了 ST-EP-PC 联用技术对 VOCs 有机污染物处理性能的中试研究。不同采样期间内，电视机线路板拆解车间内产生的有机废气中 VOCs 的总离子流图如图 8-41 所示。从图中可以看出，有机废气 VOCs 中最主要的成分（一溴甲烷、一氯乙烷、一溴乙烷、乙酸乙酯、苯、1,2-二氯丙烷和甲苯）经过 ST-EP-PC 联用设备后色谱峰（浓度）逐渐降低，这表明该联用设备能高效地去除废气中的 VOCs 组分。此外，在整个单一或联用技术降解 VOCs 过程中，并没有出现其他的毒害性反应中间化合物，绝大部分 VOCs 转化为无毒的无机小分子。

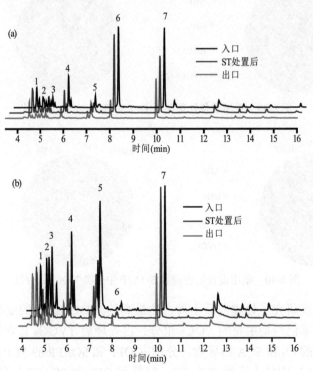

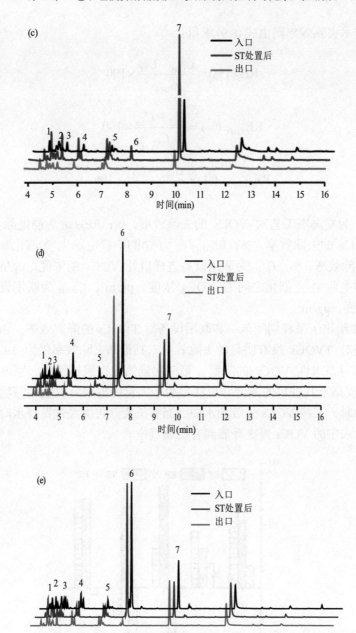

图 8-41 中试期间入口处、喷淋塔处理后及光催化处理后有机废气的总离子流图
(a)~(e)不同的采样期间:1—溴甲烷;2—一氯乙烷;3—一溴乙烷;4-乙酸乙酯;5-苯;6-1,2-二氯丙烷;7-甲苯

ST 对 VOCs 的去除效率(RE)由式(8.7)求得,EP-PC 去除效率由式(8.8)求

得，联用技术去除效率则由式(8.9)求得：

$$\mathrm{RE_{ST}}(\%) = \frac{C_{\text{inlet}} - C_{\text{ST}}}{C_{\text{inlet}}} \times 100 \tag{8.7}$$

$$\mathrm{RE_{EP\text{-}PC}}(\%) = \frac{C_{\text{ST}} - C_{\text{out}}}{C_{\text{ST}}} \times 100 \tag{8.8}$$

$$\mathrm{RE_{ST\text{-}EP\text{-}PC}}(\%) = \frac{C_{\text{inlet}} - C_{\text{out}}}{C_{\text{inlet}}} \times 100 \tag{8.9}$$

式中，$\mathrm{RE_{ST}}$ 为喷淋塔工艺对 VOCs 的去除效率，%；$\mathrm{RE_{EP\text{-}PC}}$ 为静电除尘-光催化工艺对 VOCs 的去除效率，%；$\mathrm{RE_{ST\text{-}EP\text{-}PC}}$ 为喷淋塔-静电除尘-光催化联用工艺对 VOCs 的去除效率，%；C_{inlet} 为联用设备进样口处 VOCs 的浓度，μg/m³；C_{ST} 为喷淋塔与静电除尘-光催化之间的 VOCs 浓度，μg/m³；C_{outlet} 为联用设备出口处 VOCs 的浓度，μg/m³。

图 8-42 给出了采样期间单一和联用设备对 TVOCs 的降解效率。由图可以看出，喷淋塔对 TVOCs 没有明显的去除效果，其最高去除效率仅为 19.8%，主要是对一些相对亲水性 VOCs 的降解。第三次到第五次采样期间，喷淋塔出口的 TVOCs 浓度高于其进口的浓度。可能是由于大量的拆解有机废气与喷淋塔内的循环水接触时，某些 VOCs 分散到水中，当有机废气再次经过循环水时，由于剧烈的扰动，水中的 VOCs 再次释放到有机废气中。

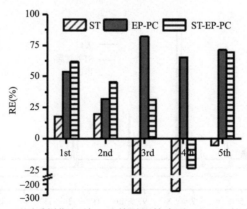

图 8-42　不同采样期间单一及其联用技术对 TVOCs 的去除效率

由于拆解工人技术的熟练程度以及所拆解的线路板数量有所差异，造成 VOCs 进口浓度在不同采样期间呈现较大的波动，会对 VOCs 的降解效率造成一定的影响。尽管如此，经过 ST-EP-PC 设备后，有机废气中 TVOCs 的去除效率依然比较高,维持在 31.9%~82.0%,这表明 ST-EP-PC 联用设备可长期有效地去除电子垃圾拆解过程排放的 VOCs 废气。首先,ST 和 EP 可有效地去除有机废气中的颗粒物,避免长时间大流量处理有机废气造成颗粒物掩蔽光催化剂与紫外灯,为后面光催化降解 VOCs 提供预处理。经过前期预处理后的 VOCs 进入光催化单元,在多孔 TiO_2 泡沫镍和真空紫外灯协同作用下,高效稳定地降解,其平均去除效率为 60.8%。已有研究报道证实,光催化过程可产生高氧化能力的氧活性物种,有效地降解气相 VOCs[85,86]。与此同时,光催化过程中真空紫外灯还会产生强氧化剂 O_3,进一步促进 VOCs 的降解。另外,使用的催化剂在真空紫外灯的照射下可以重生,确保了此体系长期的催化活性,避免明显的催化剂失活现象。

进一步考察了不同采样期间单一及其联用技术对卤代烃、芳香烃和乙酸乙酯的去除效率。由图 8-43 可以看出,ST 对 HHs 和 AHs 没有明显的去除效果,其最大去除率分别为 31.2%和 17.3%。由于乙酸乙酯具有较高的辛醇/水分配系数及

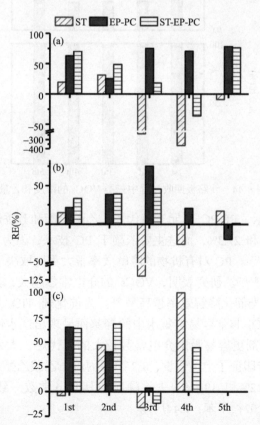

图 8-43 不同采样期间单一及联用技术对卤代烃(a)、芳香烃(b)和乙酸乙酯(c)的去除效率

较大的溶解度，具有良好的水溶性，因此喷淋塔对其去除效率明显增加。尤其是在第四次采样期间，乙酸乙酯的去除率达到 100%。这可能是此次采样期间排放的 VOCs 中乙酸乙酯的浓度最低，仅为 27 μg/m³，说明进样口浓度和溶解性的不同造成了 VOCs 种类之间去除率的差异。如图 8-44 所示，乙酸乙酯在三类 VOCs 中浓度最低(45 μg/m³)，其次为 HHs(6.9×10^2 μg/m³)和 AHs(2.3×10^3 μg/m³)。此外，乙酸乙酯在水中的溶解度(乙酸乙酯：83 g/L)远远大于卤代烃(1,2-二氯丙烷：2.6 g/L)和芳香烃(甲苯：0.53 g/L)[15]。综合上述因素，喷淋塔对乙酸乙酯的去除率最高，其次为 HHs 和 AHs。

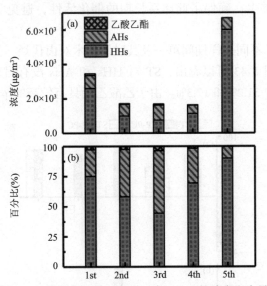

图 8-44 线路板回收过程中三类 VOCs 的浓度和含量

如图 8-43 所示，EP-PC 单元对卤代烃、乙酸乙酯和芳香烃的平均去除率分别为 62.7%、36.4%和 27.3%，而这主要依赖于 PC 技术，因为 EP 设备没有明显的 VOCs 去除能力[19]。PC 对有机物的降解效率很大程度取决于 VOCs 在催化剂表面的有效吸附[11, 87-89]。研究表明，VOCs 的介电常数越大，其对亲水性光催化剂的亲和力越好，光催化降解效率越高[87, 90]。光催化剂 TiO_2 采用水溶液配置，在配置过程中，TiO_2 非常容易富集水中的羟基而呈现出亲水性能。因此，具有亲水性表面的催化剂更容易吸附介电常数较大的卤代烃，大大地提高其去除效率。图 8-45 进一步印证了上述假设，1,2-二氯丙烷、乙酸乙酯和甲苯的去除效率分别为 76.3%、65.2%和 19.4%，去除顺序与其介电常数一致(1,2-二氯丙烷：8.93；乙酸乙酯：6.02；甲苯：2.44)[91]。

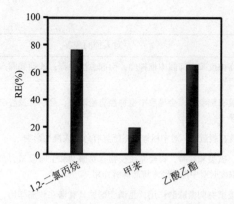

图 8-45 EP-PC 对 1,2-二氯丙烷、甲苯和乙酸乙酯的去除率

在 60 天中试期间内，TVOCs 的最高去除效率为 69.5%。ST 对亲水性的 VOCs 以及大颗粒有较好的去除率，EP 可有效地去除粒径较小的颗粒物，而 PC 对 VOCs 具有较好的去除效率。即该联用设备对有机废气中 VOCs 有一定净化作用。

8.2.2 VOCs 的健康风险消减

1. VOCs 健康风险

VOCs 是令世人关注的一类重要污染物，其对人体健康具有一定的危害作用。许多 VOCs 为有毒有害物，某些 VOCs 更被证实具有致癌、致畸和致突变作用[92]。例如 VOCs 会对人体的中枢神经系统、呼吸系统尤其是免疫系统和器官造成伤害。暴露于 VOCs 被认为是导致大量哮喘和哮喘相关症状如气喘、支气管反应和肺功能抑制等的首要诱因[93]。Zock 等研究发现经常使用能释放各种 VOCs 的家居清洗喷雾剂，发生气喘和哮喘概率会分别增加 40%和 50%[94]。Ware 等报道了 7~13 岁小孩患哮喘与 VOCs 环境浓度（2 mg/m^3）密切相关[95]。大量研究表明暴露于较高浓度（20~50 μg/m^3）VOCs（包含甲醛）中的小孩会有中度甚至重度呼吸道疾病和过敏性反应[96]。美国加州环保署的一项综述中有报道，通过室内暴露与动物毒理学机理研究证实了甲醛与哮喘类呼吸道疾病患病率的增加密切相关[97]。VOCs 中常见有毒有害物及其危害如表 8-10 所述[98]。

表 8-10 VOCs 中常见毒害性污染物的危害

VOCs	对人体的危害
甲醛	有明显的刺激性气味，可导致流泪、头晕、头痛、乏力、视物模糊等，检查可见结膜、咽部明显充血，部分患者听诊呼吸音粗糙。较重者可有持续咳嗽、声音嘶哑、胸痛、呼吸困难

续表

VOCs	对人体的危害
苯	致癌性,暴露高浓度苯会抑制中枢神经,引起急性中毒;长期暴露造成再生不良性贫血、血球减少症、白血病等
甲苯	对黏膜和皮肤具有刺激性,会导致中枢神经官能障碍,肝、肾及心脏损伤,与皮肤接触导致血管扩张、红斑等
乙苯	对皮肤和黏膜具有刺激性,对中枢神经系统具有急性或慢性影响
二甲苯	导致皮肤干燥、破裂和红肿,神经系统受损。对眼睛和上呼吸道有刺激作用,对肝脏及肾脏有轻微毒性,高浓度时对中枢神经系统起麻醉作用
苯乙烯	对眼睛和上呼吸道有刺激麻醉作用,是诱变剂,具有潜在致癌作用,暴露接触可能导致失忆,大脑和肝损伤
三氯甲烷	致癌物,导致肝、肾损伤
四氯乙烯	可引起接触者神经、肾脏和肝脏等多器官系统损伤,影响中央枢神经系统
乙酸乙酯	对眼、鼻、咽喉有刺激作用,高浓度吸入可用于麻醉

2. VOCs 的健康风险消减

致癌和非致癌风险评估:采用 USEPA 的标准方法对通过呼吸暴露在 VOCs 中产生的致癌和非致癌风险进行了评估。其中非致癌风险通过将每种 VOCs 的日浓度(C_i,μg/m³)除以其参考浓度(RfC$_i$,μg/m³)来计算,得出危险比(HR$_i$),如式(8.10)所示。VOCs 的参考浓度从 USEPA 综合风险信息系统获得[99]。终生致癌风险(LCR$_i$)是通过每种 VOCs 的日浓度(C_i,μg/m³)乘以每种化合物的单位风险(UR$_i$)来计算的,如式(8.11)所示,单位风险数据来自 USEPA 风险评估系统(http://www.epa.gov/iris/)。

$$HR_i = \frac{C_i}{RfC_i} \tag{8.10}$$

$$LCR_i = C_i \times UR_i \tag{8.11}$$

职业暴露风险评价:参照美国政府工业卫生医师协会(ACGIH)制定的方法对工人通过呼吸、皮肤接触以及非饮食摄入 VOCs 的职业暴露风险进行了评估。各 VOCs 组分的职业暴露风险(E_i)通过式(8.12)计算得到,即用该目标物的阈限值-时间加权平均值(TLV-TWA)除以其浓度即得。总职业暴露风险计算如式(8.13)所示。需要特别说明的是本次研究采样车间内的工人每天至少上班 10 个小时,每周工作 6 天,而 ACGIH 在制定时间加权平均值时默认工人每天上班 8 小时,每周工作 40 小时。因此,通过式(8.14)将时间加权平均值(TWA$_i$)校正为 TWA$_i'$。

$$E_i = \frac{C_i}{TWA_i'} \tag{8.12}$$

$$\sum E = \sum_i E_i \tag{8.13}$$

$$\text{TWA}_i'(\mu g/m^3) = \text{TWA}_i(\text{ppm}) \times \frac{\text{M.W.}}{24.45} \times 10^3 \times \frac{5 \times 8}{6 \times 10} \tag{8.14}$$

1) 案例应用一

基于 8.2.1 章节中 "2. 电子垃圾拆解过程 VOCs 减排控制应用" 中的案例应用一，静电除尘-光催化-臭氧深度氧化一体化工艺（EPO）净化 VOCs 的中试研究中，总共选择了 16 种 VOCs 进行非致癌风险评估，6 种 VOCs 进行致癌风险评估。使用式（8.10）和式（8.11），选择主要基于可用参考浓度和单位风险值，如表 8-11 所示。

表 8-11 VOCs 的非致癌参考浓度、致癌单位风险以及国际致癌研究机构的致癌分类

VOCs	非致癌风险		致癌风险		
	参考浓度（μg/m³）	来源	IARC 分类	单位风险（μg/m³）⁻¹	来源
苯	30	IRIS	1	6.0×10^{-6}	WHO
甲苯	5000	IRIS	—	—	
乙苯	1000	IRIS	2B	2.5×10^{-6}	OEHHA
二甲苯	100	IRIS	—	—	
苯乙烯	1000	IRIS	—	—	
异丙苯	400	IRIS	—	—	
正丙苯	1000	PPTRV	—	—	
三甲苯	6	PPTRV	—	—	
氯甲烷	90	IRIS	—	—	
溴甲烷	5	IRIS	—	—	
三氯甲烷	98	ATSDR	2B	2.3×10^{-5}	IRIS
1,2-二氯乙烷	2430	ATSDR	2B	2.6×10^{-5}	IRIS
氯苯	1000	OEHHA	—	—	
对二氯苯	—		2B	1.1×10^{-5}	OEHHA
丙烯腈	2	IRIS	2A	6.8×10^{-5}	IRIS
甲基丙烯酸甲酯	700	IRIS			
甲基异丁基甲酮	3000	IRIS			

"—"：Not available. ATSDR（Agency for Toxic Substances and Disease Registry）（ATSDR, 2011 IRIS）；（Integrated Risk Information System）；PPTRV（Provisional Peer Reviewed Toxicity Values of IRIS）；OEHHA（Office of Environmental Health Hazard Assessment）；WHO（World Health Organization）

首先对处理前排放的 VOCs 的非致癌风险进行了评估。如图 8-46 所示，第 1 天正丙苯、第 15 天甲苯和乙苯、第 30 天和第 45 天甲基异丁基酮、第 1、15、30 和 45 天异丙苯和氯苯和第 60 天 1,2-二氯乙烷的 HRs 均小于 0.1，低于关注水平[99]。第 45 天 (0.88) 氯甲烷、第 60 天 (0.13) 氯苯、第 15 天 (0.29) 和第 45 天 (0.72) 苯乙烯、第 30 天 (0.26) 和第 45 天 (0.19) 乙苯、第 30 天 (0.36) 和第 45 天 (0.37) 氯仿；第 1 天 (0.34)、第 30 天 (0.22)、第 45 天 (0.18) 和第 60 天 (0.68) 甲苯的 HRs 在 0.1 到 1 之间。这些 VOCs 的 HRs 高于 0.1，表明风险值高于潜在关注水平[100]。

此外，第 1 天 (1.94) 乙苯、第 30 天 (1.10) 氯甲烷、第 15 天 (26.35)、30 天 (2.91)、45 天 (2.66) 和 60 天 (1.03) 甲基丙烯酸甲酯以及第 1 天 (29.89 和 13.01)、第 30 天 (3.84 和 1.19) 和第 60 天 (36.93 和 11.16) 二甲苯和苯乙烯的 HR 超过 1，表明它们的风险值高于关注水平。而溴甲烷 (从 96.04 到 383.07)、丙烯腈 (从 26.35 到 171.40)、苯 (从 30.70 到 168.37) 和三甲基苯 (从 2.62 到 86.80) 这四种 VOCs 的 HR 远远超过 1，说明它们具有更高的非致癌风险[100]。

图 8-47 给出了六种 VOCs 的致癌风险。根据参考文献[101]，风险值 $>10^{-4}$ 被标记为"确定风险"，风险值介于 10^{-5} 和 10^{-4} 之间被标记为"可能风险"，风险值介于 10^{-5} 和 10^{-6} 之间被标记为"有可能风险"。其中 6 种选择的 VOCs 的致癌风险值都很高：丙烯腈 ($>3.58\times10^{-3}$)、苯 ($>5.53\times10^{-3}$)、乙苯 ($>1.11\times10^{-4}$)、1,2-二氯乙烷 (2.95×10^{-3})、氯仿 ($>8.21\times10^{-4}$) 和对二氯苯 (2.02×10^{-4})，说明它们具有明显的致癌风险，与之前的研究结果一致[26]。

这些结果表明，电子垃圾拆解过程排放的 VOCs 暴露导致的非致癌和致癌风险不容忽视，因此，有必要考虑使用控制技术来消减它们的风险。如图 8-46 和图 8-47 所示，EP 单元处理后 VOCs 的非致癌和致癌风险没有明显降低，表明该技术降低 VOCs 风险的能力较差。相比之下，PC-臭氧氧化组合技术显示出优异的风险消减性能。经过 PC 单元处理后，所有 VOCs 的非致癌和致癌风险都大大降低，尤其是苯乙烯 (从 0.20~16.84 降低到 0.07~2.08)、丙烯腈 (从 72.11~182.86 降低到 12.27~130.69)、苯第 1 天和第 15 天 (从 7.94×10^{-3} 降低到 4.53×10^{-3}) 以及丙烯腈 (从 7.73×10^{-3} 降低到 1.63×10^{-3})。虽然经过 PC 处理后 VOCs 的风险显著降低，但是大多数 VOCs 的 HRs 仍高于 0.1，而 LCR 均高于 10^{-4}。在随后的臭氧氧化处理后，VOCs 的非致癌和致癌风险得到进一步的消减，基本上降至 0.1 和 10^{-4} 以下，表明 PC-臭氧氧化组合技术具有很高的 VOCs 风险消减能力。但仍然有部分 VOCs 的 HRs 超过 1，如苯 (1.67~11.92)、溴甲烷 (3.29~65.79) 和丙烯腈 (2.74~33.79)，且 LCR 高于 10^{-4}，如苯 (3.0×10^{-4}~2.2×10^{-3}) 和丙烯腈 (3.7×10^{-4}~4.6×10^{-3})，表明这些 VOCs 仍然会对拆解工人构成非致癌和致癌风险。

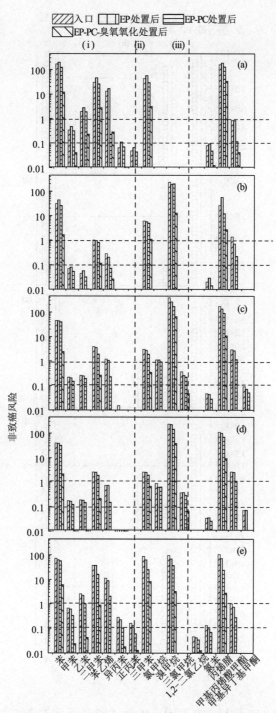

图 8-46 第 1 天(a), 15 天(b), 30 天(c), 45 天(d)和 60 天(e)处理前后 VOCs 的非致癌风险
(i): AHs; (ii): HHs; (iii): NAOCCs

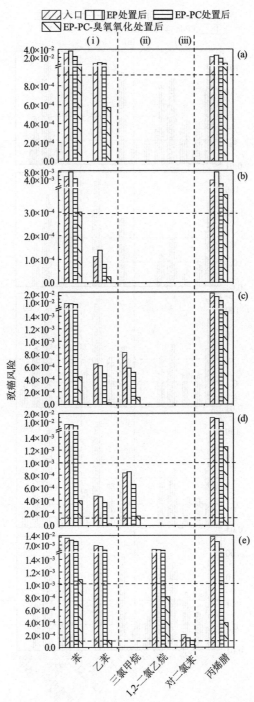

图 8-47 第 1 天(a)，15 天(b)，30 天(c)，45 天(d)和 60 天(e)处理前后 VOCs 的致癌风险
(i)：AHs；(ii)：HHs；(iii)：NAOCCs

同时，对 17 种 VOCs 的职业暴露致癌风险进行了评估(表 8-12)。如图 8-48 所示，在 EP 处理前，苯的 E_i 值(分类为 A1 确定的人类致癌物)在第 1 天、第 30 天、第 45 天和第 60 天分别为 4.74、1.27、1.10 和 2.04，所有这些值都超过 1，表明苯对电子垃圾拆解工人会构成致癌威胁。其他主要污染物如溴甲烷(第 1、30、45 和 60 天分别为 0.45、0.74、0.45 和 0.19)、丙烯腈(第 1 和 30 天分别为 0.11 和 0.12)和苯乙烯(第 1 和 60 天分别为 0.23 和 0.20)属于 A3 或 A4 组(不能视为致癌物)，其 E_i 值均在 0.1 以上，表明可能对拆解工人造成的有害影响。

表 8-12　17 种 VOCs 的 TWA 和 E_i

VOCs	Cancer level	TWA (ppm)	TWA′(μg/m³)
苯	A1	0.5	1.1
甲苯	A4	20	50.2
乙苯	A3	100	289.5
二甲苯	A4	100	289.5
苯乙烯	A4	20	56.8
异丙苯	—	50	163.9
三甲苯	—	25	81.9
氯甲烷	A4	50	68.8
溴甲烷	—	1	2.6
三氯甲烷	—	10	32.6
1,2-二氯乙烷	—	10	27.0
氯苯	—	10	30.7
1-己烯	—	50	114.7
正庚烷	—	400	1093.0
丙烯腈	A3	2	2.9
甲基丙烯酸甲酯	A4	50	136.5
甲基异丁基甲酮	A3	5	13.7

与非致癌和致癌风险评估结果一样，经过 EP 处理后 VOCs 的职业暴露致癌风险没有明显降低；而 PC-臭氧氧化组合技术处理后风险显著降低。其中 PC 单元对苯乙烯具有良好的风险消减能力，经 PC 处理后苯乙烯的 E_i 值均小于 0.1；然而，苯的 E_i 值在 PC 处理后仍然超过 1，在随后的臭氧处理后显著降低。这表明经臭氧处理后苯没有明显的致癌风险，说明 PC-臭氧氧化组合技术具有较强的消减 VOCs 风险的能力。

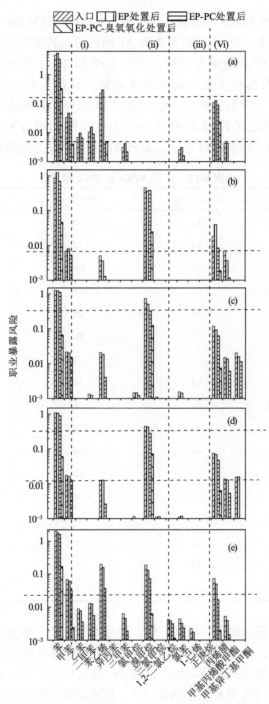

图 8-48 第 1 天(a), 15 天(b), 30 天(c), 45 天(d)和 60 天(e)处理前后 VOCs 的职业暴露风险
(i): AHs; (ii): HHs; (iii): NAOCCs

综上，在净化处理前车间内 VOCs 污染较为严重，具有非常高的致癌和非致癌风险，对暴露于其中的车间工人身体有很大危害。使用该 EC-PC-臭氧氧化一体化技术处理后，电子垃圾拆解工人暴露在 VOCs 的非致癌和致癌风险以及职业暴露致癌风险均得到显著降低。但是虽然 EPO 设备对 VOCs 有较好的处理效果，由于 VOCs 初始浓度过高，导致经处理后的尾气中某些 VOCs 的浓度仍超过可接受的安全范围，仍然存在致癌和非致癌风险。因此，为了确保车间工人的健康，有必要改善车间内旋转灰化炉的密封性以防止产生的 VOCs 废气直接逸进入车间。

另外，在 BTF-EFO 联用组合工艺净化拆解车间 VOCs 研究中，对 VOCs 处理前后致癌与非致癌风险评估分析如图 8-49 所示。

中试现场废气中 VOCs 各组分在 BTF-EPO 处理前后的非致癌风险对比如图 8-49 所示。从图可知，废气中 VOCs 的最主要非致癌危害物为 2-甲基丙烯酸甲酯和一溴甲烷，它们的非致癌风险分别高达 118.7~398.9 和 53.8~365.5；其次为 (1,3,5/1,2,4/1,2,3)-三甲苯和苯，它们的非致癌风险也均为确定致癌风险的数倍以至数十倍。另外，对/间二甲苯也是一类不容忽视的非致癌危害物。经 BTF-EPO 联用工艺处理后，废气中各组分的非致癌风险均大幅下降，其中原先最主要的非致癌危害物 2-甲基丙烯酸甲酯更是能够被完全去除，出口废气中也只有一溴甲烷(5.62~40.96) 和苯(3.44~4.16) 还存在确定非致癌风险。这主要是因为苯和一溴甲烷性质相对比较稳定，BTF-EPO 工艺对一溴甲烷和苯去除率相对偏低，导致出口废气中仍有一定含量的一溴甲烷和苯所致。

图 8-50 将废气经 BTF-EPO 联用工艺处理前后 VOCs 组分的致癌风险进行了对比。从图中可以看出，在处理前的车间内废气中所有已检测到的具有致癌风险的 VOCs 均具有确定致癌风险。尽管经 BTF-EPO 处理后，它们的致癌风险均有一定程度下降，如第 1 天和第 7 天气样中乙苯风险已降到确定致癌风险值以下，第 7 天气样中的对-二氯苯及第 15 天气样中的氯仿也没有了确定致癌风险；但大部分 VOCs 均还具有确定致癌风险。经净化处理后废气中 VOCs 的总致癌风险仍然达 $2.20~2.93\times10^{-3}$，仍是确定致癌风险值的 22~29 倍。其中需要特别指出的是 1,2-二氯乙烷在处理后其致癌风险值没有降低反而上升，这是因为在经过 BTF 中微生物分解氯代化合物时生成了中间产物 1,2-二氯乙烷，导致其浓度反而升高所致。

总之，在经 BTF-EPO 工艺净化后，尽管有机废气中的某些 VOCs 仍存在确定的致癌与非致癌风险，但是致癌与非致癌 VOCs 种类有所减少，并且其致癌与非致癌风险也大幅下降(下降 1~2 个数量级)。其中经 BTF-EPO 净化处理后的废气中引起非致癌风险的危害物主要为苯与一溴甲烷；产生致癌风险的最主要物质为苯与 1,2-二氯乙烷。

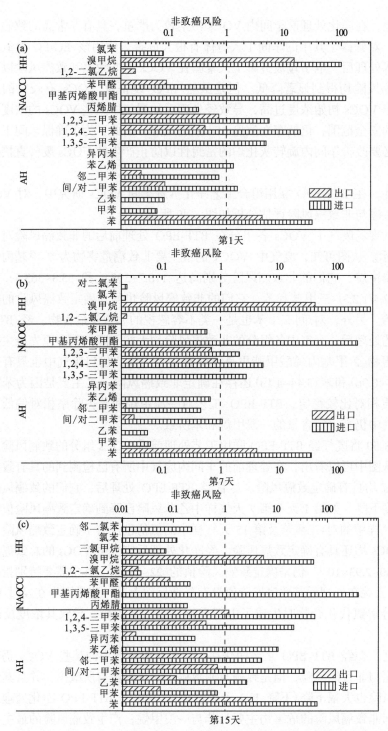

图 8-49 废气中 VOCs 各组分在 BTF-EPO 工艺处理前后非致癌风险对比

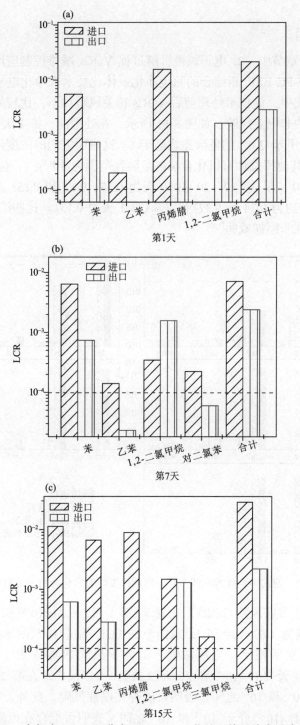

图 8-50 废气中 VOCs 组分在 BTF-EPO 工艺处理前后致癌风险对比

2) 案例应用二

基于 8.2.1 章节中 "2. 电子垃圾拆解过程 VOCs 减排控制应用" 中的案例应用二，利用 BF-PC-BTF 相结合的除尘净化一体化技术来净化电子垃圾拆解车间内 VOCs 的研究中，该技术处理前后 VOCs 相关风险如下：比较处理前后车间中与 VOCs 相关的非癌症风险。如图 8-51 所示，在处理前，第 1 天的苯 (3.46) 和第 10 天的间/对二甲苯 (1.07) 的危险比均高于 1。这表明它们的浓度超过了关注的参考水平[99, 102]。其他挥发性有机化合物的危险比范围为 0.1 至 1，包括间/对二甲苯 (第 1 天为 0.79)、苯 (第 10、20 和 30 天为 0.55、0.12 和 0.15)、邻二甲苯 (第 1 天和第 10 天分别为 0.34 和 0.32) 和甲苯 (第 1 天为 0.13)。这些结果表明，VOCs 对工人存在潜在非癌症威胁[100]。

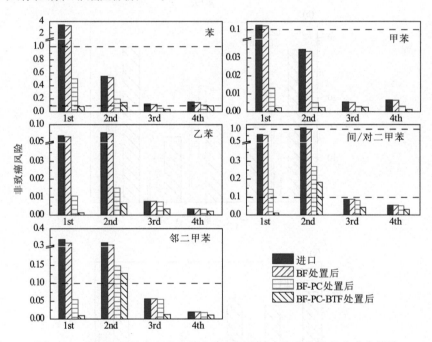

图 8-51 使用综合技术在四个样中降低挥发性有机化合物的非致癌风险

而对于终生致癌风险 (LCR)，发现苯 (第 1 天：6.24×10^{-4}) 和乙苯 (第 1 天和第 10 天：分别为 1.69×10^{-4} 和 1.92×10^{-4}) 的 LCR 超过 10^{-4} (图 8-52)。这些结果表明，与参考水平相比，当前浓度下的癌症风险非常高[101]。苯在第 10 天 (9.85×10^{-5})、20 天 (2.13×10^{-5}) 和 30 天 (2.71×10^{-5}) 以及乙苯在第 20 天 (1.89×10^{-5}) 的 LCR 介于 10^{-5} 和 10^{-4} 之间，表明可能存在癌症风险。此外，第 30 天与乙苯相关的风险 (8.43×10^{-6}) 介于 10^{-5} 和 10^{-6} 之间，表明可能存在风险。总的来说，

接触 VOCs 的工人存在暴露风险，在排放废气之前，应考虑这些风险并选择适当的处理方法。

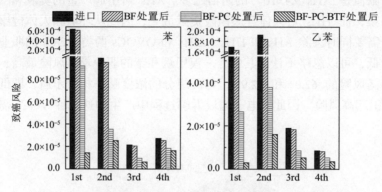

图 8-52　使用综合技术在四个样事件中降低 VOCs 的致癌风险

图 8-51 和图 8-52 表明 BF 处理后非致癌和致癌风险没有明显降低，表明除尘技术没有显著降低 VOCs 风险，即除尘技术对 VOCs 的去除贡献不大[19]。相反，PC-BTF 技术具有显著的风险降低能力，可以显著降低所研究 VOCs 的非致癌和致癌风险指数。例如，苯的致癌风险指数在第 1 天从 3.34 降至 0.51，在第 20 天从 0.11 降至 0.05。甲苯的非致癌风险指数在第 1 天从 0.12 降至 0.01。对于非致癌风险，苯从第 1 天的 $6.01×10^{-4}$ 降至 $9.15×10^{-5}$，第 20 天从 $2.04×10^{-5}$ 降至 $9.53×10^{-6}$），与乙苯相关的致癌风险也有所下降（第 1 天、第 10 天和第 20 天分别从 $1.63×10^{-4}$、$1.84×10^{-4}$ 和 $1.83×10^{-5}$ 降至 $2.65×10^{-5}$、$3.77×10^{-5}$ 和 $8.49×10^{-6}$）。

经 PC 处理后，所选 VOCs 的 LCR 均低于 10^{-4}，表明 PC 技术具有显著的降低风险的能力。由于在 PC 处理前 VOC 浓度很高，苯（第 1、10 和 30 天分别为 0.51、0.20 和 0.10）和甲苯（第 1 和 10 天分别为 0.14 和 0.27）的危害比在处理后仍高于 0.1。然而，这些挥发性有机化合物的危害指数在随后的 BTF 处理后大多低于 0.1，说明 PC-BTF 综合净化技术的在降低 VOCs 风险消减方面的成功。

3）案例应用三

基于 8.2.1 章节中"2. 电子垃圾拆解过程 VOCs 减排控制应用"中的案例应用三，在利用 ST-EP-PC 联用中试设备净化电子垃圾拆解回收过程中释放的 VOCs 的研究中，进一步对其风险消减能力进行评估。根据前期研究的评估方法中的职业暴露风险评估模型[19, 29]及相应参考值和式(8.12)、式(8.13)和式(8.14)评估了电子垃圾拆解有机废气中 VOCs 对人体健康的影响。如图 8-53 所示，进口处 TVOCs 的职业暴露风险及四类 VOCs 的平均职业暴露风险均小于 0.1，说明电子垃圾拆解工人在拆解电子垃圾过程中不存在职业暴露风险。其中 HHs 造成的职业暴露风险最高,为 2.0×10^{-2},占总的职业暴露风险值的 62.6%,其职业暴露风险贡献值是 AHs(9.4×10^{-3}) 的两倍之多。AHs 可引起严重的臭氧和二次有机气溶胶污染,而 HHs 对人体健康风险贡献最大,因此从环境和人体健康角度综合考虑,需要同时去除 AHs 和 HHs。AlHs 和 OVOCs 两类 VOCs 的职业暴露风险值非常低,可以忽略不计。其中,一溴甲烷和苯的职业暴露风险最高,分别占总职业暴露风险的 62.0% 和 32.9%,与两组分的浓度顺序相互矛盾,即低浓度的 VOCs 可引起高风险。因此,电子垃圾拆解过程中产生的这两种 VOCs 需引起高度重视。

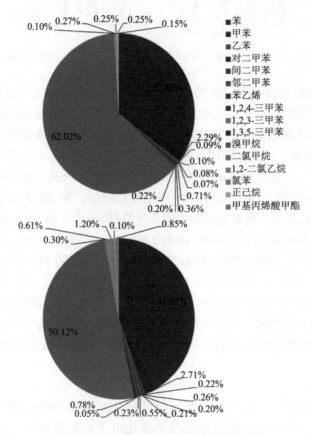

图 8-53　综合技术处理前后职业癌症风险

研究发现,所有 VOCs 在经过治理后其职业暴露风险均呈现不同程度的降低。如图 8-53 所示,职业暴露风险降低最多的是 AlHs,其次是 HHs、AHs 和 OVOCs,分别降低 70.7%、64.3%、52.2% 和 34.4%。虽然 HHs 和 AHs 仍是主要的职业暴露风险贡献者,但两者之间的差距显著减小,从 25.7%降到 5.48%(图 8-54)。造成这种现象可能的原因是苯的致癌风险百分比增加了 8.76%,而一溴甲烷的百分比则降低 11.9%。大量研究表明,含支链较多的 AHs 在光催化降解过程会生成苯[84,103,104],导致苯的百分含量从 3.0%增加到 3.3%,其相对应的职业暴露风险贡献百分比也显著增加(从 32.9%增加到 41.6%)。虽然联用设备显著降低了 VOCs 产生的总职业暴露风险,但是某些低浓度但高毒性的 VOCs 组分(例如苯)仍会在反应器内部积聚并随着高速的有机废气释放到大气环境,威胁工人身体健康。因此,需要进一步研究联用技术处理过程中产生的毒性中间产物,综合评估此联用设备的工程应用性。

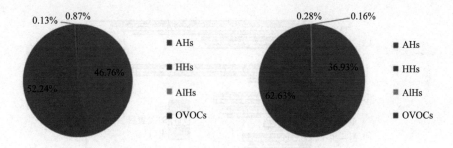

图 8-54 联用设备处理前后致癌风险中四类 VOCs 百分比

为进一步了解该设备工艺在工程应用中的性能,对该设备在电子垃圾拆解中试现场(电烤锡炉加热处理回收电视机线路板)车间的中试研究中处理前后有机废气中存在的健康风险进行了评估。从图 8-55 中可以看出,进口处单一 VOCs 或 TVOC 的职业暴露风险指数均小于 1.0,说明该车间采样期间产生的 VOCs 对操作工人不会造成危害。但是,职业暴露危害最大的一溴甲烷和苯,职业暴露风险超过了 0.1,存在可能的职业暴露风险。此外,第一次、第二次和第五次采样期间的职业暴露风险值均大于 0.1,说明该浓度的 VOCs 对车间工人造成可能的职业暴露风险。经过 ST 处理后,单一 VOCs 和 TVOC 产生的职业暴露风险没有明显的降低,反而有增加的趋势。比如,在第一次采样期间,一溴甲烷和 TVOC 的职业暴露风险值分别从 0.11 和 0.13 降至 0.06 和 0.08,而在第三次采样期间,其职业暴露风险值分别从 0.07 和 0.09 增加到 0.29 和 0.35。经过 EP-PC 处理之后,其职业暴露风险值均降到可能的职业暴露风险(0.1)以下(第五次采样除外)。其中需要特别指出的是 1,2-二氯丙烷在有机废气中比重很大,但是由于缺乏相应的参考值,从而造成总的职业暴露风险被低估。

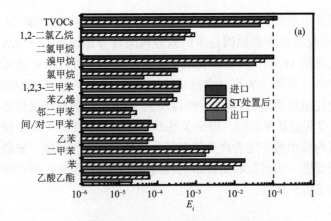

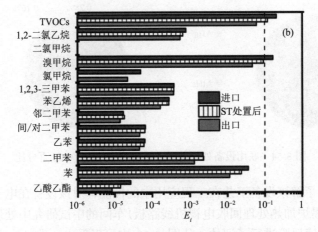

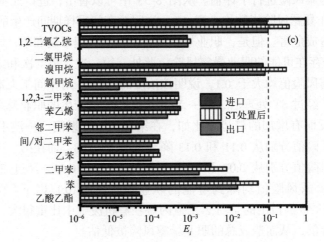

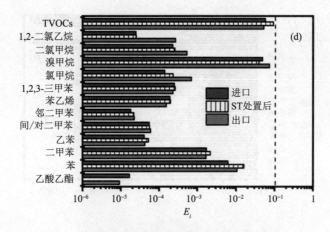

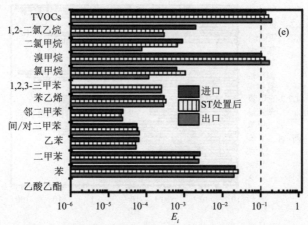

图 8-55 联用设备处理前后 VOCs 组分的职业暴露风险

(a)~(e)分别为第 1~5 次采样

13 种检测到 VOCs 被 USEPA 认定具有非致癌风险,其非致癌指数如图 8-56 所示。从图中可以看出,拆解废气中 VOCs 导致的总的非致癌风险为 64.5~1230.8,其平均值为 422.0,比确定非致癌风险值(HR=1.0)高出三个数量级。由此可见,电视机线路板拆解有机废气中的 VOCs 可引起明显的非致癌风险,必须采取相应的治理和减排措施。其中,废气中最主要的 VOCs 非致癌组分为 1,2-二氯丙烷和一溴甲烷,其非致癌风险值分别高达 9.3~1148.6 和 36.6~124.5,远远超过评估标准 1.0[99]。此外,苯和间/对二甲苯的非致癌风险均为可能非致癌风险值(0.1)的数倍至数十倍[100]。而在整个采样周期内,苯乙烯等某些 VOCs 的非致癌风险值均低于 0.1,即非致癌风险可以忽略。

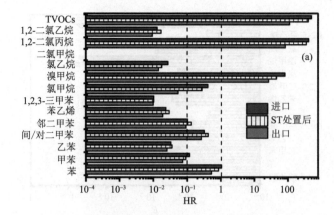

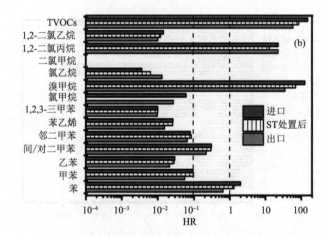

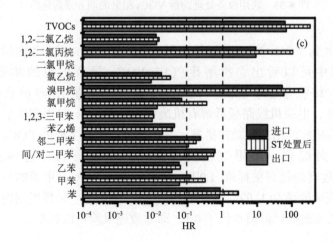

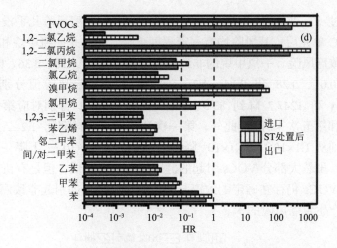

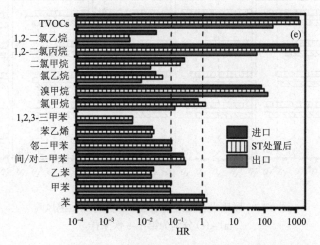

图 8-56 联用设备处理前后 VOCs 组分的非致癌暴露风险
(a)~(e)分别为第 1~5 次采样

上述 13 种具有非致癌风险的 VOCs 中，乙苯、苯、二氯甲烷和 1,2-二氯乙烷 4 种 VOCs 被认定为致癌或可疑致癌物，具有一定的致癌风险。图 8-57 给出了处理前后四种物质及 TVOCs 的致癌风险指数。拆解废气中 VOCs 导致的总的致癌风险为 1.6×10^{-4}~2.6×10^{-3}，其平均值为 1.3×10^{-3}，比确定致癌风险值（1.0×10^{-4}）高出一个数量级。由此可见，电视机线路板拆解产生的 VOCs 可引起明显的致癌风险，必须采取相应的治理和减排措施。处理前的有机废气中，除了二氯甲烷，其他三种 VOCs 均具有可疑的致癌风险（$>10^{-6}$）。其中苯和 1,2-二氯乙烷的 LCR 值分别为：2.0×10^{-4}~4.7×10^{-4} 和 8.1×10^{-4}~2.3×10^{-3}，超过了确定致癌风险值（1.0×10^{-4}）（除了第五次采样期间）[105]。

尽管经过 ST 处理后，VOCs 所导致的致癌和非致癌风险几乎没有降低，但是经过 EP-PC 单元后，其风险均有一定程度下降，尤其是一氯甲烷和 1,2-二氯丙烷引起的非致癌风险。一氯甲烷的非致癌风险值分别从 0.28、0.36、0.85 和 1.31 降到 0.05、0.01、0.28 和 0.14；1,2-二氯丙烷的非致癌风险值分别从 352.2、110.8、1052.4 和 1247.7 降到 83.5、9.3、243.2 和 56.1（分别对应第一次、第三次、第四次和第五次采样）。此外，第一次和第四次采样期间 1,2-二氯乙烷的致癌风险值分别从 1.1×10^{-3} 和 3.1×10^{-4} 降至 5.9×10^{-4} 和 2.8×10^{-5}，下降一个数量级。经过处理后，虽然大部分 VOCs 引起的健康风险有所降低，但是有机废气中的某些高浓度的 VOCs 仍存在确定的致癌与非致癌风险，比如引起非致癌风险的 1,2-二氯丙烷和产生致癌风险的苯。

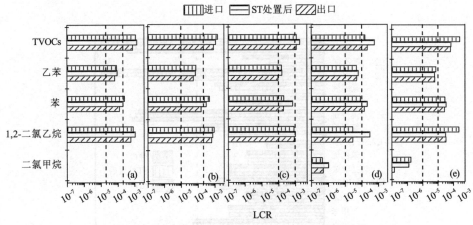

图 8-57 联用设备处理前后 VOCs 组分的致癌风险
(a)~(e) 分别为第 1~5 次采样

综上，经联用设备净化后，有机废气中 VOCs 所导致的职业暴露风险、致癌和非致癌风险均显著降低。

8.3 持久性毒害有机物的控制与健康风险消减

电路板作为电子产品的核心部件，含有大量的金属和非金属材料，具有很高的回收再利用价值，但是在废电路板的回收处理，尤其是加热熔融焊锡、拆解去除元器件的过程中，将会产生大量的大气颗粒物和有机污染物等毒害气体物质，不仅对区域环境造成严重的污染，危害当地群体健康，并且废气中的持久性毒害污染物（PTS）也会随着大气长距离的迁移扩散造成更大范围的空气污染，而吸入高浓度的 PTS 会对车间内的工人造成重大的健康风险。因此，电子垃圾拆解活动中 PTS 的控制和减排以及对其健康风险消减的研究也受到广泛关注。本节首先分析比较了废电路板加热、人工或机械拆解去除元器件过程中产生烟气中 PBDEs 等 PTS 的排放特征及控制，然后采用 8.2.1 章节中"2. 电子垃圾拆解过程 VOCs 减排控制应用"案例应用二所采用的"袋式除尘器-光催化反应器-生物滴滤塔"（BF-PC-BTF 组合技术，图 8-30）一体化气体净化反应器为例，详细介绍了 BF-PC-BTF 组合技术来净化 PTS，并对其控制前后的健康风险消减能力进行评估。

8.3.1 持久性毒害有机物的排放

1. 废电路板加热-人工拆解排放 PBDEs 特征及控制

Guo 等[106]对废电路板加热-人工拆解处理工艺所排放烟气及 PBDEs 的污染特征进行了现场监测研究，具体的处理工艺、气体及灰尘样品的采样信息如图 8-58 所示。研究得出，由于工艺采用负压操作台，大部分加热废电路板所产生的烟气由集气罩收集进入尾气处理系统，前期研究表明，负压操作台可以收集去除 95.4%的烟气颗粒物[8]，但仍然存在逸出集气罩的烟气污染车间内的空气。研究表明车间内颗粒物的质量浓度和数量浓度均明显高于车间外，其中车间内 TSP 和 $PM_{2.5}$ 的Σ_8PBDEs 浓度分别为 20.3 ng/m^3 和 16.1 ng/m^3，说明车间内存在气体 PBDEs 污染现象[106]。废电路板加热释放的烟气中 PBDEs 可以吸附或沉降于灰尘样品中，4 种灰尘样品中Σ_8PBDEs 的浓度水平由高到低依次为：溶液浮渣（68000 ng/g）>集气罩内部（20200 ng/g）>集气罩外部（10 300 ng/g）>车间地面灰尘（2680 ng/g）。碱液吸收所产生的溶液浮渣成为吸附 PBDEs 的重要载体和汇，其 PBDEs 浓度最高，同系物 BDE-28、BDE-47、BDE-99 和 BDE-209 的浓度分别为 16300 ng/g（24.0%）、17700 ng/g（26.0%）、15400 ng/g（22.6%）和 12100 ng/g（17.8%），说明一定量的 PBDEs 被碱液吸收或吸附去除。另外，后续的活性炭吸附塔对烟气中 PBDEs 也具有一定的吸附去除效果。通过测试分析 3 个不同采样点位置处 PUF 膜中 PBDEs 的浓度水平，发现集气罩内部、集气管道内和排放口处 PUF 中 Σ_6PBDEs 的平均浓度分别为 7440 pg/m^3、1310 pg/m^3 和 416 pg/m^3，且具有类似的 BDE 同系物组成，这说明尾气处理装置对气体中 PBDEs 的排放浓度有一定的消减作用，其 PBDEs 的控制机制主要是物理去除。另外，根据采样时间所处理的废电路板质量、烟气气流流速、采样时间及排放口处 PUF 膜中 PBDEs 的浓度水平等参数，可以估算出每加热处理 1 千克废电路板产生烟气中 PBDEs 的排放系数为 47.3 $ng\Sigma_6$PBDEs。综合比较处理车间、烟气管道及排放口处气体中 PBDEs 的浓度水平和同系物组成，以及碱液浮渣中含有高浓度的 PBDEs 等现场实验数据，可以得出，"碱液吸收+活性炭吸附"的气体控制技术可以一定程度截留去除烟气中 PBDEs 等 PTS，从而降低废电路板加热产生烟气污染物对环境的排放量。但是，碱液吸收对 PTS 的截留去除机理，以及活性炭对 PTS 的吸附效能及活性炭的饱和再生等问题还需后续研究。

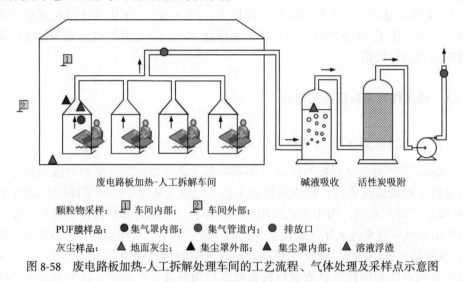

图 8-58　废电路板加热-人工拆解处理车间的工艺流程、气体处理及采样点示意图

颗粒物采样：⬜ 车间内部；⬜ 车间外部；
PUF膜样品：● 集气罩内部；● 集气管道内；● 排放口
灰尘样品：▲ 地面灰尘；▲ 集尘罩外部；▲ 集尘罩内部；▲ 溶液浮渣

2. 废电路板加热-机械拆解过程排放 PBDEs 特征及控制

Tan 等[107]利用自制的废电路板加热-机械拆解装置对电路板上元器件进行加热拆解实验，借助安德森八级撞击式采样器采集不同温度条件下（100 ℃、150 ℃、200 ℃、260 ℃）排放的烟气样品，分析气态和不同粒径颗粒物（10~9.0 μm、9.0~5.8 μm、5.8~4.7 μm、4.7~3.3 μm、3.3~2.1 μm、2.1~1.1 μm、1.1~0.7 μm、0.7~0.4 μm 和 0.4~0 μm）中 PBDEs 的化学组成、浓度水平及排放因子等特征。随着加热温度的升高，加热废电路板所排放烟气中气态和颗粒态 PBDEs 的质量浓度也随之增加（图 8-59）。在 260 ℃的加热拆解条件下，排放颗粒物 PM_{10} 的质量浓度为 2240.4 μg/m³，其中 $PM_{2.5}$ 的浓度达到 1418.3 μg/m³，占比颗粒物总质量的 63.2%，且颗粒物质量主要分布在 0.4~0.7 μm 与 5.8~9.0 μm 两个粒径段。加热废电路板排放烟气中 PBDEs 以气态和颗粒态两种形式存在，当拆解温度由 100 ℃升到 260 ℃时，烟气中总 PBDEs（气态+颗粒态）的排放因子从 6.14 ng/g 增加到 674.08 ng/g，其中气态 $\sum_{39}$PBDEs 的排放因子范围为 1.72~25.00 ng/g，而颗粒结合态 $\sum_{39}$PBDEs 的排放因子范围为 4.44~649.08 ng/g。

在相同温度条件下，颗粒态 $\sum_{39}$PBDEs 的排放因子都比气态 $\sum_{39}$PBDEs 的排放因子大得多。当加热温度为 260 ℃时，颗粒态 PBDEs 的最大排放因子值为

649.08 ng/g，约是气态 PBDEs 最大排放因子值(25.00 ng/g)的 25 倍。在不同温度下，颗粒态Σ_{39}PBDES 呈现不同的粒径分布规律。随着加热温度的升高，细颗粒中Σ_{39}PBDEs 的丰度增加，而粗颗粒中的丰度降低。当在 100 ℃处理时，Σ_{39}PBDEs 的质量分布集中在 5.8~9 μm 和>9 μm 的两个粒径段中，而在 200℃或 260℃时，颗粒态的Σ_{39}PBDEs 主要在 0.7~1.1 μm 的粒径范围并呈现单峰分布。

当温度从 100 ℃提高到 260 ℃时，Σ_{39}PBDEs 的质量中位数空气动力学直径（MMAD）从 2.51 μm 减小到 0.66 μm，即说明随着加热温度的升高，颗粒态 PBDEs 容易向小粒径颗粒物聚集。在各个温度条件下，六溴和七溴等高溴代二苯醚的 MMAD 值均大于 2.5 μm，而一溴至五溴等低溴代二苯醚的 MMAD 值小于 2.5 μm。由于 PBDEs 的半挥发特性，其同系物沸点在 50~300 ℃之间，与颗粒物中有机碳（OC1 和 OC2）的温度范围比较吻合，从而解释了不同加热温度产生烟气中 PBDEs 浓度水平与有机碳（OC1 和 OC2）具有较好的相关性。

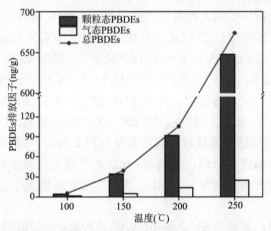

图 8-59 废电路板加热-机械拆解装置产生烟气中 PBDEs 的排放因子(ng/g)

Guo 等[32]对废电路板加热-机械拆解车间的大气和灰尘等环境介质中的 PBDEs 污染特征进行研究发现，车间 PM_{10} 和 TSP 中Σ_8PBDEs 的浓度分别为 479 ng/m³ 和 1670 ng/m³，其中 TSP 以 BDE-47、BDE-99 和 BDE-209 为主要同系物，其浓度分别为 746 ng/m³、487 ng/m³ 和 70.6 ng/m³。同时，采集车间 4 种类型灰尘样品，得出灰尘样品中Σ_8PBDEs 的浓度水平依次为：旋风除尘器集灰（317000 ng/g）>加热拆解设备内部灰尘（84700 ng/g）>车间地面灰尘（31100 ng/g）>电路板表面积灰（15600 ng/g）。结果表明，废电路板加热过程排放烟气被旋风除尘器收集分离，得到的旋风除尘器集灰成为 PBDEs 的主要汇，其中同系物 BDE-47、BDE-99 和 BDE-209 的质量占比分别为 45.7%、32.8%和 5.5%。也就是说，利用旋风除尘器去除截留烟气颗粒物的同时，颗粒物内部或表面可以吸附或吸收一定量的 PBDEs 等 PTS 污染物，再通过对旋风除尘器集灰的安全处置可以减少废电路板加热产生 PTS 等污染物对环境的排放。

8.3.2 持久性毒害有机物的控制

前文阐述了废电路板加热-拆解过程中 PBDEs 等 PTS 的排放特征，实验监测也表明负压集气罩可以收集去除大部分的烟气组分，且后续的碱液吸收和活性炭吸附可以一定程度减少 PTS 的排放浓度水平。但是，综合考虑电子垃圾拆解活动排放 PTS 的污染控制和风险减排等因素，还需对排放烟气的净化去除进行深入研究。本节以 8.2.1 章节中"2. 电子垃圾拆解过程 VOCs 减排控制应用"中的案例应用二所采用的"袋式除尘器-光催化反应器-生物滴滤塔"（BF-PC-BTF 组合技术，图 8-30）一体化气体净化反应器为例，对气体净化组合技术处理前后 PTS 的健康风险消减能力进行评估。

样品收集与分析：使用大容量空气采样器，在 0.15~0.25 m^3/min 的流量下，废气先后通过聚氨酯泡沫（PUF）(密度为 0.03 g/cm^3，直径 6.5 cm × 厚度 10.5 cm)和 GFF (20.3 cm × 25.4 cm)吸入 108~180 m^3 的空气，持续 12 h。PUF 事先使用二氯甲烷索氏提取 72 h，GFF 在 450 ℃下烘烤 12 h。采样后，将 GFF 包裹在预焙铝箔中，密封在双层聚乙烯袋中，并将 PUF 储存在带铝箔内衬盖的溶剂清洗玻璃罐中，所有样品在-20 ℃下储存。用 200 mL 二氯甲烷对 PUF 和 GFF 进行索氏提取 72 h。通过旋转蒸发器将提取物蒸发至约 2 mL，浓缩的萃取物在装有硅胶和氧化铝的柱上清洗和分馏。使用 70 mL 正己烷和二氯甲烷(7∶3)混合溶剂洗脱柱子，收集的洗涤液蒸发至 1 mL，然后氮吹浓缩至 200 μL，采用气相 GC-MS 测定有机物。

PTS 净化效率：如图 8-60(a)和(b)所示，ΣPAHs 和ΣPBDEs 的平均去除率分别为 87.2%和 94.1%，这证实了选用的组合技术能够高效地去除 PTS。其中 PC 和 BF 分别对 PAHs 和 PBDEs 的去除贡献最大，分别占到 63.2%和 80.5%，造成这种差异的原因是 PAHs 和 PBDEs 不同的气相-颗粒相分布规律。

大多数 PAHs 以气相为主，PC 技术可有效去除这些气相中的 PAHs，平均去除效率为 69.9%(图 8-61)，而 BF 能高效去除颗粒相中的少量 PAHs，平均去除效率为 93.3%(图 8-62)。相反，大约 90.2%的 PBDEs 分布在颗粒相上，BF 技术很好地捕获了这些颗粒(图 8-63)。与气态 PAHs 相似，PC 对气态 PBDEs 的去除率相对比较低，只有 48.9%。残留的 PAHs 和 PBDEs 可通过后续的 BTF 技术去除。

对于不同的 PAHs，组合技术对 Nap 的平均去除效率最高(95.7%)，其次是 Phe(85.9%)、Flua(81.3%)和 Pyr(77.7%)。这表明环数越少的 PAHs 越容易被 PC

去除[108]。此外，与 BF(10.4%~22.9%)和 BTF(1.8%~6.6%)相比，PC 在消减这四种 PAHs 的贡献最大(51.8%~82.7%)(图 8-63)。

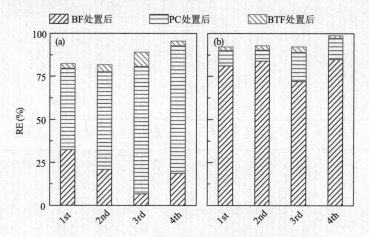

图 8-60　组合技术对 ΣPAHs(a)和 ΣPBDE(b)的去除效率

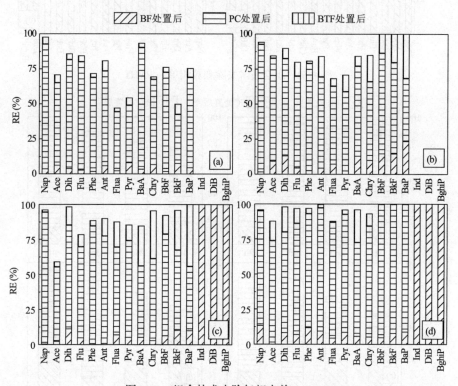

图 8-61　组合技术去除气相中单一 PAHs

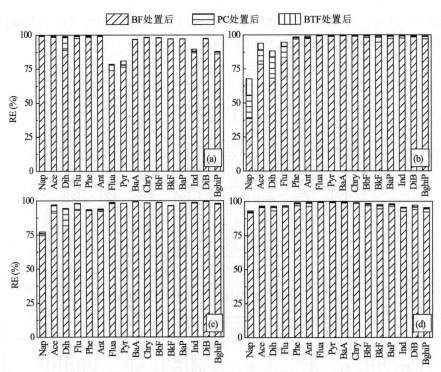

图 8-62 组合技术去除颗粒相单一 PAHs

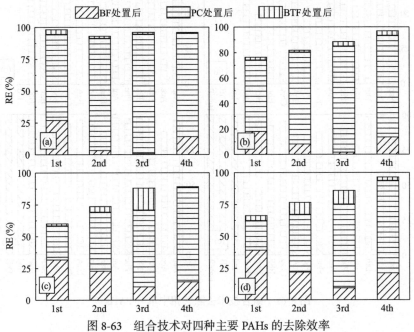

图 8-63 组合技术对四种主要 PAHs 的去除效率
(a) Nap;(b) Phe;(c) Flua;(d) Pyr

而组合技术对于三种主要的 PBDEs 同系物(BDE-47、BDE-99 和 BDE-209)的去除效率分别为 94.9%、95.9%和 88.6%(图 8-64)。BF 的贡献最大(平均去除效率 78.5%),其次是 PC(平均去除效率 11.8%),BTF 对这三种 PBDEs 的去除性能最差(平均去除效率 2.8%)。

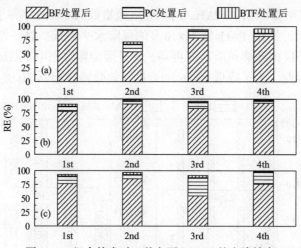

图 8-64 组合技术对三种主要 PBDEs 的去除效率
(a) BDE-209; (b) BDE-99; (c) BDE-47

综上所述,典型的 PTS(如 PAHs 和 PBDEs)一旦释放到大气中,以气态和颗粒两种形式分布,通过 BF-PC-BTF 组合技术可以有效地消除 PTS 污染。首先,BF 有效地拦截颗粒物,从而降低颗粒物上 PS 的浓度。例如 BF 对 TSP 的平均去除率为 86.0%(从 2.20 mg/m³ 降低到 0.31 mg/m³),有效地去除了颗粒态 PAHs 和 PBDEs,保护了 PC 和 BTF 反应器免受颗粒物的负面影响;而无颗粒污染的废气随后进入 PC 反应器,PAHs 和 PBDEs 被有效分解,甚至矿化;最后,BTF 中的微生物也能有效地降解有机物。采用 BF-PC-BTF 组合技术可以有效地消除电子垃圾拆解过程排放的 PAHs 和 PBDEs 等典型 PTS。

8.3.3 持久性毒害有机物的健康风险消减

1. PAHs 和 PBDEs 风险评估方法

根据 8.1.2 中给出的方法评估 PAHs 的致癌风险,职业暴露时间为 25 年。通过式(8.15)和式(8.16)计算 PBDEs 的人体每日平均摄入量和非致癌风险。

$$\sum \text{Exposure} = \frac{\text{CTR}_R}{24\text{BW}} \tag{8.15}$$

$$\text{Hazard Index} = \frac{\sum \text{Exposure}}{\text{RfD}} \qquad (8.16)$$

2. 健康风险消减评估

图 8-65 表明颗粒相中 PAHs 的致癌风险指数(1.47×10^{-3})高于显著风险水平(10^{-3})[109];然而,气相 PAHs(1.74×10^{-4})的风险水平很低。这一结果说明含量更低的颗粒相 PAHs 的致癌风险反而更高,这是因为颗粒相中 PAHs 具有更高的毒性当量因子。因此,为了降低 PAHs 的毒性,消除颗粒相 PAHs 是关键。

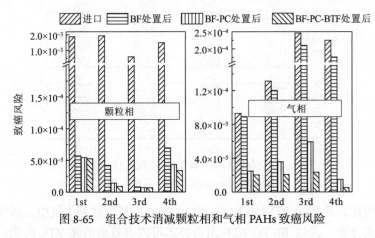

图 8-65 组合技术消减颗粒相和气相 PAHs 致癌风险

经过 BF 处理后,颗粒相和气相中 PAHs 的致癌风险指数分别下降到 4.41×10^{-5} 和 1.49×10^{-4}。相比而言,BF 技术能更有效地降低颗粒相 PAHs 的风险。而经 PC 处理后,气态 PAHs 的致癌风险指数显著降低至 3.36×10^{-5},进一步经 BTF 处理后,颗粒相和气相 PAHs 的风险指数分别降低至 2.54×10^{-5} 和 1.73×10^{-5}。这些风险指数远低于处理前,并且低于显著风险水平(10^{-3}),说明了组合技术能够有效地消减 PAHs 的致癌风险。

图 8-66 和图 8-67 给出了组合技术处理前后 PBDEs 的暴露风险及其消减效率。如图 8-66 所示,工人吸入颗粒相 PBDEs 的平均暴露指数为 639.43 ng/(kg·d),是气相的 9 倍(73.35 ng/(kg·d))。这表明前者对工人会造成更高的暴露风险。经过 BF 处理后,颗粒相 PBDEs 的暴露风险指数迅速降至 55.55 ng/(kg·d),再次证明 BF 高效消减分布在颗粒相上的 PBDEs 的暴露风险。进一步经过 PC 处理后,气相 PBDEs 的暴露风险指数降至 15.11 ng/(kg·d),表明 PC 可以有效降低气态 PBDEs 的暴露风险。总的来说,使用 BF-PC-BTF 组合技术,颗粒相和气相中 PBDEs 的暴露风险消减效率分别达到了 95.0%和 84.6%(图 8-67),说明该技术可以有效地降低 PBDEs 呼吸暴露的风险。

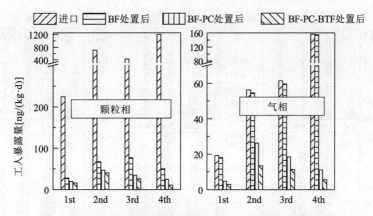

图 8-66 组合技术处理后颗粒相和气相中 PBDEs 的暴露风险

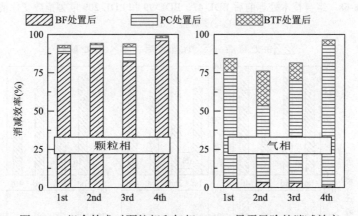

图 8-67 组合技术对颗粒相和气相 PBDEs 暴露风险的消减效率

进一步对三种主要的 PBDEs 同系物 BDE-47、BDE-99 和 BDE-209 的危害指数进行了研究。如图 8-68 所示,颗粒相中 BDE-47、BDE-99 和 BDE-209 的平均危害指数为 1.04×10^{-2}、1.33×10^{-2} 和 1.50×10^{-2},高于气相(2.86×10^{-1}、5.82×10^{-2} 和 4.55×10^{-5})。这表明颗粒相中的 BDE-47、BDE-99 和 BDE-209 对人体构成的威胁更大。与暴露风险相似,经过 BF 处理后颗粒相中 BDE-47、BDE-99 和 BDE-209 的危害指数的平均消减效率为 95.7%、90.6%和 76.6%,占总危害指数消减的 97.5%、94.0%和 86.5%(图 8-69)。然而 BF 在消减气相 BDE-47、BDE-99 和 BDE-209 风险方面表现不佳,仅占总危险指数的 2.2%、7.7%和 23.2%。而经过后续 PC-BTF 处理后,三种 PBDEs 的危害指数消减效率提高到 85.2%以上。综上所述,采用 BF-PC-BTF 组合技术在有效地消除电子垃圾拆解过程排放的 PAHs 和 PBDs 等典型 PTS 的浓度的同时也实现其健康风险的高效消减。

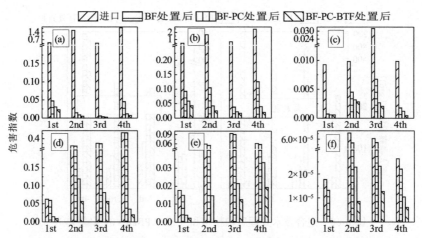

图 8-68 组合技术处理前后 BDE-47、BDE-99 和 BDE-209 危害指数变化情况
(a~c)颗粒相;(d~f)气相

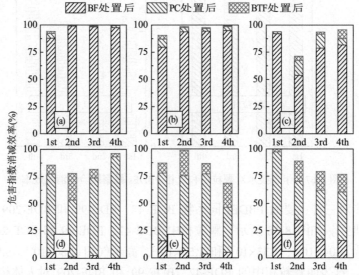

图 8-69 组合技术对 BDE-47、BDE-99 和 BDE-209 危害指数的消减效率
(a~c)颗粒相;(d~f)气相

通过上述实际案例可知,采用各种不同的处理技术组合在电子垃圾拆解排放大气污染物的控制和健康风险消减方面都有显著的作用,通过负压收集、前处理、光催化、生物、臭氧氧化等多种处理技术联用组合的方法,对电子垃圾拆解过程中排放的废气进行处理,使得排放的废气中颗粒物和有机物的浓度均有明显降低,其致癌和非致癌风险以及职业暴露风险也有明显降低,可见多种技术联用的方法在电子垃圾拆解过程的大气污染控制和健康风险消减方面有较好的研究价值和应用前景,为电子垃圾拆解过程有机废气净化提供了重要的理论和技术支

持，同时也为制定合理可行的控制手段和治理措施提供了可靠的原始数据与基础理论。

（陈江耀　郭　杰　安太成）

参 考 文 献

[1] Luo P, Bao L J, Li S M, et al. Size-dependent distribution and inhalation cancer risk of particle-bound polycyclic aromatic hydrocarbons at a typical e-waste recycling and an urban site [J]. Environmental Pollution, 2015, 200: 10-15.

[2] Degrendele C, Okonski K, Melymuk L, et al. Size specific distribution of the atmospheric particulate PCDD/Fs, dl-PCBs and PAHs on a seasonal scale: Implications for cancer risks from inhalation [J]. Atmospheric Environment, 2014, 98: 410-416.

[3] Bi X H, Sheng G Y, Peng P, et al. Size distribution of n-alkanes and polycyclic aromatic hydrocarbons (PAHs) in urban and rural atmospheres of Guangzhou, China [J]. Atmospheric Environment, 2005, 39(3): 477-487.

[4] Chrysikou L P, Gemenetzis P G, Samara C A. Wintertime size distribution of polycyclic aromatic hydrocarbons (PAHs), polychlorinated biphenyls (PCBs) and organochlorine pesticides (OCPs) in the urban environment: Street- vs rooftop-level measurements [J]. Atmospheric Environment, 2009, 43(2): 290-300.

[5] Kawanaka Y, Matsumoto E, Sakamoto K, et al. Estimation of the contribution of ultrafine particles to lung deposition of particle-bound mutagens in the atmosphere [J]. Science of the Total Environment, 2011, 409(6): 1033-1038.

[6] Yu H, Yu J Z. Polycyclic aromatic hydrocarbons in urban atmosphere of Guangzhou, China: Size distribution characteristics and size-resolved gas-particle partitioning [J]. Atmospheric Environment, 2012, 54: 194-200.

[7] Zhang K, Zhang B-Z, Li S-M, et al. Calculated respiratory exposure to indoor size-fractioned polycyclic aromatic hydrocarbons in an urban environment [J]. Science of the Total Environment, 2012, 431: 245-251.

[8] Guo J, Ji A, Wang J, et al. Emission characteristics and exposure assessment of particulate matter and polybrominated diphenyl ethers (PBDEs) from waste printed circuit boards de-soldering [J]. Science of the Total Environment, 2019, 662: 530-536.

[9] Almuhanna E A, Maghirang R G, Murphy J P, et al. Laboratory scale electrostatically assisted wet scrubber for controlling dust in livestock buildings [J]. Applied Engineering in Agriculture, 2009, 25(5): 745-750.

[10] Mohan B R, Jain R K, Meikap B C. Comprehensive analysis for prediction of dust removal efficiency using twin-fluid atomization in a spray scrubber [J]. Separation and Purification Technology, 2008, 63(2): 269-277.

[11] An T C, Sun L, Li G Y, et al. Photocatalytic degradation and detoxification of o-chloroaniline in the gas phase: Mechanistic consideration and mutagenicity assessment of its decomposed gaseous intermediate mixture [J]. Applied Catalysis B-Environmental, 2011, 102(1-2): 140-146.

[12] Kocik M, Dekowski J, Mizeraczyk J. Particle precipitation efficiency in an electrostatic precipitator [J]. Journal of Electrostatics, 2005, 63(6-10): 761-766.

[13] Tokarek S, Bernis A. An exemple of particle concentration reduction in Parisian subway stations by electrostatic precipitation [J]. Environmental Technology, 2006, 27(11): 1279-1287.

[14] 黄勇, 陈江耀, 李建军, 等. 生物滴滤塔耦合光催化氧化技术处理电子垃圾拆解车间排放废气的中试研究 [J]. 生态环境学报, 2014, 23(5): 817-823.

[15] He Z G, Li J J, Chen J Y, et al. Treatment of organic waste gas in a paint plant by combined technique of

[16] Deng W J, Louie P K K, Liu W K, et al. Atmospheric levels and cytotoxicity of PAHs and heavy metals in TSP and PM2. 5 at an electronic waste recycling site in southeast China [J]. Atmospheric Environment, 2006, 40(36): 6945-6955.

[17] Li G Y, Sun H W, Zhang Z Y, et al. Distribution profile, health risk and elimination of model atmospheric SVOCs associated with a typical municipal garbage compressing station in Guangzhou, south China [J]. Atmospheric Environment, 2013, 76: 173-180.

[18] Koo J, Hong J, Lee H, et al. Effects of the particle residence time and the spray droplet size on the particle removal efficiencies in a wet scrubber [J]. Heat and Mass Transfer, 2010, 46(6): 649-656.

[19] Chen J Y, Huang Y, Li G Y, et al. VOCs elimination and health risk reduction in e-waste dismantling workshop using integrated techniques of electrostatic precipitation with advanced oxidation technologies [J]. Journal of Hazardous Materials, 2016, 302: 395-403.

[20] Chen S-C, Liao C-M. Health risk assessment on human exposed to environmental polycyclic aromatic hydrocarbons pollution sources [J]. Science of the Total Environment, 2006, 366(1): 112-123.

[21] Kang Y, Man Y B, Cheung K C, et al. Risk Assessment of Human Exposure to Bioaccessible Phthalate Esters via Indoor Dust around the Pearl River Delta [J]. Environmental Science & Technology, 2012, 46(15): 8422-8430.

[22] Wu C-C, Bao L-J, Guo Y, et al. Barbecue Fumes: An Overlooked Source of Health Hazards in Outdoor Settings? [J]. Environmental Science & Technology, 2015, 49(17): 10607-10615.

[23] Lei B, Zhang K, An J, et al. Human health risk assessment of multiple contaminants due to consumption of animal-based foods available in the markets of Shanghai, China [J]. Environmental Science and Pollution Research, 2015, 22(6): 4434-4446.

[24] Wang Y, Hu J, Lin W, et al. Health risk assessment of migrant workers' exposure to polychlorinated biphenyls in air and dust in an e-waste recycling area in China: Indication for a new wealth gap in environmental rights [J]. Environment International, 2016, 87: 33-41.

[25] Chen J-W, Wang S-L, Hsieh D P H, et al. Carcinogenic potencies of polycyclic aromatic hydrocarbons for back-door neighbors of restaurants with cooking emissions [J]. Science of the Total Environment, 2012, 417: 68-75.

[26] Luo P, Ni H-G, Bao L-J, et al. Size distribution of airborne particle-bound polybrominated diphenyl ethers and its implications for dry and wet deposition [J]. Environmental Science & Technology, 2014, 48(23): 13793-13799.

[27] Weschler C J, Beko G, Koch H M, et al. Transdermal uptake of diethyl phthalate and di(n-butyl) phthalate directly from air: Experimental verification [J]. Environmental Health Perspectives, 2015, 123(10): 928-934.

[28] Wu C-C, Bao L-J, Tao S, et al. Dermal uptake from airborne organics as an important route of human exposure to e-waste combustion fumes [J]. Environmental Science & Technology, 2016, 50(13): 6599-6605.

[29] An T C, Huang Y, Li G Y, et al. Pollution profiles and health risk assessment of VOCs emitted during e-waste dismantling processes associated with different dismantling methods [J]. Environment International, 2014, 73: 186-194.

[30] An T C, Zhang D L, Li G Y, et al. On-site and off-site atmospheric PBDEs in an electronic dismantling workshop in south China: Gas-particle partitioning and human exposure assessment [J]. Environmental Pollution, 2011, 159(12): 3529-3535.

[31] Guo J, Lin K F, Deng J J, et al. Polybrominated diphenyl ethers in indoor air during waste TV recycling process [J]. Journal of Hazardous Materials, 2015, 283: 439-446.

[32] Guo J, Zhang R, Xu Z M. PBDEs emission from waste printed wiring boards during thermal process [J]. Environmental Science & Technology, 2015, 49(5): 2716-2723.

[33] 郝吉明, 马广大. 大气污染控制工程(第二版) [M]. 北京: 高等教育出版社, 2002.

[34] Parmar G R, Rao N N. Emerging control technologies for volatile organic compounds [J]. Critical Reviews in Environmental Science and Technology, 2009, 39(1): 41-78.

[35] 陆豪, 吴祖良, 高翔. 吸附法净化挥发性有机物的研究进展 [J]. 环境工程, 2013, 31(3): 93-97.

[36] 李婕, 关宁. 挥发性有机物(VOC_S)活性炭吸附回收技术综述 [J]. 四川环境, 2007, (6): 101-105, 111.

[37] 王祥云, 谭念华, 储政. 混合气体中挥发性有机物的回收技术 [C]. 中国化工学会 2003 年石油化工学术年会论文集, 2003.

[38] 牛茜, 李兵, 徐校良, 等. 催化燃烧法处理挥发性有机化合物研究进展 [J]. 现代化工, 2013, 33(11): 19-23.

[39] Haque F, De Visscher A, Sen A. Biofiltration for BTEX removal [J]. Critical Reviews in Environmental Science and Technology, 2012, 42(24): 2648-2692.

[40] 杜波, 陈晓燕. 光化学催化氧化技术研究进展 [J]. 内蒙古环境科学, 2007, (2): 52-54.

[41] Mo J, Zhang Y, Xu Q, et al. Photocatalytic purification of volatile organic compounds in indoor air: A literature review [J]. Atmospheric Environment, 2009, 43(14): 2229-2246.

[42] Mo J H, Zhang Y P, Xu Q J. Effect of water vapor on the by-products and decomposition rate of ppb-level toluene by photocatalytic oxidation [J]. Applied Catalysis B-Environmental, 2013, 132: 212-218.

[43] Weon S, Choi W. TiO_2 nanotubes with open channels as deactivation-resistant photocatalyst for the degradation of volatile organic compounds [J]. Environmental Science & Technology, 2016, 50(5): 2556-2563.

[44] Al-Sabahi J, Bora T, Al-Abri M, et al. Efficient visible light photocatalysis of benzene, toluene, ethylbenzene and xylene (BTEX) in aqueous solutions using supported zinc oxide nanorods [J]. PLoS One, 2017, 12(12): e0189276.

[45] Dhada I, Nagar P K, Sharma M. Challenges of TiO_2-based photooxidation of volatile organic compounds: designing, coating, and regenerating catalyst [J]. Industrial & Engineering Chemistry Research, 2015, 54(20): 5381-5387.

[46] Tang S, Lu N, Li J, et al. Improved phenol decomposition and simultaneous regeneration of granular activated carbon by the addition of a titanium dioxide catalyst under a dielectric barrier discharge plasma [J]. Carbon, 2013, 53: 380-390.

[47] Hall W J, Williams P T. Fast pyrolysis of halogenated plastics recovered from waste computers [J]. Energy & Fuels, 2006, 20(4): 1536-1549.

[48] Duan H B, Li J H. Thermal degradation behavior of waste video cards using thermogravimetric analysis and pyrolysis gas chromatography/mass spectrometry techniques [J]. Journal of the Air & Waste Management Association, 2010, 60(5): 540-547.

[49] Quan C, Li A M, Gao N B, et al. Characterization of products recycling from PCB waste pyrolysis [J]. Journal of Analytical and Applied Pyrolysis, 2010, 89(1): 102-106.

[50] Chen J, Zhang D, Li G, et al. The health risk attenuation by simultaneous elimination of atmospheric VOCs and POPs from an e-waste dismantling workshop by an integrated de-dusting with decontamination technique [J]. Chemical Engineering Journal, 2016, 301: 299-305.

[51] Huang A, Wang N, Lei M, et al. Efficient oxidative debromination of decabromodiphenyl ether by TiO_2-mediated photocatalysis in aqueous environment [J]. Environmental Science & Technology, 2013, 47(1): 518-525.

[52] Wen S, Zhao J C, Sheng G Y, et al. Photocatalytic reactions of phenanthrene at TiO_2/water interfaces [J]. Chemosphere, 2002, 46(6): 871-877.

[53] Lei M, Wang N, Zhu L, et al. Peculiar and rapid photocatalytic degradation of tetrabromodiphenyl ethers over Ag/TiO_2 induced by interaction between silver nanoparticles and bromine atoms in the target [J]. Chemosphere, 2016, 150: 536-544.

[54] Nunez C M, Ramsey G H, Ponder W H, et al. Corona destruction: An innovative control technology for

VOCs and air toxics [J]. Air & Waste, 1993, 43(2): 242-247.
[55] 张莉莉. 臭氧-催化协同脱除挥发性有机物的试验研究[D]. 杭州: 浙江大学, 2012.
[56] Zhong L, Haghighat F, Lee C-S, et al. Performance of ultraviolet photocatalytic oxidation for indoor air applications: Systematic experimental evaluation [J]. Journal of Hazardous Materials, 2013, 261: 130-138.
[57] Kim Y M, Kim S, Lee J Y, et al. Pyrolysis reaction pathways of waste epoxy-printed circuit board [J]. Environmental Engineering Science, 2013, 30(11): 706-712.
[58] Grause G, Furusawa M, Okuwaki A, et al. Pyrolysis of tetrabromobisphenol-A containing paper laminated printed circuit boards [J]. Chemosphere, 2008, 71(5): 872-878.
[59] Blazso M, Czegeny Z, Csoma C. Pyrolysis and debromination of flame retarded polymers of electronic scrap studied by analytical pyrolysis [J]. Journal of Analytical and Applied Pyrolysis, 2002, 64(2): 249-261.
[60] Kim H J, Han B, Kim Y J, et al. Submicrometer particle removal indoors by a novel electrostatic precipitator with high clean air delivery rate, low ozone emissions, and carbon fiber ionizer [J]. Indoor Air, 2013, 23(5): 369-378.
[61] Demeestere K, Dewulf J, Van Langenhove H. Heterogeneous photocatalysis as an advanced oxidation process for the abatement of chlorinated, monocyclic aromatic and sulfurous volatile organic compounds in air: State of the art [J]. Critical Reviews In Environmental Science and Technology, 2007, 37(6): 489-538.
[62] Huang H B, Leung D Y C, Li G S, et al. Photocatalytic destruction of air pollutants with vacuum ultraviolet (VUV) irradiation [J]. Catalysis Today, 2011, 175(1): 310-315.
[63] Kim J, Zhang P Y, Li J G, et al. Photocatalytic degradation of gaseous toluene and ozone under UV254+185 (nm) irradiation using a Pd-deposited TiO_2 film [J]. Chemical Engineering Journal, 2014, 252: 337-345.
[64] Guo T, Bai Z, Wu C, et al. Influence of relative humidity on the photocatalytic oxidation (PCO) of toluene by TiO_2 loaded on activated carbon fibers: PCO rate and intermediates accumulation [J]. Applied Catalysis B: Environmental, 2008, 79(2): 171-178.
[65] Jeong J, Sekiguchi K, Sakamoto K. Photochemical and photocatalytic degradation of gaseous toluene using short-wavelength UV irradiation with TiO_2 catalyst: comparison of three UV sources [J]. Chemosphere, 2004, 57(7): 663-671.
[66] Zhao W R, Shi Q M, Liu Y. Performance, deactivation and regeneration of SnO_2/TiO_2 nanotube composite photocatalysts [J]. Acta Physico-Chimica Sinica, 2014, 30(7): 1318-1324.
[67] Paca J, Klapkova E, Halecky M, et al. Interactions of hydrophobic and hydrophilic solvent component degradation in an air-phase biotrickling filter reactor [J]. Environmental Progress, 2006, 25(4): 365-372.
[68] Li G Y, Zhang Z Y, Sun H W, et al. Pollution profiles, health risk of VOCs and biohazards emitted from municipal solid waste transfer station and elimination by an integrated biological-photocatalytic flow system: A pilot-scale investigation [J]. Journal of Hazardous Materials, 2013a, 250: 147-154.
[69] 马兴元, 牛艳芳, 吕凌云, 等. 轻质陶粒滤料生态滤床的挂膜与启动研究[J]. 生态环境学报, 2009, 18(6): 2118-2121.
[70] 陈江耀, 张德林, 李建军, 等. 光催化与生物技术联用工艺处理油漆废气中试研究 [J]. 环境工程学报, 2010, (6): 1389-1393.
[71] Liu R R, Chen J Y, Li G Y, et al. Cutting down on the ozone and SOA formation as well as health risks of VOCs emitted from e-waste dismantlement by integration technique [J]. Journal of Environmental Management, 2019, 249.
[72] Liu R R, Chen J Y, Li G Y, et al. Using an integrated decontamination technique to remove VOCs and attenuate health risks from an e-waste dismantling workshop [J]. Chemical Engineering Journal, 2017, 318: 57-63.
[73] Shao M, Lu S H, Liu Y, et al. Volatile organic compounds measured in summer in Beijing and their role in ground-level ozone formation [J]. Journal of Geophysical Research-Atmospheres, 2009, 114: D00G06.
[74] Liu Y, Yuan B, Li X, et al. Impact of pollution controls in Beijing on atmospheric oxygenated volatile organic

compounds (OVOCs) during the 2008 Olympic Games: observation and modeling implications [J]. Atmospheric Chemistry and Physics, 2015, 15(6): 3045-3062.

[75] Ou J, Zheng J, Li R, et al. Speciated OVOC and VOC emission inventories and their implications for reactivity-based ozone control strategy in the Pearl River Delta region, China [J]. Science of the Total Environment, 2015, 530-531: 393-402.

[76] Guo S, Hu M, Guo Q, et al. Primary sources and secondary formation of organic aerosols in Beijing, China [J]. Environmental Science & Technology, 2012, 46(18): 9846-9853.

[77] Wu W, Zhao B, Wang S, et al. Ozone and secondary organic aerosol formation potential from anthropogenic volatile organic compounds emissions in China [J]. Journal of Environmental Sciences, 2017, 53: 224-237.

[78] Wu R R, Xie S D. Spatial distribution of ozone formation in China derived from emissions of speciated volatile organic compounds [J]. Environmental Science & Technology, 2017, 51(5): 2574-2583.

[79] Zheng J Y, Shao M, Che W W, et al. Speciated VOC emission inventory and spatial patterns of ozone formation potential in the Pearl River Delta, China [J]. Environmental Science & Technology, 2009, 43(22): 8580-8586.

[80] Liu C, Ma Q X, Chu B W, et al. Effect of aluminium dust on secondary organic aerosol formation in m-xylene/NOx photo-oxidation [J]. Science China-Earth Sciences, 2015, 58(2): 245-254.

[81] Wang S Y, Wu D W, Wang X M, et al. Relative contributions of secondary organic aerosol formation from toluene, xylenes, isoprene, and monoterpenes in Hong Kong and Guangzhou in the Pearl River Delta, China: an emission-based box modeling study [J]. Journal of Geophysical Research-Atmospheres, 2013, 118(2): 507-519.

[82] Einaga H, Futamura S, Ibusuki T. Heterogeneous photocatalytic oxidation of benzene, toluene, cyclohexene and cyclohexane in humidified air: comparison of decomposition behavior on photoirradiated TiO_2 catalyst [J]. Applied Catalysis B-Environmental, 2002, 38(3): 215-225.

[83] Lewandowski M, Ollis D F. A Two-Site kinetic model simulating apparent deactivation during photocatalytic oxidation of aromatics on titanium dioxide (TiO_2) [J]. Applied Catalysis B-Environmental, 2003, 43(4): 309-327.

[84] Sun L, Li G Y, Wan S G, et al. Mechanistic study and mutagenicity assessment of intermediates in photocatalytic degradation of gaseous toluene [J]. Chemosphere, 2010, 78(3): 313-318.

[85] Fu P, Zhang P, Li J. Photocatalytic degradation of low concentration formaldehyde and simultaneous elimination of ozone by-product using palladium modified TiO_2 films under UV254+185nm irradiation [J]. Applied Catalysis B: Environmental, 2011, 105(1-2): 220-228.

[86] Jeong J, Sekiguchi K, Lee W, et al. Photodegradation of gaseous volatile organic compounds (VOCs) using TiO_2 photoirradiated by an ozone-producing UV lamp: decomposition characteristics, identification of by-products and water-soluble organic intermediates [J]. Journal of Photochemistry and Photobiology A: Chemistry, 2005, 169(3): 279-287.

[87] Chen J Y, Li G Y, He Z G, et al. Adsorption and degradation of model volatile organic compounds by a combined titania-montmorillonite-silica photocatalyst [J]. Journal of Hazardous Materials, 2011, 190(1-3): 416-423.

[88] Zhang M L, An T C, Fu J M, et al. Photocatalytic degradation of mixed gaseous carbonyl compounds at low level on adsorptive TiO_2/SiO_2 photocatalyst using a fluidized bed reactor [J]. Chemosphere, 2006, 64(3): 423-431.

[89] Bhatkhande D S, Pangarkar V G, Beenackers A A C M. Photocatalytic degradation for environmental applications—A review [J]. Journal of Chemical Technology & Biotechnology, 2002, 77(1): 102-116.

[90] An T C, Zhang M L, Wang X M, et al. Photocatalytic degradation of gaseous trichloroethene using immobilized ZnO/SnO_2 coupled oxide in a flow-through photocatalytic reactor [J]. Journal of Chemical Technology and Biotechnology, 2005, 80(3): 251-258.

[91] Maryott A A, Smith E R. Table of Dielectric Constants of Pure Liquids [M]. 514. U.S. Department of Commerce. National Bureau of Standards Circular, 1951.

[92] 钱华, 戴海夏. 室内空气污染与人体健康关系探讨 [J]. 上海环境科学, 2006, 25(1): 33-38, 42.

[93] Rumchev K, Spickett J, Bulsara M, et al. Association of domestic exposure to volatile organic compounds with asthma in young children [J]. Thorax, 2004, 59(9): 746-751.

[94] Zock J-P, Plana E, Jarvis D, et al. The use of household cleaning sprays and adult asthma: an international longitudinal study [J]. American Journal of Respiratory and Critical Care Medicine, 2007, 176(8): 735-741.

[95] Ware J H, Spengler J D, Neas L M, et al. Respiratory and irritant health-effects of ambient volatile organic-compounds - the Kanawha County health study [J]. American Journal of Epidemiology, 1993, 137(12): 1287-1301.

[96] Mendell M J. Indoor residential chemical emissions as risk factors for-respiratory and allergic effects in children: a review [J]. Indoor Air, 2007, 17(4): 259-277.

[97] California Environmental Protection Agency. Formaldehyde Reference Exposure Levels [R]. 2007.

[98] 周裕敏, 郝郑平, 王海林. 北京城乡结合地空气中挥发性有机物健康风险评价 [J]. 环境科学, 2011b, (12): 3566-3570.

[99] Ramirez N, Cuadras A, Rovira E, et al. Chronic risk assessment of exposure to volatile organic compounds in the atmosphere near the largest Mediterranean industrial site [J]. Environment International, 2012, 39(1): 200-209.

[100] McCarthy M C, O'Brien T E, Charrier J G, et al. Characterization of the chronic risk and hazard of hazardous air pollutants in the United States using ambient monitoring data [J]. Environmental Health Perspectives, 2009, 117(5): 790-796.

[101] Sexton K, Linder S H, Marko D, et al. Comparative assessment of air pollution-related health risks in Houston [J]. Environmental Health Perspectives, 2007, 115(10): 1388-1393.

[102] He Z G, Li G Y, Chen J Y, et al. Pollution characteristics and health risk assessment of volatile organic compounds emitted from different plastic solid waste recycling workshops [J]. Environment International, 2015, 77: 85-94.

[103] Pei C C, Leung W W F. Photocatalytic oxidation of nitrogen monoxide and o-xylene by $TiO_2/ZnO/Bi_2O_3$ nanofibers: Optimization, kinetic modeling and mechanisms [J]. Applied Catalysis B-Environmental, 2015, 174: 515-525.

[104] Dhada I, Sharma M, Nagar P K. Quantification and human health risk assessment of by-products of photo catalytic oxidation of ethylbenzene, xylene and toluene in indoor air of analytical laboratories [J]. Journal of Hazardous Materials, 2016, 316: 1-10.

[105] Li G, Zhang Z, Sun H, et al. Pollution profiles, health risk of VOCs and biohazards emitted from municipal solid waste transfer station and elimination by an integrated biological-photocatalytic flow system: a pilot-scale investigation [J]. Journal of Hazardous Materials, 2013, 250-251: 147-154.

[106] Guo J, Patton L, Wang J, et al. Fate and migration of polybrominated diphenyl ethers in a workshop for waste printed circuit board de-soldering [J]. Environmental Science and Pollution Research, 2020, 27(24): 30342-30351.

[107] Tan S, Chen Z, Wang R, et al. Emission characteristics of polybrominated diphenyl ethers from the thermal disassembly of waste printed circuit boards [J]. Atmospheric Environment, 2020, 226: 117402.

[108] Gonzalez D, Ruiz L M, Garralon G, et al. Wastewater polycyclic aromatic hydrocarbons removal by membrane bioreactor [J]. Desalination and Water Treatment, 2012, 42(1-3): 94-99.

[109] Rodricks J V, Brett S M, Wrenn G C. Significant risk decisions in federal regulatory agencies [J]. Regulatory Toxicology and Pharmacology: RTP, 1987, 7(3): 307-320.

第 9 章 电子垃圾健康风险管控：发展趋势与挑战

随着经济社会和信息技术的不断发展，全球电子电器产品的产量还将持续增加，也必将导致电子垃圾产生量的逐年递增。由于电子垃圾存在跨区域间转移、处理处置过程中释放毒害污染物质所引起的全球环境污染及风险问题，鉴于各国国情、经济发展水平及电子垃圾的管理水平等方面存在较大差异性，导致不同国家的电子垃圾污染现状、法律法规等管理制度、资源化与污染控制技术及污染物健康暴露风险等领域具有不同特征，从而给电子垃圾的全球共治带来了诸多困难和挑战。本章概述了电子垃圾治理相关的全球性公约及多个国家相应的管理政策，为电子垃圾的环境治理提供经验借鉴参考，并提出电子垃圾污染治理及风险管控的发展趋势和所面临的挑战。

9.1 全球的政策

电子垃圾的全球管理涉及的领域包括与电子垃圾相关的国际性公约、产生量的统计、收集管理体系、越境转移信息及各国的相关法律行政规范等。目前，与电子垃圾相关的国际性公约主要有《控制危险废物越境转移及其处置巴塞尔公约》(简称《巴塞尔公约》)、《关于持久性有机污染物的斯德哥尔摩公约》(简称《斯德哥尔摩公约》)、《关于在国际贸易中对某些危险化学品和农药采用事先知情同意程序的鹿特丹公约》(简称《鹿特丹公约》)和《关于汞的水俣公约》(简称《水俣公约》)等，其中《巴塞尔公约》对电子垃圾的越境转移进行相关控制，《斯德哥尔摩公约》主要对电子产品相关的持久性有机污染物进行禁限用，而《水俣公约》与电子垃圾中的汞污染有一定关联。2012 年，《巴塞尔公约》、《斯德哥尔摩公约》以及《鹿特丹公约》三项公约的秘书处环境署部分合并为一个秘书处，为三项公约的联合协同增效提供服务。

电子垃圾相关的管理体系主要包括相关立法、标准及指南导则、生产者责任延伸制(EPR)等管理制度。截至 2019 年 10 月，在国家层面拥有电子垃圾相关的政策、立法或法规的国家有 78 个，已覆盖全球 71%的人口，其中中国和印度两个人口大国已经制定了电子垃圾相关的法律条例，使人口覆盖率较高。EPR 是一项重要的环境管理政策，对解决电子垃圾回收处理产生环境污染问题具有显著优势，也是国际上普遍采纳的环境管理制度。EPR 是将生产者对其产品承担的资源环境责任从生产环节延伸到产品设计、流通消费、回收利用、废物处置等全生命周期的制度。EPR 的核心是通过引导生产者承担产品废弃后的回收和资源化利用责任，激励生产者推行产品源头控制、绿色生产，从而在产品全生命周期中最大限度提升资源利用效率。

电子垃圾的跨境转移是全球长期关注的一个问题，当电子垃圾流入没有环保处理能力或缺乏环保管理的国家时，会对人类健康和生态环境产生不利影响。1989 年通过并于 1992 年生效的《巴塞尔公约》，是各国政府为控制《巴塞尔公约》规定的危险废物越境转移所作全球努力的主要成果之一。这项由 190 个缔约方批准的多边条约专注于防止有害于环境和社会的危险废物贸易模式，包括与电子垃圾有关的贸易。2006 年《内罗毕宣言》（Nairobi Declaration）和 2011 年《卡塔赫纳决定》（Cartagena Decisions）提出了关于电子垃圾相关的《巴塞尔公约》后续增编。2006 年在巴塞尔公约缔约方大会第 8 次会议上，通过了关于为电子垃圾环境无害化管理的创新解决方案《内罗毕宣言》。签署国注意到电子垃圾的跨界转移在全世界迅速扩大，并对人类健康构成风险，特别是在没有能力安全管理这类废物的国家，宣布迫切需要促进公众对风险、技术发展和关于最佳管理实践的信息交流的认识，并加强《巴塞尔公约》的规定执行，制止电子垃圾非法贩运和指导这类废物安全管理的全球文书[1]。但是，《巴塞尔公约》是否是合适的框架来处理包括就业、技术转让、通信、经济发展、环境保护和对人类健康的影响在内的多方面问题？针对这些问题，《巴塞尔公约》秘书处在 2012 年 12 月制定了一份关于电子废物跨境转移的技术准则草案，重点是区分废物和非废物（例如，旨在回收或翻新二手电子产品）。该指南承认，政府当局在评估和区分用于修理、翻新、转售或人道主义援助再利用的二手电子设备与用于资源回收的电子垃圾时将面临困难。新的准则将需要关于销售或转让功能齐全的二手电子设备的发票和合同文件，以及对准备修理或翻新设备进行评估和测试的证据，以及防止运输过程中损坏的适当保护措施的证明，包括装卸过程中的保护性包装。这些准则应使跨越国界转移电子垃圾变得更加困难，成本效益也更低，特别更难转移到劳动力廉价、采用不环保方法回收电子垃圾中少量贵金属的地区。2019 年，《巴塞尔公约》秘书处临时通过了《关于电子和电气废物以及废旧电气和电子设备的越境转移——尤其是关于依照〈巴塞尔公约〉对废物和非废物加以区别的技术准则》，虽然这只是指导方针，并不是强制性文件，但可以一定程度上为二手电子产品和电子垃圾的区别判定提供参考。在 2022 年 6 月《巴塞尔公约》缔约方第十五次会议，通过了《巴塞尔公约》附件二、八和九的修正案，从而将所有电子垃圾纳入《事先知情同意控制程序》，即《电子垃圾修正案》。因此，从 2025 年 1 月 1 日起，所有跨越《巴塞尔公约》缔约方国际边界的电子垃圾都将受到严格的控制程序，各国政府将能够决定是否从其他国家进口电子垃圾。《电子垃圾修正案》将在法律上约束《巴塞尔公约》下的国家严格控制电子垃圾的跨境转移，并确保其环境无害管理。

由于全球数据有限和信息不完善，目前很难完全准确估计电子垃圾的跨境移动。监测这种流动对于各国更好地准备控制危险废物的越境转移和推进对这类废物的环境无害化管理至关重要。《巴塞尔公约》缔约方根据第13条授权采集的数据为分析电子垃圾的跨境流动和数量提供了一些信息，但这些信息不足以进行综合分析，原因如下：①报告不完整。许多缔约方不提交国家报告，或不每年提交报告，只有不到50%的缔约方在2012年提交报告。②定义不明确。各缔约方对定义的解释不同，导致数据不规范，阻碍了数据的汇总和分析。③不正确的分类。《巴塞尔公约》附件一、附件八和附件九分别提供了需要控制的废物类别、需要控制的危险废物清单和非危险废物清单，但是各缔约方对废物的分类是不同的。④报告中的差异。各国报告中的危险废物越境转移量可能不准确，因为通知和转移文件中描述的数额通常不同。⑤数据不准确。同一越境货物可能按出口国和进口国进行描述，可能包含不同数量的危险废物，有时项目的代码和描述不匹配。

许多《巴塞尔公约》缔约国并没有记录和统计电子垃圾越境转移情况，特别是在美洲、亚洲、大洋洲和非洲。由于报告数据缺失、数据质量较低以及缺乏对电子垃圾越境转移的控制，从而对电子垃圾的环保管理构成了威胁，并可能助长该地区的非法越境转移。越境转移既发生在各大洲之间，也发生在大洲内部。北美、欧洲和大洋洲的高收入国家是净出口国，而非洲、亚洲和拉丁美洲等地则是净进口国。数据显示，越来越多来自西欧的电子垃圾流入东欧和东南亚。亚洲仍然是电子垃圾的进出口中心，因为亚洲是电子产品的制造基地，并且有一定量的区域内电子垃圾贸易。另外，区域内处理电子垃圾设施数量的增加也相应增加了区域内的电子垃圾流通量。登记合法的越境转移电子垃圾的单位质量价值最低，因为环境无害化处理和管理需要花费更多成本。所有电子垃圾的越境转移中，只有9%的电子垃圾发生在各大洲之间的合法越境转移。废电路板是电子垃圾中价值较高的一类产品，51%的废电路板越境转移发生在各大洲之间，而38%的二手电子产品和非法电子垃圾越境转移发生在各大洲之间。也就是说，电子垃圾的单位质量价值越高，它可以被运输得越远。

9.2 各大洲的电子垃圾管控

全球各大洲电子垃圾的正规收集、环境友好处理及跨境管理水平等具有不同特点，具体各大洲的电子垃圾产生量、正式收集及跨国流通量详见表1-1。高收入国家或地区，如北美、欧洲和东亚等地，拥有较好的电子垃圾处理设施和较高的收集回收率，并进口一些价值高的电子垃圾（如废电路板等）进行资源回收处理。在欧洲，42%的企业被记录为环保型管理，其次是亚洲（12%），然后是美洲（9.2%）和大洋洲（8.6%）。高收入国家已经具备了一定数量的电子垃圾环保处理设施，这些设施也不能处理产生的所有电子垃圾，其主要处理对象是价值高的电子垃圾，包括进口的废电路板等。废电路板含有高含量的金、银、钯和铜等有价金属，回收价值高。每年产生电子垃圾中含有 120 万吨左右的废电路板，其中 41.5 万吨（34%）的废电路板从电子垃圾中分离出来并得到正规处理，包括约 36 万吨废电路板通过跨境运输到达欧盟、北美、韩国和日本等发达国家中的环保型冶炼厂。相比 17%的电子垃圾总体正规处理率，34%的废电路板得到正规合理的回收处理，也就是说，大约有一半的废电路板来自非正规处理渠道，且还有 66%的废电路板没有记录在环保回收设施中。低收入和中等收入国家或地区，如非洲、拉丁美洲以及南亚和东南亚等地，他们本身就存在电子垃圾处理基础设施短缺和管理立法不完善等问题，而进口电子垃圾又加重了其国内电子垃圾的处理负担。以非洲为例，其是全球范围内有正式记录的收集回收率最低的大洲，只有 1%的工厂被记录为符合环保管理要求。2019 年，除非洲本土产生 290 万吨电子垃圾外，还有 55 万吨进口的二手电子产品或电子垃圾。进口电子垃圾通过集装箱中二手电子产品和电子垃圾混合运输，或被塞在报废或二手车辆的方式被交易到非洲大陆。另外，非洲大陆拆解得到的废电路板等高价值电子垃圾，部分会被出口到具有处理设施或冶炼能力的西欧等地[2]。

针对电子垃圾管理的全球挑战，联合国国际电信联盟和《巴塞尔公约》秘书处于 2012 年签署了一项协议，将开展联合项目和方案，以制定信息和通信技术标准，提高各国对电子垃圾管理的认识，从而减少电子垃圾对环境的负面影响。2018 年 3 月 21 日，在信息社会世界首脑会议论坛上，国际电信联盟和其他六个联合国实体组织签署了一份意向书，为在电子垃圾管理领域加强合作，建立电子垃圾联盟铺平了道路。此外，在 2019 年世界峰会论坛上，三个新的联合国实体组织也签署了意向书。该意向书的目标包括签署方承诺加强合作、建立伙伴关系和支持会员国应对全球电子垃圾挑战，其联盟的三个核心职能为"宣传、知识、联合实施"。该联盟的具体目标包括：①支持各国减少和管理电子垃圾，以创造就业机会，同时保护工人、人类健康和环境；②加强各国制定和实施废电器电子产品综合管理政策和实际措施的能力；③联盟实体采取"统一行动"，为现有方案、伙伴关系和项目创造协同增效和增值，避免资源和工作的重复；④在全球、区域、国家和地方等多层次范围内提高主要电子垃圾利益攸关方的认识和参与度；⑤利用现有的国际专业知识，支持电子产品循环经济的发展；⑥防止电子产品的跨境非法贩运，确保其符合国际要求；⑦促进非国家职能部门（如行业）参与解决电子垃圾问题。

9.2.1 欧美地区

1. 欧洲

为了促进可持续性生产与消费,欧盟早在 2003 年就公布了关于电子垃圾的《废旧电气电子设备指令》(WEEE 指令,2002/96/EC)和《关于在电子电气设备中限制使用某些有害物质的指令》(RoHS 指令,2002/95/EC)两个指令,以此来推动在欧盟成员国从电子垃圾灰色产业到绿色经济转变的进程,并于 2012 年颁布了修订后的 WEEE 指令(2012/19/EU)和 RoHS(2011/65/EU)。WEEE 指令的目标是实现电子垃圾的再利用、再循环和其他形式的回收,以减少废弃物的处理。同时努力改进涉及电子电气设备生命周期的所有操作人员,如生产者、销售者、消费者,特别是直接涉及报废电子电气设备处理人员的环保行为。WEEE 指令对电子产品的设计、电子垃圾的分类、收集、处理、再利用或循环利用、资金支付及信息颁布等内容进行了规定阐述。修订的 WEEE 指令于 2012 年 8 月 13 日生效,指令更新了不同类型电子电器产品的回收率、再利用/循环利用率的分阶段目标值,主要包括以下 3 个方面:①一个雄心勃勃的欧洲目标,将 4 千克/(人·年)改为目标收集率,即到 2016 年后,成员国每年电子垃圾的最低收集率应为前三年投放市场电子电气设备的 65%,或者是在成员国境内产生电子垃圾的 85%;②到 2016 年后,无论电器产品的类别,回收率均提高 5%;③旨在第三方国家重复使用的二手设备的国际越境转移的可追溯性要求,以打击未申报废物管理的隐藏贸易。RoHS 指令主要对 6 类毒害物质在电子电器产品中的最大重量含量进行了限制,具体包括:镉(0.01%)、铅(0.1%)、汞(0.1%)、六价铬(0.1%)、多溴联苯(0.1%)、多溴联苯醚(0.1%)。修订后的 RoHS 指令条款与欧盟《化学品的注册、评估、授权和限值》(REACH)的条款保持了一致,并提出重点关注 REACH 法规列为对健康和环境存在风险的物质。2015 年 3 月,欧盟正式发布了 RoHS2.0 修订指令(EU)2015/863,在原来 6 种管控物质基础上,新增 4 项管控物质,包括邻苯二甲酸二(2-乙基己基)酯(DEHP)、邻苯二甲酸基丁酯(BBP)、邻苯二甲酸二丁基酯(DBP)和邻苯二甲酸二异丁酯(DIBP),其重量含量限值均为 0.1%。

2002 年 4 月,欧洲 6 国(奥地利、比利时、荷兰、挪威、瑞典和瑞士)运营的 WEEE 生产者责任组织成立了电子垃圾论坛(WEEE Forum)。电子垃圾论坛是公开的、非营利的、代表生产者或生产者协会的组织,以提供环保及高效的电子垃圾收集、物流及处理解决方法,并努力形成全球范围内电子垃圾回收处理最佳实践方案。目前,电子垃圾论坛由全球 46 个 WEEE 生产者责任组织组成,其成员代表了整个制造业领域,其中三分之二的成员是市场领导者。电子垃圾论坛组织在电子垃圾的收集、物流和处理技术等方面获得了大量的专业知识，论坛成员们已经累计收集回收、环境无害化处理了 2850 万吨电子垃圾。2018 年，WEEE 论坛首次提出将 10 月 14 日定为"国际电子垃圾日(International E-Waste Day)"，旨在提高消费者对电子垃圾回收需求的认识，该活动也获得了《巴塞尔公约》秘书处的支持。电子垃圾论坛拟定了相关标准和研发了软件平台，允许生产者责任组织对其运营进行基准测试，并获得关键数据信息。针对电子垃圾的收集、储存、运输、回收准备、处理处置等相关事宜，电子垃圾论坛制定了统一的泛欧洲标准(WEEE Laboratory of Excellence，WEEELABEX)，并形成了由欧洲电工标准化委员会(CENELEC)托管的欧洲 EN50625 标准。电子垃圾论坛研发了一个数据库应用程序——废物流报告工具(waste flow reporting tool，WF-RepTool)。WF-RepTool 以透明、可追溯的方式确定电子垃圾处理结果，符合 WEEELABEX 和 CENELEC 50625 等标准要求。WF-RepTool 的主要用户为电子垃圾生产者责任组织和电子垃圾处理运营商，并允许用户使用统一的系统收集电子垃圾材料组成的数据，计算电子垃圾的回收和回收率，建立自己的流程图，监控下游的输出分量直到垃圾状态结束。该数据库平台覆盖电子垃圾处理过程的整个链条，可以用于政府、审计师和专业人士等第三方报告和记录处理绩效和结果。

荷兰作为电子垃圾论坛早期的成员国之一，基于国际公认的测算框架，整合了所有可用的统计数据，包括实地研究、家庭和企业调查、合规方案的内部数据以及来自荷兰国家电子垃圾登记平台(National WEEE Register，NWR)的数据，比较 2018 年和 2010 年荷兰的电子垃圾流通数据后得出：2010 年，WEEELABEX 认证在荷兰还不是强制性的，39%的电子垃圾得到了合规收集，34%的电子垃圾由金属废料部门管理，剩下的 27%被丢弃在垃圾箱中，出口重复使用，或无法确定。自 2015 年 7 月起，WEEELABEX 认证已被法律强制执行，即要求在 NWR 中登记符合 WEEELABEX 标准要求的电子垃圾收集数据。2018 年，荷兰的电子垃圾产生量为 36.6 万吨，其中合规收集并在 NWR 中登记的电子垃圾重量为 18.4 万吨(50%)，27%的记录为不符合规定的回收，余下 23%电子垃圾的去向为垃圾箱或出口重复使用，或无法确定。对比可知，荷兰电子垃圾的合规收集率由 2010 年的 39%增加到 2018 年的 50%。这说明法律要求的 WEEELABEX 认证使荷兰电子垃圾的回收管理更加透明，也一定程度增加了电子垃圾的合规收集率。另外，根据前三年投放市场电子产品数量基准核算，荷兰 2018 年的电子垃圾回收率为 45%。但是，根据修订后 WEEE 指令(2012/19/EU)第 7 条规定(最低收集率应为前三年投放市场电子电气设备的 65%，或者是在成员国境内产生电子垃圾的 85%)，荷兰 2019 年很难达到电子垃圾的最低收集率目标。

据德国联邦环境署公布的数据显示，2005 年德国电子垃圾总量为 4.9 万吨，随后一路攀升，在 2009 年达到了 60.5 万吨，2016 年和 2019 年的电子垃圾产生量则分别为 188.4 万吨和 160.7 万吨。尽管电子垃圾数量庞大，德国却能进行有效的回收和循环利用。德国积极响应欧盟 2003 年出台的 WEEE 指令，2005 年就制定了相应的《电子电气设备法》。为了适应电子垃圾总量不断攀升等新形势，德国在 2015 年对法案进行修订，于同年 10 月正式实施。新的《电子电气设备法》规定，大型电子电器产品销售商卖出等价电子电器产品的同时，将有义务免费回收相应的废旧设备；而周长小于 25 厘米的小型电子电器产品如手机、电吹风机等则必须无条件回收。根据《电子电气设备法》，德国联邦环境署成立了电子垃圾注册基金会。基金会协调废弃物的流向、分配生产商的责任份额并监控其回收情况。生产厂商在电子产品进入德国市场前，必须向基金会提交注册信息，并且需要汇报每月产品在德销售情况及每年的回收情况。生产商有义务免费向公共垃圾中心提供电子垃圾回收容器。目前德国共设立了 1500 多个公共电子垃圾回收中心。

2. 美国

2019 年美国电子垃圾的产生量为 691.8 万吨（全球第二），人均 21.0 千克，其 2017 年记录在案的收集电子垃圾量为 102 万吨。与欧盟不同，美国没有全国统一的电子垃圾相关法律，但是美国有 25 个州已经颁布了州的电子垃圾政策，另外一半的州还没有相关立法，这就导致美国碎片化的拼凑管理。在 25 个州的电子垃圾法律中，23 个州（加州和犹他州除外）包含了生产者责任延伸制（EPR）。美国联邦政府对电子垃圾的政策制定进行了多次尝试，却收效甚微，其对电子垃圾的约束性限制也比较匮乏。美国出台了联邦资源保护与回收法案（the federal Resource Conservation and Recovery Act，RCRA），其只针对少部分的电子垃圾产品，如由于含铅等重金属而被列为危险废物的阴极射线管（CRT）显示器，法案禁止填埋 CRT 显示器。虽然 RCRA 禁令在一定程度可以预防健康和环境危害，但是它忽略了更大的电子垃圾来源，因为它只是适用于大型企业和公共机构，而家庭或小型机构所产生的废旧家电可以被豁免。加利福尼亚州（简称"加州"）在 2003 年率先通过了《电子垃圾回收法》（Electronic Waste Recycling Act），且加州的填埋禁令范围比联邦 RCRA 禁令更广。从 2005 年 1 月 1 日起，加州的消费者在购买电子设备产品时要缴纳电子垃圾收集和回收的费用，也是美国地方电子垃圾法律中唯一要求消费者购买时需要预先支付回收费用。自《电子垃圾回收法》颁布以来，加州建立了 600 多个回收站点，30 多个经批准的回收商。加州的电子垃圾的回收量从 2005 年的 2.72 万吨增加到 2008 年的 9.71 万吨，且 2008 年所覆盖电子垃圾的回收率达到了 58%。另外，加州也颁布了有关手机（AB 2901）、充电电池（AB 1125）、CRT 面板玻璃（AB 1419）等电子垃圾产品的相关政策法规，在一定程度上促进了加州电子垃圾的回收和再循环。数据显示，2010 年加州售出的 1800 万部手机中，370 万部被回收，回收率为 21%，比 2007 年 17%的回收率提高了 4%，而 2010 年美国的手机回收率只有 10%。

美国在国家层面没有类似欧盟 RoHS 指令的法规，但有些州颁布了类似的法规。例如，加州的健康与安全规范（Health and Safety Code Sections 25214.9-25214.10.2）在一定程度上借鉴参考了欧盟的 RoHS 指令，其规定了特定电子屏幕及其显示设备中有害物质的限量标准。与欧盟 RoHS 不同的是，加州的 RoHS 主要针对显示屏幕对角线长度大于 4 英寸的电子设备，如电视、电脑显示屏等。2007 年加州还通过了《加州灯具功率和毒害降低法》（California Lighting Efficiency and Toxics Reduction Act），规定任何人不得在加州范围内生产、销售或供应有害物质含量超过欧盟 RoHS 指令规定限值的普通型灯具。

为了解电子垃圾回收处理人员所面临的职业危害，美国职业安全与卫生研究所（NIOSH）于 2012 年对美国数个电子垃圾拆解处理企业场所进行过多次调查采样，采集了粉尘、大气、金属表面擦拭样品及工人的血样及尿样等样品，并发布了相关的健康暴露危害评估报告。报告指出：回收场所处理的电子垃圾主要是废电视机和废旧的 CRT 显示器，工人接触的空气环境存在铅和镉含量超标，2 名工人血铅水平超过了 10 μg/dL。不良的工作习惯，如不洒水就进行清扫作业等，会使有毒的金属进入到回收场所其他作业区。另外，NIOSH 官员认为电子垃圾回收行业的雇主们不应该参照当前的职业安全与健康署的铅标准，因为这些标准对工人们没有提供任何保护措施，并建议电子垃圾回收商应该参照由其他认证机构制定的更严格的推荐性标准，如由电子垃圾回收组织（e-Stewards）制定的《负责任的电子设备循环及再利用标准》等，这些标准可以为电子垃圾回收作业人员进行更好的保护指导。e-Stewards 总部位于美国加州西雅图市，其源自非营利组织巴塞尔行动网络（Basel Action Network，BAN）。2003 年，BAN 启动了电子垃圾"管家承诺计划"（e-Steward Pledge program），致力于阻止有毒物质和废物流入发展中国家。当时，在美国和加拿大拥有 100 个网点和 40 多家合格的电子垃圾回收商，他们承诺只使用负责任的、安全的方法来处理电子垃圾。2006 年，BAN 将"管家承诺计划"转变为独立审核的认证计划，致力于建立和维护具有安全、道德和全球责任的电子垃圾回收和翻新标准，并通过共享和使用电子垃圾回收和再利用的实践标准和最佳实践方案。

9.2.2 非洲地区

非洲大多数国家的电子垃圾回收处理，主要还是以非正规的行业收集者和回收者为主。各国政府对于电子垃圾回收处理行业的控制比较少、效率也很低，相关的管理问题包括：缺乏政府的政策和立法、缺乏有效的回收/收集系统和 EPR 系统、缺乏充分的公众意识、缺乏足够的回收设施以及缺乏危险废物管理所需的经费、回收部门由不受控制与装备不良且污染环境的非正规行业主导等。因此，电子垃圾的粗放式回收处理，必将排放大量的毒害物质，从而对非洲相关人群和环境构成了风险。但是，电子垃圾如果处理得当，也会产生经济和社会效益，包括为回收贵金属、为电子垃圾从业者提供就业机会和经济利益、帮助清洁环境、翻新出售廉价电子和电气设备，回收金属减少矿产开采等好处。

1. 加纳

加纳人口超过 2900 万，位于撒哈拉沙漠以南，2019 年的电子垃圾产生量为 5.3 万吨，人均 1.8 千克。加纳是电子垃圾的主要接收国，尤其是来自欧洲、北美等区域产生的电子垃圾。随着西非国家进口的廉价二手电器电子产品和设备的寿命缩短，估计15%的设备完全不能使用，这使加纳接收的电子垃圾数量进一步增加。以加纳首都阿克拉(Accra)郊区的阿博布罗西(Agbogbloshie)为例，该区域是非洲最大的电子垃圾回收站之一。20 世纪 90 年代初，由于种族冲突及内乱，加纳国家北部居民逃到南部的城市，包括首都阿克拉。阿克拉郊区的阿博布罗西靠近加纳最大的消费市场、临海、空地大、有大量劳动力但失业问题严重，后来当地失业青年发现可以通过暴力拆解、焚烧等粗暴方法得到电线中的铜，从而获得几美元的收入。据估计，有 3 万~5 万人生活工作在阿博布罗西，焚烧电子垃圾排放的黑烟吞噬了整个阿博布罗西，给从业人员及附近居民造成严重的空气污染及毒害物质暴露问题。从事电子垃圾拆解的工人属于弱势群体，受到社会的排斥，且报酬很低，并忍受着环境和工作所带来的伤害。2016 年 8 月，加纳颁布了第 917 号法案《危险和电子废物控制和管理法》，随后签署了《有害、电子和其他废物(分类)、控制和管理条例》。这两项法律条例共同为加纳可持续管理电子垃圾的创新战略奠定了基础。2018 年 2 月，加纳环境科学技术和创新部等部门发布了《电子垃圾环境无害化管理技术指南》，为电子垃圾的可持续管理和回收方面又迈出了重要一步。基于国际标准认证框架和加纳当地的业务需求，瑞士资助的可持续回收行业项目(Sustainable Recycling Industries，SRI)项目支持了相关准则的制定，包括一个由决策者、行业以及非正规拆解行业代表、民间社会组织和科学家参与的为期两年的利益攸关方参与进程。项目涵盖了活跃在电子垃圾回收链中的各种经济行为体的具体要求，如收集者、回收/回购中心、运输商、处理者和最终处置场。2021 年，当地政府推倒、拆除了阿博布罗西的电子垃圾从业人员所居住的简易木屋和工棚，并安排工人到距离阿克拉 1 小时车程的郊区定居工作生活，这标志着阿克拉市非正规电子垃圾处理的终结[4-5]。

2. 尼日利亚

2019 年尼日利亚电子垃圾产生量为 46.1 万吨，人均 2.3 千克。近年来，由于中国、印度和巴基斯坦等早期电子垃圾接收国加强了对电子垃圾的进口监管，导致流向尼日利亚的电子垃圾数量逐渐增加，成为发达国家产生电子垃圾的新目的地。2015 年，尼日利亚接收了 6.6 万吨的所谓二手电脑、电视机和显示器等电子产品，其中约 1.69 万吨的二手电子产品已经处于不能工作的状态。数据显示，尽管欧盟对电子垃圾交易有严格的法律，但尼日利亚约 70% 的电子垃圾来自欧洲。在尼日利亚计算机及相关产品经销商协会(CAPDAN，一个协调 IT 行业事务的监管机构)的保护下，电子垃圾通过中间商或进口商从事的非法入境通道而抵达尼日利亚，主要是拉各斯海港(Lagos seaport)。进口商按重量购买集装箱，经简单修理或翻新后，这些二手电子产品在 Ikeja 电脑城和 Alaba 国际市场进行出售，不能工作的电子产品直接被送到城市的各个垃圾填埋场，包括拉各斯的 Olusosun、Igodun 和 Ikorodu 垃圾场等[6]。数千名工人，包括未成年儿童，住在垃圾场或垃圾场附近，从事收集、分类、拆解和从电子垃圾中提取金属的工作。他们要么直接与冶炼厂打交道，要么将拆解产物卖给中间商。这种非正规的处理方法给当地环境和人体健康造成了伤害，并在各个拾荒者群体之间造成纠纷。尽管尼日利亚在电子垃圾管理方面没有强有力的具体法规，但该国正在积极实施《国家环境条例(电子部门)》，明确禁止无法使用电子产品的贸易。尼日利亚也已经批准在拉各斯州的 Ojota 开设第一家由 Hinckley recycling 公司运营的正规电子垃圾回收工厂。

9.2.3 拉美地区

巴西是拉丁美洲地区第一大电子垃圾产生国，2019 年其电子垃圾产生量为 214.3 万吨(人均 10.2 千克)。据估计，2001 年至 2030 年期间，巴西需要累计处理 2500 万吨电子垃圾。巴西专门针对电子垃圾的国家立法还不是很完善，其国家颁布的《垃圾法》("Law of the Garbage")规定了个人对电子垃圾的责任，也适用于制造商、零售商、政府组织和官员，也适用于最终用户。2010 年 8 月国家第 12.305 号联邦法律有关国家固体废物的政策中规定电子垃圾应得到适当处理。2010 年 7 月，圣保罗州颁布的州法(第 13.576 号)确立了电子垃圾的正式处理程序，包括回收、管理、处置及 EEE 生命周期的共同责任原则，从而在监管框架层面确保可回收利用固废的再利用。但是，在圣保罗和联邦层面，相关立法引起了制造商的反对。此外，在联邦和州法律的保护下，由于电子垃圾能够创造就业机会而被视为具有社会价值的经济资产，巴西有专门的电子垃圾管理、处理公司，如专门收集和回收废旧家电的 Ecobraz、专门为电子垃圾提供回收的 Reciclagem Brasil、与圣保罗普雷菲图拉市政府合作提供电子垃圾回收管理的 Coopermiti、提供大规模的收集和回收服务的 Descarte Certo 等。电子垃圾一旦收集起来，就会经过拆解分类，然后将不同部件分别送入回收公司进行处理，回收塑料、金属、电线、电缆等原材料[6]。

在"加强国家倡议和区域合作，对废弃电子电器产品中的持久性有机污染物进行环境无害化管理"项目框架内，由全球环境基金资助，联合国工业发展组织协调的"UNIDO-GEF 5554"项目，对拉丁美洲不包含巴西的 13 个国家(阿根廷、玻利维亚、智利、哥斯达黎加、厄瓜多尔、危地马拉、洪都拉斯、尼加拉瓜、巴拿马、秘鲁、萨尔瓦多、乌拉圭、委内瑞拉)的电子垃圾统计、立法和管理基础设施进行了区域监测，并取得了如下一些有关电子垃圾的信息成果[3]：

(1)国际公约及法律法规管理方面：①该地区所有 13 个参与国都有一些关于废物管理的法律和监管框架，其中 5 个国家(玻利维亚、智利、哥斯达黎加、厄瓜多尔和秘鲁)有针对电子垃圾和 EPR 等较为完善的立法框架和具体立法。阿根廷、萨尔瓦多、危地马拉、洪都拉斯、尼加拉瓜、巴拿马、乌拉圭和委内瑞拉既没有 EPR，也没有明确的电子垃圾收集目标，其电子垃圾管理主要是在一般废物或危险废物的法律法规中体现的。所有国家都有包括持久性有机污染物在内的危险废物法规，但没有一个国家有专门针对电子垃圾中持久性有机污染物的立法。②上述 13 个拉丁美洲国家都批准了《巴塞尔公约》，并颁布了禁止电子垃圾进口的国家禁令，但这些措施的执行仍然是一个重大挑战。该区域的许多国家没有向《巴塞尔公约》秘书处提交越境转移报告，使得很难监测区域内外电子垃圾和持久性有机污染物的越境转移。玻利维亚、智利、厄瓜多尔和巴拿马没有 2016~2019 年电子垃圾进出口的官方数据，且一些有关该区域进出口越境转移的电子垃圾数量没有记录在向《巴塞尔公约》秘书处提交的报告中。例如，尼加拉瓜、洪都拉斯等国收集的电子垃圾被拆解，一些有价值的部件(如铝、铁、金等)在国内市场上出售，而其他部件(如印刷电路板、电池等)则储存在集装箱中，积累到足够数量时就出口到其他国家进行处理。为了避免这种情况，萨尔瓦多正在开发一个跟踪危险废物进出口的电子平台，该平台将协助当局监视相关公司参与废物进出口的授权，控制和监控不同国家当局之间的信息交流，将拉丁美洲多个国家联系起来，以改进区域合作。13 个国家都不限制出口危险废物和用于最后处置或回收的其他废物。另外，二手电子电器产品进口导致接受国生成更多的电子垃圾，并给现有的电子垃圾管理带来负担。③上述 13 个拉丁美洲国家都批准了《斯德哥尔摩公约》，但没有一个国家有管理电子垃圾塑料中持久性有机污染物的具体法律文书。智利和哥斯达黎加目前正在制定本国电子垃圾塑料的持久性有机污染物的法律框架，在该项目框架的组成部分 2(加强废物拆解和回收设施/

基础设施的国家能力)中,大多数国家目前正在制定电子垃圾相关的最低标准,以管理和最终处置电子垃圾塑料中所含的持久性有机污染物。其他国家也有一些持久性有机污染物的相关条例规范。例如,2013年,洪都拉斯达成了《关于"对含有多氯联苯或受多氯联苯污染的设备和废物进行无害环境管理"》的部长级协议,它确立了环境无害化管理的程序、措施、条款和责任,以防止污染和保护环境,但是还缺少针对电子垃圾管理和持久性有机污染物的环境健康安全标准。2017年,哥斯达黎加批准了《多氯联苯的识别和环境无害化消除条例》,要求在持久性有机污染物信息系统中注册。厄瓜多尔有与化学品和危险废物有关的环境健康安全标准,但没有专门针对电子垃圾中的持久性有机污染物的标准。2021年,萨尔瓦多通过了一项与危险废物管理有关的新法规——第41号法令《危险物质、残留物和废物特别法规》。

(2)电子垃圾相关数据信息方面:①电子产品市场投放量与电子垃圾产生量。2010年至2019年期间,投入市场上的电子电器产品总量出现波动。2010年进入市场的电子产品总量为170万吨(人均8.9千克),2017年增加到190万吨,2019年又下降到170万吨(人均8.1千克)。只有阿根廷、哥斯达黎加和智利在国内生产电子电器产品,其他10个国家完全依赖进口。同期,该地区的电子垃圾产生量增加了49%,从2010年的90万吨(人均4.7千克)增加到2019年的130万吨(人均6.7千克)。电子垃圾人均产生量最高的是哥斯达黎加(人均13.2千克),最低的是尼加拉瓜(人均2.5千克)。电子垃圾的小型设备(V类)、温度交换设备(Ⅰ类)和大型设备(Ⅳ类)的所占比例较高,累计占该地区总量的75%。各个类型的电子垃圾(屏幕显示器除外)均保持正增长,但其年增长率有下降的趋势。②电子产品塑料和电子垃圾废塑料。投放市场的电子产品塑料从2010年的47万吨(人均2.49千克)下降到2019年的46万吨(人均2.22千克)。产生的电子垃圾塑料从2010年的24万吨(人均1.29千克)稳步增长到2019年的38万吨(人均1.85千克)。投放市场电子产品塑料中的溴化阻燃剂(BFR)从2010年的4万吨(人均0.20千克)波动变化到2019年的3万吨(人均0.17千克)。而废塑料中含有的BFR从2010年的2万吨(人均0.12千克)稳步增加到2019年的3万吨(人均0.15千克)。按照欧盟对电子垃圾的分类,调查发现小型设备中BFR含量占比最大,为1.6万吨(人均0.08千克),其次是小型IT设备的1万吨(人均0.05千克)和显示器0.5万吨(人均0.02千克)。③电子垃圾的正规收集处理。相比2019年全球电子垃圾17%的正规收集率,拉丁美洲国家的正规收集率很低(不到10%),13个国家的官方正式收集电子垃圾总量为3.6万吨(人均0.21千克),收集率仅为2.7%。按重量计,收集量较高的国家是智利(6.84千吨)、哥斯达黎加(5.10千吨)、秘鲁(3.02千吨)和厄瓜多尔(3.0千吨)。按正规收集率计,哥斯达黎加(8%)和智利(5%)的收集率最高,其次是阿根廷(4%)、玻利维亚(4%)、厄瓜多尔

(4%)、乌拉圭(3%)、秘鲁(2%)、萨尔瓦多(1%)、洪都拉斯(1%)、委内瑞拉(0.4%)、尼加拉瓜(0.4%)和巴拿马(0.4%),危地马拉的收集率还正在评估收集阶段。由于电子垃圾的正规收集率与国家立法、执法、电子垃圾管理基础设施之间存在很强的相关性。而拉丁美洲这些国家的电子垃圾相关立法处于起步阶段,立法框架与执法策略不一致,且尚未建立必要的电子垃圾管理基础设施和收集点,这些原因导致该区域电子垃圾的正规收集率较低。

(3)面对电子垃圾资源和污染属性的双重挑战。电子垃圾的管理是拉丁美洲的经济机遇,13 国 2019 年产生的电子垃圾中含有 7 吨黄金、0.31 吨稀土金属、59.1 万吨铁、5.4 万吨铜和 9.1 万吨铝等二次资源,总价值约为 17 亿美元。拉丁美洲地区 97%以上的电子垃圾没有被正式收集或送往环境无害化处理设施进行管理。在非正规处理方式中,当挑出电子垃圾中有价值的材料后,大多数电子垃圾最终都被扔进了垃圾填埋场。该区域内电子垃圾中的有害物质,包括至少 2.2 吨汞、0.6 吨镉、4.4 吨铅、4 吨溴化阻燃剂和 5.6 吨温室气体当量(以制冷剂核算),由于管理不善,电子垃圾中的毒害物质得不到环境无害化管理,对生态环境和人体健康产生潜在风险。有关电子垃圾管理和持久性有机污染物的评估、统计数字、立法和现有挑战研究表明,上述 13 个国家的电子垃圾和持久性有机污染物的管理系统存在较大差异,可通过以下两点措施来改进:①建立一个强有力的法律和政策框架,以电子垃圾的环境无害化管理为重点;②加强现有的监测管理系统,使其更加高效和实用,并鼓励所有利益攸关方参与并支持监测系统的开发合作以确保电子垃圾管理的相关政策能够顺利实施执行。

9.2.4 亚洲地区

1. 日本

日本是世界上最早实施 EPR 国家之一,拥有先进的电子垃圾回收系统以及完善的回收、再利用基础设施。从 20 世纪 60 年代开始,日本经历经济高度增长期,日本政府先后制定并完善了一系列相关的法律法规,相继出台了《废弃物处理法》、《家电循环利用法》和《食品循环利用法》等法规。2001 年 4 月 1 日,日本正式实施《家电循环利用法》,主要内容包括:以电视机、电冰箱、空调和洗衣机四种电器为立法对象,采用由零售商进行回收、制造商进行回收利用的方式。日本在 2006 年 4 月修订了《资源有效利用促进法》,要求电器电子产品制造商或进口商根据工业标准《电器电子设备特定化学物质标识标准》(JIS C 0950 title "The marking for presence of the specific chemical substances for electrical and electronic equipment",J-MOSS)对 7 种电器电子产品(个人计算机、家用空调、电视机、微波炉、烘衣机、电冰箱、洗衣机)中的 6 种特定有害物质(铅、汞、镉、六价铬、多溴联苯和多溴联苯醚)进行标识,并鼓励企业对产品中有害物质进行管理并使用更环保的材料。日本的 J-MOSS 也被称为日本的 RoHS,其对 6 种有害物质在电器电子产品中设定了参考限值,其质量限值与欧盟 RoHS 的限值一样,且对以下 3 种情况必须在产品、包装和目录上作"内容标记(橙色标记)":①没有豁免申请,且含量超过参考限值;②有豁免申请(超过参考限值),且其他部分的含量超过参考限值;③有豁免申请(低于参考限值),且其他部分的含量超过参考限值。

2013 年 4 月 1 日,日本施行《小型家电回收法》,目的是对废旧小型电子产品中使用的金属等有用资源进行回收利用。列入可回收范围的小型家电包括手机、电脑、数码相机等 21 个品类的 104 种产品,电子燃气灶和吸尘器也在回收范围内。据《小型家电回收法》规定,由日本地方政府或指定的企业进行回收利用,将回收来的小型家电被再次变成数码相机、手机等产品,重新回到消费者手中。日本从 2017 年 4 月起,在全国的政府机构、邮局、商业设施等地设置了约 7 万个电子垃圾回收箱,用于回收智能手机、计算器、游戏机、相机等小型电子产品。据日本环境省公布的数据,2017 年根据《家电循环利用法》规定的回收范围在日本全国指定场所共回收约 1189 万台。另外,日本明确规定,对电视机、洗衣机、空调、冰箱必须进行回收再利用,其回收资源再商品化率方面,空调为 92%、显像管式电视为 73%、液晶等离子电视为 88%、电冰箱为 80%、洗衣机/烘干机为 90%。2017 年 2 月至 2019 年 3 月,日本东京奥组委执行了"利用城市矿山制作奖牌"项目,呼吁国民广泛参与,捐出不用的旧手机和家电,从这些电子零部件中提取贵金属,满足制造 5000 枚奖牌所需的 30.3 千克黄金、4100 千克白银和 2700 千克铜。根据东京奥组委官网公布的统计数据显示,日本最大的移动运营商(NTT DoCoMo)已经回收了 575 万部手机,全国地方政府共回收小家电 67180 吨。至 2019 年,日本电子垃圾的产生量为 256.9 万吨,人均 20.4 千克,其 2017 年记录在案的收集回收电子垃圾量为 57 万吨。

2. 印度

随着印度经济的快速增长和中等收入人群的扩大,其社会购买能力不断增强,也促使了电子垃圾的指数级增长。Pathak 等[7]采用预测数学模型,量化了印度市场上最畅销的计算机和手机两种电子产品。研究发现,计算机的电子垃圾量将持续增加到 2022 年,并在 2028 年慢慢达到一个饱和点;而手机电子垃圾则没有达到饱和点。2019 年印度电子垃圾产生量为 323 万吨,人均 2.4 千克,其 2016 年记录在案的电子垃圾收集回收量仅为 3 万吨。十个印联邦产生了印度 70%的电子垃圾,主要包括马哈拉施特拉邦、泰米尔纳德邦、安得拉邦等。印度 65 个城市产生的电子垃圾占印度电子垃圾总量的60%以上，产生电子垃圾数量前十的城市为，孟买、德里、班加罗尔、金奈、加尔各答、艾哈迈达巴德、海德拉巴、浦那、苏拉特和那格浦尔。考虑到电子垃圾对可持续发展的不利影响，印度已经制定了针对电子垃圾的立法——《2016年电子垃圾管理规则》，并对制造商和翻新商实行生产者责任延伸制度。电子垃圾立法的出台提高了人们对电子垃圾管理的意识，并涌现了一些收集中心、有组织的回收公司，如Recyclekaro.com、E-Parisaraa等。总体来说，印度还是比较缺乏大规模有组织的电子垃圾回收设施，90%的电子垃圾回收处理仍属于无组织的回收行为[6]。

3. 越南

2019年越南的电子垃圾产生量为25.7万吨（人均2.7千克）[8]，且以每年10万吨的速度增长[9]。越南的河内和胡志明市为人口最稠密的城市，根据越南国家统计局的数据估计，2020年河内和胡志明市的居民分别丢弃了16.1万和70万台电视机、9.7万和29万台个人电脑、17.8万和42.4万台冰箱、13.6万和33.9万台洗衣机、9.7万和33万台空调[9]。另外，由于中国自2018年起实施了废物进口禁令，一定规模的电子垃圾拆解处理业务从中国转移到越南、泰国等东南亚国家，使越南成为新兴的电子垃圾拆解转移国家之一。

越南的大部分电子垃圾都是由非正规部门处理。在收集方面，成千上万的小贩骑着摩托车、自行车或步行挨家挨户地从终端用户那里购买废弃的电子电气设备和其他可回收材料，然后以更高的价格出售给私人商店或贸易商，而收集得到的电子垃圾将在非正规废物回收站进行拆解处理。越南约有30个村庄参与电子垃圾的拆解处理，主要位于越南北部城市和省份的农村或郊区，如河内，兴安省（Tinh Hung Yen）和永福省（Tinh Hai Phong）等地。其中，兴安省的Bui Dau村是在回收活动和环境影响方面研究最广泛的地区之一。当地拆解工人使用原始技术处理电子垃圾，以回收铜、铝和塑料等可回收材料，主要包括6类处理流程：①收集和运输；②在室内和室外堆放处理前和处理后的电子垃圾；③人工拆解和分离材料，将其分为金属、玻璃、塑料等；④塑料回收；⑤露天焚烧电缆和电线，回收金属铜；⑥以露天倾倒和焚烧的方式处置价值较低的材料。研究表明，上述原始加工活动，如手工拆解、露天燃烧和塑料回收，已被确定为环境排放和人类暴露于有毒元素（特别是砷、镍、铅和锌等）和有机污染物（如阻燃剂、多环芳烃、多氯联苯和二噁英类化合物）的重要贡献者[10]。

近年来，越南不断出台相应的法律政策来加强电子垃圾的管理。2013年，越南颁布了第50/QD-TTg号关于规定回收和处置废弃产品的总理决议，被认为是第一个将电子垃圾视为特定废物的立法，也被认为是建立电子垃圾管理制度的第一个法律基础。该制度是基于完整的环境保护责任政策，规定回收和处置废弃产品的所有责任都由生产者/进口商/分销商承担。2014年，越南的《环境保护法》明确指出，生产者/销售者有责任收集和处理废弃产品(第87条)，消费者有责任将废弃产品带到适当的地方(第88条)。这些术语在2015年的关于回收和处理废气产品的第16/2015/QD-TTg号总理决议(这是越南的EPR制度)中进行了详细的迭代，该决议取代了第50/QD-TTg号决议[11]。2019年，关于进口二手机械设备和工艺线的第18/2019/QD-TTg号总理决议，其中规定了设备使用年限不应超过10年。根据《对外贸易管理法》指导方针的第69/2018/ND-CP号法令，其中罗列了禁止出口和进口的二手货物清单(附件1)，包括电子产品、制冷设备、家用电器、信息技术产品等；禁止临时进口用于再出口和边境口岸转移的二手货物清单(附件6)，包括蓄电池、铅酸蓄电池等；临时进口用于再出口的二手货物贸易清单(附件9)，包括电风扇、空调、冰箱、冰柜/制冷机组、洗碗机、洗衣机、个人电脑、吸尘器、电话、扬声器、相机等，其中贸易公司必须支付再出口保证金。2020年，《环境保护法》(72/2020/QH14)严禁以任何形式从境外进口、暂进口、转移过境废物(第6条)。

在越南，电子垃圾管理最大的障碍之一是缺乏关于废弃物生成和当前管理信息的数据库。随着电子垃圾数量的快速增长，现有的不恰当的电子垃圾管理已经造成了许多环境和人类健康问题，这导致了对以下问题的迫切需求：①更全面、具体且定量的关于电子废物产生和管理系统的信息；②防止有毒气体排放，有效管理废物流中的宝贵物质资源的全面、实用的解决办法。Tran等[12]采用物质流分析得出：在1966~2035年间，越南的报废电视机的数量逐渐增加。在2012年的废旧电视中，66%被直接重复使用或在维修/翻新后重复使用，3%在国内回收或开放燃烧以回收有价值的材料，9%被非法出口，其余22%为露天倾倒。废旧电视机中的75%的铜、铝、铁等贱金属被重复利用或回收利用；其余25%被出口或排放。而对于金、银、钯等贵金属以及塑料和玻璃，则有更大的材料损失：大约34%被非法出口(贵金属和塑料)或露天倾倒(玻璃)。研究还得出，非正规部门大量参与废电视机管理系统，使废弃物处理情况更加复杂和难以控制，且会对环境和人类健康造成潜在风险。

4. 中国

2019年我国电子垃圾的产生量为1012.9万吨(全球第一)，人均7.2千克，其2018年记录在案的收集回收电子垃圾量为154.6万吨。面对数量巨大的电子垃圾，我国相关的政府职能部门承担着较大的法律制度保障、处理能力建设的压力。目前，我国电子垃圾的管理制度体系主要基于《中华人民共和国固体废物污染环境防治法》(以下简称《固废法》)、《中华人民共和国清洁生产促进法》和《中华人民共和国循环经济促进法》三部法律。电子垃圾作为固体废物的重要组成部分，对其管理应该遵循《固废法》。《固废法》于 1995 年 10 月通过，并自 1996 年 4 月 1 日施行，其对防治固体废物污染环境，保障人体健康，维护生态安全，促进经济社会可持续发展起了纲领性作用。《固废法》历经多次修改，最近的一次修改通过时间为 2020 年 4 月。修订后的《固废法》于 2020 年 9 月 1 日起施行，并对我国电子垃圾相关的管理工作提出新的要求，其中第六十六条、六十七条明确提出：国家将建立电器电子、铅蓄电池等产品的生产者责任延伸制度，鼓励产品的生态设计和资源回收利用；国家将对废弃电器电子产品实行多渠道回收和集中处理制度，拆解、利用、处置废弃电器电子产品等，应当遵守有关法律法规的规定，采取防止污染环境的措施。2020 年底，生态环境部、商务部、发展改革委、海关总署联合发布《关于全面禁止进口固体废物有关事项的公告》，公告要求自 2021 年 1 月 1 日起，禁止以任何方式进口固体废物，禁止我国境外的固体废物进境倾倒、堆放、处置。这就意味着我国从法律层面杜绝了各类电子垃圾等固体废物的入境问题，以后我国电子垃圾的处理对象主要来自国内产生。

根据《固废法》等相关法律的相关规定，国务院于 2009 年 2 月颁布了《废弃电子产品回收处理管理条例》（国务院第 551 号令），并于 2011 年 1 月 1 日起实施。《废弃电子产品回收处理管理条例》对我国电子垃圾相关的目录制度、基金制度、行业规划制度、处理企业资格许可制度、信息管理制度等进行了阐述和规定，并陆续出台了《废弃电子产品回收处理管理条例》的配套政策，如《废弃电器电子产品处理目录》《废弃电器电子产品处理基金征收使用管理办法》《废弃电器电子产品规范拆解作业及生产管理指南》《废弃电器电子产品拆解处理情况审核工作指南》和《废弃电器电子产品处理企业资格审查和许可指南》等，从而形成我国以废弃电器电子产品处理基金为核心内容的生产者责任延伸制度，建成符合我国国情的电子垃圾相关的环境综合管理体系（图 9-1）[13]。

我国已经形成了废旧家电处理基金征收及补贴的制度体系（图 9-2）[14]，其中，电子垃圾正规处理企业是享受基金补贴的第一责任人，对废弃电器电子产品拆解处理的规范性和基金补贴申报的真实性、准确性承担责任；省级生态环境主管部门负责组织本地区处理企业废弃电器电子产品拆解处理情况的审核工作，对审核结论负责；生态环境部负责组织基金制度相关的规范、指南、审核办法等政策的制定、实施等，并对各省级生态环境主管部门报送的审核情况汇总确认后提交财政部。2012~2020 年，国家累计发放 219 亿元的基金用于补贴处理企业，引导约 6 亿台废旧电视机、电冰箱、洗衣机、空调和微型计算机进入正规企业处理，使上述废旧家电的规范回收处理率超 40%，其中电视机回收率高达 94%以上，电冰箱的回收率达到 77%以上，处于国际领先水平[13]。国家生态环境主管部门对基金补贴名单内企业进行全过程和无死角监控，严控处理企业出现环境违法或者污染防控不到位。目前，基金补贴名单内处理企业管理比较规范，配备必要的环境污染防治设施，污染排放监测开展情况较好，对环境污染较小，有效消减了电子垃圾可能造成的环境风险。

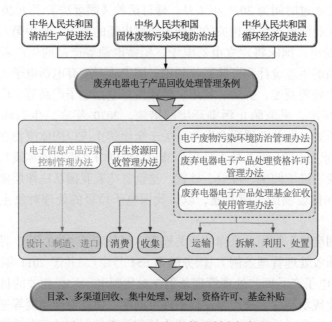

图 9-1　我国电子垃圾的管理制度框架[13]

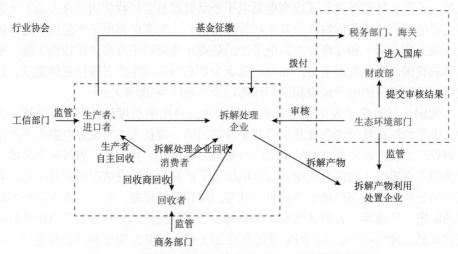

图 9-2　我国废弃电器电子产品处理基金征收使用机制[14]

我国电子垃圾处理技术和设备以国产和自主研发为主，欧洲和日本等进口技术和设备占据少量的市场份额。处理方式以手工拆解与机械处理相结合的方式，属于劳动密集型行业，但拆解工艺和处理流程逐渐向高效化方向发展。随着我国拆解处理废旧家电类型的不断变化，大多数正规处理企业已经开始技术和装备的升级改造，以满足不同处理对象的新要求。另外，电子垃圾拆解产物的高值化深加工，如稀贵金属的提取净化、废塑料的深加工及高值化利用等，逐渐受到各个企业的重视。

如上所述，我国针对电子垃圾的管理已经颁布了较为完善的法律制度和配套条例，形成了覆盖全国的电子垃圾正规处理企业的合理布局，有效遏制了电子垃圾拆解回收所产生的环境污染，显著降低了相关的环境健康风险，初步探索出一条适合我国国情的电子垃圾回收之路。

9.3 发展趋势及健康风险挑战

9.3.1 发展趋势

随着全球经济社会的不断发展，电子电器产品的需求将不断增长，也将促使全球电子垃圾产生量的不断增加，鉴于各国国情不一样，各国所面临的电子垃圾收集回收、制度管理、处理处置等问题也不尽相同，虽然现在还很难针对电子垃圾形成一个放诸四海而皆准的最佳实践方案，但可以从以下几个方面进行尝试努力，从而形成国际组织、国家、企业、社会等多层级多元素共治电子垃圾问题的良好局面。

(1) 国际层面：基于国际公约准则，加强国际组织间、区域各国之间的通力合作，鼓励各缔约国向《巴塞尔公约》/《斯德哥尔摩公约》秘书处提交国家报告、并定期更新，如非法越境转移的二手电子产品和电子垃圾的数据和路径，电子垃圾中持久性有机物等毒害物质的信息，高质量的数据信息将有助于掌握不同区域或国家的电子垃圾污染源现状，并有助于进一步制定防治电子垃圾非法贩运的行动方案和形成禁限用持久性有机物的履约行动计划。

(2) 国家层面：完善电子垃圾相关的法律法规、标准指南、技术规范等制度体系；通过政府监管、市场调节、社会监督，形成了电子垃圾全流程管理监控模式；推动电子垃圾等可再生资源的规模化、规范化、清洁化利用，促进电子垃圾产业的集聚发展，高水平建设资源回收再利用基地，形成电子垃圾的"回收-拆解-处理-处置"的全产业链循环利用模式；完善和改进电子垃圾的收集回收网络，将电子垃圾回收体系融入废旧物质回收体系，借助废旧物资回收网点与生活垃圾分类网点等基础设施，积极推广"互联网+回收"的线上线下相融合的回收模式；规范电子垃圾回收行业的经营秩序，提升行业经营管理水平，不断提高电子垃圾的交投意愿和便利性，从而让更多电子垃圾能够进入正规拆解处理的渠道。

(3)企业层面：电子电器产品的生产制造商，应该强化社会环境责任意识，采用环保、低毒的绿色材料，鼓励产品的绿色设计，让产品的回收利用更加便捷，其材料的使用寿命更加持久，从源头上减少毒害材料的使用；企业应该积极参与电子垃圾的回收处理工作或提供信息指导，即将生产者责任延伸制落实到生产企业，企业是产品的设计开发者，对其构造功能及有毒害物质的使用更加清楚，有了生产企业的参与，将有助于电子垃圾的绿色环保处理。电子垃圾拆解处理企业要密切结合国家相应的发展策略，如循环经济发展规划、无废城市建设等，实现企业与国家发展的协同发展。

(4)社会层面：每个人都是电子产品的消费者，也是电子垃圾的制造者。普通百姓要养成绿色环保的生活方式，不以快速淘汰更新电子产品为荣，而要让电子产品满足要求的条件下，让其物尽其用，从而在一定程度上减缓电子垃圾人均产生量的增长。消费者在处理淘汰或损坏的电子电器产品时，应该具备相应的环境保护意识，不要随意丢弃电子垃圾，应该将废旧电子电器产品交投或丢弃到指定回收站点，从源头上减少电子垃圾对环境的潜在污染。电子垃圾相关的从业人员，包括行业管理者、科研人员、收集回收者、处理处置者等，一方面要各司其职，不断完善符合国情的法律政策来促进行业的健康发展，研发绿色环保的处理技术来减少回收过程的二次污染，采用满足环保要求的工艺方法来实现可回收资源的最大化利用。另一方面，电子垃圾各利益攸关方要加强沟通联动，形成合力共同促进电子垃圾行业的绿色可持续发展。

9.3.2 健康风险挑战

针对电子垃圾的收集回收、处理处置及健康风险管控等研究领域，还将面临如下一些挑战：

(1)新产生电子垃圾的回收处理体系。随着人民生活水平的提高和电子信息技术的更新，将迎来更多各式各样的废弃电器电子产品，给我国的电子垃圾的收集、规范化处理处置提出了新的要求。为避免新产生电子垃圾流入社会面，形成潜在污染源，国家可以通过制定相应的政策方案，健全线上线下相融合的回收网络，打通优化逆向物流回收渠道，持续增强企业的环保处理能力，促使新产生的电子垃圾尽量流入正规处置企业，并得到环保规范处理处置，从而降低电子垃圾对环境的污染和对人体健康的危害。

(2)电子垃圾的资源化技术及污染排放控制新技术研究。虽然我国电子垃圾的回收处理模式实现了家庭作坊式分散拆解向正规企业或园区集中处理的转变，但是电子垃圾资源化处理及污染控制技术还有待升级改造，集中处理过程将导致污染物的局部高浓度释放，形成重金属和有机污染物的复合污染特征，故还需要对排放大气等环境介质污染物进行专门研究，研发出绿色环保、经济可行、处理效率高的污染排放控制工程应用技术。

（3）关注电子垃圾相关的新污染物问题。电器电子产品的社会需求量还会持续增加，其中电子产品所需材料的种类及组分也不断更新迭代。以阻燃剂添加剂为例，随着多溴联苯醚等溴系阻燃剂的禁用限用，致使其他阻燃剂（如有机磷阻燃剂）和新型阻燃剂的使用量将不断增加，当使用新型阻燃剂的电器电子产品淘汰废弃后，电子垃圾的处理处置将遇到新型污染物的环境污染及健康风险问题。废塑料是电子垃圾的重要组成部分，其中含有多种类型的助剂或添加剂，还需对废塑料中新污染物的化学组分及含量水平进行深入研究。另外，对废塑料进行资源化利用过程中，是否引起微塑料污染等问题还需进一步探讨。

（4）从业人员及周边居民的暴露风险评估及防护研究。随着人们健康防护意识的不断提高，园区或企业管理人员、一线操作工人及周边居民对污染物的暴露风险有了一定防范意识。面对电子垃圾处理企业车间或园区车间的高浓度污染物的暴露风险，应该加强对车间污染物的控制消减和从业人员的职业健康防护，降低车间的环境暴露浓度和减少污染物进入人体的暴露剂量，从而将电子垃圾拆解回收处理所产生的健康风险降至最低。

（安太成　郭　杰）

参 考 文 献

[1] Ogunseitan O A. The Basel Convention and e-waste: translation of scientific uncertainty to protective policy [J]. Lancet Global Health, 2013, 1(6): E313-E314.
[2] Baldé C P, D'Angelo E, Luda V, et al. Global transboundary e-waste flows monitor—2022, United Nations Institute for Training and Research (UNITAR), Bonn, Germany[R].
[3] Seidu F, Kaifie A. The end of informal e-waste recycling in Accra, Ghana? [J]. Annals of Work Exposures and Health, 2022, 66(8): 1091-1093.
[4] Akon-Yamga G, Daniels C U, Quaye W, et al. Transformative innovation policy approach to e-waste management in Ghana: Perspectives of actors on transformative changes [J]. Science and Public Policy, 2021, 48(3): 387-397.
[5] Srivastava R R, Pathak P. Handbook of Electronic Waste Management. Policy issues for efficient management of E-waste in developing countries [M]. Elsevier, 2020.
[6] Wagner M, Baldé C P, Luda V, et al. Regional E-waste Monitor for Latin America: Results for the 13 countries participating in project UNIDO-GEF 5554, Bonn (Germany), 2022[R].
[7] Pathak P, Srivastava R R, Ojasvi. Assessment of legislation and practices for the sustainable management of waste electrical and electronic equipment in India [J]. Renewable & Sustainable Energy Reviews, 2017, 78: 220-232.

[8] Forti V, Baldé C P, Kuehr R, et al. The Global E-waste Monitor 2020: Quantities, flows and the circular economy potential. United Nations University (UNU)/United Nations Institute for Training and Research (UNITAR)-co-hosted SCYCLE Programme, International Telecommunication Union (ITU) & International Solid Waste Association (ISWA), Bonn/Geneva/Rotterdam[R].
[9] Brindhadevi K, Barcelo D, Chi N T L, et al. E-waste management, treatment options and the impact of heavy metal extraction from e-waste on human health: Scenario in Vietnam and other countries [J]. Environmental Research, 2023, 217: 114926.
[10] Hoang A Q, Tue N M, Tu M B, et al. A review on management practices, environmental impacts, and human exposure risks related to electrical and electronic waste in Vietnam: Findings from case studies in informal e-waste recycling areas [J]. Environmental Geochemistry and Health, 2022: (doi.org/10.1007/s10653-10022-01408-10654).
[11] Hai H T, Hung H V, Quang N D. An overview of electronic waste recycling in Vietnam [J]. Journal of Material Cycles and Waste Management, 2017, 19(1): 536-544.
[12] Tran H P, Schaubroeck T, Nguyen D Q, et al. Material flow analysis for management of waste TVs from households in urban areas of Vietnam [J]. Resources Conservation and Recycling, 2018, 139: 78-89.
[13] 中国电子废物环境综合管理(2012—2021) [R]. 生态环境部固体废物与化学品管理技术中心, 2021.
[14] 2020 年全国废弃电器电子产品拆解处理产业形势分析报告 [R]. 生态环境部固体废物与化学品管理技术中心, 2021.